Based on the 1999
NATIONAL ELECTRICAL CODE®

House Wiring
with the National
Electrical Code®

RAY C. MULLIN

Delmar Publishers

an International Thomson Publishing company

Albany • Bonn • Boston • Cincinnati • Detroit • London • Madrid
Melbourne • Mexico City • New York • Pacific Grove • Paris • San Francisco
Singapore • Tokyo • Toronto • Washington

NOTICE TO THE READER

Cover Design: Courtesy of Brucie Rosch

Delmar Staff:
Publisher: Alar Elken
Acquisitions Editor: Mark Huth
Developmental Editor: Jeanne Mesick
Project Editor: Megeen Mulholland
Production Coordinator: Toni Hansen
Art and Design Coordinator: Cheri Plasse
Editorial Assistant: Dawn Daugherty
Production Manager: Mary Ellen Black
Marketing Manager: Mona Caron

COPYRIGHT © 1999
By Delmar Publishers
an International Thomson Publishing company I(T)P®

The ITP logo is a trademark under license
Printed in the United States of America

Online Services

Delmar Online

To access a wide variety of Delmar products and services on the World Wide Web, point your browser to:

http://www.delmar.com
or email: info@delmar.com

A service of I(T)P®

For more information, contact:

Delmar Publishers
3 Columbia Circle, Box 15015
Albany, New York 12212-5015

International Thomson Publishing Europe
Berkshire House 168-173
168-173 High Holborn
London, WC1V 7AA
United Kingdom

Nelson ITP, Australia
102 Dodds Street
South Melbourne,
Victoria, 3205 Australia

Nelson Canada
1120 Birchmont Road
Scarborough, Ontario
M1K 5G4, Canada

International Thomson Publishing France
Tour Maine-Montparnasse
33 Avenue du Maine
75755 Paris Cedex 15, France

International Thomson Editores
Seneca 53
Colonia Polanco
11560 Mexico D. F. Mexico

International Thomson Publishing GmbH
Königswinterer Strasße 418
53227 Bonn
Germany

International Thomson Publishing Asia
60 Albert Street
#15-01 Albert Complex
Singapore 189969

International Thomson Publishing Japan
Hirakawa-cho Kyowa Building, 3F
2-2-1 Hirakawa-cho, Chiyoda-ku,
Tokyo 102, Japan

ITE Spain/Parninfo
Calle Magallanes, 25
28015-Madrid, Espana

   2   3   4   5   6   7   8   9   10   XXX   04   03   02   01   00   99

Library of Congress Cataloging-in-Publication Data

Mullin, Ray C.
   House wiring with the NEC / Ray C. Mullin.
     p. cm.
   Includes index.
   ISBN 0-8273-8350-9 (alk. paper)
   1. Electric wiring—Insurance requirements. 2. Electric wiring,
Interior. 3. National Fire Protection Association. National
Electrical Code (1996) I. Title.
TK3275.M854   1998
621.319'24'021873—dc21
                        98-4515
                        CIP

CONTENTS

PREFACE

You are about to embark on a journey. A journey that will take you through the major *National Electrical Code®* issues involved in house (residential) wiring. As you already know, there are many books available on house wiring for the homeowner and/or handyman. Generally, these books do not cover *in detail* the important *National Electrical Code®* requirements that will result in a safe electrical installation. ***House Wiring with the NEC®*** fills this need.

To set the stage, here are a few definitions.

Dwelling Unit: *One or more rooms for the use of one or more persons as a housekeeping unit with space for eating, living, and sleeping, and permanent provisions for cooking and sanitation.*

One-Family Dwelling: *A building consisting solely of one dwelling unit.*

Two-Family Dwelling: *A building consisting solely of two dwelling units.*

Building: *A structure that stands alone or which is cut off from adjoining structures by fire walls with all openings therein protected by approved fire doors.*

The terms *house*, *dwelling*, and *residence* are used throughout this book, and have the same meaning.

What this book covers: The material found in ***House Wiring with the NEC®*** focuses on how to "Meet Code" when doing house wiring. It covers **all** of the key *National Electrical Code®* rules relating to house wiring.

What this book does not cover: ***House Wiring with the NEC®*** does not cover topics such as "how to fish a cable through the wall," "how to tape a joint," "how to hang a fixture," "how to fasten an outlet box to the wall," "how to pull wires through conduit," "how to cut a hole in the drywall to fasten a box," etc. Let the other "handyman" books cover these topics. You are probably already pretty handy at these things. But do you really understand the Code? A little? A lot? None?

Use this book to learn the *National Electrical Code®* rules so that your wiring installation will be safe. Remember, **safety isn't everything—it's the only thing.** To set things straight as to what ***House Wiring with the NEC®*** is all about, let us take a look at the first sentence in the *National Electrical Code®*

> **Section 90-1. Purpose.**
>
> (a) **Practical Safeguarding.** *The purpose of this Code is the practical safeguarding of persons and property from hazards arising from the use of electricity.*

As you study ***House Wiring with the NEC®*** and the *National Electrical Code®* discussing house wiring, never for a moment forget that *everything* presented is based on safety. There can be no compromise when it comes to the safety of people and property. That is why the *National Electrical Code®* exists.

House Wiring with the NEC® is loaded with diagrams, illustrations, calculations, and examples, as well as references to specific sections of the *National Electrical Code®* You will get more out of ***House Wiring with the NEC®*** if you also have a copy of the *National Electrical Code®* to refer to. Copies of the *National Electrical Code®* NFPA 70 can be obtained from:

Delmar Publishers
P.O. Box 15015
Albany, New York 12214-5511

Another text available from Delmar Publishers is ***Electrical Wiring—Residential***. This book contains much information about house wiring not found in the *National Electrical Code®* such as electronic "time-of use" watt-hour meters, "off-peak" metering for energy savings, how electric range heating elements and controls work, how electric dryer heating elements and controls work, exhaust fans and speed controls, dimmers for lighting, door chime wiring, telephone wiring, cable and conventional television wiring, suggested wiring and lighting design for typical house wiring, baseboard electric heating, heating and air-conditioning control wiring, and electric water heaters (connections, recovery time charts, time to use up the hot water). The book comes complete with a set of full-size electrical plans (called "blue-prints" in the old days) of a typical residence.

Many thanks go to my good friend, Ron O'Riley. Ron watched over my shoulders as I developed the material for this text. He made suggestions as to what should be included and what should not be included. He checked the material as it was being created and refined.

Ron is a great educator. He has written a considerable amount of electrical apprenticeship training material, and is the author of *Illustrated Changes in The National Electrical Code®* and another book on the subject of *Electrical Grounding.* These texts are widely used in electrical apprenticeship training programs across the country.

Thanks to the National Fire Protection Association for their permission to quote certain material directly from the *National Electrical Code.®*

ABOUT THE AUTHOR

This text was written by Ray C. Mullin, former electrical instructor for the Wisconsin Schools of Vocational, Technical, and Adult Education. He is a former member of the International Brotherhood of Electrical Workers. He is a member of the International Association of Electrical Inspectors, the Institute of Electrical and Electronic Engineers, and the National Fire Protection Association, Electrical Section, and has served on Code Making Panel 4 of the *National Electrical Code.®*

Mr. Mullin completed his apprenticeship training and worked as a journeyman and supervisor for residential, commercial, and industrial installations. He has taught both day and night electrical apprentice and journeyman courses, has conducted engineering seminars, and has conducted many technical Code workshops and seminars at International Association of Electrical Inspectors Chapter and Section meetings, and has served on their Code panels.

He has written many technical articles that have appeared in electrical trade publications. He has served as a consultant to electrical equipment manufacturers regarding conformance of their products to industry standards, and on legal issues relative to personal injury lawsuits resulting from the misuse of electricity and electrical equipment. He has served as an expert witness.

Mr. Mullin presents his knowledge and experience in this text in a clear-cut manner that is easy to understand. This presentation will help students to fully understand the essentials required to pass the residential licensing examinations and to perform residential wiring that "meets Code."

Mr. Mullin is the author of *Electrical Wiring—Residential*, that focuses entirely on the *National Electrical Code®* requirements for house wiring, plus coverage of most of the additional topics that electricians need to know about electrical ranges, water heaters, clothes dryers, exhaust and air circulating ceiling paddle fans, indoor and outdoor lighting, garage door openers, and similar topics not directly covered in the *NEC.®*

He is co-author of *Electrical Wiring—Commercial, Illustrated Electrical Calculations,* and *The Smart House.* He has contributed technical material to Delmar's *Electrical Grounding,* and to the International Association of Electrical Inspectors texts *Soares Book on Grounding* and *Ferm's Fast Finder.*

He served on the Executive Board of the Western Section of the International Association of Electrical Inspectors, on their Code Clearing Committee, and served as past Secretary/Treasurer of the Indiana Chapter of the IAEI. He presently serves on the Electrical Commission in his hometown.

Mr. Mullin is past Director and Technical Liaison for a major electrical manufacturer. In this position, he was deeply involved in electrical codes and standards as well as contributing and developing technical training material for use by this company's field engineering personnel.

Mr. Mullin attended the University of Wisconsin, Colorado State University, and the Milwaukee School of Engineering.

National Electrical Code® and *NEC®* are registered trademarks of the National Fire Protection Association, Inc., Quincy, MA 02269.

DELMAR PUBLISHERS IS YOUR ELECTRICAL BOOK SOURCE!

Whether you're a beginning student or a master electrician, Delmar Publishers has the right book for you. Our complete selection of proven best-sellers and all-new titles is designed to bring you the most up-to-date, technically accurate information available.

NATIONAL ELECTRICAL CODE

National Electrical Code® 1999/NFPA
Revised every three years, the *National Electrical Code®* is the basis of all U.S. electrical codes.
> Order # 0-8776-5432-8
> Loose-leaf version in binder
> Order # 0-8776-5433-6

National Electrical Code® Handbook 1999/NFPA
This essential resource pulls together all the extra facts, figures, and explanations you need to interpret the 1999 *NEC.®* It includes the entire text of the Code, plus expert commentary, real-world examples, diagrams, and illustrations that clarify requirements.
> Order # 0-8776-5437-9

Illustrated Changes in the 1999 National Electrical Code®/O'Riley
This book provides an abundantly illustrated and easy-to-understand analysis of the changes made to the 1999 *NEC.®*
> Order # 0-7668-0763-0

Understanding the National Electrical Code,® 3E/Holt
This book gives users at every level the ability to understand what the *NEC®* requires, and simplifies this sometimes intimidating and confusing code.
> Order # 0-7668-0350-3

Illustrated Guide to the National Electrical Code®/Miller
Highly detailed illustrations offer insight into Code requirements, and are further enhanced through clearly written, concise blocks of text that can be read very quickly and understood with ease. Organized by classes of occupancy.
> Order # 0-7668-0529-8

Interpreting the National Electrical Code,® 5E/Surbrook
This updated resource provides a process for understanding and applying the *National Electrical Code®* to electrical contracting, plan development, and review.
> Order # 0-7668-0187-X

Electrical Grounding, 5E/O'Riley
Electrical Grounding is a highly illustrated, systematic approach for understanding grounding principles and their application to the 1999 *NEC.®*
> Order # 0-7668-0486-0

ELECTRICAL WIRING

Electrical Raceways and Other Wiring Methods, 3E/Loyd
The most authoritative resource on metallic and non-metallic raceways, provides users with a concise, easy-to-understand guide to the specific design criteria and wiring methods and materials required by the 1999 *NEC.®*
> Order # 0-7668-0266-3

Electrical Wiring—Residential, 13E/Mullin
Now in full color! Users can learn all aspects of residential wiring and how to apply them to the wiring of a typical house from this, the most widely used residential wiring book in the country.
> Softcover Order #0-8273-8607-9
> Hardcover Order # 0-8273-8610-9

House Wiring with the NEC®/Mullin
The focus of this new book is the applications of the *NEC®* to house wiring.
> Order # 0-8273-8350-9

Electrical Wiring—Commercial, 10E/Mullin and Smith
Users can learn commercial wiring in accordance with the *NEC®* from this comprehensive guide to applying the newly revised 1999 *NEC.®*
> Order # 0-7668-0179-9

Electrical Wiring—Industrial, 10E/Smith & Herman
This practical resource has users work their way through an entire industrial building—wiring the branch-circuits, feeders, service entrances, and many of the electrical appliances and subsystems found in commercial buildings.
> Order # 0-7668-0193-4

Cables and Wiring, 2E/AVO
This concise, easy-to-use book is your single-source guide to electrical cables—it's a "must-have" reference for journeyman electricians, contractors, inspectors, and designers.
> Order # 0-7668-0270-1

ELECTRICAL MACHINES AND CONTROLS

Industrial Motor Control, 4E/Herman and Alerich
This newly revised and expanded book, now in full color, provides easy-to-follow instructions and essential information for controlling industrial motors. Also available are a new lab manual and an interactive CD-ROM.
> Order # 0-8273-8640-0

Electric Motor Control, 6E/Alerich and Herman
Fully updated in this new sixth edition, this book has been a long-standing leader in the area of electric motor controls.
Order # 0-8273-8456-4

Introduction to Programmable Logic Controllers/
Dunning
This book offers an introduction to Programmable Logic Controllers.
Order # 0-8273-7866-1

Technician's Guide to Programmable Controllers, 3E/
Cox
Uses a plain, easy-to-understand approach and covers the basics of programmable controllers.
Order # 0-8273-6238-2

Programmable Controller Circuits/Bertrand
This book is a project manual designed to provide practical laboratory experience for one studying industrial controls.
Order # 0-8273-7066-0

Electronic Variable Speed Drives/Brumbach
Aimed squarely at maintenance and troubleshooting, *Electronic Variable Speed Drives* is the only book devoted exclusively to this topic.
Order # 0-8273-6937-9

Electrical Controls for Machines, 5E/Rexford
State-of-the-art process and machine control devices, circuits, and systems for all types of industries are explained in detail in this comprehensive resource.
Order # 0-8273-7644-8

Electrical Transformers and Rotating Machines/Herman
This new book is an excellent resource for electrical students and professionals in the electrical trade.
Order # 0-7668-0579-4

Delmar's Standard Guide to Transformers/Herman
Delmar's Standard Guide to Transformers was developed from the best-seller *Standard Textbook of Electricity* with expanded transformer coverage not found in any other book.
Order # 0-8273-7209-4

DATA AND VOICE COMMUNICATION CABLING AND FIBER OPTICS

Complete Guide to Fiber Optic Cable System Installation/Pearson
This book offers comprehensive, unbiased, state-of-the-art information and procedures for installing fiber optic cable systems.
Order # 0-8273-7318-X

Fiber Optics Technician's Manual/Hayes
Here's an indispensable tool for all technicians and electricians who need to learn about optimal fiber optic design and installation as well as the latest troubleshooting tips and techniques.
Order # 0-8273-7426-7

A Guide for Telecommunications Cable Splicing/
Highhouse
A "how-to" guide for splicing all types of telecommunications cables.
Order # 0-8273-8066-6

Premises Cabling/Sterling
This reference is ideal for electricians, electrical contractors, and inspectors needing specific information on the principles of structured wiring systems.
Order # 0-8273-7244-2

ELECTRICAL THEORY

Delmar's Standard Textbook of Electricity, 2E/Herman
This exciting full-color book is the most comprehensive book on DC/AC circuits and machines for those learning the electrical trades.
Order # 0-8273-8550-1

Industrial Electricity, 6E/Nadon, Gelmine and Brumbach
This revised, illustrated book offers broad coverage of the basics of electrical theory and is perfect for those who wish to be industrial maintenance technicians.
Order # 0-7668-0101-2

EXAM PREPARATION

Journeyman Electrician's Exam Preparation, 2E/Holt
This comprehensive exam prep guide includes all of the topics on the journeyman electrician competency exams.
Order # 0-7668-0375-9

Master Electrician's Exam Preparation, 2E/Holt
This comprehensive exam prep guide includes all of the topics on the master electrician's competency exams.
Order # 0-7668-0376-7

REFERENCE
ELECTRICAL REFERENCE SERIES

This series of technical reference books is written by experts and designed to provide the electrician, electrical contractor, industrial maintenance technician, and other electrical workers with a source of reference information about virtually all of the electrical topics that they encounter.

Electrician's Technical Reference—Motor Controls/Carpenter
Electrical Reference—Motor Controls is a source of comprehensive information on understanding the controls that start, stop, and regulate the speed of motors.
    Order # 0-8273-8514-5

Electrician's Technical Reference—Motors/Carpenter
Electrician's Technical Reference—Motors builds an understanding of the operation, theory, and applications of motors.
    Order # 0-8273-8513-7

Electrician's Technical Reference—Theory and Calculations/Herman
Electrician's Technical Reference—Theory and Calculations provides detailed examples of problem-solving for different kinds of DC and AC circuits.
    Order # 0-8273-7885-8

Electrician's Technical Reference Transformers/Herman
Electrician's Technical Reference—Transformers focuses on the theoretical and practical aspects of single-phase and 3-phase transformers and transformer connections.
    Order # 0-8273-8496-3

Electrician's Technical Reference—Hazardous Locations/Loyd
Electrician's Technical Reference—Hazardous Locations covers electrical wiring methods and basic electrical design considerations for hazardous locations.
    Order # 0-8273-8380-0

Electrician's Technical Reference—Wiring Methods/Loyd
Electrician's Technical Reference—Wiring Methods covers electrical wiring methods and basic electrical design considerations for all locations, and shows how to provide efficient, safe, and economical applications of various types of available wiring methods.
    Order # 0-8273-8379-7

Electrician's Technical Reference—Industrial Electronics/Herman
Electrician's Technical Reference—Industrial Electronics covers components most used in heavy industry, such as silicon control rectifiers, triacs, and more. It also includes examples of common rectifiers and phase-shifting circuits.
    Order # 0-7668-0347-3

RELATED TITLES

Common Sense Conduit Bending and Cable Tray Techniques/Simpson
Now geared especially for students, this manual remains the only complete treatment of the topic in the electrical field.
    Order # 0-8273-7110-1

Practical Problems in Mathematics for Electricians, 5E/Herman
This book details the mathematics principles needed by electricians.
    Order # 0-8273-6708-2

Electrical Estimating/Holt
This book provides a comprehensive look at how to estimate electrical wiring for residential and commercial buildings with extensive discussion of manual versus computer-assisted estimating. The residential estimating portion of this book is based upon *Electrical Wiring—Residential*, making both books a great combination for studying residential wiring and residential estimating.
    Order # 0-8273-8100-X

Electrical Studies for Trades/Herman
Based on Delmar's *Standard Textbook of Electricity*, this new book provides non-electrical trades students with the basic information they need to understand electrical systems.
    Order # 0-8273-7845-9

General Information

OBJECTIVES

After studying this unit, you will be able to:

- **Understand some of the hazards involved when working with electricity.**
- **Realize that there are federal laws governing safety in the workplace.**
- **Know more about the *National Electrical Code*,® and how it is arranged.**
- **Discuss some of the key words used in the *National Electrical Code*.®**
- **Realize that there are codes other than the *National Electrical Code*® that can affect an installation.**
- **Understand the basics of licensing and permits.**
- **Know that there are recognized testing laboratories that test and "list" electrical equipment.**

INTRODUCTION

All types of work involve some type of hazard. Personal injuries can be costly. This unit discusses safety in the workplace. To do electrical work, you need to know about electrical codes and standards, licensing requirements, permits, and recognized testing agencies.

SAFETY IN THE WORKPLACE

Electricity is dangerous! Before venturing into the study of residential wiring, let us discuss safety. **Wiring is a skilled trade!** A considerable amount of "on-the-job training" and "book-learning" is needed before one can say "I am an electrician." Working with electricity is not for the novice. Electricity is safe when handled properly. However, electricity is not forgiving. There is no room for error. There may not be a second chance. A question often asked **after** someone is injured or killed by electricity is "Would the injury or fatality have occurred had the power been shut off?" The answer is "Probably not."

Whether the project involves new construction where the only electrical power is temporary power, or in remodel work where the home already has electrical power, serious injury death can result if the power is not turned off.

Safety starts with the individual. The individual must accept and be aware of this responsibility. It is the **personal responsibility** of the **individual** to make sure the power is off. Do not rely on others. When **you** turn off the circuit breaker or disconnect switch, remove the fuses, lock the switch or circuit breaker in the OFF position, and attach an explanatory tag—**you** are reasonably assured that someone else will not turn the power back on (Figure 1–1).

The voltage level in a home is 120 volts measured between one "hot" conductor and the grounded "neutral" conductor, or between one "hot" conductor and a grounded surface such as a metal water pipe. This is the *line-to-ground* voltage. Between the two "hot" conductors, the voltage is 240 volts. This is the *line-to-line* voltage.

From basic electrical theory, *line voltage appears across an open in a series circuit.* Getting caught "in series" with a 120-volt circuit will give

1

Figure 1–1 A typical disconnect switch with a lock and a tag attached to it.

you a 120-volt shock. Likewise, getting caught "in series" with a 240-volt circuit will give you a 240-volt shock. A second thing that theory tells us is that the *current in a series circuit is the same in all parts of the series circuit.* Once a person is caught in a series circuit, the current flowing in the circuit will be the same in all parts of the circuit—including the individual who has become part of the series circuit. It is important to know that it is the current that kills, not the voltage; 120 volts can be just as deadly as 240 volts. The current we are talking about is measured in milliamperes (thousandths of an ampere). This is covered in detail in Unit 7.

For example, open-circuit voltage across the two terminals of a single-pole switch on a lighting circuit is 120 volts when the switch is in the OFF position and the lamp(s) are in place. When working on electrical equipment with the power turned on, the outcome can be death, burns, or serious injury, either as a direct result of electricity, or from an indirect secondary reaction such as falling off a ladder or jerking away from the "hot" conductor into moving parts of equipment, such as into the turning blades

of a furnace fan. Dropping a metal tool onto live parts, allowing metal shavings from a drilling operation to fall onto live parts of electrical equipment, cutting into a "live" conductor and a "neutral" conductor at the same time, or touching the "live" wire and the "neutral" wire or a grounded surface at the same time can cause injury.

A short circuit or ground fault can result in an *arc blast* that can cause serious injury or death. The heat of an electrical arc has been determined to be hotter than the sun. Tiny hot "balls" of copper can fly into your eye, or onto your skin. Clothing can be ignited. It cannot be said often enough—working on switches, receptacles, fixtures, appliances, panels, or other electrical equipment with the power turned on is dangerous.

Dirt, debris, and moisture can also set the stage for equipment failures and personal injury. Neatness and cleanliness in the workplace are a must.

Safety and the Law

Not only is it a good idea to use proper safety measures as you work on and around electrical systems, it is **required** by law. As an electrician or electrical contractor, you need to be aware of these regulations.

The Federal Regulations in the Occupational Safety and Health Act (OSHA) Number 29, Subpart S, in Part 1910.332 discusses the training needed for those who face a risk of electrical injury. Proper training means trained in and familiar with the safety-related work practices required by paragraphs 1910.331 through 1910.360. Numerous texts are available on OSHA requirements.

QUALIFICATIONS NEEDED TO DO ELECTRICAL WORK

The *National Electrical Code®* defines a *qualified person* as *"one familiar with the construction and operation of the equipment and the hazards involved."* Merely telling someone or being told "be careful" does not meet the definition of proper training, and does not make the person qualified.

Only qualified persons are permitted to work on or near exposed energized equipment. To become qualified, a person must have the skill and techniques necessary to distinguish exposed live parts from the other parts of electrical equipment, must be able to determine the voltage of exposed live parts, and must be trained in the use of special precautionary techniques such as personal protective equipment, insulation, shielding material, and insulated tools.

In the OSHA regulations, Subpart S, 1910.333 requires that safety-related work practices shall be employed to prevent electric shock or other injuries resulting from either direct or indirect electrical contact. Live parts to which an employee may be exposed shall be de-energized before the employee works on or near them, unless the employer can demonstrate that de-energizing introduces additional or increased hazards.

Working on equipment "live" is acceptable only if there would be a greater hazard if the system was de-energized. Examples of this would be hospital life support systems, some alarm systems, certain ventilation systems in hazardous locations, and the power for critical illumination circuits. None of these **excuses** to work on "hot" electrical equipment are found in residential wiring. Working on energized equipment requires proper insulated tools, proper nonflammable clothing, rubber gloves, protective shields and goggles, and in some cases, rubber blankets.

OSHA regulations allow only qualified personnel to work on or near electrical circuits or equipment that have not been de-energized. The OSHA regulations provide rules regarding "lockout and tagging" to make sure that the electrical equipment being worked on will not inadvertently be turned on while someone is working on the supposedly "dead" equipment. As the OSHA regulations state, "a lock and a tag shall be placed on each disconnecting means used to de-energize circuits and equipment. ..."

Some electricians' work agreements require that as a safety measure, two or more qualified electricians must work together when working on energized circuits. They do not allow apprentices in training to work on "live" equipment. These work agreements usually allow apprentices to stand back and observe.

The National Fire Protection Association publications *Electrical Safety Requirements for Employee Workplaces, NFPA 70E,* and *Electrical Equipment Maintenance NFPA 70B* present much of the same material regarding electrical safety as does the OSHA regulation.

Safety cannot be compromised! **The rule is turn off and lock-off the power, then properly tag the disconnect with a description as to exactly what that particular disconnect controls.**

GROUND-FAULT PROTECTION

One of the most common ways to prevent serious electrical ground-fault shocks when using portable electrical tools and extension cords is to plug into a *ground-fault circuit-interrupter (GFCI)* device. GFCIs are available as wall receptacles, circuit breakers in a panel, or as portable cord sets where you plug the male end (line) of the GFCI portable device into any 120-volt wall receptacle, then plug the electrical tool into the female end (load) of the GFCI device. GFCIs are discussed and illustrated in detail in Unit 7.

NATIONAL ELECTRICAL CODE®
NFPA 70

The *National Electrical Code® NFPA 70* is published by the National Fire Protection Association, and is quite often just referred to as the *NEC.* The *National Electrical Code®* is available in soft back, hard back, CD-ROM, and $3\frac{1}{2}$" computer disk for Windows, MacIntosh, or DOS (Figure 1–2).

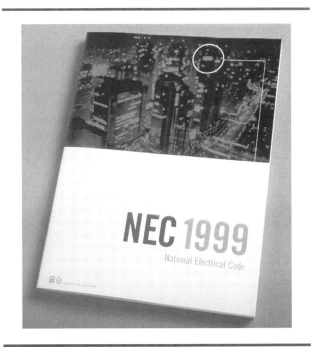

Figure 1–2 The *National Electrical Code®* book. (*Courtesy of National Fire Protection Association*)

Because of the ever-present danger of fire or shock hazard through some failure of the electrical system, the electrician and the electrical contractor must not only make the installation in accordance with the requirements contained in the *National Electrical Code*, they must also use "listed" materials and perform all work in accordance with recognized standards. The term "listed" refers to products and equipment that have been tested and evaluated by a recognized testing laboratory such as Underwriters Laboratories, and have passed the tests.

The *National Electrical Code* is the basic standard that governs electrical work. In *Section 90-1(a)* we find that the purpose of the Code is "*the practical safeguarding of persons and property from hazards arising from the use of electricity.*" In *Section 90-1(b)* we find that if the *NEC* rules are followed, the installation will be "*essentially free from hazard but not necessarily efficient, convenient, or adequate for good service or future expansion of electrical use.*" *Section 90-1(c)* tells us that the Code "*is not intended as a design specification nor an instruction manual for untrained persons.*"

The *NEC* is a minimum standard for the *NOW* installation and really does not provide for the future. It is the electrician's responsibility to ensure that the installation meets these present requirements. Allowing additional capacity for the future is permitted by the *NEC*.

The material in this book is based on the *National Electrical Code*. You must also consider local and state codes. Contact your municipality–talk to the people in the electrical inspection department. Before starting an electrical project, find out what edition of the electrical code is enforced. Learn about special requirements in your municipality. You will save yourself a lot of time, grief, and money if you do.

All *National Electrical Code* references presented throughout this text are printed in *italics*.

CODE ARRANGEMENT

It is important to understand the arrangement of the *NEC*.

- Introduction   Explains the purpose of the *NEC*, what is covered, what is not covered, who enforces the *NEC*.

- Chapters 1, 2, 3, and 4   These chapters contain general requirements that cover most situations.

- Chapters 5, 6, and 7   These chapters pertain to special occupancies, special equipment, and special conditions. The requirements in these chapters are in addition to, or are changes to (modify or amend) the general rules found in Chapters 1, 2, 3, and 4

- Chapter 8   Covers communication, telephone, fire and burglar alarm, radio, television, cable systems, etc.

- Chapter 9   This chapter contains examples and tables

- Appendix A   Lists text that has been extracted from other NFPA codes. These are identified throughout the *NEC* in particular code sections with the superscript letter "x".

- Appendix B   Contains useful information not actually enforceable as a *NEC* rule.

- Appendix C   Contains 24 tables showing the number of conductors permitted in various types of raceways.

- Appendix D   Examples of load calculations.

- Appendix E   A cross-reference to *Article 250* showing the new numbering system compared to the old numbering system.

- Index   The alphabetical index for the *National Electrical Code*.

Chapters are divided into Articles. Articles are divided into Sections. Sections are divided into subsections and exceptions.

SPECIAL TERMS IN THE *NEC*

The electrical trade uses many terms that are unique. In some ways, one might say that the *National Electrical Code* has a language of its own. These unique terms need definitions to enable the electrician to understand the meaning intended by the Code.

Article 100 of the *National Electrical Code®* is a "dictionary" of these special words. *Article 90* also provides further clarification of words used in the *NEC®.* Refer to *Article 100* whenever you come across a word that appears to be special to the electrical trade.

Here are just a few examples.

Ampacity: *The current in amperes that a conductor can carry continuously under the conditions of use without exceeding its temperature rating.*

Approved: *Acceptable to the authority having jurisdiction.* It is wrong to say that Underwriters Laboratories *approves* a product. UL does not approve anything. UL "lists" a product. See the definition of "Listed" below.

Authority Having Jurisdiction: An organization, office, or individual responsible for "approving" equipment, an installation, or a procedure, *Section 90-4.* Nothing is approved until the authority having jurisdiction approves it. In most instances, the AHJ is the electrical inspector.

Listed: *Equipment, materials, or services included in a list published by an organization that is acceptable to the authority having jurisdiction and concerned with evaluation of products or services, that maintains periodic inspection of production of listed equipment or materials or periodic evaluation of services, and whose listing states that either the equipment, material, or services meets identified standards or has been tested and found suitable for a specified purpose.*

Note: The term "listed" appears in the *National Electrical Code®* more than 400 times. For obvious reasons, using equipment and materials that have been tested and listed by a recognized testing laboratory makes for a safer installation, rather than using equipment and materials that have not been tested. The "listing" mark will usually appear on the label of the product or on the carton.

Throughout the Code, the word *"shall"* indicates a mandatory rule, *Section 90-5.* As you study the *National Electrical Code,®* think of *"shall"* as meaning *"must."* Some examples found in the *NEC®* where the word "shall" is used in combination with other words are:

shall be, shall have, shall not

A permissive rule might use words such as:

shall be permitted, shall not be required

Fine Print Notes (FPN) are also found throughout the Code. FPNs are "explanatory" notes and contain a wealth of information. FPNs are not mandatory. FPNs might reference other sections of the Code, refer to other codes, offer cautions or warnings, make recommendations, call attention to a potential electrical problem, offer the Code intent of the material that is located above the FPN, or be used to define something where further description is necessary.

HISTORY OF THE *NATIONAL ELECTRICAL CODE®*

The first *National Electrical Code®* was published in 1897. It is revised and updated every three years by changing, editing, adding technical material, adding sections, adding articles, adding tables, and so on, so as to be as up-to-date as possible relative to electrical installations. For example, the Code added requirements pertaining to circuits supplying computers and similar electronic equipment. Upscale homes utilizing the latest electronic technology for controlling lighting, receptacle outlets, and appliances come under the jurisdiction of the *NEC®* insofar as conventional wiring methods are concerned. Home automation systems such as X10™, CEBus™, AmpOnQ, Echelon™, Home-Star,® and Elan® are being installed in these homes. Components for home automation systems are available through electrical and electronic supply houses, home centers, and through the mail. *Article 780* in the *NEC®* entitled *Closed-Loop and Programmed Power Distribution* is involved with this unique type of wiring. Home automation is discussed in Unit 14.

As material, equipment, and technologies change, the National Fire Protection Association solicits proposals from individuals in the electrical industry or others interested in electrical safety. The process for making changes to the *National Electrical Code®* is fair. Anyone can send in a proposal to change something that they feel needs to be revised, added, or deleted. Instructions, deadline dates for sending in proposals, and a suggested proposal form are found in the back of the *National Electrical Code®* book.

There are 20 Code-Making Panels. Code-Making Panel (CMP) members meet on specific dates to review, accept, accept in principle, accept in part, accept in principle in part, hold for further study, or reject each of the many proposals. There is also a Correlating Committee that sees to it that proposals are referred to the proper CMP, and that the actions of one CMP are not in conflict with the actions of other CMPs. A *Report of Proposals (ROP)* explaining the reason for the CMP's action on each proposal is published. This provides the means for public review of the CMP actions. The CMPs meet again to review, accept, accept in principle, accept in part, accept in principle in part, hold for further study, or reject the comments that are sent in based on the *ROP*. A *Report on Comments* is then published, showing all of the CMP actions on the comments. Final action (voting) on these proposals is taken at the National Fire Protection Association's annual meeting.

Code-Making Panels consist of individuals representing over 65 organizations such as the International Association of Electrical Inspectors, the Institute of Electrical and Electronic Engineers, the National Electrical Contractors Association, the Edison Electric Institute, the Independent Electrical Contractors Association, Inc., Underwriters Laboratories, Inc., Consumer Product Safety Commission, International Brotherhood of Electrical Workers, and many others.

The *National Electrical Code®* does not become "law" until adopted by official action of the legislative body of a city, county, or state.

National Electrical Code® Edition

This text is based on the 1999 edition of the *National Electrical Code®* However, be aware that some municipalities, counties, and states have not yet adopted the 1999 edition, and might still be using the 1996 or earlier editions. Always check on this with the local authority having jurisdiction before starting a wiring project to make sure your installation complies with the proper electrical code.

A condensed version of the *National Electrical Code® NFPA 70* is available that mainly covers Code regulations that apply to house wiring. It is

the *Electrical Code for One- and Two-Family Dwellings, NFPA 70A.*

Copies of the *National Electrical Code® NFPA 70* and *NFPA 70A* may be ordered from:

> Delmar Publishers
> P.O. Box 15015
> Albany, New York, 12212-5015
> 1-800-347-7707
>
> or
>
> National Fire Protection Association
> 1 Batterymarch Park
> Quincy, Massachusetts 02269-9101
>
> or
>
> International Association of
>   Electrical Inspectors
> 901 Waterfall Way
> Suite 602
> Richardson, TX 75080-7702

or check your local electrical supply houses and book stores. There is a good possibility that they may sell the *National Electrical Code®*

RESIDENTIAL ELECTRICAL MAINTENANCE CODE FOR ONE- AND TWO-FAMILY DWELLINGS (NFPA 73)

The *NEC®* is not a retroactive code, other than a few rules for replacing receptacles. This is covered in Unit 7. The National Fire Protection Association developed the *Residential Electrical Maintenance Code for One- and Two- Family Dwellings NFPA 73* code. It is a rather brief seven-page code that provides requirements for evaluating already installed electrical systems within and associated with **existing** one- and two-family dwellings to identify safety, fire, and shock hazards, such as improper installations, overheating, physical deterioration, and abuse. This code lists most of the electrical things that should be checked out in an existing dwelling that could result in a fire or shock hazard if not corrected. It only points out things to look for that are visible, and does not get into examining concealed wiring that would require removal of permanent parts of the structure. It does not get into calculations, location requirements, and complex topics as does the *National Electrical Code® NFPA 70* and *NFPA 70A.*

NFPA 73 uses simple statements such as: *The service shall be adequate to serve the connected load, overcurrent protective devices shall be prop-*

erly rated for conductor ampacities; conductors shall be properly terminated and supported at panelboards, boxes, and devices; raceways shall be securely fastened in place; cables and cable assemblies shall be properly secured and supported, fixture canopies shall be in place and properly secured; covers shall be in place and properly secured, and so forth.

NFPA 73 can be an extremely useful guide for electricians doing remodel work, and for electrical inspectors wanting to bring an existing dwelling to a reasonably safe condition. Many localities require that when a home changes ownership, the wiring must be brought up to some minimum standard, but not necessarily as extensively as would be the case for new construction. For example, a local code might require that a minimum 100-ampere, 120/240-volt service be installed—that the service ground be brought up to the requirements of the latest edition of the *National Electrical Code®*—that 20-ampere appliance circuits be installed in the kitchen and dining room—that a separate 20-ampere branch circuit be installed for the laundry equipment (clothes washer)—and that GFCI receptacles be installed in bathrooms and other specified areas.

Other Building Codes

The majority of building departments across the country have for the most part adopted the *National Electrical Code® NFPA 70* rather than developing their own electrical codes. In a community, the electrical inspector may be full-time or part-time. He may also have responsibility for other trades such as plumbing or heating and air-conditioning. The heads of building departments generally are called the Building Commissioner or Director of Development. Regardless of title, they are responsible to make sure that the building codes in their communities are enforced.

Most communities belong to one of the major building officials organizations, and have adopted the building codes of that particular organization. These organizations are:

- **BOCA** Building Official & Code Administrators International, Inc. 4051 W. Flossmoor Road Country Club Hills, Illinois 60478-5795 Phone: 708-799-2300

- **ICBO** International Conference of Building Officials 5360 Workman Mill Road Whittier, California 90601-2298 Phone: 213-699-0541

- **SBCCI** Southern Building Code Congress International, Inc. 900 Montclair Road Birmingham, Alabama 35212-1206 Phone: 205-591-1853

- **IAPMO** International Association of Plumbing and Mechanical Officials 5032 Alhambra Avenue Los Angeles, CA 90032-3490 Phone: 213-223-1471

- **ICC** International Code Council 5203 Leesburg Pike, Suite 708 Falls Church, VA 22041 Phone: 703-931-4533

ICC is an organization that provides technical, educational, and informational products and services in support of international codes. BOCA, ICBO, and SBCCI are members of this organization. Chapters 39 through 46 of the ICC *One- and Two-Family Dwelling Code* contain the basic electrical requirements for one- and two-family dwellings based on the *NEC.®* The remaining chapters cover plumbing, heating, and construction requirements.

In summary, the following electrical codes are available:

- *National Electrical Code® NFPA 70*—needed for all types of electrical installations.

- *Electrical Code for One- and Two- Family Dwellings and Mobile Homes NFPA 7OA*—can be used when residential wiring only is involved.

- *Residential Electrical Maintenance Code for One- and Two- Family Dwellings NFPA 73.*

- *ICC's One- and Two- Family Dwelling Code*—contains residential electrical, plumbing, heating, and construction requirements.

Local electrical code requirements may differ from those found in the *National Electrical Code.®* Local requirements usually are more stringent that the *National Electrical Code.®* They may have additions to the *National Electrical Code,®* or they might delete certain parts of the *National Electrical Code.* Check with the local "authority having jurisdiction" to determine which electrical code is enforced, and what local amendments to this code might take precedence.

AMERICAN NATIONAL STANDARDS INSTITUTE (ANSI)

The American National Standards Institute is an organization that coordinates the efforts and results of the various standards-developing organizations, such as those mentioned in previous paragraphs. Through this process, ANSI approves standards that then become recognized as American national standards. One will find much similarity between the technical information found in ANSI standards, the Underwriters standards, the International Electronic and Electrical Engineers standards, and the *National Electrical Code.* Many of the *Fine Print Notes* in the *National Electrical Code*® reference ANSI Standards.

LICENSING AND PERMITS

Most communities, counties, and/or states require electricians and electrical contractors to be licensed. This usually means they have taken and have passed a test. To maintain a valid license, many states require electricians and electrical contractors to attend and satisfactorily complete approved continuing education courses consisting of a specified number of classroom hours over a given period of time. Quite often, a community will have a "Residential Only" license for electricians and contractors that limit their activity to house wiring.

Permits are a means for a community to permanently record electrical work to be done, who is doing the work, and to schedule inspections during and after the "rough-in" stage, and in the "final" stages of construction. Usually, permits must be issued prior to starting an electrical project. In most cases, homeowners are allowed to do electrical work in their own home where they live, but not in other properties they might own.

Figure 1–3 is a simple application for electrical permit form. Some permit application forms are much more detailed.

If you are not familiar with licensing and permit requirements in your area, it makes sense to check this out with your local electrical inspector or building department before starting an electrical project. Not to do so could prove to be very costly. Many questions can be answered You will find out what tests if any must be taken, what permits are needed, what electrical code is enforced in your community, minimum size electrical service, and so on. Generally, the electrical permit is taken out by an electrical contractor who is licensed and registered as an electrical contractor in the jurisdictional area.

For new construction or for a main electrical service change, you will also need to contact the electric utility.

NATIONALLY RECOGNIZED TESTING AGENCIES

Before reading this next section, review the definitions in *Article 100* of Listed - Labeled - Identified.

How does one know if a product is safe to use? Manufacturers, consumers, regulatory authorities, and others recognize the importance of independent, "third-party" testing of products in an effort to reduce safety risks. The surest way to know if a product is safe is to accept the findings of a "third-party" testing agency. In other words, "Look For The Label." Nationally recognized testing laboratories (NRTL) have the knowledge and wherewithal to test and evaluate products for safety. Make sure the product has a listing marking on it. If the product is too small to have a listing mark on the product itself, then look for the marking on the carton the product came in. The listing or labeling may also include the identified use of the product as to its suitability for a specific purpose, function, use, environment, application, or other specific use.

You should always install "listed" products. Look for the label. To ensure safety, the products must also be used and installed properly. This basic rule is found in *Section 110-3(b)* of the *National Electrical Code*® which states that *"Listed or labeled equipment shall be installed and used in accordance with any instructions included in the listing or labeling."* The true meaning of *Section 110-3(b)* is that any instructions and/or specifications furnished by the manufacturer of the product are an enforceable part of the *National Electrical Code.*

APPLICATION FOR ELECTRICAL PERMIT
VILLAGE OF ANYWHERE, USA   1-234-567-8900, EXT. 1234

Date_____   Permit No. _____

Owner_____   Job Address_____

Telephone No._____   Job Start Date _____

CONTRACTOR INFORMATION AND SIGNATURE

Electrical Contractor_____   Tel. No. _____

Address_____ City_____ State_____ Zip_____

Registration No._____ City of Registration _____

Supervising Electrician (Please Print Name)_____

Supervising Electrician's Signature _____

Insurance Bond_____ Village Business License _____

SERVICE INSPECTIONS OR REVISIONS

Existing Service Size: Amps_____ Volts_____ No. of Circuits_____ No. Added _____

New Service Size: Amps_____ Volts_____ No. of Circuits_____

Type: Overhead_____ Underground_____

Service Installation Fees: 100-200 amps $50_____ over 200 amps $75_____

MOTORS AND AIR CONDITIONING EQUIPMENT

No. of Motors: up to 1 HP @ $50_____   Over 1-10 HP @ $50_____

                11-25 HP @ $25_____   Over 25 HP @ $25_____

Air Conditioner/Heat Pump:

No. of Tons_____ ($20 for first ton, $5 for each additional ton)   _____

Furnace (electric):   kW_____ Amps_____ ($25)   _____

Dryer (electric):   kW_____ Amps_____ ($10)   _____

Range, oven, cooktop (electric): Total kW_____ Amps_____ ($10 each)   _____

Water Heater (electric): kW_____ Amps_____ ($10)   _____

TYPE OF MISCELLANEOUS ELECTRIC WORK

_____

_____

                Minimum Inspection Fee: $40.00   _____

                Escrow Deposit, if applicable   _____

TOTAL DUE   _____

Figure 1–3   A typical form of an Application for Electrical Permit.

There are a number of nationally recognized laboratories. The following laboratories do a considerable amount of testing and listing of electrical equipment.

Underwriters Laboratories Inc. (UL)

Underwriters Laboratories Inc. (UL), founded in 1894, is a highly qualified, nationally recognized testing laboratory. UL develops standards, and performs tests to these standards.

Many reputable manufacturers of electrical equipment submit their products to UL where the equipment is subjected to numerous tests. These tests determine if the product can perform safely under normal and abnormal conditions to meet published standards. After UL tests, evaluates, and determines that a product complies with the specific

Figure 1–4 Underwriters Laboratories (UL) logo.

standard, the manufacturer is then permitted to **label** its product with the UL Mark. The products are then **listed** in a UL Directory. Figure 1–4 shows the typical UL logo.

It should be emphasized that UL **does not approve** products. Rather, UL **lists** products. A UL Listing Mark on a product means that representative samples of the product have been tested and evaluated to nationally recognized safety standards with regard to fire, electric shock, and related safety hazards.

Useful UL publications are:

Electrical Construction Materials Directory (Green Book)

Electrical Appliances and Utilization Equipment Directory (Orange Book)

General Information for Electrical Equipment (White Book)

These directories are extremely useful for looking up specific requirements, permitted uses, or limitations for a given product that cannot be found in the *National Electrical Code.*

The Green and Orange Directories provide technical information regarding a particular product, as well as list the names and addresses of manufacturers and the manufacturers' identification numbers. The White Book provides the technical information by which a product is tested, but does not show manufacturers' names and addresses. If you must choose only one directory, the White Book is probably the best companion to the *National Electrical Code.*

The directories mentioned previously can be obtained by writing or calling:

Underwriters Laboratories Inc.
333 Pfingsten Road
Northbrook, Illinois 60062-2096
(847) 272-8800

Intertek Testing Services (ITS)

Formerly known as Electrical Testing Laboratories, ITS is a nationally recognized testing laboratory. They provide testing, evaluation, labeling, listing, and follow-up service for the safety testing of electrical products. This is done in conformance to nationally recognized safety standards, or to specifically designated requirements of jurisdictional authorities.

Information can be obtained by writing to:

Intetek Testing Services
3933 U.S. Route 11
Cortland, New York 13045-0950
Phone: 1-800-345-3851

Canadian Standards Association (CSA)

Many electrical products used in this country bear the CSA label. This is the label of the Canadian Standards Association. Some electrical products may have both the CSA and UL label on them.

The Canadian Standards Association is the Canadian counterpart of Underwriters Laboratories Inc. in the United States. CSA is the source of the *Canadian Electrical Code (CEC),* which is very different from the *National Electrical Code.* Many UL standards and Canadian standards have been "harmonized." This harmonization of standards enables a manufacturer to have a product tested, evaluated, and listed by either organization because the standards are similar. Where there is a difference in the standard, such differences are tested and evaluated by the testing laboratory.

Copies of the *Canadian Electrical Code (CEC)* and information regarding CSA standards can be obtained by contacting:

Canadian Standards Association
178 Rexdale Blvd.
Rexdale, Ontario, Canada
M9W 1R3

PREFIX	SYMBOL	MULTIPLIER	SCIENTIFIC NOTATION (POWERS OF TEN)	VALUE
tera	T	1 000 000 000 000	10^{12}	one trillion (1 000 000 000 000/1)
giga	G	1 000 000 000	10^{9}	one billion (1 000 000 000/1)
mega	M	1 000 000	10^{6}	one million (1 000 000/1)
kilo	k	1 000	10^{3}	one thousand (1 000/1)
hecto	h	100	10^{2}	one hundred (100/1)
deka	da	10	10^{1}	ten (10/1)
unit		1	—	one (1)
deci	d	0.1	10^{-1}	one tenth (1/10)
centi	c	0.01	10^{-2}	one hundredth (1/100)
milli	m	0.001	10^{-3}	one thousandth (1/1 000)
micro	μ	0.000 001	10^{-6}	one millionth (1/1 000 000)
nano	n	0.000 000 001	10^{-9}	one billionth (1/1 000 000 000)
pico	p	0.000 000 000 001	10^{-12}	one trillionth (1/1 000 000 000 000)

Figure 1–5   Metric prefixes, symbols, multipliers, powers, and values.

METRICS AND THE *NATIONAL ELECTRICAL CODE*®

You might be surprised to learn that in addition to English measurements, the *National Electrical Code*® also shows metric measurements.

The metric system is a "base-10" or "decimal" system in that values can be easily multiplied or divided by "ten" or "powers of ten." The metric system as we know it today is known as the International System of Units (SI) derived from the French term "le Système International d'Unités." The metric system is used by all industrial nations in the world but only to a limited extent in the United States. In this country, we say "meter" (mee-tur) and we spell it "meter." Most other countries say "meter" but spell it "metre."

Metric measurements are not shown for conduit size, box size, wire size, horsepower designation for motors, and other "trade sizes" that do not reflect actual measurements.

Guide to Metric Usage

By assigning a name to a measurement, such as a *watt*, the name becomes the unit. Adding a prefix to the unit, such as *kilo*, forms the new name *kilowatt*, meaning 1,000 watts. In the metric system, units increase or decrease in multiples of 10, 100, 1,000, and so on. For example, one megawatt (1,000,000 watts) is 1 000 times greater than one kilowatt (1,000 watts). Refer to Figure 1–5 for prefixes used in the metric system.

The prefixes used most commonly are *centi, kilo,* and *milli*. Consider that the basic unit is a meter (one). Therefore, a centimeter is 0.01 meter, a kilometer is 1,000 meters, and a millimeter is 0.001 meter.

Some common measurements of length and equivalents are shown in Figure 1–6.

Figure 1–7 shows conversion factors, English to metric and metric to English.

A comprehensive metric conversion chart is included in Delmars' text *Electrical Wiring–Commercial.*

one inch	=	2.54	centimeters
	=	25.4	millimeters
	=	0.025 4	meter
one foot	=	12	inches
	=	0.304 8	meter
	=	30.48	centimeters
	=	304.8	millimeters
one yard	=	3	feet
	=	36	inches
	=	0.914 4	meter
	=	914.4	millimeters
one meter	=	100	centimeters
	=	1 000	millimeters
	=	1.093	yards
	=	3.281	feet
	=	39.370	inches

Figure 1–6   Some common measurements of length and their equivalents.

inches (in) × 0.025 4	=	meters (m)
inches (in) × 0.254	=	decimeters (dm)
inches (in) × 2.54	=	centimeters (cm)
centimeters (cm) × 0.393 7	=	inches (in)
inches (in) × 25.4	=	millimeters (mm)
millimeters (mm) × 0.039 37	=	inches (in)
feet (ft) × 0.304 8	=	meters (m)
meters (m) × 3.280 8	=	feet (ft)
square inches (in^2) × 6.452	=	square centimeters (cm^2)
square centimeters (cm^2) × 0.155	=	square inches (in^2)
square feet (ft^2) × 0.093	=	square meters (m^2)
square meters (m^2) × 10.764	=	square feet (ft^2)
square yards (yd^2) × 0.836 1	=	square meters (m^2)
square meters (m^2) × 1.196	=	square yards (yd^2)
kilometers (km) × 1 000	=	meters (m)
kilometers (km) × 0.621	=	miles (mi)
miles (mi) × 1.609	=	kilometers (km)

Figure 1–7   Useful conversions—English to metric and metric to English.

REVIEW QUESTIONS

1. Safety in the workplace is of utmost importance. Do not take chances. When working on an electrical circuit, you should:
   a) Turn off the power yourself.
   b) Leave the power on, but be careful.
   c) Depend upon your supervisor to turn the power off.

2. In a typical home, the voltage measured between a "hot" conductor and the grounded neutral conductor is _____.
   a) 120 volts
   b) 240 volts

3. In a typical home, the voltage measured between the two "hot" conductors is _____.
   a) 120 volts
   b) 240 volts

4. To reduce electrical hazards, there is a federal regulation that applies to workers doing electrical work. This regulation discusses safety in the workplace, including issues such as turning off the power, locking off switches and circuit breakers in the OFF position, tagging the switches, etc. What is this regulation?

_____

5. What electrical code is most commonly used when referring to electrical installations?

_____

6. Electrical equipment that bears the label of a recognized testing laboratory, such as Underwriters Laboratories, should always be used. This ensures that the equipment has been:

    a) tested to specific standards, then "Approved."

    b) tested to specific standards, then "Listed."

7. What Chapters of the *National Electrical Code*® cover general requirements for electrical installations?

_____

8. There are many words in the *National Electrical Code*® that are unique to the electrical industry. The meanings of these special words are found in what article of the *National Electrical Code*®?

    a) Article 100

    b) Article 210

    c) Article 250

9. What edition of the *National Electrical Code*® is enforced in your community?

_____

10. Does your community have an electrical code in addition to the *National Electrical Code*®? If "yes," what are some of the significant requirements that differ from those of the *National Electrical Code*®?

_____

_____

_____

_____

11. Does the community where you live require that a permit be taken out for electrical work?

_____

12. Does the community where you live require electricians and electrical contractors to be "licensed?"

_____

13. Electrical equipment must be installed in conformance to the requirements of the *National Electrical Code*.® Is the following statement true or false? *Section 110-3(b)* states that *"Listed or labeled equipment shall be installed and used in accordance with any instructions included in the listing or labeling."*

    a) True

    b) False

Useful Electrical Formulas

OBJECTIVES

After studying this unit, you will be able to:

- **Perform electrical calculations for determining amperes, wattage, kilowatts, power factor, and other calculations necessary for house wiring.**

- **Perform voltage drop calculations to determine proper size conductors to use.**

- **Understand multiwire circuits and how they are used in residential wiring.**

- **Understand the hazards of multiwire circuits as a result of open neutrals.**

- **Know how to calculate your electric utility bill.**

INTRODUCTION

Doing electrical wiring involves not only physical work, but requires a working knowledge of electrical theory as well. This unit covers most of the more important electrical formulas that you will need to know when doing residential wiring.

This text focuses on the *National Electrical Code.* It is not intended to be a text on electrical theory. Qualified electricians have studied and developed a working knowledge of basic electrical theory. For quick reference, this unit presents most of the electrical formulas that electricians need to know. In some cases, only the formulas are shown, with little or no explanation of the theory behind the formulas. In other cases such as voltage drop, multiwire branch circuits, and open neutrals, a minimum discussion is included to make the subject easier to understand. This unit should be considered as a review for the qualified electrician. Full technical information relative to electrical theory and electrical calculations can be found in any good electrical theory book, such as Delmar's *Illustrated Electrical Calculations,* and Delmar's *Standard Textbook of Electricity.*

Figure 2–1 shows many useful formulas for calculating watts, amperes, kilovolt-amperes, etc. House wiring does not involve three-phase or direct current, but the formulas are included for reference.

Figure 2–2 is the so-called "Pie" chart showing equations for calculating watts, amperes, ohms, and volts. These equations are based on Ohm's law, which states that in a *direct current circuit, current is directly proportional to voltage and inversely proportional to resistance.* For example, if the voltage is doubled, the current will double. If the voltage is cut in half, the current will be cut in half. If the resistance is doubled, the current will be cut in half. If the resistance is cut in half, the current will double.

To use these formulas, you need to know two of the three factors involved. For example, to find **P** (power measured in watts), select any of the three formulas shown in the outer ring of the upper left quarter (slice) of the pie. They are:

$P = I^2R$ this means: amperes $\times$ amperes $\times$ ohms

$P = EI$ this means: volts $\times$ amperes

$P = E^2/R$ this means: (volts $\times$ volts) $\div$ ohms

These equations are all right to use with resistive loads such as incandescent lamps and electric heaters. For inductive loads such as motors, transformers, and fluorescent ballasts, it is best to use the formulas found in Figure 2–1 as they take into consideration power factor and efficiency.

TO FIND	SINGLE PHASE	THREE PHASE	DIRECT CURRENT
AMPERES when kVA is known	$\dfrac{kVA \times 1000}{E}$	$\dfrac{kVA \times 1000}{E \times 1.73}$	not applicable
AMPERES when horsepower is known	$\dfrac{HP \times 746}{E \times \% \text{ eff.} \times pf}$	$\dfrac{HP \times 746}{E \times 1.73 \times \% \text{ eff.} \times pf}$	$\dfrac{HP \times 746}{E \times \% \text{ eff.}}$
AMPERES when kilowatts are known	$\dfrac{kW \times 1000}{E \times pf}$	$\dfrac{kW \times 1000}{E \times 1.73 \times pf}$	$\dfrac{kW \times 1000}{E}$
HORSEPOWER	$\dfrac{I \times E \times \% \text{ eff.} \times pf}{746}$	$\dfrac{I \times E\ 1.73 \times \% \text{ eff.} \times pf}{746}$	$\dfrac{I \times E \times \% \text{ eff.}}{746}$
KILOVOLT AMPERES	$\dfrac{I \times E}{1000}$	$\dfrac{I \times E \times 1.73}{1000}$	not applicable
KILOWATTS	$\dfrac{I \times E \times pf}{1000}$	$\dfrac{I \times E \times 1.73 \times pf}{1000}$	$\dfrac{I \times E}{1000}$
WATTS	$E \times I \times pf$	$E \times I \times 1.73 \times pf$	$E \times I$

$$\text{ENERGY EFFICIENCY} = \frac{\text{Load Horsepower} \times 746}{\text{Load Input kVA} \times 1000}$$

$$\text{POWER FACTOR (pf)} = \frac{\text{Power Consumed}}{\text{Apparent Power}} = \frac{W}{VA} = \frac{kW}{kVA} = \cos\varnothing$$

I = Amperes HP = Horsepower	E = Volts % eff. = Percent Efficiency e.g., 90% eff. is 0.90	kW = Kilowatts	kVA = Kilovolt-amperes pf = Power Factor e.g., 95% pf is 0.95

Figure 2–1 Useful electrical formulas.

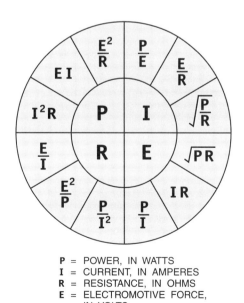

P = POWER, IN WATTS
I = CURRENT, IN AMPERES
R = RESISTANCE, IN OHMS
E = ELECTROMOTIVE FORCE,
      IN VOLTS

Figure 2–2 Equations based on Ohm's law.

WATTS, AMPERES, VOLT-AMPERES. WHAT SHOULD I USE?

Some electrical equipment is marked in watts. Some is marked in amperes, Others are marked in volt-amperes. What is the difference, and what should be used?

The whole idea behind selecting equipment such as conductors, panels, switches, controllers, etc. is to make sure that they are capable of safely carrying the current that will flow through them. Let's look at some examples.

VOLTS × AMPERES = VOLT-AMPERES

Yet, many times we say that:

VOLTS × AMPERES = WATTS

What we really mean to say is that:

VOLTS × AMPERES × POWER FACTOR
= WATTS

For a pure resistive load such as an incandescent light bulb, a toaster, a flat iron, or a resistance electric heating element, the power factor is 100%, in which case:

VOLTS × AMPERES × 1 = WATTS

When we are involved with transformers, motors, ballasts, and other "inductive" loads, wattage is not necessarily the same as volt-amperes.

Example No. 1: Calculate the wattage and volt-amperes of a 120-volt, 10-ampere resistive load.

Solutions:

1. VOLTS × AMPERES × 1 = WATTS
2. 120 × 10 × 1 = 1200 watts
3. 120 × 10 = 1200 volt-amperes

Note that watts and volt-amperes are the same.

Example No. 2: Calculate the wattage and volt-amperes of a 120-volt, 10-ampere motor load at 80% power factor.

Solutions:

1. VOLTS × AMPERES × 1 = WATTS
2. 120 × 10 × 0.8 = 960 watts
3. 120 × 10 = 1200 volt-amperes

Note that watts and volt-amperes are not the same. Using the wattage value to calculate amperes would result in a current value of 8 amperes rather than the actual ampere rating of 10 amperes.

Example No. 3: A recreation room will have six recessed fluorescent fixtures. Each fixture has two 2-lamp ballasts. In total, there are twenty four 40-watt fluorescent lamps. Two types of ballasts are being considered. Low cost, low efficiency ballasts are rated 1.70 amperes, 204 volt-amperes, 120-volts and have an efficiency of 39% (0.39). Higher cost, high efficiency ballasts are rated 0.70 amperes, 84 volt-amperes, 120-volts and have an efficiency of 95% (0.95).

Calculation using wattage of lamps only:

Calculation using the low efficiency magnetic (with core, coils & capacitors) ballasts:

I = 1.70 × 12 = 20.4 amperes

OR: Calculation using the high efficiency electronic ballasts that have the letter "E" in a circle marked on the label:

I = 0.70 × 12 = 8.4 amperes

Why such a big difference? Because of the watts loss (heat) wasted in the ballast. In the case of the low efficiency ballasts, only a small fraction of the total energy consumption is converted to light. The rest is consumed by the ballast.

Obviously, the correct choice is to use the high efficiency ballasts where only a small amount of energy is lost in the ballast. One 15-ampere branch circuit would be adequate for the load of 8.4 amperes. The low efficiency ballasts would require two branch circuits, and would result in much higher energy (bigger light bills) costs.

Watts equals volt-amperes **only** in a pure resistive circuit.

Therefore, to be sure that adequate ampacity is provided for in branch-circuit wiring, feeder sizing, and service-entrance calculations, we use **volt-amperes**. This allows us to ignore power factor, and address the true current draw that will enable us to determine the correct ratings of electrical equipment.

In some instances the terms **watts** and **volt-amperes** can be used interchangeably without creating any problems. For instance, the Code, in *Section 220-19*, recognizes that for electric ranges and other cooking equipment, the kVA and kW ratings shall be considered to be equivalent for the purpose of branch-circuit and feeder calculations.

Examples No. 2(a) and (b) in Chapter 9 of the Code indicate that for wall-mounted ovens, counter-mounted cooking units, water heaters, dishwashers, and combination clothes dryers, their kW ratings are equivalent to kVA values. Throughout this text, the terms **wattage** and **volt-amperes** are appropriately used when calculating and/or estimating loads.

VOLTAGE DROP

Watch out when the branch-circuit wiring is long! Electrical conductors have a natural, but unwanted, built-in resistance that opposes the flow of current. The result is that part of the voltage (electrical pressure) is used up when forcing the current through the conductors. This loss is measured in volts, and is referred to as *voltage drop*.

Throughout this text, the proper sizing of conductors for given loads is explained. However, when the conductors are relatively long, such as the "home run" from the main panel to the connected load, conductor sizing based only on load calculations may not be large enough. According to the *National Electrical Code®* voltage drop must also be taken into consideration. When long runs are encountered, first determine the minimum conductor size based on the connected load, then double-check the conductor size by doing a voltage drop calculation.

Low-voltage conditions can cause lamps to burn dim or "dip," some television pictures to "shrink," motors to run hot, electric heaters to not produce their rated heat output, and appliances may not operate properly. Under severe voltage drop conditions, personal computers might shut off, the result being lost data if the data had not been saved.

Low-voltage conditions in a home can be caused by:

- conductors that are too small for the load being served.
- a circuit that is too long.
- poor connections at the terminals or lugs.
- conductors operating at high temperatures such as in attics. Under these conditions, the resistance of the conductors is higher than when operating at lower temperatures.
- very long service-entrance conductors and/or service-drop conductors.
- electric utility company problems at transformer, fuse cutout, or somewhere on their lines.

Code Reference to Voltage Drop

The *Fine Print Note* to *Section 310-15* tells us that selecting conductor sizes from the allowable ampacity tables, such as the most often used *Table 310-16,* does not take voltage drop into consideration. This *FPN* refers us to *Fine Print Note No. 4* to *Section 210-19(a)* for branch circuits, and *Fine Print Note No. 2* to *Section 215-2(b)* for feeders, which tells us that the voltage drop to the farthest outlet on the circuit or feeder should not exceed 3% (Figure 2–3). Where a branch circuit and a feeder are both involved, the total voltage drop to the farthest outlet should not exceed 5% (Figure 2–4). Although *Fine Print Notes* are not mandatory as stated in *Section 90-5,* the recommended voltage drop percentages are good numbers to follow.

For most residential voltage drop calculations, a simple formula can be used. The formula shown below assumes that the load is resistive (100% power factor). Although the power factor for inductive loads such as motors and ballasts is less than 100%, the results obtained using the simple formula in most instances will be acceptable. Should the solution of a calculation seem to "sit on the fence," go to the next larger size conductor. The simple formula works reasonably well with conductors No. 1 AWG and smaller, and gets less accurate as conductor size increases. This is because the simple formula considers only the conductor's resistance.

Where more accuracy is needed, formulas should be used that include type of conductor (copper, copper-clad, or aluminum), resistance, reactance, temperature, spacing, and if the conductors are installed in metal or nonmetallic raceways. This data is found in *Tables 8* and *9* of *Chapter 9, National Electrical Code.®* *Table 8* is shown in Figure 2–5. *Table 9* is not included in this text as it is not used often in residential wiring. This subject is covered in great detail in *Electrical Wiring — Commercial.*

To find voltage drop in single-phase circuits:

$$E_d = \frac{K \times I \times L \times 2}{}$$

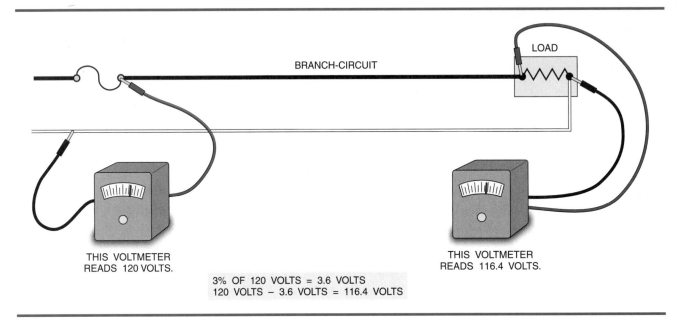

3% OF 120 VOLTS = 3.6 VOLTS
120 VOLTS − 3.6 VOLTS = 116.4 VOLTS

Figure 2–3 Maximum recommended voltage drop on a branch circuit is 3%, *Section 210-19(a), FPN No. 4.*

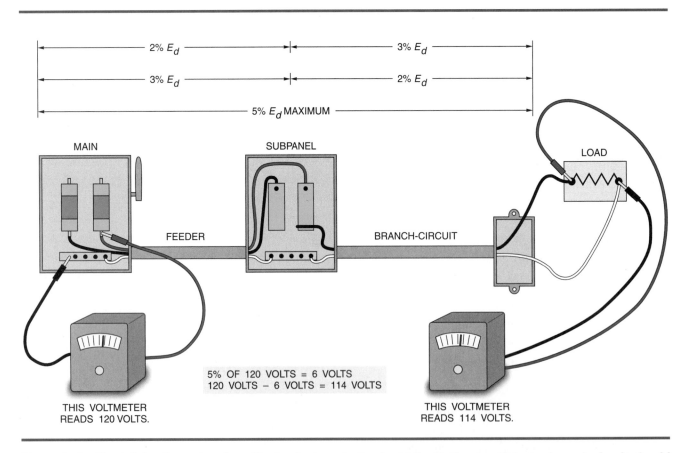

5% OF 120 VOLTS = 6 VOLTS
120 VOLTS − 6 VOLTS = 114 VOLTS

Figure 2–4 The total voltage drop from the beginning of a feeder to the farthest outlet on a branch circuit should not exceed 5%. If the voltage drop in the feeder is 3%, then do not exceed 2% voltage drop in the branch circuit. If the voltage drop in the feeder is 2%, then do not exceed 3% voltage drop in the branch circuit. See *Section 210-19(a), FPN No. 4* and *Section 215-2, FPN No. 2.*

Table 8. Conductor Properties

Size (AWG or kcmil)	Area (Circular Mils)	Conductors				Direct-Current Resistance at 75°C (167°F)		
		Stranding		Overall		Copper		Aluminum
		Quantity	Diameter (in.)	Diameter (in.)	Area (in.²)	Uncoated (ohm/1000 ft)	Coated (ohm/1000 ft)	(ohm/1000 ft)
18	1620	1	—	0.040	0.001	7.77	8.08	12.8
18	1620	7	0.015	0.046	0.002	7.95	8.45	13.1
16	2580	1	—	0.051	0.002	4.89	5.08	8.05
16	2580	7	0.019	0.058	0.003	4.99	5.29	8.21
14	4110	1	—	0.064	0.003	3.07	3.19	5.06
14	4110	7	0.024	0.073	0.004	3.14	3.26	5.17
12	6530	1	—	0.081	0.005	1.93	2.01	3.18
12	6530	7	0.030	0.092	0.006	1.98	2.05	3.25
10	10380	1	—	0.102	0.008	1.21	1.26	2.00
10	10380	7	0.038	0.116	0.011	1.24	1.29	2.04
8	16510	1	—	0.128	0.013	0.764	0.786	1.26
8	16510	7	0.049	0.146	0.017	0.778	0.809	1.28
6	26240	7	0.061	0.184	0.027	0.491	0.510	0.808
4	41740	7	0.077	0.232	0.042	0.308	0.321	0.508
3	52620	7	0.087	0.260	0.053	0.245	0.254	0.403
2	66360	7	0.097	0.292	0.067	0.194	0.201	0.319
1	83690	19	0.066	0.332	0.087	0.154	0.160	0.253
1/0	105600	19	0.074	0.372	0.109	0.122	0.127	0.201
2/0	133100	19	0.084	0.418	0.137	0.0967	0.101	0.159
3/0	167800	19	0.094	0.470	0.173	0.0766	0.0797	0.126
4/0	211600	19	0.106	0.528	0.219	0.0608	0.0626	0.100
250	—	37	0.082	0.575	0.260	0.0515	0.0535	0.0847
300	—	37	0.090	0.630	0.312	0.0429	0.0446	0.0707
350	—	37	0.097	0.681	0.364	0.0367	0.0382	0.0605
400	—	37	0.104	0.728	0.416	0.0321	0.0331	0.0529
500	—	37	0.116	0.813	0.519	0.0258	0.0265	0.0424
600	—	61	0.099	0.893	0.626	0.0214	0.0223	0.0353
700	—	61	0.107	0.964	0.730	0.0184	0.0189	0.0303
750	—	61	0.111	0.998	0.782	0.0171	0.0176	0.0282
800	—	61	0.114	1.030	0.834	0.0161	0.0166	0.0265
900	—	61	0.122	1.094	0.940	0.0143	0.0147	0.0235
1000	—	61	0.128	1.152	1.042	0.0129	0.0132	0.0212
1250	—	91	0.117	1.289	1.305	0.0103	0.0106	0.0169
1500	—	91	0.128	1.412	1.566	0.00858	0.00883	0.0141
1750	—	127	0.117	1.526	1.829	0.00735	0.00756	0.0121
2000	—	127	0.126	1.632	2.092	0.00643	0.00662	0.0106

Notes:

1. These resistance values are valid **only** for the parameters as given. Using conductors having coated strands, different stranding type, and, especially, other temperatures changes the resistance.

2. Formula for temperature change: $R_2 = R_1[1 + \alpha(T_2 - 75)]$ where: $\alpha_{cu} = 0.00323$, $\alpha_{AL} = 0.00330$

3. Conductors with compact and compressed stranding have about 9 percent and 3 percent, respectively, smaller bare conductor diameters than those shown. See Table 5A for actual compact cable dimensions.

4. The IACS conductivities used: bare copper = 100%, aluminum = 61%.

5. Class B stranding is listed as well as solid for some sizes. Its overall diameter and area is that of its circumscribing circle.

FPN: The construction information is per NEMA WC8-1992. The resistance is calculated per National Bureau of Standards Handbook 100, dated 1966, and Handbook 109, dated 1972.

Figure 2–5 The data in this table is extremely useful because it contains the information needed to do voltage drop calculations, watts loss calculations, equipment grounding conductor, grounding electrode conductor sizing, and other calculations that require circular mil area and conductor resistance values. (*Reprinted with permission from NFPA 70-1999*)

CMA

To find conductor size for single-phase circuits:

$$CMA = \frac{K \times I \times L \times 2}{E_d}$$

In the preceding formulas:

E_d = allowable voltage drop in volts

K = approximate resistance in ohms per circular-mil foot at 167°F (75°C)
- for uncoated copper wire use 12 ohms*
- for aluminum wire use 20 ohms*

I = current in amperes flowing through the conductors

L = length in feet from beginning of circuit to the load

CMA = cross-sectional area of the conductors in circular mils*

We use the factor of 2 for single-phase circuits because there is voltage drop in both conductors, to and from the connected load.

For three-phase voltage drop calculations, use a factor of 1.73 instead of 2 in the preceding formulas.

Example 1: Finding the voltage drop. What is the approximate voltage drop on a 120-volt, single-phase circuit consisting of No. 14 AWG copper conductors where the load is 11 amperes, and the length of the circuit is 85 feet? Look up the circular mil area in Figure 2–5 or in *Table 8, Chapter 9, NEC.*®

Answer:

$$E_d = \frac{K \times I \times L \times 2}{CMA}$$

$$= \frac{12 \times 11 \times 85 \times 2}{4,110}$$

$$= 5.46 \text{ volts drop}$$

The permitted voltage drop is 3% of 120 volts, which is 3.6 volts. The 5.46 volts drop exceeds the recommended maximum voltage drop of 3.6 volts, and would not be in conformance with the intent of *Section 210-19(a), Fine Print Note No. 4.* The No. 14 AWG copper conductors are too small.

Let's try it again using No. 12 AWG conductors.

$$E_d = \frac{K \times I \times L \times 2}{CMA}$$

* Values derived from Figure 2–5, or *Table 8, Chapter 9, NEC.*®

$$= \frac{12 \times 11 \times 85 \times 2}{6,530}$$

$$= 3.44 \text{ volts drop}$$

A voltage drop of 3.44 volts is less than the permitted 3.6 volts drop. Therefore, using No. 12 AWG copper conductors is all right for this example.

Example 2: Finding the voltage drop. A single-phase residential type 240-volt air-conditioner's-nameplate is marked "Minimum circuit ampacity 40 amperes." The circuit originates at the main panel located approximately 65 feet from the air conditioner unit. Two "hot" conductors are needed. A neutral is not required. What is the minimum size Type THHN copper conductor needed to keep the voltage drop to no more than 3%?

Answer: Checking *Table 310-16, NEC,*® we find conductor ampacities as follows:

WIRE SIZE	60°C	75°C	90°C
No. 10	30	35	40
No. 8	40	50	75
No. 6	55	65	75

We might quickly select No. 10 conductors from the 90°C column since that is the temperature rating of the Type THHN conductors we are using. **This is incorrect.**

In Unit 8 we will discuss conductors. We will learn that *Section 110-14(c)* requires that unless the equipment is marked otherwise, we must use the 60°C column in *Table 310-16* for circuits rated 100 amperes or less, or marked for No. 14 through No. 1 conductors. We use the 75°C column for circuits rated over 100 amperes or marked for conductors larger than No. 1. In our example, there is no mention that the equipment is marked with a temperature rating. Therefore, we must use the 60°C column in conformance with *Section 110-14(c)* even though the Type THHN conductors we are using are rated for 90°C.

The correct size conductor is No. 8.

This is why many electricians and electrical contractors make it a habit to install 90°C insulated conductors, and read the ampacity from the 60°C column of *Table 310-16.*

The permitted voltage drop on this circuit is:

$$E_d = 240 \times 0.03 = 7.2 \text{ volts}$$

Then:

$$E_d = \frac{K \times I \times L \times 2}{CMA}$$
$$= \frac{12 \times 40 \times 65 \times 2}{16,510}$$
$$= 3.78 \text{ volts drop}$$

This is well below the permissible 7.2 volts drop. The No. 8 copper conductors are acceptable.

Example 3: Finding the conductor size. Find the minimum size copper conductor needed to supply a submersible water pump motor load. The pump is marked 120 volts, 16 amperes, single-phase. The circuit is approximately 150-feet long. Keep the voltage drop to not more than 3%.

Answer: The voltage drop is not to exceed: $120 \times 0.03 = 3.6$ volts.

$$CMA = \frac{K \times I \times L \times 2}{E_d}$$
$$= \frac{12 \times 16 \times 150 \times 2}{3.6}$$
$$= 16,000 \text{ circular mils}$$

Checking Figure 2–5, which is *Table 8, Chapter 9* of the *NEC®* we find that a No. 10 has a circular mil area of 10,380 circular mils. A No. 8 has a circular mil area of 16,510 circular mils. Therefore, No. 8 conductors would be the proper size to use. These No. 8 conductors could be rated 60, 75, or 90°C as listed in *Table 310-16*.

BRANCH-CIRCUIT WIRING IN A HOME

The *NEC®* in *Article 100* defines a *branch circuit* as *"the circuit conductors between the final overcurrent device protection the circuit and the outlet(s)."*

The electrical service in a home is 120/240 volt, 3-wire, single-phase.

Typically, 15- and 20-ampere branch circuits start out at the main service panel and run to some point in the house, such as to a ceiling or wall electrical outlet or device box. From this point, the branch circuit spreads out to feed other lighting outlets or receptacle outlets. Figure 2–6 illustrates a 2-wire branch circuit supplying a hypothetical 10- ampere load.

Figure 2–7 illustrates two 2-wire branch circuits, each supplying a hypothetical 10-ampere load.

MULTIWIRE BRANCH CIRCUITS

The *NEC®* recognizes multiwire branch circuits. A *multiwire branch circuit* is defined as a *"branch circuit consisting of two or more ungrounded conductors having a potential difference between them, and a grounded conductor having equal potential difference between it and each ungrounded conductor of the circuit and that is connected to the neutral or grounded conductor of the system."* Figure 2–8 shows how two individual 2-wire branch circuits can be replaced with one 3-wire multiwire branch circuit. Note that both branch circuits share a common neutral.

Why use a multiwire branch circuit? Here are a couple of reasons. There are savings in material and

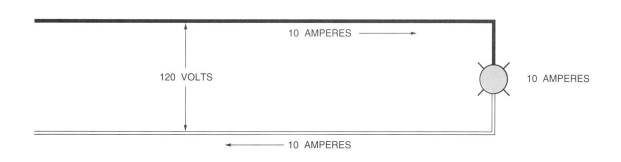

Figure 2–6 A 120-volt, 2–wire branch circuit. Note that the current flowing through the black conductor is the same as the current returning through the white conductor.

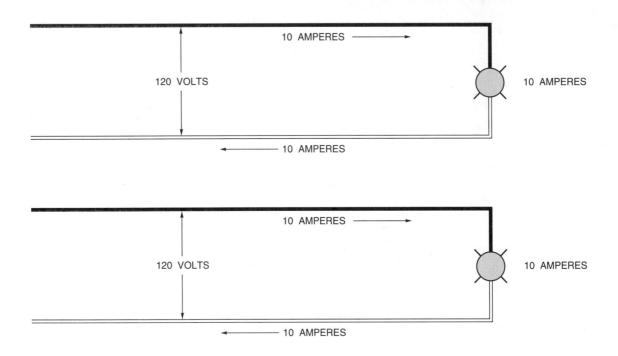

Figure 2–7   Two separate 120-volt, 2–wire branch circuits. Each branch circuit consists of its own black and white conductors.

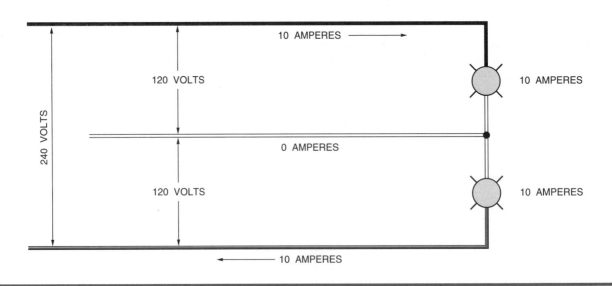

Figure 2–8   A 120/240-volt, 3-wire multiwire branch circuit. The black conductor of one branch circuit and the red conductor of the other branch circuit share the white neutral conductor. The white neutral conductor is *common* to both branch circuits. The white neutral conductor carries the unbalanced current, which is the difference between the current flowing in the black conductor and the current flowing in the red conductor. The unbalanced current is 10 - 10 = zero ampere. A 120/240-volt, 3-wire multiwire branch circuit does the same job as the two separate 120-volt, 2–wire branch circuits in Figure 2–7, but with less total watts loss in the conductors, as discussed in the text.

labor by installing one 3-wire cable instead of two 2-wire cables. For the same size and type, 3-wire nonmetallic-sheathed cable costs roughly 1.7 times more than a 2-wire nonmetallic-sheathed cable. An electrical contractor's recent **actual** estimate of a 100-foot "home run" of one No. 14/3 nonmetallic-sheathed cable from a panel to a junction box came out to be $82.00. For two runs of No. 14/2 nonmetallic-sheathed cable, the result was $124.00. These **total installed cost** estimates included material, labor, overhead, taxes, benefits, inflation, adjustment for wear and tear of tools, and profit.

Other advantages are savings in power consumption and less voltage drop. Because the grounded neutral conductor carries the unbalanced current of the two "hot" phase conductors, there is less power lost (watts loss) in the conductors. This translates into energy savings. Watts loss and voltage drop calculations are discussed further on.

By now you might be saying to yourself *"Yah, I know all about 3-wire circuits because that's exactly what the electric utility company used for their incoming power lines to my house, and that is how*

the service-entrance conductors and main panel are connected in my house." You are absolutely right!

Warning! When installing a 3-wire multiwire branch circuit, be absolutely sure that the "hot" conductors are connected to opposite phases in the panel. The grounded neutral conductor carries the unbalanced current between the two "hot" conductors. This current is the difference between the current in the black "hot" wire and the red "hot" wire. For example, if one load connected to the black wire is 12 amperes and the other load connected to the red wire is 10 amperes, the neutral current in the white wire is the difference between these loads—2 amperes. This is clearly illustrated in Figure 2–9.

Let us see what happens if the black and red conductors of this 3-wire multiwire branch circuit are improperly connected to the same phase in the panel (Figure 2–10). Note that because the black and red conductors are connected to the same phase, the white **grounded** conductor is no longer carrying the unbalanced current of 2 amperes. It is no longer a neutral conductor and will now carry the total return current of both branch circuits, which is

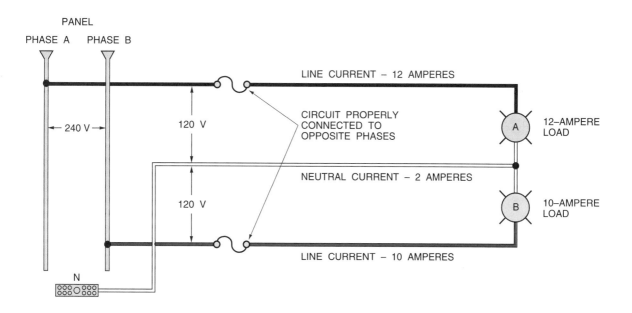

Figure 2–9   The correct wiring for a 120/240-volt, 3-wire multiwire branch circuit. Note that the black and red conductors are connected to the *opposite* phases in the panel. The black and red conductors share a *common* neutral. The white neutral conductor carries the unbalanced current. The unbalanced current is 12 − 10 = 2 amperes. A 120/240-volt, 3-wire multiwire branch circuit does the same job as the two separate 120-volt, 2-wire branch circuits in Figure 2–7, but with less total watts loss in the conductors as discussed in the text.

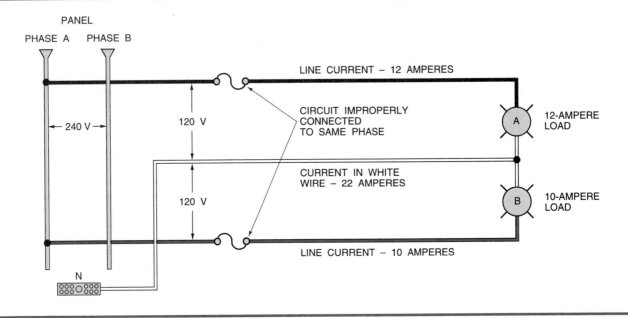

Figure 2–10 An improperly connected 120/240-volt, 3-wire multiwire branch circuit. Note that the black and red conductors are connected to the *same* phases in the panel. This results in the white wire having to carry the total return current from both phase conductors, which is 12 + 10 = 22 amperes. In this diagram, the white wire is greatly overloaded and will run hot. Over time, the insulation on this overloaded white conductor will be destroyed.

12 + 10 = 22 amperes. As a result, the white **grounded** conductor will be severely overloaded and will run hot. Over time, the insulation will be destroyed, and this could lead to possible fire and loss of life. All single-phase, 120/240-volt panels are clearly marked to help prevent an error when making the connections.

See Unit 4 for details relating to connecting receptacles when multiwire branch circuits are installed.

WATTS LOSS AND VOLTAGE DROP COMPARISONS

Let us take a look at the energy losses in the conductors and do a voltage drop calculation for all of the above diagrams. Assume that all of the conductors are No. 14 AWG copper, and the length of the circuit is 50 feet, supply to load. In the following examples, use the formula $E = I \times R$ for the voltage drop calculations, and $W = I^2R$ for the watts loss calculations.

In Figure 2–6, the watts loss in each current-carrying conductor is:

$$\text{Watts} = I^2R = 10 \times 10 \times 0.154 = 15.4 \text{ watts}$$

The watts loss in both conductors is:

$$15.4 + 15.4 = 30.8 \text{ watts}$$

The resistance value of 0.154 ohms is derived from Figure 2–5. The resistance of 1,000 feet of a No. 14 AWG copper conductor is 3.07 ohms. The resistance of a conductor is directly proportional to its length. Therefore, the resistance of 50 feet is:

$$\frac{3.07}{1,000} \times 50 = 0.154 \text{ ohms}$$

The voltage drop in each conductor is:

$$\begin{aligned} E_d &= IR \\ &= 10 \times 0.154 \\ &= 1.54 \text{ volts} \end{aligned}$$

The voltage drop for both conductors is:

$$2 \times 1.54 = 3.08 \text{ volts}$$

In Figure 2–7 we have four conductors, each with a watts loss of 15.4 watts.

Therefore, we have a total watts loss of:

$$15.4 \times 4 = 61.6 \text{ watts}$$

The voltage drop in each of the 2-wire branch circuit is 3.08 volts.

In Figure 2–8 we have a 3-wire multiwire branch circuit.

The watts loss in the black conductor is 15.4 watts.

The watts loss in the red conductor is also 15.4 watts.

There is no watts loss in the white neutral conductor since it is carrying no unbalanced current. The loads are perfectly balanced.

The total watts loss is:

$$15.4 + 15.4 = 30.8 \text{ watts}$$

The voltage drop across the black conductor is:

$$10 \times 0.154 = 1.54 \text{ volts}$$

The voltage drop across the red conductor is:

$$10 \times 0.154 = 1.54 \text{ volts}$$

There is no voltage drop across the white grounded neutral conductor.

In Figure 2–9 we have a 3-wire multiwire branch circuit but this time the loading is unbalanced.

The watts loss in the black conductor is:

$$12 \times 12 \times 0.154 = 22.2 \text{ watts}$$

The watts loss in the red conductor is:

$$10 \times 10 \times 0.154 = 15.4 \text{ watts}$$

The watts loss in the white conductor is:

$$2 \times 2 \times 0.154 = 0.6 \text{ watts}$$

The total watts loss is 38.2 watts.

The voltage drop across the black conductor is:

$$12 \times 0.154 = 1.85 \text{ volts}$$

The voltage drop across the red conductor is:

$$10 \times 0.154 = 1.54 \text{ volts}$$

The voltage drop across the white grounded neutral conductor is:

$$2 \times 0.154 = 0.308 \text{ volts}$$

In Figure 2–10, we have an improperly connected 3-wire multiwire branch circuit.

The watts loss in the black conductor is:

$$12 \times 12 \times 0.154 = 22.2 \text{ watts}$$

The watts loss in the red conductor is:

$$10 \times 10 \times 0.154 = 15.4 \text{ watts}$$

The watts loss in the white conductor is:

$$22 \times 22 \times 0.154 = 74.5 \text{ watts}$$

The total watts loss is 112.1 watts.

The voltage drop across the black conductor is:

$$12 \times 0.154 = 1.85 \text{ volts}$$

The voltage drop across the red conductor is:

$$10 \times 0.154 = 1.54 \text{ volts}$$

The voltage drop across the white grounded neutral conductor is:

$$22 \times 0.154 = 3.39 \text{ volts}$$

OPEN NEUTRAL

When the white grounded conductor of a 2-wire circuit opens, for whatever reason, the power to the connected load goes off. The danger with this situation is that part of the circuit is still "live." The branch circuit breaker will not trip on an open circuit. Someone might think that the circuit has been turned off, and this presents a real hazard. Touching any grounded surface and the "hot" conductor at the same time will result in an electrical shock. Also, when the load is still connected, touching any grounded surface and the white grounded conductor at the same time anywhere downstream from the open will result in an electrical shock. Why? Because you are putting yourself in series with the circuit, and line voltage appears across an open!

An open neutral in a 3-wire multiwire branch circuit is not good for additional reasons! If an open neutral occurs on a 3-wire branch circuit, some of the connected load can burn out if operated at voltages higher than their rated voltage. Motors operated at low voltage will run hot and may burn out. If the voltage is so low that the motor cannot start, it is possible that the motor might "hum." Most motors are able to operate at 10% of their rated voltage.

Figure 2–11 shows a properly connected 3-wire multiwire branch circuit, but with an open neutral. With the open neutral, Loads A and B are now connected in series. Load A "sees" 109.1 volts. Load B "sees" 130.9 volts.

An open neutral is easily noticed in a home because just as soon as the neutral opens, some incandescent lamps will burn brightly and some will burn dim. If you turn on or off some of the lamps, the dimness or brightness of the other lamps will change. How dim or how bright the lamps burn depends upon what lamps were burning at the time the open neutral situation occurred. Clocks, television sets, videocassette recorders (VCRs), stereo

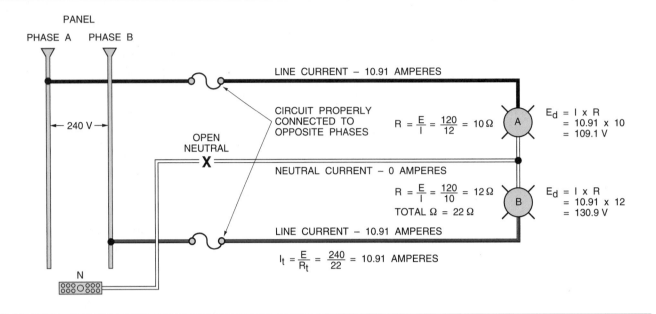

Figure 2–11 Example of an open neutral situation. With the open neutral, Loads **A** and **B** are now connected in series. Load **A** is subjected to 109.1 volts. Load **B** is subjected to 130.9 volts. This condition is easily recognized in a home because some incandescent lamps will be burning dim and others will be burning very bright. Some sensitive electronic equipment might immediately burn out. The "dimness" or "brightness," and the voltage drop across the different loads depends upon what lamps and/or other loads were on at the time the neutral opened.

equipment, computers, and other sensitive electronic equipment may instantly burn out when an open neutral takes place.

An open neutral can occur in the branch circuit wiring, at the main electrical panel, in the electric meter base outside the home, in the electric utilities incoming power lines (overhead or underground), or at the electric utilities transformer connections.

Let us look at a real situation. An example of what can occur should the neutral of a 120/240-volt, 3-wire multiwire branch circuit open is shown in Figure 2–12. Trace the flow of current from Phase A through the television set, through the toaster, then back to Phase B, thus completing the circuit. The following simple calculations show why the television set, stereo, home computer, or other sensitive electronic equipment can be expected to burn up, and why the toaster will not toast when operating at 22.9 volts.

$$R_t = 8.45 + 80 = 88.45 \text{ ohms}$$

$$I = \frac{E}{R} = \frac{240}{88.45} = 2.71 \text{ amperes}$$

The voltage that the toaster "sees":

$$IR = 2.71 \times 8.45 = 22.9 \text{ volts}$$

The voltage that the television "sees":

$$IR = 2.71 \times 80 = 216.8 \text{ volts}$$

As mentioned earlier, an open neutral situation can arise when the neutral of the utility company's incoming service-entrance conductors (underground or overhead) opens. However, the problem is minimized because the neutral of the service is solidly grounded to the metal water piping system within the building and is supplemented in most cases by a driven ground rod. This stabilizes the voltage between the "hot" phase conductors and the grounded neutral conductor, and between the "hot" phase conductors and ground. That is why proper grounding and bonding of the main service is so important. See Unit 12 for more information on the purpose of grounding.

EFFECT OF VOLTAGE VARIATION

As previously discussed, long runs can result in excessive voltage drop. Low voltage can cause all sorts of problems, like lights to burn dim, heaters to not heat as much as they should, and some TV pictures to "shrink."

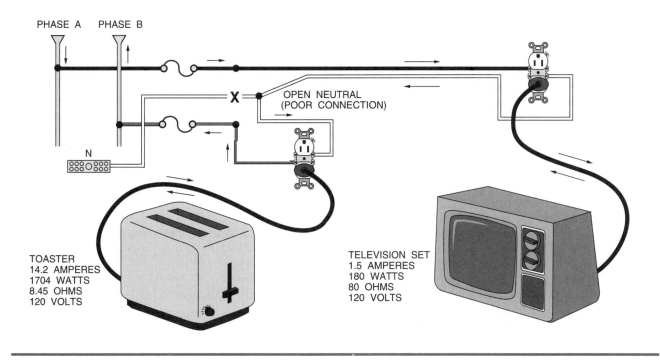

PHASE A   PHASE B

X   OPEN NEUTRAL
(POOR CONNECTION)

N

TOASTER
14.2 AMPERES
1704 WATTS
8.45 OHMS
120 VOLTS

TELEVISION SET
1.5 AMPERES
180 WATTS
80 OHMS
120 VOLTS

Figure 2–12   Diagram of a properly connected 3-wire multiwire branch circuit. A television set is plugged into a receptacle fed by one circuit. A toaster is plugged into a receptacle fed by another circuit. The neutral opens. The toaster is subjected to 22.9 volts. It will not toast your bread. The television set is subjected to 216.8 volts. It will burn out.

Here are some simple formulas to use to determine changes in a circuit due to voltage variation. Consider a 3,000-watt, 240-volt resistive heating element in an electrical water heater.

At rated voltage, the resistance is:

$$R = \frac{E \times E}{W} = \frac{240 \times 240}{3,000} = 19.2 \text{ ohms}$$

At rated voltage, the current is:

$$I = \frac{E}{R} = \frac{240}{19.2} = 12.5 \text{ amperes}$$

At 220 volts, the wattage is:

$$W = \frac{E \times E}{R} = \frac{220 \times 220}{19.2} = 2,521 \text{ watts}$$

At 220 volts, the current is:

$$I = \frac{W}{E} = \frac{2,521}{220} = 11.46 \text{ amperes}$$

Another way to calculate the effect of voltage variation is:

$$\text{Correction Factor} = \frac{\text{Applied Voltage Squared}}{\text{Rated Voltage Squared}}$$

Using the previous example:

$$\text{Correction Factor} = \frac{220 \times 220}{240 \times 240} = 0.8403$$
(0.84 is close enough)

Therefore, 3000 watts × 0.84 = 2520 watts

The above formulas only work for resistive circuits. The formulas for inductive circuits such as an electric motor are much more involved than this book is intended to cover. For typical electric motors, the following information will suffice.

VOLTAGE VARIATION	FULL-LOAD CURRENT	STARTING CURRENT
110%	7% Decrease	10–12% Increase
90%	11% Increase	10–12% Decrease

Electric motors are designed to operate at ±10% of their nameplate voltage. This is a NEMA standard.

WATT HOUR METERS

Electrical energy used in a home is measured by the watt-hour meter on the outside of the house. Typically, the electrician installs all of the service-entrance equipment except the actual watt-hour meter, which is furnished and installed by the electrical utility. The watt-hour meter is sometimes referred to as the "utility's cash register."

Figure 2–13 shows typical single-phase watt-hour meters. The first is a conventional five-dial meter. The second is a five-dial meter where one set of dials registers total kilowatt-hours and the second set of dials registers premium time kilowatt-hours. The third is a programmable electronic digital display watt-hour meter that can register kilowatt-hour consumption for up to four different rate schedules. Utilities that offer special "time-of-use" rates install meters of the second and third types.

Before installing the main electrical service to a residence, check with your electric utility to find out what rate schedules are available to homeowners. Incentives in the form of lower rates for electricity used during specified hours or for specific appliances can be very cost effective. Special rates for electric heating, electric water heaters, air conditioners, and heat pumps might be available. "Time-of-use" rate schedules are becoming more and more popular across the country.

Reading a Watt-hour Meter

A digital watt-hour meter is easy to read because the actual numbers are shown.

Figure 2–14 shows the dials of a conventional meter. This type of meter is read as follows:

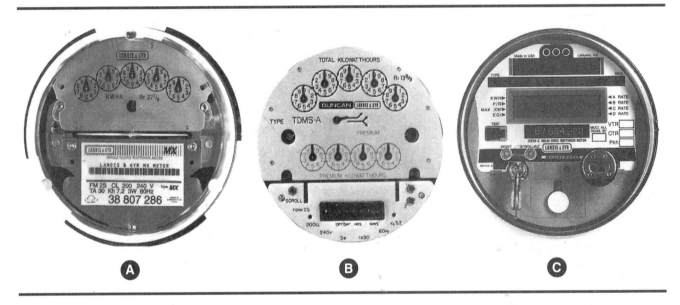

Figure 2–13 Photo A shows a conventional single-phase watt-hour meter. Photo B shows a single-phase watt-hour meter that has two sets of dials. One set registers the total kilowatt-hours. The second set registers premium time kilowatt-hours. Photo C shows a programmable electronic digital display watt-hour meter that is capable of registering four different "Time-of-Use" rates. (*Courtesy of Landis & Gyr Energy Management, Inc.*)

Figure 2–14 The reading of this five-dial watt-hour meter is 18,672 kilowatt-hours.

Figure 2–15 One month later, the watt-hour meter reads 18,975 kilowatt-hours, indicating that 303 kilowatt-hours of electricity were used during the month.

Starting with the first dial, record the last number the pointer has passed. Continue doing this until the readings of all five dials are obtained. The reading on the five-dial meter in Figure 2–14 is 18,672 kilowatt-hours.

If the meter reads 18,975 one month later as shown in Figure 2–15, subtract the previous reading of 18,672 from 18,975. The customer will be billed for 303 kilowatt-hours of electricity used for that month.

THE COST OF USING ELECTRICAL ENERGY

The kilowatt (kW) is a convenient unit of electrical power. One thousand watts (W) is equal to one kilowatt. The watt-hour meter measures and records both wattage and time.

For residential metering, most utilities have rate schedules based upon "cents per kilowatt-hour." Stated another way—how much wattage is being used and for how long?

Burning a 100-watt light bulb for 10 hours is the same as using a 1,000-watt electric heater for 1 hour. Both equal one kilowatt-hour (kWh).

$$\text{kWh} = \frac{\text{watts} \times \text{hours}}{1,000} = \frac{100 \times 10}{1,000} = 1\text{kWh}$$

$$\text{kWh} = \frac{\text{watts} \times \text{hours}}{1,000} = \frac{1000 \times 1}{1,000} = 1\text{kWh}$$

In the preceding examples, if the electric rate is 8 cents per kilowatt-hour, the cost to operate the 100-watt light bulb for 10 hours and the cost to operate the electric heater for 1 hour are the same—8 cents. Both loads use one kilowatt-hour of electricity.

Calculating the Cost of Using Electricity

The cost of electrical energy used can be calculated as follows:

$$\text{Cost} = \frac{\text{watts} \times \text{hours used} \times \text{cost per kWh}}{1,000}$$

Example: Find the cost of operating a color television set for 8 hours. The label on the back of the television set indicates 175 watts. The electric rate is $0.10494 per kilowatt-hour.

$$\text{Cost} = \frac{175 \times 8 \times \$0.10494}{1,000} = \$0.1469$$
(approximately 15 cents)

Example: Find the approximate cost of operating a central air conditioner per day, that on average runs 50% of the time during a 24-hour period on a typical hot summer day. The unit's nameplate is marked 240 volts, single-phase, 23 amperes. The electric rate is $.09 cents per kilowatt-hour. The steps are:

1. The time the air conditioner operates each day is: $24 \times 0.50 = 12$ hours.

2. Convert the nameplate data to use in the calculations:

Watts = volts × amperes = $240 \times 23 = 5,520$ watts

3. $\text{Cost} = \dfrac{\text{watts} \times \text{hours used} \times \text{cost per kWh}}{1,000}$

$$= \frac{5,520 \times 12 \times \$0.09}{1,000} = \$5.96$$

Note: The answers to the preceding examples are approximate because they were based on watts. Power factor and efficiency factors were not included as they generally are unknown when coming up with rough estimates of an electric bill. For all practical purposes, the answers are acceptable.

Here is an example of what a typical monthly electric bill might look like (see Figure 2–16).

GENERIC ELECTRIC COMPANY

Anyplace, USA

Days of Service:	From: 02-01-99	To: 02-28-99	Due Date: 03-25-99

Present reading	84,980
Previous reading	83,655
Kilowatt-hours used	1,325
Rate/kWh: 1st 400 kWh @ $0.10494	$41.98
remaining 925 kWh @ $0.06168	57.05
Energy charge	99.03
Basic service charge (single-dwelling)	8.91
State tax	4.24
Total current charges	$112.18
Total amount due by 03-25-99	$112.18
Total amount due after 03-25-99	$113.30

Figure 2–16 An example of a typical monthly electric bill.

Some utilities increase their rates during the hot summer months when the air-conditioning load is high. Other utilities provide a second meter for specific loads such as electric water heaters, air-conditioners, heat pumps, or total electric heat. Other utilities use electronic watt-hour meters that have the capability of registering kilowatt-hour consumption during specific "time-of-use". These electronic watt-hour meters might have up to four different "time-of-use" periods, each period having a different "cents per kilowatt-hour" rate.

Other charges that might appear on a "light bill" might be a fuel adjustment charge based on a "per kWh" basis. Such charges enable a utility to recover from the consumer extra expenses it might incur for fuel costs used in generating electricity. Fuel charges can vary with each monthly bill without the utility having to apply to the regulatory agency for a rate change.

REVIEW QUESTIONS

Refer to Figures 2–1 and 2–2 to solve the following problems. show your work.

1. A toaster is marked 1,000 watts, 120 volts. How much current does the toaster draw?

2. Calculate the resistance (in ohms) of the toaster in question 1.

3. An electric baseboard heater is marked 1,500 watts and 6.25 amperes. The voltage rating on the nameplate is scratched so badly that it is illegible. With the information given, determine the heater's voltage rating.

4. A two-lamp fluorescent ballast is marked 0.90 amperes at 120 volts. The fluorescent lamps are marked 40 watts. What is the power factor of this ballast?

5. The nameplate on a pad-mount transformer located at the rear lot line of a house provides the following data: 240/120 volts, single-phase, 100 kVA. What is the transformer's 240-volt full-load current rating?

6. A tool shed is located approximately 100 feet from the house. A two-wire, 20-ampere, 120-volt branch circuit is run underground from the main electrical panel in the house to the tool shed. The branch circuit is intended to supply two 100-watt incandescent lamps and a 120-volt receptacle that will be used to plug in portable tools. The largest portable tool is a ¼-horsepower, 115-volt, 5.8-ampere table saw. There will also be a 1,000-watt, 120-volt electric heater. Under worst case conditions with everything turned on:

    a) Calculate the voltage drop using No. 12 AWG copper conductors.

    b) Calculate the voltage drop using No. 10 AWG copper conductors.

    c) Calculate the voltage drop using No. 8 AWG copper conductors.

7. Watts loss in a conductor results in heat in the conductor. This is wasted power. A 12-ampere load is located 100 feet from the main electric panel. The conductors are Type THHN. Referring to Figure 2–5 (*Table 8, NEC®*), using the DC resistance values at 75°C for copper conductors, calculate the approximate watts loss in this branch circuit:

    a) Using No. 14 AWG *solid* conductors.

    b) Using No. 12 AWG *solid* conductors.

    c) Using No. 10 AWG *stranded* conductors.

8. Multiwire branch circuits can be advantageous in certain situations for house wiring. For example, a 3-wire nonmetallic-sheathed cable could be run from the main panel to an outlet box located a considerable distance from the main panel. In this outlet box, the multiwire circuit can be separated into two 2-wire branch circuits. When a multiwire branch circuit is used, be sure that:

   a) the ungrounded "hot" conductors are connected to the same phase in the main panel.

   b) the ungrounded "hot" conductors are connected to opposite phases in the main panel.

   c) it does not make any difference to what phases the "hot" conductors are connected.

9. One of the "negatives" of using multiwire branch circuits is what happens when the neutral opens. Let us assume that a 3-wire multiwire branch circuit has been installed. The system is a 120/240-volt system as found in a typical home. For some reason, the neutral opens. At the time the neutral opens, a 1,400-watt toaster-oven was operating and was connected to a branch circuit supplied by Phase A. A personal computer (4 amperes), monitor (2 amperes), and printer (0.5 amperes) were operating and were connected to Phase B. Calculate the voltage across the toaster-oven and across the computer, monitor, and printer after the neutral opened. To keep the calculations simple, assume all loads to be resistive.

10. Most electric motors are designed to operate at _____ of their rated voltage.

   a) ±5%

   b) ±10%

   c) ±15%

11. Two 750-watt, 120-volt resistance heating cables were laid in the rain gutters to keep ice from accumulating in the gutters. Snow and sleet had been falling for two days. The heating elements had been turned on for 48 hours. How much did it cost to operate these cables continuously for this 48-hour period of time? The electric rate is $0.08 per kilowatt-hour.

Electrical Specifications, Plans, and Symbols

OBJECTIVES

After studying this unit, you will be able to:

- Understand the meaning and importance of "Plans and Specifications."
- Understand the symbols used on residential blueprints (plans) to represent electrical outlets, switches, and special purpose outlets.
- Learn about notations that are found on plans.
- Learn about symbols that represent nonelectrical items on architectural plans.

INTRODUCTION

"A picture tells a million words." Whether you are working with a complete and detailed set of structural plans, electrical plans, and specifications, or are working with hand draw sketches, most of the items can be represented by symbols. Instead of writing detailed words for an item, a symbol that represents the item is used. This unit covers virtually all of the symbols found on today's electrical, architectural, plumbing, and heating plans.

SPECIFICATIONS, PLANS, AND SYMBOLS

When wiring a new home, the electrician will usually have two documents to work with. One is the **Specifications** and the other the **Plans**.

Specifications are generally in book form and may consist of a few pages to many pages, depending on the magnitude of the project. Specifications contain information detailing who is responsible for specific items, general conditions, schedule of drawings, etc. It is a good idea to read the entire specifications first, then the electrical plans, then the nonelectrical plans. Doing this will keep conflicts between the various building trades to a minimum. It is amazing how much information can be found on the nonelectrical plans that are of particular interest to the electrician. When there is a difference between the plans and specifications, the specifications take preference. Specifications are discussed in much greater detail in the *Electrical Wiring— Commercial* book.

Construction plans are still referred to as **blueprints**, a term carried over from many years ago when the plans were blue with white lines. Today, plans are white paper with black lines. Many are computer-generated as opposed to drafting by hand. Electrical symbols are used on electrical plans to show the location and type of electrical device, fixture, or appliance to be installed at that particular location. Most architects, consulting engineers, electricians, contractors, and home designers use the American National Standards Institute (ANSI) symbols. Symbols might be called "a secret code," the "short hand" of the industry, and sometimes they are called "downright confusing." Regardless, all building plans and electrical plans contain many, many symbols, so it is necessary for anyone working "with the tools" to become familiar with electrical and other architectural symbols.

Figure 3–1 is a portion of a typical electrical plan showing a ceiling fixture, two 3-way switches, and two wall receptacles. When a "dashed" line connects a switch or switches and an outlet, it indicates that that particular outlet is controlled by the switch or switches. The "dashed" line is curved so it will not be confused with building construction lines. Outlets shown without curved "dashed" lines are independent outlets and have no switch control.

Figure 3–2 shows symbols commonly used to indicate wall and ceiling lighting outlets.

Figure 3–3 shows symbols commonly used to indicate conventional receptacle outlets, as well as most other types of outlets,

Figure 3–4 shows symbols commonly used to indicate various types of switches.

Figure 3–5 shows symbols commonly used to indicate circuiting, conduit and cable runs, number of conductors in a given raceway or cable, and other details found on a good set of electrical plans.

Figure 3–6 shows many miscellaneous symbols commonly found on electrical plans.

Occasionally plans may contain symbols that are not found in the ANSI standards. Whenever a "nonstandard" symbol appears on a plan, check the plan for a legend or notation that explains the meaning of the symbol. Notations are quite often used with a specific symbol to call attention for such things as a variation, type, size, quantity, or to make reference to a schedule that might be common to a number of items on the plans. Lighting fixtures are a good example of this.

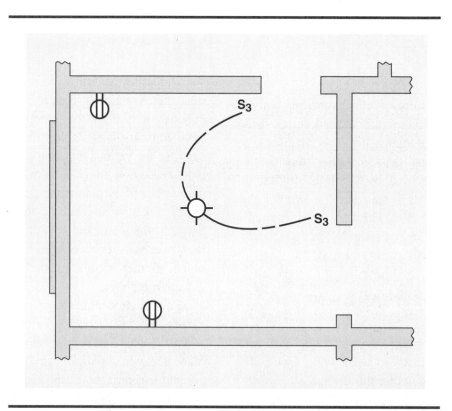

Figure 3–1   A portion of a typical electrical plan. The center ceiling light is controlled by the two 3-way switches as indicated by the curved dashed line. The wall receptacle outlets are not controlled by any switches. They are on at all times.

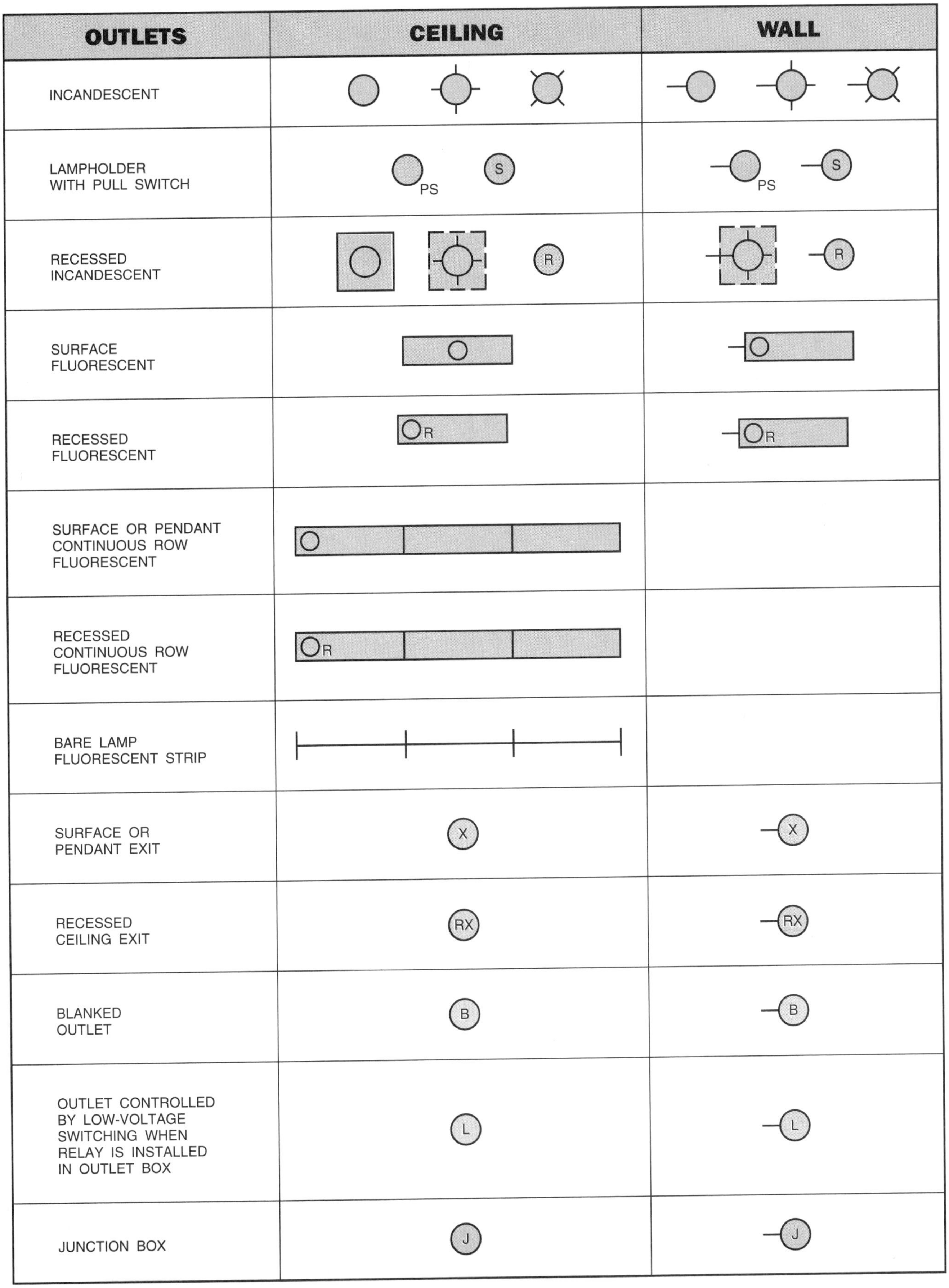

OUTLETS	CEILING	WALL
INCANDESCENT		
LAMPHOLDER WITH PULL SWITCH	PS S	PS S
RECESSED INCANDESCENT	R	R
SURFACE FLUORESCENT		
RECESSED FLUORESCENT	R	R
SURFACE OR PENDANT CONTINUOUS ROW FLUORESCENT		
RECESSED CONTINUOUS ROW FLUORESCENT	R	
BARE LAMP FLUORESCENT STRIP		
SURFACE OR PENDANT EXIT	X	X
RECESSED CEILING EXIT	RX	RX
BLANKED OUTLET	B	B
OUTLET CONTROLLED BY LOW-VOLTAGE SWITCHING WHEN RELAY IS INSTALLED IN OUTLET BOX	L	L
JUNCTION BOX	J	J

Figure 3–2 Lighting outlet symbols.

RECEPTACLE OUTLETS

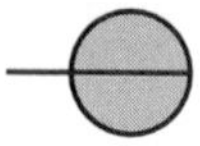 SINGLE RECEPTACLE OUTLET	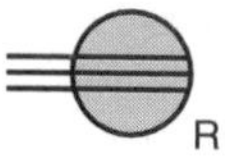 RANGE OUTLET
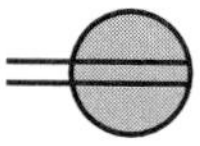 DUPLEX RECEPTACLE OUTLET	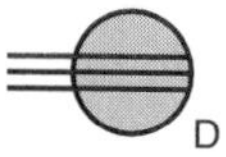 CLOTHES DRYER OUTLET
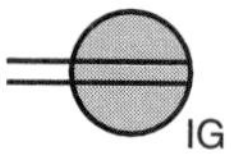 INSULATED (ISOLATED) GROUND RECEPTACLE OUTLET	 FAN OUTLET
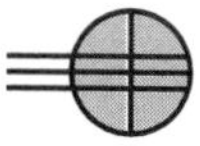 TRIPLEX RECEPTACLE OUTLET	 CLOCK OUTLET
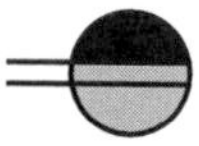 DUPLEX RECEPTACLE OUTLET, SPLIT-CIRCUIT	 FLOOR OUTLET
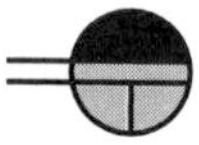 TRIPLEX RECEPTACLE OUTLET, SPLIT-CIRCUIT	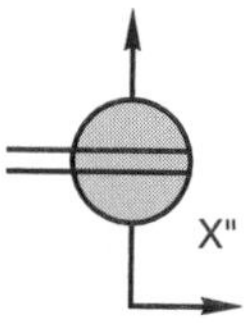 MULTIOUTLET ASSEMBLY; ARROW SHOWS LIMIT OF INSTALLATION. APPROPRIATE SYMBOL INDICATES TYPE OF OUTLET. SPACING OF OUTLETS INDICATED BY "X" INCHES.
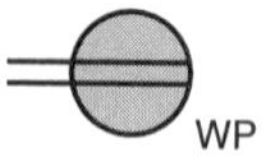 WEATHERPROOF RECEPTACLE OUTLET	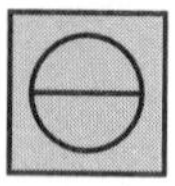 FLOOR SINGLE RECEPTACLE OUTLET
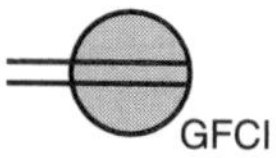 GROUND-FAULT CIRCUIT INTERRUPTER RECEPTACLE OUTLET	 FLOOR DUPLEX RECEPTACLE OUTLET
 SPECIAL-PURPOSE OUTLET (SUBSCRIPT LETTERS INDICATE SPECIAL VARIATIONS: DW = DISHWASHER. ALSO a, b, c, d, ETC. ARE LETTERS KEYED TO EXPLANATION ON DRAWINGS OR IN SPECIFICATIONS).	 FLOOR SPECIAL-PURPOSE OUTLET

Figure 3–3 Receptacle outlet symbols.

SWITCH SYMBOLS

S   SINGLE-POLE SWITCH

S_2 DOUBLE-POLE SWITCH

S_3 THREE-WAY SWITCH

S_4 FOUR-WAY SWITCH

S_D DOOR SWITCH

S_{DS} DIMMER SWITCH

S_K KEY SWITCH

S_L LOW-VOLTAGE SWITCH

S_{LM} LOW-VOLTAGE MASTER SWITCH

S_P SWITCH WITH PILOT LAMP

S_R VARIABLE-SPEED SWITCH

S_T TIME SWITCH

S_{WP} WEATHERPROOF SWITCH

Figure 3–4   Switch symbols.

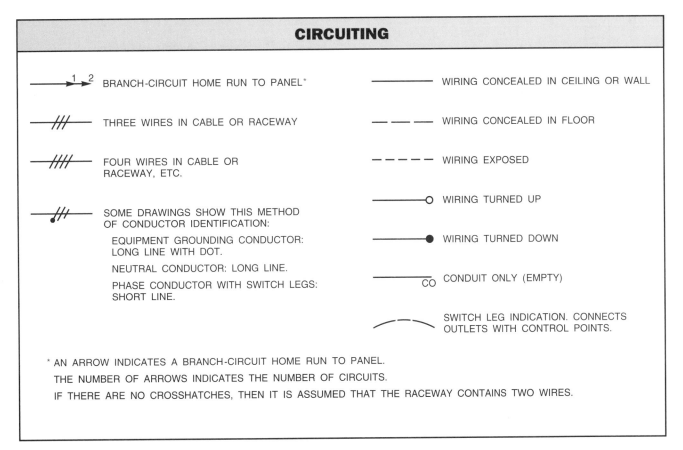

Figure 3–5   Circulating symbols.

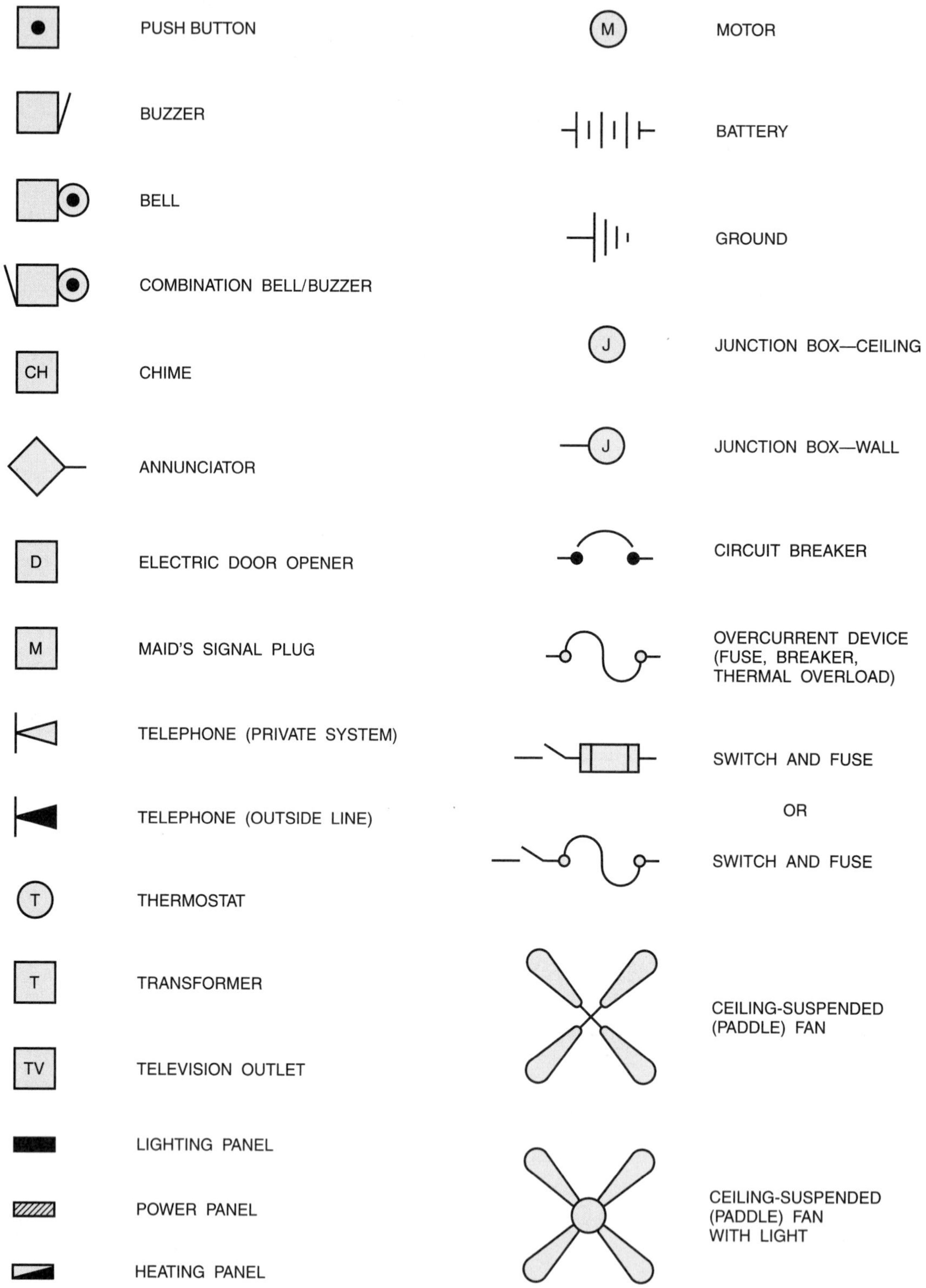
PUSH BUTTON
BUZZER
BELL
COMBINATION BELL/BUZZER
CH
CHIME
ANNUNCIATOR
D
ELECTRIC DOOR OPENER
M
MAID'S SIGNAL PLUG
TELEPHONE (PRIVATE SYSTEM)
TELEPHONE (OUTSIDE LINE)
T
THERMOSTAT
T
TRANSFORMER
TV
TELEVISION OUTLET
LIGHTING PANEL
POWER PANEL
HEATING PANEL
M
MOTOR
BATTERY
GROUND
J
JUNCTION BOX—CEILING
J
JUNCTION BOX—WALL
CIRCUIT BREAKER
OVERCURRENT DEVICE (FUSE, BREAKER, THERMAL OVERLOAD)
SWITCH AND FUSE
OR
SWITCH AND FUSE
CEILING-SUSPENDED (PADDLE) FAN
CEILING-SUSPENDED (PADDLE) FAN WITH LIGHT

Figure 3–6 Miscellaneous symbols.

SYMBOL	NOTATION
1	PLUGMOLD ENTIRE LENGTH OF WORKBENCH. OUTLETS 18" (457 mm) O.C. INSTALL 48" (1.2 m) TO CENTER FROM FLOOR. GFCI PROTECTED.
2	TRACK LIGHTING. PROVIDE 5 LAMPHOLDERS.
3	TWO 40-WATT RAPID START FLUORESCENT LAMPS IN VALANCE. CONTROL WITH DIMMER SWITCH.

Figure 3–7 Examples of notations for items that do not have a standard symbol, where there are space limitations on the plans, or where the architect or engineer merely wants to explain in more detail the intent and meaning of the special symbol.

Notations are also used to provide additional information about an item on the plan when there is insufficient room on the plan itself. Figure 3–7 shows examples of how notations might be used on electrical plans for items that do not have a standard symbol, or where it is necessary because of space limitations.

Typical Electrical Plans

In the Appendix of this text you will find two fold-out electrical plans, one for a First Floor and one for a Basement. Take a look at these plans now. You will get a better idea of how symbols, legends, and notations are used on residential electrical plans. Many home floor plans are not as detailed as these because the electrical symbols are included on the construction plans.

House wiring involves planning for many outlets. This term is used rather loosely by electricians, yet the *National Electrical Code®* clearly defines an *outlet* as *"a point on a wiring system where current is taken to supply utilization equipment."* The term *outlet* can be broken down even further.

- A *receptacle outlet* is *"an outlet where one or more receptacles are installed."*

  In Figure 3–8, (A) is where the branch-circuit wiring enters the box. Because the intent is to install a receptacle in the box, it is called a *receptacle outlet.* (B) shows a *receptacle outlet* with a single receptacle installed. This is one receptacle. (C) shows a *receptacle outlet* with a multiple (duplex) receptacle installed. This is two receptacles. (D) shows a *receptacle outlet* with two multiple (duplex) receptacles installed. This is four receptacles.

- A *lighting outlet* is *"an outlet intended for the direct connection of a lampholder, a lighting fixture, or a pendant cord terminating in a lampholder"* (Figure 3–9).

  Receptacles and toggle switches are **not** outlets. They are wiring devices. A wiring device carries current, but does not consume current. The Code defines a *device* as follows:

- A *device* is *"a unit of an electrical system which is intended to carry but not utilize electric energy."*

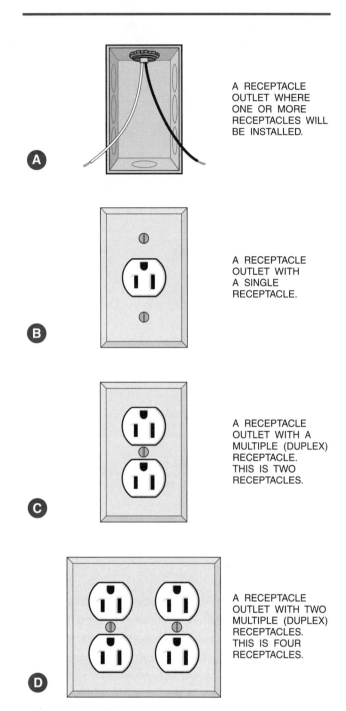

Figure 3–8 A *receptacle outlet* is where one or more receptacles will be installed. You plug into a receptacle. As defined by the *NEC*, a receptacle is a contact device installed at the outlet for the connection of a single contact device. Each receptacle on one strap (yoke) is defined as a receptacle. On the same yoke, you could have a single receptacle (A) or a multiple receptacle (B). A multiple receptacle might have two (duplex) or three receptacles (triplex).

A RECEPTACLE OUTLET WHERE ONE OR MORE RECEPTACLES WILL BE INSTALLED.

A RECEPTACLE OUTLET WITH A SINGLE RECEPTACLE.

A RECEPTACLE OUTLET WITH A MULTIPLE (DUPLEX) RECEPTACLE. THIS IS TWO RECEPTACLES.

A RECEPTACLE OUTLET WITH TWO MULTIPLE (DUPLEX) RECEPTACLES. THIS IS FOUR RECEPTACLES.

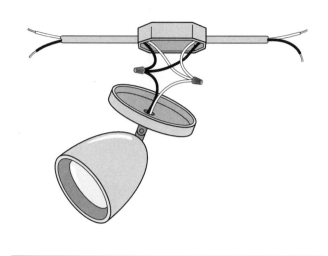

Figure 3–9 A *lighting outlet* is where a lighting fixture is intended to be installed.

Figure 3–10 offers a visual correlation of an electrical symbol and the type of fixture or receptacle that the symbol represents.

To select the size and type of box to be installed for a given wiring device, fixture, or appliance, you will need to know how many conductors will enter that particular box. Unit 5 covers the connecting of switches and receptacles. Unit 10 covers outlet and switch boxes, and the maximum number of conductors permitted in a given box.

ARCHITECTURAL SYMBOLS (NONELECTRICAL)

Electricians must work together with individuals of the other building trades. Each of these building trades has its own unique symbols. It is necessary to become familiar with these symbols. Adopted by the American Association of Architects, Figures 3–11, 3–12, and 3–13 show many nonelectrical symbols found on building plans and construction drawings.

RECOMMENDED HEIGHT FOR SWITCHES, RECEPTACLES, AND WALL LIGHTING OUTLETS

For the most part, the *National Electrical Code*® does not specify the height for switches, receptacles, and wall lighting outlets. You will find different recommended heights in different parts of the country

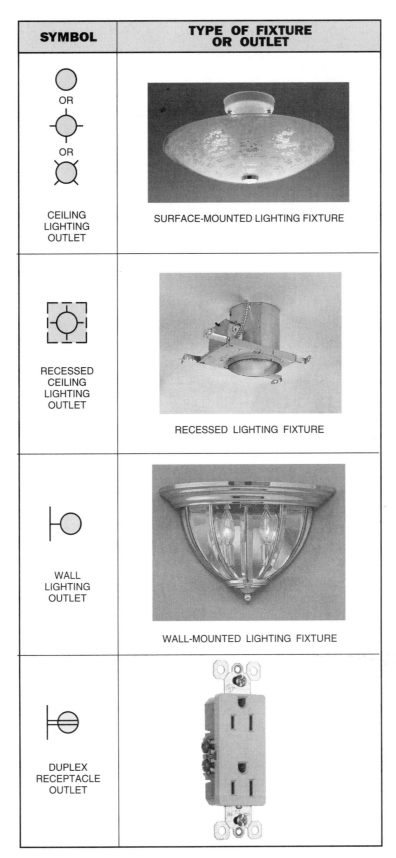

SYMBOL	TYPE OF FIXTURE OR OUTLET
CEILING LIGHTING OUTLET	SURFACE-MOUNTED LIGHTING FIXTURE
RECESSED CEILING LIGHTING OUTLET	RECESSED LIGHTING FIXTURE
WALL LIGHTING OUTLET	WALL-MOUNTED LIGHTING FIXTURE
DUPLEX RECEPTACLE OUTLET	

Figure 3–10 An illustration showing some electrical symbols and the type of fixture or receptacle represented by that symbol. (*Photos Courtesy of Progress Lighting*)

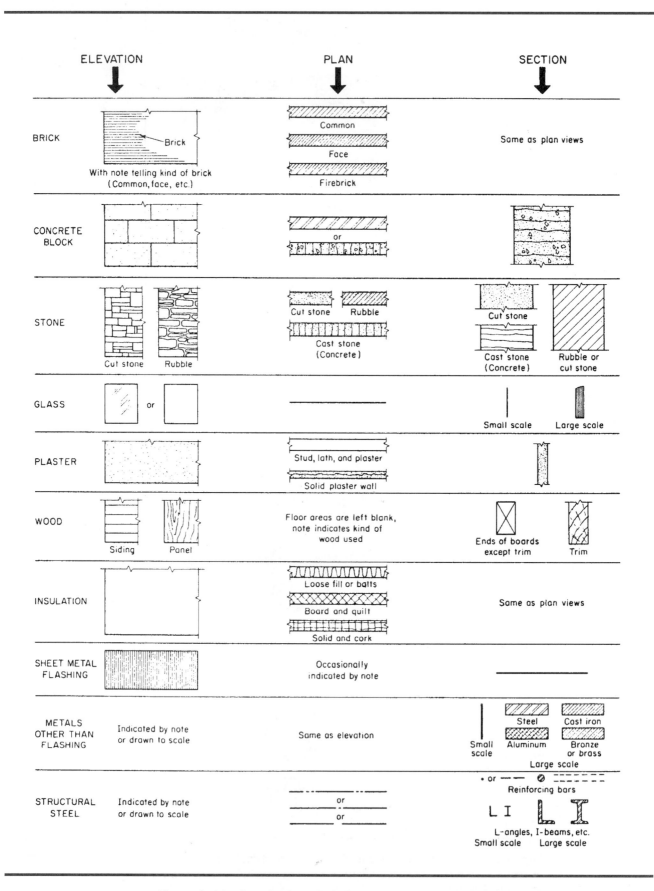

Figure 3–11   Standard symbols for architectural drawings.

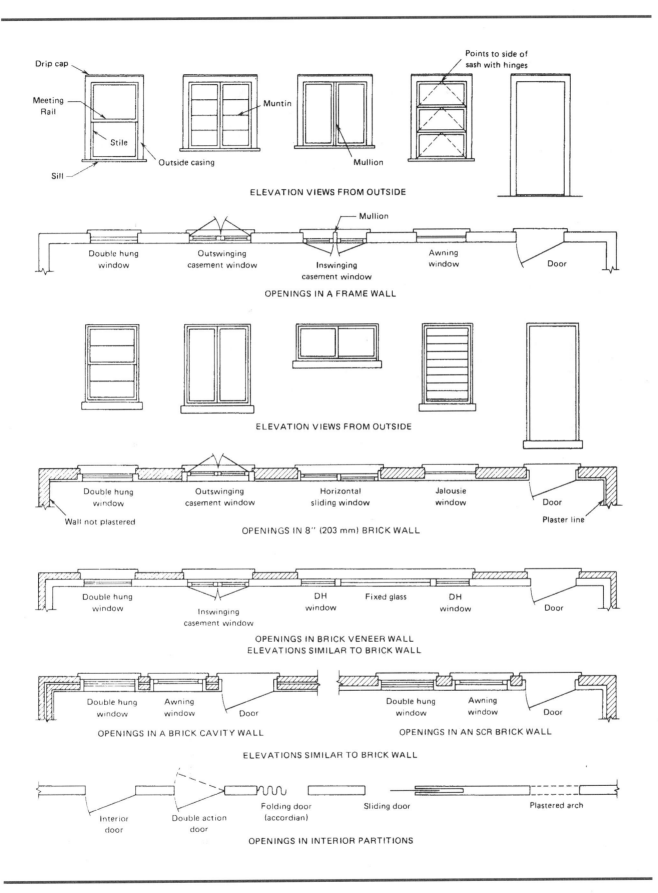

ELEVATION VIEWS FROM OUTSIDE

OPENINGS IN A FRAME WALL

ELEVATION VIEWS FROM OUTSIDE

OPENINGS IN 8″ (203 mm) BRICK WALL

OPENINGS IN BRICK VENEER WALL
ELEVATIONS SIMILAR TO BRICK WALL

OPENINGS IN A BRICK CAVITY WALL   OPENINGS IN AN SCR BRICK WALL

ELEVATIONS SIMILAR TO BRICK WALL

OPENINGS IN INTERIOR PARTITIONS

Figure 3–11 (Continued)

PIPING

Piping, in general (Lettered with name of material conveyed)	————
Non-intersecting pipes	
Steam	————
Condensate	— — — —
Cold water	————
Hot water	—··—··—
Air	—•—•—•—
Vacuum	—o—o—o—
Gas	— — — —
Refrigerant	—+—+—+—
Oil	—···—···—

PIPE FITTINGS

For welded or soldered fittings, use joint indication shown below	Screwed	Bell and spigot
Joint		
Elbow - 90 deg		
Elbow - 45 deg		
Elbow - turned up		
Elbow - turned down		
Elbow - long radius		
Side outlet elbow - outlet down		
Side outlet elbow - outlet up		
Base elbow		
Double branch elbow		
Single sweep tee		
Double sweep tee		
Reducing elbow		
Tee		
Tee - outlet up		
Tee - outlet down		
Side outlet tee outlet up		
Side outlet tee outlet down		
Cross		
Reducer		
Eccentric reducer		

PIPE FITTINGS (continued)

For welded or soldered fittings, use joint indication shown below	Screwed	Bell and spigot
Lateral		
Expansion joint flanged		

VALVES

For welded or soldered fittings, use joint indication shown below	Screwed	Bell and spigot
Gate valve		
Globe valve		
Angle globe valve		
Angle gate valve		
Check valve		
Angle check valve		
Stop cock		
Safety valve		
Quick opening valve		
Float opening valve		
Motor operated gate valve		

PLUMBING

Corner bath	
Recessed bath	
Roll rim bath	
Sitz bath	SS
Foot bath	FB
Bidet	B
Shower stall	
Shower head	(Plan) (Elev)
Overhead gang shower	(Plan) (Elev)
Pedestal lavatory	PL
Wall lavatory	WL
Corner lavatory	Lav
Manicure lavatory Medical lavatory	ML
Dental lavatory	Dental lav

PLUMBING (continued)

Plain kitchen sink	S
Kitchen sink, R & L drain board	
Kitchen sink, L H drain board	
Combination sink & dishwasher	
Combination sink & laundry tray	S&T
Service sink	SS
Wash sink (Wall type)	
Wash sink	
Laundry tray	LT
Water closet (Low tank)	
Water closet (No tank)	
Urinal (Pedestal type)	
Urinal (Wall type)	
Urinal (Corner type)	
Urinal (Stall type)	
Urinal (Trough type)	TU
Drinking fountain (Pedestal type)	DF
Drinking fountain (Wall type)	DF
Drinking fountain (Trough type)	DF
Hot water tank	HWT
Water heater	WH
Meter	M
Hose rack	HR
Hose bibb	HB
Gas outlet	G
Vacuum outlet	
Drain	D
Grease separator	
Oil separator	
Cleanout	
Garage drain	
Floor drain with backwater valve	
Roof sump	

Types of joints

Flanged   Screwed   Bell & spigot   Welded   Soldered

Figure 3–12   Standard symbols for plumbing, piping, and valves.

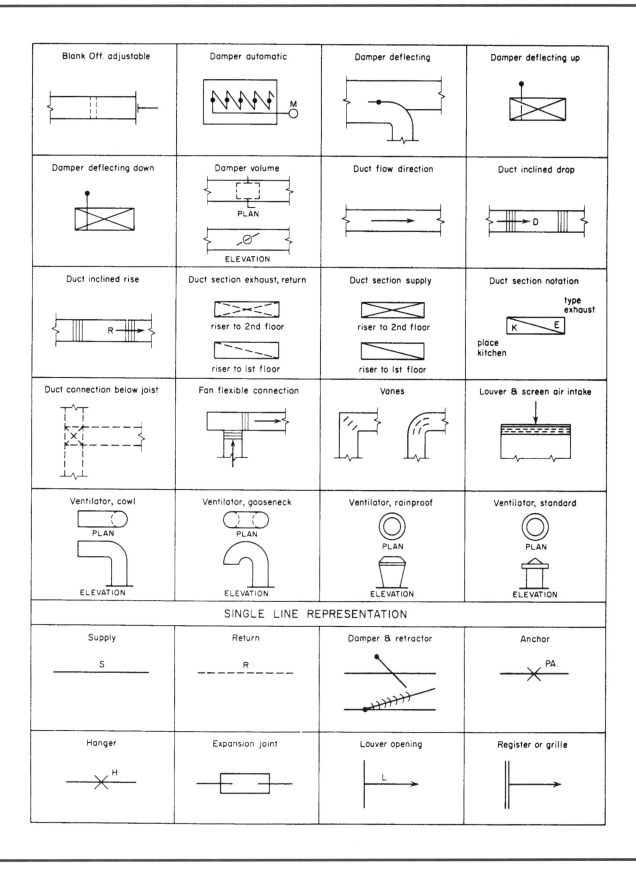

Figure 3–13   Standard symbols for sheet metal ductwork.

that came about over the years which have become somewhat of a tradition.

The same holds true for the positioning of wall receptacles. In most parts of the country, receptacles are mounted in the vertical position because it is easy to secure single-gang switch boxes to a stud. In other areas, such as the Greater Chicago Area, all house wiring is done with electrical metallic tubing (EMT), and most wall boxes are 4 inches square, making it easy to attach a 4-inch square single-gang plaster ring to the box in either the vertical or horizontal position. It is, therefore, a matter of choice.

Figure 3–14 provides some recommended heights that can be followed. Where physically challenged individuals are the intended occupants, switch heights may need to be lowered, and wall receptacle heights may need to be raised.

Some electrical contractors are installing wall switches 30 to 36 inches above the finished floor. At first pass, this might seem strange. But think about it! Children can reach the switches without messing up the wall. Adults carrying packages can easily turn the switches on and off with arms extended.

SWITCHES	
Regular	46" (1.17 m). Some homeowners want wall switches mounted 30–36" so as to be easy for children to reach and easy for adults to reach when carrying something. The choice is yours!
Between counter and kitchen cabinets. Depends on backsplash.	44"–46" (1.12 m–1.17 m)

RECEPTACLE OUTLETS	
Regular	12" (305 mm)
Between counter and kitchen cabinets. Depends on backsplash.	44"–46" (1.12 m–1.17m)
In garages	46" (1.17 m)
In unfinished basements	46" (1.17 m)
In finished basements	12" (305 mm)
Outdoors (above grade or deck)	18" (457 mm)

WALL LIGHTING FIXTURE OUTLETS	
Outside entrances. Depends on fixture.	66" (1.68 m). If fixture is "upward" from box on fixture, mount wall box lower. If fixture is "downward" from box on fixture, mount wall box higher.
Inside wall brackets	60" (1.52 m)
Side of medicine cabinet	60" (1.52 m)
Above medicine cabinet	You need to know the measurement of the medicine cabinet. Check the rough-in opening. Mount box approx. 6" to center of box above rough-in opening.

Note: All dimensions are from finished floor to center of the electrical box. If possible, try to mount wall boxes for lighting fixtures based upon the type of fixture to be installed. Verify all dimensions before "roughing in." If wiring for physically handicapped, the above heights may need to be lowered in the case of switches, and raised in the case of receptacles.

Figure 3–14 Recommended typical heights for switches, receptacles, and wall lighting outlets.

REVIEW QUESTIONS

1. The two items generally provided for all construction projects are the

   _____ and the _____

2. In the space provided, draw the symbol the words represent.

   a) Duplex receptacle   _____

   b) Split-circuit duplex receptacle   _____

   c) GFCI receptacle   _____

   d) Electric range outlet   _____

   e) Special purpose outlet   _____

   f) Single-pole switch   _____

   g) Three-way switch   _____

   h) Switch leg   _____

   i) Chime   _____

   j) Television outlet   _____

   k) Lighting panel   _____

3. Although somewhat confusing to the layperson, an electrician knows that the *NEC*® defines a *receptacle outlet* as a point on the wiring system where _____

   _____

4. A *duplex receptacle* is actually:

   a) one receptacle.

   b) two receptacles.

5. Recommended mounting heights for switches and receptacles in homes are shown in Figure 3–14. If you are accustomed to different mounting heights, indicate the measurements.

   Receptacles (centerline to finished floor)   _____ inches

   Switches (centerline to finished floor)   _____ inches

   Receptacles above kitchen countertops
      (centerline to finished floor)   _____ inches

   Do you mount wall receptacles vertically
      or horizontally?   _____

Switches and Receptacles

OBJECTIVES

After studying this unit, you will be able to:

- **Understand some key words as defined in the *National Electrical Code*® for electrical wiring.**

- **Learn about the many types of switches used for wiring houses.**

- **Learn about the many types of receptacles used for wiring houses.**

- **Understand the different types of terminals on wiring devices, and how to make connections that "Meet Code."**

- **Learn about the problems associated with aluminum conductors when not properly connected and terminated.**

INTRODUCTION

Switches of all types are installed in homes for the control of lighting and other electrical equipment. Receptacles are installed for plugging in cord-connected lamps, electrical equipment, and appliances. This unit covers switches and receptacles of the types generally used for residential wiring, the proper use of terminals on wiring devices, and some of the problems that might be encountered when using aluminum conductors.

IMPORTANT DEFINITIONS

For a better understanding of some of the terms found in the *National Electrical Code*, following are some definitions.

- An outlet is *"a point on a wiring system at which current is taken to supply utilization equipment."*

- A *receptacle* is *"a contact device installed at the outlet for the connection of an attachment plug. A single receptacle is a single contact device with no other contact device on the same yoke. A multiple receptacle is a single device containing two or more receptacles"* (refer back to Figure 3–8).

- A *receptacle outlet* is *"an outlet where one or more receptacles are installed"* (refer back to Figure 3–8).

- A *lighting outlet* is *"an outlet intended for the direct connection of a lampholder, a lighting fixture, or a pendant cord terminating in a lampholder"* (refer back to Figure 3–9).

- A *device* is *"a unit of an electrical system that is intended to carry but not utilize electric energy."* Toggle switches and a receptacles are good examples of devices. That is why they are listed in catalogs as "wiring devices." They certainly do carry current, but they do not consume current.

- A *general-use snap switch* is *"a form of general-use switch constructed so that it can be installed in device boxes or on box covers, or otherwise used in conjunction with wiring systems recognized by this code."*

SWITCHES

Unit 5 contains wiring diagrams most often used in house wiring.

This unit discusses switches and receptacles of the types generally used in homes. There are literally hundreds of styles and types of switches and receptacles to choose from.

The most frequently used switch for residential use is the *snap switch*. These quite often are referred to as toggle switches, but the UL designation is *snap switch*. Webster's dictionary defines a toggle switch as "a switch consisting of a projecting lever moved back or forth through a small arc to open or close an electric circuit." The *National Electrical Code®* in *Article 380* merely refers to them as *switches*. Manufacturers' catalogs show many varieties of snap switches such as toggle switches, rocker switches, slider switches, tap action switches, lock-type switches, quiet switches, etc. They all are one and the same.

These switches are available in single-pole switches (control from one point), three-way switches (control from two points), four-way switches (control from three or more points), switches with pilot lights in their toggle handle, dimmer switches, interchangeable switches and receptacles, combinations switches and receptacles on one yoke, key switches, and switches with spring-wound and electronic timers. Other types of control can be obtained with motions sensors (that

sense motion to turn on a light), photocells (that sense darkness to turn on a light), and low-voltage remote control switches.

When mounted in a flush switch (device) box, the switch is concealed in the box, with only the insulated handle protruding. The color of the wiring devices most commonly used in homes in white, brown, and ivory, although other colors are available. Switches are attached to switch (device) boxes and to plaster rings used with 4-inch square outlet boxes with No. 6-32 screws. These No. 6-32 screws are usually held captive to the yoke of a switch or receptacle by a small fiber or cardboard washer. This keeps you from dropping and losing the screw. Figure 4–1 is an assortment of typical toggle (snap) switches.

SWITCH RATINGS

Underwriters Laboratories lists switches that are used for general purpose lighting control as *general-use snap switches*. The UL requirements are a mirror image of *Section 380-14(a)* and *(b)* of the *National Electrical Code®*.

AC General-Use Snap Switches

AC general-use snap switches are for use on:

- alternating current (AC) only.

- resistive and inductive loads not to exceed the ampere rating of the switch at rated voltage. This includes electric discharge lamps such as fluorescent lamps.

- tungsten-filament lamp loads not to exceed the ampere rating of the switch at 120 volts.

- motor loads not to exceed 80% of the ampere rating of the switch at rated voltage (300 volts or less). An additional requirement is that the motor load shall not exceed 2 horsepower, *Section 430-83(c)(2)*.

AC general-use snap switches are marked "AC only" in addition to identifying their current and voltage rating. A typical switch marking is 15A, 120-277 V AC. The 277-volt rating is required when the switches are used in commercial buildings where the system voltage might be 277/480 volts, in which case the voltage from a "hot" phase conductor to the grounded white neutral conductor is 277 volts. In a home, the voltage from a "hot" phase conductor to the grounded white neutral conductor is 120 volts.

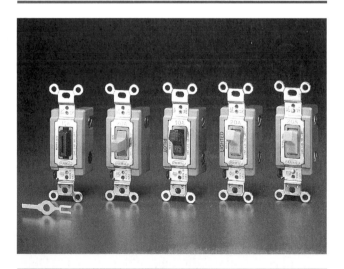

Figure 4–1 An assortment of typical toggle switches. (*Courtesy Hubbell, Incorporated*)

AC/DC General-Use Snap Switches

AC/DC general-use snap switches are for use on:

- alternating-current (AC) or direct-current (DC) circuits.

- resistive loads not to exceed the ampere rating of the switch at rated voltage.

- inductive loads not to exceed one-half the ampere rating of the switch at rated voltage unless otherwise marked.

- tungsten filament lamp loads not to exceed the ampere rating of the switch at 125 volts when marked with the letter "T."

- for switches marked with a horsepower rating, a motor load shall not exceed the rating of the switch at rated voltage.

- AC/DC general-use snap switches normally are not marked AC/DC. However, it is always marked with the current and voltage rating, such as 10A-125V, or 5A-250V-T.

Why are some switches required to have a "T?" The letter "T" stands for tungsten. Tungsten is the metal used to make the filament of most conventional incandescent lamps. Tungsten filament lamps draw a very high momentary inrush current at the instant the circuit is energized, thus the switches contacts are subjected to substantial amount of arcing when the switch contacts close.

The cold resistance of a typical 100-watt incandescent lamp is approximately 9.5 ohms. This same lamp has a hot resistance of approximately 144 ohms when operating at 100% of the lamp's rated voltage. The filament resistance increases very rapidly, going from the cold resistance of 9.5 ohms to the hot resistance of 144 ohms in about 1/240 second (one-quarter of a cycle).

Normal operating current for this 100-watt lamp is:

$$I = \frac{E}{R} = \frac{120}{144} = 0.83 \text{ amperes}$$

But the momentary instantaneous inrush current could be as high as:

$$I = \frac{E}{R} = \frac{170 \text{ (peak voltage)}}{9.5} = 17.9 \text{ amperes}$$

The instantaneous inrush current in the preceding example is about 21 times normal operating current. This inrush current drops off to normal operating current in about 6 cycles (0.10 second). Conductors in the circuit are not overheated by this momentary high value of inrush current because the time period is extremely short.

On very rare occasions, a circuit breaker might trip off when it supplies a rather large incandescent lamp load that is controlled by a dimmer switch. If the dimmer switch is turned to a preset (low) setting, and the circuit is turned on, the low voltage cannot heat up the lamp filament fast enough. The circuit breaker's instant trip mechanism goes into action, shutting off the breaker. This nuisance tripping problem is generally experienced on heavy dimming load applications, such as in restaurants, libraries, and similar commercial buildings. It is not a common occurrence in residential installations.

Refer back to Figure 3–14 for recommended heights to mount switches.

TYPES OF SWITCHES

The following discusses many types of switches that are available for use in residential installations.

Interchangeable Type

This term is applied to wiring devices that are small in size, allowing up to three devices to be mounted on one mounting bar in a single-gang box. Single-gang faceplates are available with one, two, or three openings. The devices can be switches, receptacles, or pilot lights. Figure 4–2 shows a three-hole mounting bar, a grounding-type receptacle, and a switch. Many electricians continue to refer to this type of wiring device as a Despard switch, receptacle, or pilot light.

Door Switches

A door switch is mounted into the door frame on the hinged side of a door. The junction box is furnished with the switch. The plunger is adjustable. Door switches are available in two types: "light on - door open," and "light on - door closed." Door switches are a nice feature for use on closet doors to

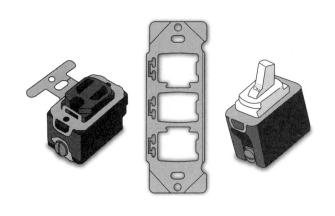

Figure 4–2 Interchangeable wiring devices and a three-hole mounting bar. Any combination of three interchangeable wiring devices such as switches, receptacles, or pilot lights can be mounted on the mounting bar.

automatically turn on the closet light when the door is opened. When the door is closed the plunger is pushed inward. This is the OFF position. When the door is opened, the plunger comes outward. This is the ON position. The wiring of a door switch is easy because it is a simple single-pole switch, connected in series with the closet light. Because the junction box is small, the wiring must be arranged so that only the two-wire switch loop cable is run to the junction box. Figure 4–3 shows a typical door switch.

Figure 4–3   A door switch.

Key Switches

Key switches, sometimes referred to as lock switches, are used when you do not want "just anyone" turning on or turning off the particular load controlled by that switch. Only those who have the key can operate the switch. Key switches operate mechanically the same as conventional snap switches. The toggle handle has been replaced by a slot. The key must be inserted to operate the switch. Key (lock) switches are available in single-pole, double-pole, three-way, and four-way types. Figure 4–4 shows a key switch with the key inserted into the slot.

Switches in Wet Locations

Figure 4–5 is a switch protected by a weather-proof cover, used when the switch is mounted outdoors, or in other wet locations. Switches are not permitted in wet locations in tub and shower spaces. See *Section 380-4.* The only switches permitted in tubs and shower areas are those switches that are part of a "listed" tub or shower assembly.

Low-Voltage Switches

For residential application, low-voltage systems generally operate on 24 volts through a step-down

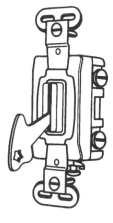

Figure 4–4   A key-operated switch. The key is inserted into the slot on the switch to operate the switch, then removed. (*Courtesy Pass & Seymour Legrand*)

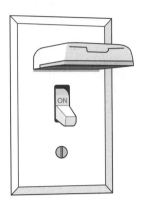

Figure 4–5 A switch protected by a weatherproof cover. This is required by *Section 380-4* for switches that are located outdoors or in wet locations. Switches are not permitted to be installed in wet locations in tub or shower spaces.

transformer having a primary input of 120 volts and a secondary output of 24 volts. The step-down transformer is incorporated in the system. Because of this low voltage, switch boxes are not required although they can be used. Most often, plaster covers are nailed to the studs, with the low-voltage cable merely brought out through the opening, connected to the low-voltage switch, then pushed back into the open space. The faceplates fasten to the plaster cover with No. 6-32 screws the same as conventional switches.

The low-voltage switches control a relay, which in turn switches the 120-volt circuit. Low-voltage switches are discussed in Unit 14.

Electronic Dimming Switches

Dimmers operate on the principle that reducing the voltage will dim the lighting. Let us take an example to show how the wattage of an incandescent lamp changes when the voltage to it is reduced. We will use simple formulas to show this principle.

Example: A 100-watt incandescent lamp draws approximately 0.833 ampere at 120 volts.

The lamp's wattage is:

$$W = E \times I = 120 \times 0.883 = 100 \text{ watts}$$

The lamp's hot resistance is:

$$R = \frac{E}{I} = \frac{120}{0.833} = 144 \text{ ohms}$$

The lamp's wattage at 60 volts is:

$$W = \frac{E \times E}{R} = \frac{60 \times 60}{144} = 25 \text{ watts}$$

These calculations can be done for other voltages. Other formulas can be used. See Unit 2.

The most common type of dimmer for residential use is the electronic dimmer rated 600-watt maximum, 120 volts. However, electronic dimmers are also available in 1,000-, 1,500-, and 2,000-watt capability. These higher wattage dimmers need to be derated if ganged together in one box. The manufacturer's instruction will specify the derating values. Electronic dimmers for controlling incandescent lamps only are marked in *watts*. Electronic dimmers for controlling inductive loads are marked in *volt-amperes*. An example of this would be lighting systems that incorporate a step-down transformer, as in low-voltage outdoor decorative lighting. Another example is low-voltage track lighting that also includes a step-down transformer. The instructions will clearly indicate if the dimmer is suitable for use on an inductive load.

Be sure to turn off the power when hooking up an electronic dimmer. The possibility of causing a short circuit or a ground fault while working the circuit "hot" would burn out the dimmer instantly. Also, repetitive "make and breaks" before the actual final splicing is completed will cause the dimmer's internal electronic circuitry to heat up beyond its capability.

Figure 4–6 shows some typical electronic dimmers used in homes. Dimmer switches are available in single-pole and three-way types.

The higher quality dimmers will reduce or eliminate the annoying buzzing picked up by radios.

Sometimes the filament in an incandescent lamp will vibrate when connected to an electronic dimmer. This hum usually can be eliminated or greatly reduced by installing "rough service" lamps, or changing the dimmer.

Most instructions furnished with dimmers recommend that the dimmer *not* be used to control a receptacle. The logic is that more than the intended lamp might be plugged into the receptacle, leading to the possibility of serious overloading of the

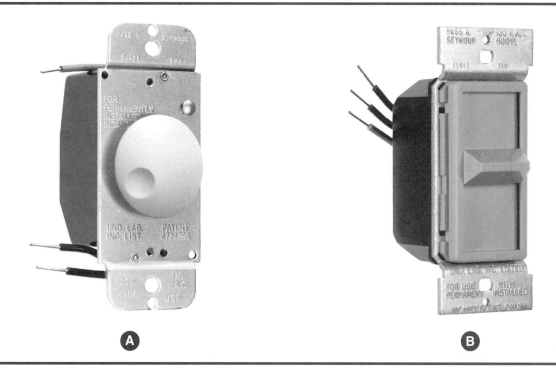

(A)   (B)

Figure 4–6   Some typical electronic dimmers found in homes. All perform the same function by varying the voltage to the connected incandescent lamp load. (A) rotating knob, (B) slider knob. (*Courtesy Pass & Seymour Legrand*)

dimmer. Furthermore, reduced voltage to a television, radio, stereo components, computers, and so forth, can cause costly damage to the appliance.

Autotransformer Dimmers

When larger loads are to be dimmed, it might be necessary to install an autotransformer (one winding) type of dimmer. Because they contain an actual autotransformer, they are physically larger and require a special wall box that is furnished with the dimmer. Since the advent of electronic dimmers, autotransformer-type dimmers are very rarely used in residential applications. Figure 4–7 is an exploded view of an autotransformer-type dimmer.

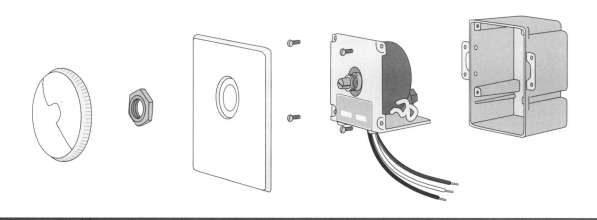

Figure 4–7   An autotransformer-type dimmer.

Dimming Fluorescent Lamps

Special dimming ballasts are required to control the light output of fluorescent lamps. **Do not** hook up a conventional fluorescent fixture to a dimmer control. It will not work, and could cause a fire! Today, rapid start 32-watt T8 lamps are generally used with electronic dimming ballasts. In the past, 40-watt F40T12 rapid start lamps were used. A wiring diagram for dimmer control of fluorescent is found in Unit 5.

Because incandescent lamps and fluorescent lamps have different characteristics, they cannot be controlled with the same dimmer control.

Switches For Speed Control of Fans

Speed control switches for ceiling fans are similar in appearance to lighting dimmer controls. The higher quality electronic speed controls are designed to reduce or eliminate the humming associated with ceiling fans when their speed is reduced. For this reason, it is unwise to use a lighting dimmer control as a speed control for a fan. Fan speed controls are connected "in series" with the fan motor. Figure 4–8 shows two types of speed controls.

Switches for Humidity Control

Ceiling paddle fans circulate the air in a room. Exhaust fans remove hot air from the room or attic, replacing it with cooler outside air through open windows, roof vents, or gable vents. In addition to speed control as discussed previously, humidity controls are available that will turn the fan on when the relative humidity reaches a certain level. Decent comfort level is about 50% relative humidity. A humidity controller can be adjusted for the desired humidity, and also manually operated. Exhaust fans can be connected so they will operate with manual control, timing control, and humidity control. Figure 4–9 is an exploded view of a line-voltage humidity control, sometimes referred to as a *humidistat*.

There are so many varieties of dimmers, speed controllers, and switches available today, that you need to think about what it is you want to do, then check with the various manufacturers' catalogs or

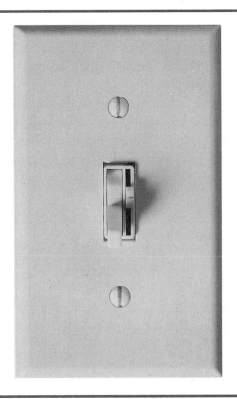 

Figure 4–8 Two types of wiring devices that combine control of lighting as well as control of a ceiling fan. (A) has a toggle switch for turning a ceiling fan on or off, as well as a "slider" that provides speed control for the fan. (B) has a toggle switch for turning a light on or off, a "slider" for dimming the light, and a second "slider" for "on" or "off" and speed control of the fan. (*Courtesy of Lutron*)

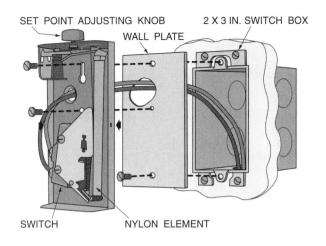

Figure 4–9 An exploded view showing the parts of a humidity control for use with exhaust fans.

visit the electrical distributor or home center to see what devices are available.

Faceplates for Switches and Receptacles

Faceplates are available in many styles and colors, such as brown, ivory, white, red, and gray. They are available in thermoplastic (will change shape and melt if overheated), thermoset material (will not change shape if overheated), chrome plated, wood (but lined with noncombustible material), and stainless steel. Oversized jumbo faceplates are also available should the opening in the wall be too large for a standard faceplate to conceal.

- Use nonmetallic faceplates when the metal wall box is ungrounded, *Section 380-9*. This could be the case in very old homes.

- Metal faceplates must be grounded, *Section 410-56(d)*. The same requirement is found in *Section 380-12*. If the wall box is properly grounded, the faceplate will be grounded through the metal yoke on the switch and No. 6-32 faceplate screws. If the wall box is nonmetallic, then use switches that have the extra equipment grounding screw of the type shown in Figure 4–10.

- Faceplates must cover the wall opening and seat against the wall surface, *Section 380-9*. The measurements of standard faceplates are:

Figure 4–10 This switch has a terminal for the connection of an equipment grounding conductor in accordance with *Section 380-9(b)*. Thus, when a metal faceplate is installed, it will be grounded in accordance with *Section 410-56(d)*. (*Courtesy Pass & Seymour Legrand*)

NO. OF GANGS	HEIGHT	WIDTH
1	4½"	2¾"
2	4½"	4⁹⁄₁₆"
3	4½"	6⅜"
4	4½"	8³⁄₁₆"
5	4½"	10"
6	4½"	11¹³⁄₁₆"

- If the box is set back from the wall surface, make sure that the plaster ears on the yoke of the switch fit tightly against the surface of the wall, *Sections 380-10(b)* and *410-56(f)(1)*.

- If the box is set flush with the wall surface, make sure that the plaster ears on the yoke of the switch fit tightly against the box, *Sections 380-10(b)* and *410-56(f)(2)*.

- Wall boxes must be flush with the surface of the wall if the wall is combustible, and set back not over ¼ inch if the wall is noncombustible, *Section 370-20*. See Unit 10 for more information about boxes.

- There is nothing worse than a loose switch or receptacle. Many electricians will put spacers (metal washers) between the yoke and the set back wall box. This tightly secures the device to the box, and does not rely on soft or crumbling plaster to keep the switch or receptacle tight.

- *Section 410-56(e)* requires that faceplates *"completely cover the opening and seat against the mounting surface."* For boxes that are set back from the finished wall surface, receptacles *"shall be installed so that the mounting yoke or strap of the receptacle is held rigidly at the surface of the wall."* For boxes that are flush with the finished wall surface, receptacles *"shall be installed so that the mounting yoke or strap of the receptacle is seated against the box or raised box cover."*

- *Section 410-56(f)(3)* prohibits supporting a receptacle to a cover by only one screw, unless a device assembly or box cover "listed" and "identified" for securing by a single screw is used.

And while we're talking about faceplates, add a little class to your work by aligning the faceplate screws so the slots are in precisely the vertical position for vertically mounted receptacles and switches, and in the horizontal position for horizontally mounted wiring devices. You will be amazed at how neat this looks.

Although this might be stretching things, you will recall *Section 110-12: "Mechanical Execution of Work. Electric equipment shall be installed in a neat and workmanlike manner."*

Timers

Timers are unique in that they provide switch control that is operated by a clock. Timers are also referred to as time clocks.

Time clocks are used where a load is to be controlled for specific ON/OFF times of the day or night. The capabilities of time clocks are endless.

Figure 4–11 shows two types of spring-loaded timers that are connected in series with the load—the same as standard wall switch. They install in a standard single-gang device box. Because they take up quite a bit of cubic inch space, make sure the wall box is large enough to meet the box-fill requirements of *Section 370-16*. A typical application is in

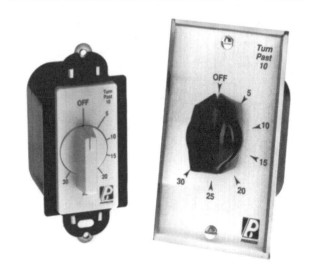

Figure 4–11 Spring-loaded timers. (*Courtesy Paragon Electric Company, Inc.*)

bathrooms for the control of an exhaust fan or an electric heater.

Figure 4–12 is a 24-hour time clock commonly used to control security lighting, decorative lighting, or energy management. This particular time clock

Figure 4–12 A 24–hour time clock. (*Courtesy Paragon Electric Company, Inc.*)

Figure 4–13 An electronic time clock. (*Courtesy Paragon Electric Company, Inc.*)

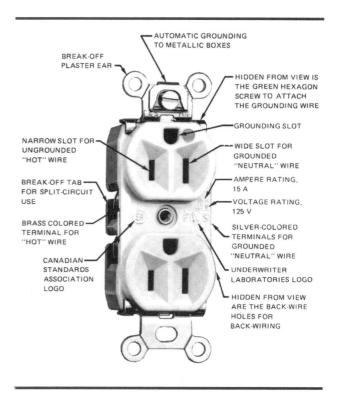

Figure 4–14 A grounding-type receptacle detailing various parts of the receptacle. (*Courtesy of Pass & Seymour, Inc.*)

has four adjustable and removable "pins" that control two ON and two OFF operations. Additional "pins" can be added. There is an "override" switch for manual control.

Figure 4–13 is an electronic 24-hour, 7-day time clock. It can be programmed to skip certain days. Up to sixteen events can be programmed. There is an "override" switch for manual control.

RECEPTACLES

The following text provides the information necessary for selecting, installing and connecting receptacles in homes.

Ratings

Receptacles installed in homes are generally 15 amperes, 125 volts of the type shown in Figure 4–14. As found in *Table 210-21(b)(3)*, 15-ampere receptacles may be used on either 15- or 20-ampere branch circuits. This is only permitted when the branch circuit supplies two or more receptacles or

outlets. General purpose lighting circuits in homes that supply receptacles are usually supplied by 15-ampere branch circuits. In some instances, 20-ampere lighting circuits are used, but this is not too common for house wiring. Of course, the small appliance branch circuits required in the kitchen, dining room, laundry, and other specific areas are required to be 20-ampere branch circuits.

Clearly illustrated in Figure 4–14, general purpose receptacles have a wide slot, a narrow slot, and a horseshoe-shaped slot. The narrow slot is for the ungrounded "hot" conductor and is internally connected to the brass-colored terminal. The wide slot is for the grounded "neutral" conductor and is internally connected to the white- (silver) colored terminal. The horseshoe-shaped slot is for the equipment grounding conductor that is internally connected to the green-colored, hexagon-shaped screw terminal.

Grounding of Receptacles

Removing the fiber or cardboard washer usually found on the No. 6-32 mounting screws gives metal-to-metal contact of the yoke to a metal box. This is great when you want to ensure that a good equipment ground continuity has been established between the grounded metal box and the yoke of the receptacle. But many times getting the metal-to-metal contact is impossible because the box is set back a small distance from the wall surface. Many electricians use metal washers and nuts as spacers to fill the small gap between the yoke and the metal box. This makes a very solid mechanical and electrical metal-to-metal connection, the receptacle will not be loose, and the No. 6-32 screws are not being depended upon as the equipment grounding means.

Excellent equipment ground continuity is obtained between the yoke of the receptacle and the metal wall box when receptacles of the type shown in Figure 4–14 are used. This particular grounding-type receptacle has a metal clip that holds the No. 6-32 screw in place on the yoke. In addition, this screw and clip assembly has passed the Underwriters Laboratories tests, making this an effective means of establishing good electrical ground continuity between the yoke and the metal box.

Receptacles rated 20 amperes are rarely used in homes. They cost more than 15-ampere receptacles. A 20-ampere receptacle can be recognized easily by the wide neutral grounded conductor slot that has the shape of a "T." When a single receptacle is installed on an individual branch circuit, the ampere rating of the receptacle must not be less than the ampere rating of the branch circuit, *Section 210-21(b)(1)*. An example of this might be a separate 20-ampere branch circuit supplying a receptacle in a basement for a sump pump. All receptacles in a basement must be GFCI protected—except a single receptacle that supplies a specific appliance, *Section 210-8(a)(5), Exception No. 2*. This topic is covered in detail in Unit 7.

The blade configuration of 15-ampere and 20-ampere receptacles is shown in Figures 4–15 and 4–16.

Split-Circuit Receptacles

Most receptacles today can be converted into a split-circuit receptacle by breaking off the tab between the two terminals (see Figures 4–14 and 4–17). Tabs can be removed from the "hot" conductor terminals, the neutral conductor terminals, or both. In house wiring, split-circuit receptacles are most often

Figure 4–15 This is a 15-ampere, 125-volt receptacle and is a NEMA 5-15R configuration. The letters NEMA stand for National Electrical Manufacturers Association.

Figure 4–16 This is a 20-ampere, 125-volt receptacle and is a NEMA 5-20R configuration. The letters NEMA stand for National Electrical Manufacturers Association.

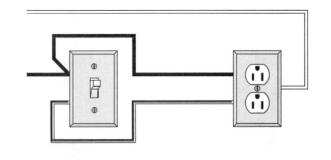

Figure 4–17 Suggested wiring and positioning for a switched split-circuit receptacle. The "hot" receptacle is on the top; the "switched" receptacle is on the bottom.

used where it is desired to have one of the receptacles "on" all of the time, and have the other receptacle controlled by a wall switch, in which case the tab is removed from between the two "hot" terminals.

Establish some sort of consistency when installing split-circuit receptacles. For vertically installed receptacles, place the switched receptacle on the bottom as illustrated in Figure 4–17. For horizontally installed receptacles, place the switched receptacle to the right.

Split-Circuit Receptacles and Multiwire Branch Circuits

Split-circuit receptacles may be connected to a multiwire branch circuit. This might be advantageous to feed areas such as in kitchens where there is always a heavy concentration of high wattage electrical appliances. There may be a cost savings (labor and material) to run one three-wire branch circuit instead of two two-wire branch circuits to serve the receptacles in these areas. When multiwire branch circuits are connected, the neutral conductor must not be broken at the receptacle. *Section 300-13(b)* states the grounded "neutral" conductor connections shall be made up independently of the connections to the receptacle itself. In other words, do not use the terminals on the receptacle as a splicing means for the grounded "neutral" conductors of a multiwire branch circuit.

Figure 4–18 is an exploded view of a grounding-type receptacle that has break-off tabs to convert the receptacle to a split-circuit type.

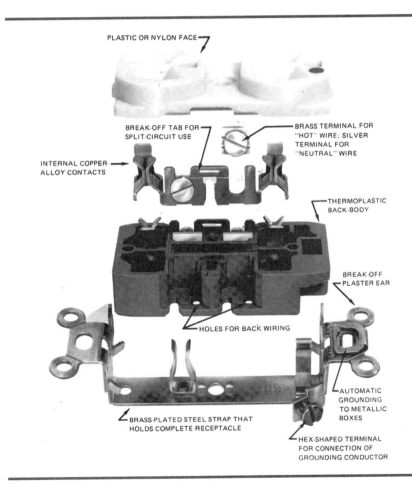

Figure 4–18 An exploded view of a grounding-type receptacle. This receptacle has both screw terminals and holes for back wiring, has an integral clip for automatic grounding to a grounded metal box, and has break-off tabs to convert the receptacle into a split-circuit type. Push-in terminals not permitted for No. 12 AWG conductors for 20-ampere receptacles. This is a U.L. requirement.

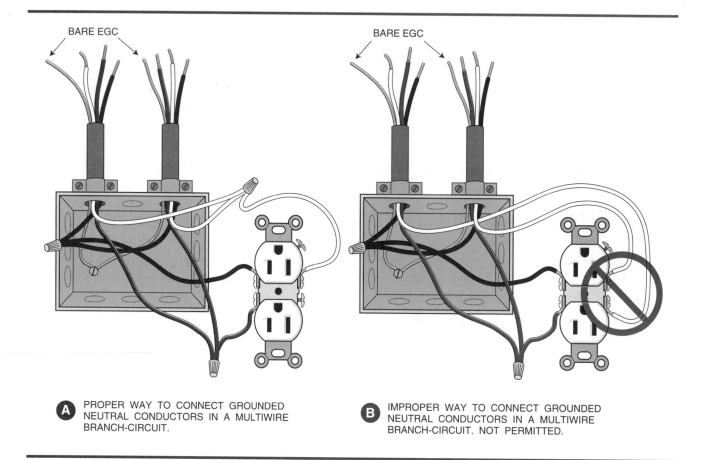

BARE EGC

BARE EGC

A   PROPER WAY TO CONNECT GROUNDED NEUTRAL CONDUCTORS IN A MULTIWIRE BRANCH-CIRCUIT.

B   IMPROPER WAY TO CONNECT GROUNDED NEUTRAL CONDUCTORS IN A MULTIWIRE BRANCH-CIRCUIT. NOT PERMITTED.

Figure 4–19   The right and wrong way to connect the neutral in a three-wire (multiwire) branch circuit, *Section 300-13(b)*.

Figure 4–19 shows the right way and the wrong way to terminate the neutral conductors on a receptacle.

To review the hazards when a neutral opens on a three-wire branch circuit, see Unit 2.

GFCI receptacles are easily identified by their *reset* and *test* buttons. GFCI receptacles are available with feed-through feature, but are not available with the split-circuit design. Figure 4–20 is a feed-through GFCI receptacle.

To use GFCI receptacles on multiwire branch circuits, the circuitry can be made up as shown in the diagram in Figure 4–21.

Section 210-4(b) requires that both of the "hot" conductors of a multiwire branch circuit be disconnected simultaneously. A 2-pole circuit breaker, or two 1-pole circuit breakers with a "listed" handle tie meets this requirement. Figure 4–22 shows how this is accomplished.

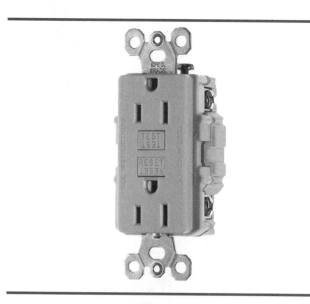

Figure 4–20   A feedthrough ground-fault circuit-interrupter (GFCI) receptacle. GFCI receptacles are not available with the split-circuit feature. (*Courtesy Pass & Seymour Legrand*)

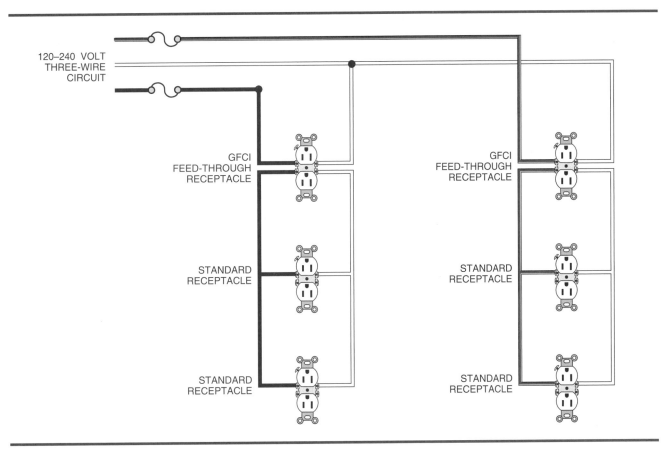

Figure 4–21 This wiring diagram shows how a multiwire branch circuit can be used to carry two circuits to one point in the system, then split the three-wire circuit to two-wire circuits. One GFCI receptacle of the feedthrough type is used for the first receptacle in each two-wire branch circuit. All receptacles downstream from the feed-through GFCI receptacles are GFCI protected.

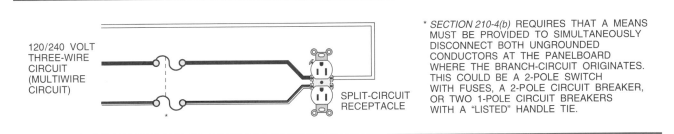

Figure 4–22 A split-circuit receptacle connected to a three-wire multiwire branch circuit. The disconnect must simultaneously disconnect both ungrounded conductors.

Positioning of Receptacles

There have been proposals made to have a requirement in the *National Electrical Code*® regarding the positioning of receptacles. These proposals have been rejected by the Code-Making Panel responsible for receptacle requirements. In the past few years, a number of manufacturers have been attaching the yoke and positioning the markings on their receptacles so that when the receptacle is installed, the grounding slot is to the top. This can clearly be seen on the receptacle illustrated in Figure 4–14.

What is the reasoning behind putting the grounding slot to the top? There is always the possibility that a metal wall plate could come loose and fall downward onto the blades of an attachment plug cap that is loosely plugged into the receptacle. This creates a potential shock and fire hazard. Let us take a look at some of the options.

- **Recommended:** Grounding hole to the top (Figure 4–23). A loose metal plate could fall onto the grounding blade of the attachment plug cap. Sparks would fly.

- **Not Recommended:** Grounding hole to bottom (Figure 4–24). A loose metal plate could fall onto both the grounded neutral and "hot" blades. Sparks would fly.

- **Recommended:** Grounded neutral blades on top (Figure 4–25). A loose metal plate could fall onto these grounded neutral blades. No sparks would fly.

- **Not Recommended:** "Hot" blades on top (Figure 4–26). A loose metal plate could fall onto these "live" blades. Sparks would fly. If this was a split-circuit receptacle fed by a 3-wire, 120/240-volt circuit, the short circuit would actually be across the 240-volt line. Sparks would really fly!

Ground-Fault Circuit-Interrupter Receptacles

The *National Electrical Code®* requires that certain receptacles in a home be ground-fault protected. A ground-fault circuit-interrupter receptacle is shown in Figure 4–20. A ground-fault circuit-interrupter circuit breaker is shown in Figure 7–3. GFCI protection is discussed in detail in Unit 7.

Figure 4–23 Recommended receptacle positioning where the grounding hole is to the top. A loose metal plate could fall onto the grounding blade of the attachment plug cap. Sparks would fly.

Figure 4–24 Not recommended receptacle positioning where the grounding hole is to the bottom. A loose metal plate could fall onto both the grounded neutral and "hot" blades. Sparks would fly.

Figure 4–25 Recommended receptacle positioning where the grounded neutral blades are on top. A loose metal plate could fall onto these grounded neutral blades. No sparks would fly.

Figure 4–26 Not recommended receptacle positioning where the "hot" blades are on top. A loose metal plate could fall onto these "live" blades. Sparks would fly.

Receptacles in Outdoor Locations

Water and electricity do not mix! Rain water and water from sprinkling must be kept out of electrical equipment.

A *damp location* is a location that is partially protected, such as under canopies, under an eave, or under roofed open porches. These areas are highly unlikely to be deluged with water from a beating rain.

A *wet location* is a location that is subject to saturation with water, such as locations exposed to the weather and not protected in any way.

Section 410-57 provides us with the guidelines for receptacles installed outdoors in damp or wet locations. The key points are:

• Receptacles shall not be installed in tub or shower spaces.

• In damp locations, the enclosure must be made weatherproof for those times when nothing is plugged into the receptacle, and the cover is closed.

• An installation suitable for wet locations is also suitable for damp locations.

• In wet locations, the enclosure must be made weatherproof even for those times when something is plugged into the receptacle. Covers that do not provide weatherproof conditions when something is plugged into the receptacle are all right to use, but only when used on a temporary basis such as for portable tools. The cord will be unplugged as soon as the person using the receptacle is finished. Examples of this might be hedge trimmers, extension cords, grills, and similar appliances. The self-closing spring-loaded cover will automatically close tightly just as soon as the plug is pulled out.

Figure 4–27 shows a flush mounted receptacle that meets the weatherproof requirements of *Section 410-57(b)*. Because the receptacle itself is set far back into the wall box, an attachment plug cap can be plugged into the receptacle, and still allow the cover to close tightly, making the installation in compliance with *Section 410-57(b)*.

Outdoor receptacles might also be installed away from the building, such as in flower beds or under trees and shrubbery for supplying 120-volt or low-voltage decorative lighting. Figure 4–28 illustrates the *NEC®* requirements relating to outdoor electrical conduit bodies that will be supported by conduits that rise up from the ground. Generally, these type of enclosures have ½-inch female threaded openings or hubs in which to secure the conduits. Any unused openings must be closed with ½-inch "plugs" that are screwed tightly into the threaded hubs.

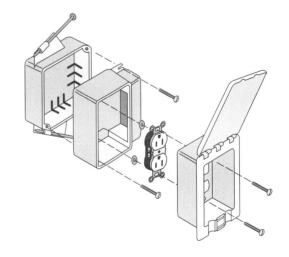

Figure 4–27   Photo of a recessed type of weatherproof receptacle that meets the requirements of *Section 410-57(b)*. This section requires that the enclosure be weatherproof while in use if unattended, where the cord will be plugged in for long periods of time. The "exploded" view illustrates one method of "roughing-in" this type of weatherproof receptacle. (*Photos Courtesy of TayMac Corporation*)

RIGID METAL, RIGID NONMETALLIC, INTER-MEDIATE METAL CONDUIT, OR ELECTRICAL METALLIC TUBING OK TO SUPPORT BOXES THAT DO NOT HAVE DEVICES OR FIXTURES. RIGID METAL OR INTERMEDIATE METAL CONDUIT OK TO SUPPORT BOXES THAT DO HAVE DEVICES OR FIXTURES. RIGID NONMETALLIC CONDUIT NOT PERMITTED TO SUPPORT LIGHTING FIXTURES OR OTHER EQUIPMENT, *SECTION 347-3(b)*.

THIS BODY IS CONSIDERED TO BE ADEQUATELY SUPPORTED.

A WHEN TWO OR MORE CONDUITS ARE TIGHTLY THREADED INTO THE HUBS OF A CONDUIT BODY, THE CONDUIT BODY IS CONSIDERED TO BE ADEQUATELY SUPPORTED, *SECTION 370-23(e)*. FOR ENCLOSURES THAT SUPPORT FIXTURES OR CONTAIN WIRING DEVICES, THE ENCLOSURE MUST BE SUPPORTED WITHIN 18 INCHES (457 mm). CONDUITS COMING OUT OF THE GROUND ARE ACCEPTABLE AS MEETING THIS SUPPORT REQUIREMENT, *SECTION 370-23(f)*.

B THIS CONDUIT BODY IS NOT ADEQUATELY SUPPORTED BY THE ONE CONDUIT THREADED INTO THE HUB. THIS CONDUIT BODY COULD TWIST VERY EASILY, RESULTING IN DAMAGED INSULATION ON THE CONDUCTORS AND A POOR GROUND CONNECTION BETWEEN THE CONDUIT AND THE CONDUIT BODY. SEE *SECTIONS 370-23(e)* AND *(f)*.

THIS BODY IS NOT CONSIDERED TO BE ADEQUATELY SUPPORTED.

C CONDUCTORS MAY BE SPLICED IN THESE CONDUIT BODIES ONLY IF THE CONDUIT BODY IS MARKED WITH ITS CUBIC-INCH CAPACITY SO THAT THE PERMISSIBLE CONDUCTOR FILL MAY BE DETERMINED USING THE CONDUCTOR VOLUME FOUND IN *TABLE 370-16(b)*.

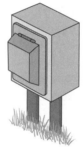

SPLICES MAY BE MADE IN CONDUIT BODIES WHEN MARKED WITH THEIR CUBIC-INCH CAPACITY, *SECTION 370-16(c)(2)*.

Figure 4–28 **Supporting threaded conduit bodies. Refer to** *Section 370-23(e)* **for supporting enclosures that will not contain devices or will not support fixtures. Refer to** *Section 370-23(f)* **for supporting enclosures that will contain devices or will support fixtures.**

Receptacles in Floors

On occasion, because of rather large sliding glass doors, or window-type walls that almost touch the floor making it impossible to install receptacles in the wall, it might be necessary to install one or two receptacles in the floor. According to *Section 210-52(a)(1)*, no point along the floor line shall be more that 6 feet from a receptacle in homes.

On other occasions, a floor receptacle might come in handy under the dining room, breakfast room, or dinette room table. Warming trays and other electrical appliances can be placed on the table, then plugged into the floor receptacle under the table instead of running extension cords across the floor from the table to the wall receptacle.

Do not install a conventional receptacle in the floor! Over time, dirt, insects, foreign objects, and water will accumulate in the receptacle and box.

Section 370-27(b) requires that floor boxes specifically be "listed" for installation in the floor. As mentioned many times in this text, "listed" means that the product has been tested by a recognized testing laboratory, such as Underwriters Laboratories, and has passed the tests. These boxes come complete with a special brass cover that has one or two removable threaded plugs, about the size of a silver dollar. These plugs can be removed by inserting a coin or screwdriver into the slot on the threaded plug. Turn counterclockwise to remove the plug—turn clockwise to replace the plug. These plugs keep dirt and water out of the receptacle when not in use.

Another type of cover has spring-loaded covers that automatically snap back tightly when the receptacle is not in use.

Figure 4–29 shows three different styles of floor receptacles.

Receptacles Above Electric Baseboard Heaters

"Listed" electric baseboard heaters must not be installed below wall receptacle outlets unless the instructions furnished with the baseboard heaters indicate that they may be installed below receptacle outlets. This requirement is found in the *Fine Print Note* following the second paragraph of *Section 210-52.* Cords that hang over and rub on an electric baseboard heater could become damaged. The insulation on the cord might melt, allowing the bared "hot" conductor to touch the metal of the heater. The resulting "ground fault" would cause arcing and sparking that could start a fire. A shock hazard is also present if the "hot" conductor in the cord is exposed because of damaged insulation (Figure 4–30).

Section 210-52(a)(1) requires that receptacles must be placed so that no point along the floor line in any wall space is more than 6 feet from an outlet. When baseboard electric heaters are permanently installed, careful consideration must be given to receptacle placement. Consider installing an electric baseboard heater that has a factory-installed receptacle on it (Figure 4–31). Such receptacles may be counted in the "6-foot rule," but must not be connected to the electric heater's branch circuit. This is permitted by *Section 210-52,* second paragraph.

Another way to solve the receptacle placement dilemma is to use "listed" blank spacers between electric baseboard heating sections. Receptacles positioned above wide spacers (12 inches or more)

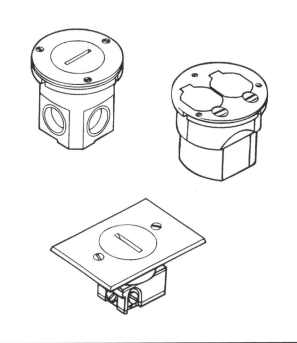

Figure 4–29 Three types of receptacles that are permittted to be installed in floors as required by *Section 370-27(b)* of the *National Electrical Code.* (*Courtesy of Hubbell Electrical Products, a Division of Hubbell Incorporated [Delaware]*)

are not a problem because it is highly unlikely that the cord plugged into the receptacle would hang over the heating section of the baseboard and thus, would not be subjected to the extreme heat. Refer to the instructions furnished with the heater. High-density electric heaters have more wattage per foot and are 'hotter to the touch" than low-density electric heaters. If there still is some doubt as to whether or not this installation is acceptable, discuss it with the electrical inspector *before* making the installation. (Figure 4–32).

Receptacles for Room Air Conditioners

The Code rules for central air conditioning equipment are found in *Article 440* of the *National Electrical Code.* Branch-circuit calculations for central air-conditioning equipment are discussed in Unit 13.

For homes that do not have central air-conditioning, window units or through-the-wall air-conditioning units are installed. These types of air conditioners are available in both 120-volt and

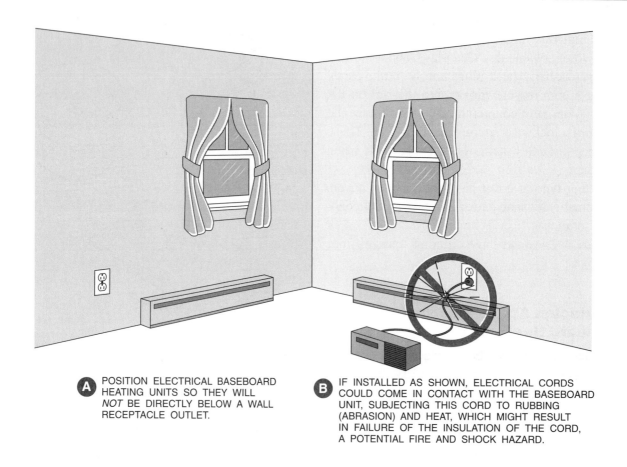

A POSITION ELECTRICAL BASEBOARD HEATING UNITS SO THEY WILL *NOT* BE DIRECTLY BELOW A WALL RECEPTACLE OUTLET.

B IF INSTALLED AS SHOWN, ELECTRICAL CORDS COULD COME IN CONTACT WITH THE BASEBOARD UNIT, SUBJECTING THIS CORD TO RUBBING (ABRASION) AND HEAT, WHICH MIGHT RESULT IN FAILURE OF THE INSULATION OF THE CORD, A POTENTIAL FIRE AND SHOCK HAZARD.

Figure 4–30   When positioning receptacles where permanently installed electric baseboard heaters are involved, do not install the receptacles above the heater. Position them to one side of the heater (A). Damage to an electrical cord could result if the receptacle were to be positioned above the heater (B). Or use factory-installed receptacles as illustrated in Figure 4–31.

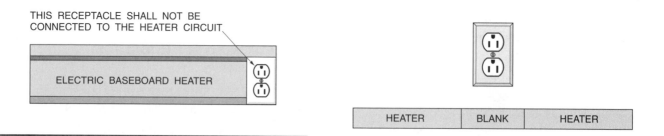

Figure 4–31   Factory-installed receptacle outlets or receptacle outlet assemblies provided by the manufacturer for use with its electric baseboard heaters may be counted as the required receptacle outlet for the space occupied by a permanently installed electric baseboard heater.

Figure 4–32   It may be all right to install a receptacle above a wide blank spacer in an electric baseboard heater installation. Check the instructions furnished with the heater.

240-volt ratings. Because room air conditioners are plug- and cord-connected, the receptacle outlet and the circuit capacity must be selected and installed according to applicable Code regulations.

The Code requirements for room air-conditioning units are found in *Sections 440-60* through *440-64*. To make it easy to calculate circuit requirements for a room air conditioner, the *National Electrical Code®* in *Section 440-62(a)* considers a room air conditioner to be a single motor unit, even though the unit contains two motors: a hermetic motor compressor and a fan.

The basic Code rules for installing these cord- and plug-connected units are as follows:

- The air conditioners must be grounded. This is accomplished through the attached cord, which contains an equipment grounding conductor (EGC). Internal to the attachment plug cap, this EGC is connected to the round blade, which is longer than the other two flat blades. Internal to the air conditioner unit, the EGC is connected to the metal frame of the unit. Grounding of appliances is covered in Units 12 and 13.

- The air conditioners must be connected using a cord and attachment plug.

- Maximum length of cord: 10 feet (3.05 m) for 120-volt units and 6 feet (1.83 m) for 240-volt units.

- The air conditioner rating may not exceed 40 amperes at 250 volts, single-phase.

- The rating of the branch-circuit overcurrent device must not exceed the branch-circuit conductor rating or the receptacle rating, whichever is less.

- The air conditioner load shall not exceed 80% of the branch-circuit ampacity if no other loads are served.

- The air conditioner load shall not exceed 50% of the branch-circuit ampacity if other loads are served.

- The attachment plug cap may serve as the disconnecting means. The unit itself has manual ON/OFF and temperature controls.

Some of the preceding requirements are already met when the unit bears the Underwriters Laboratories label. Others of the preceding require-

ments need to be met by the electrician installing the branch-circuit wiring for the air conditioner unit.

Small room air conditioners probably can be plugged into one of the receptacles that is connected to a conventional 15-ampere branch circuit without tripping the breaker or blowing a fuse. However, larger room air conditioners draw a considerable amount of current. The best solution is to install a separate 15- or 20-branch circuit. These separate branch circuits would be 120 volts or 240 volts, depending on the type of air conditioner to be installed. It would be best to install a single-receptacle device rather than a duplex so that there is no misunderstanding that the receptacle is intended for a specific use—that of supplying the air conditioner.

Figure 4–33 is a 15-ampere, 250-volt NEMA configuration 6-15R.

Figure 4–34 is a 20-ampere, 250-volt NEMA configuration 6-20R.

Figure 4–33 This is a 15-ampere, 250-volt receptacle and is a NEMA 6-15R configuration. The letters NEMA stand for National Electrical Manufacturers Association.

Figure 4–34 This is a 20-ampere, 250-volt receptacle and is a NEMA 6-20R configuration. The letters NEMA stand for National Electrical Manufacturers Association.

The tandem blade configuration in Figures 4–33 and 4–34 meets the requirements of *Section 210-7(f)*, which states that receptacle outlets for different voltage levels must be non interchangeable with each other. It would be impossible to plug in the familiar plug cap of, for example, a computer because the configuration of the 125-volt and 250-volt receptacles and plug caps are very different.

Receptacles For Electric Ranges And Dryers

Residential electric ranges, electric dryers, wall-mounted ovens, and surface-mounted cooking units may be connected with cords designed specifically for this use. Depending on the appliances' wattage rating, the receptacles are usually rated 30 amperes or 50 amperes, 125/250 volts. Cord sets rated 40 amperes or 45 amperes contain 40- or 45-ampere conductors, yet the plug cap is rated 50 amperes. It becomes a cost issue to use a full 50-ampere-rated cord set in all cases, or to match the cord ampere rating as closely as possible to that of the appliance being connected. Load calculations for ranges and dryers are covered in Unit 13.

Range and dryer receptacles are available in both surface mount and flush mount.

Blade and Slot Configuration for Plug Caps and Receptacles

Figure 4–35 shows the different configurations for receptacles commonly used for electric range and dryer hook-ups. The National Electrical Manufacturers Association (NEMA) uses the letter "R" to indicate a receptacle, and the letter "P" for plug cap.

30-ampere, 3-wire. The L-shaped slot on the receptacle (NEMA 10-30R) and the L-shaped blade on the plug cap (NEMA 10-30P) are for the white neutral wire. Frames of electric ranges, ovens, and clothes dryers installed prior to the adoption of the 1996 *National Electrical Code®* were permitted to be grounded to the neutral conductor as stated in *Sections 250-60* and *250-61* of previous editions of the *NEC.®* The neutral conductor served a dual purpose, that of the neutral and that of the appliance's required equipment ground. Since the 1996 *NEC®* this permission is for existing branch circuits only, *Section 250-140.*

30-ampere, 4-wire. The L-shaped slot on the receptacle (NEMA 14-30R) and the L-shaped blade on the plug cap (NEMA 14-30P) are for the white neutral wire. The horseshoe-shaped slot on the receptacle and the round- or horseshoe-shaped blade on the plug cap are for the equipment ground.

50-ampere, 3-wire. The wide flat slot on the receptacle (NEMA 10-50R) and the matching wide blade on the plug cap (NEMA 10-50P) are for the white neutral wire. Frames of electric ranges, ovens, and clothes dryers installed prior to the adoption of the 1996 *National Electrical Code®* were permitted to be grounded to the neutral conductor as stated in *Sections 250-60* and *250-61* of previous editions of

50-ampere, 3-pole, 3-wire, 125/250-volt NEMA 10-50R permitted prior to 1996 *NEC®*

50-ampere, 4-pole, 4-wire, 125/250-volt NEMA 14-50R required since the 1996 *NEC®*

30-ampere, 3-pole, 3-wire, 125/250-volt NEMA 10-30R permitted prior to 1996 *NEC®*

30-ampere, 4-pole, 4-wire, 125/250-volt NEMA 14-30R required since the 1996 *NEC®*

Figure 4–35 Illustrations of 30-ampere and 50-ampere receptacles used for cord- and plug connection of electric ranges, ovens, counter-mounted cooking units, and electric clothes dryers. Four-wire receptacles and four-wire cord sets are now required on new electric range and dryer branch-circuit installations. Prior to the adoption of the 1996 *NEC,®* three-wire receptacles and three-wire cord sets were permitted. See *Sections 250-140* and *142(b)* of the *National Electrical Code.®*

the *NEC.* The neutral connection served a dual purpose, that of the neutral and that of the appliance's required equipment ground. Now, this permission is for existing branch circuits only.

50-ampere, 4-wire. The wide flat slot on the receptacle (NEMA 14-50R) and the matching wide blade on the plug cap (NEMA 14-50P) are for the white neutral wire. The horseshoe shaped slot on the receptacle and the round- or horseshoe-shaped blade on the plug cap are for the equipment ground.

Figure 4-36 shows the connections required by the current *National Electrical Code*® for the cord connection of electric ranges, wall-mounted ovens, surface cooking units, and electric dryers. Four-wire cord sets and four-wire receptacles are required. The receptacle, the cord set, and the branch circuit have a separate equipment grounding conductor to serve as a means to ground the frame of the appliance in accordance with *Sections 250-114, 118, 134,* and *138* of the *National Electrical Code.*®

Figure 4-37 shows the connections permitted prior to the adoption of the 1996 *National Electrical Code*® for electric range and electric dryer receptacles. Note how the appliance is grounded to the neutral conductor. The neutral conductor is serving as both the neutral conductor and the equipment

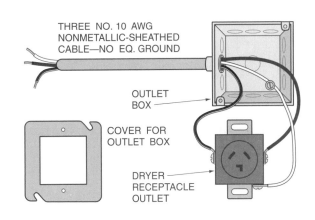

Figure 4–37 Prior to the 1996 *National Electrical Code*, it was permitted to ground the junction box to the neutral conductor, but only when the box was part of the circuit serving electric ranges, wall-mounted ovens, surface-cooking units, and electric clothes dryers. The frames of these appliances were also grounded to the neutral conductor, which served the appliance as both the neutral conductor and the equipment grounding conductor. This is no longer permitted! See *Section 250-140* of the *NEC.*®

grounding conductor. For new installations, this three-wire cord set and three-wire receptacle method is not permitted. However, for existing branch-circuit installations, nothing need be changed or rewired.

All of this is covered in *Section 250-140* of the *National Electrical Code*®.

Grounding of appliances and other electrical equipment is discussed in Units 12 and 13.

Cord Sets for Electric Ranges and Dryers

The following text explains the ratings, conductor sizes, and configurations for range and dryer cords.

A 30-AMPERE CORD SET CONTAINS:

- **3-wire:** three No. 10 AWG conductors. The attachment plug cap (NEMA 10-30P) is rated 30 amperes.

- **4-wire:** four No. 10 AWG conductors. The attachment plug cap (NEMA 14-30P) is rated 30 amperes.

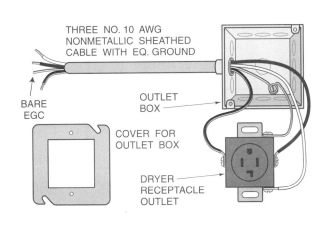

Figure 4–36 New installations of branch-circuit wiring for electric ranges, wall-mounted ovens, surface-cooking units, and electric clothes dryers require that a separate equipment grounding conductor be used. The frames of these appliances shall not be grounded to the neutral conductor! See *Section 250-140* of the *NEC.*®

A 40-AMPERE CORD SET CONTAINS:

- **3-wire:** three No. 8 AWG conductors and one No. 10 AWG conductor. The attachment plug cap (NEMA 10-50P) is rated 50 amperes.

- **4-wire:** three No. 8 AWG conductors and one No. 10 AWG conductor. The attachment plug cap (NEMA 14-50P) is rated 50 amperes.

A 45-AMPERE CORD SET CONTAINS:

- **3-wire:** two No. 6 AWG conductors and one No. 8 AWG conductor. The attachment plug cap (NEMA 10-50P) is rated 50 amperes.

- **4-wire:** three No. 6 AWG conductors and one No. 8 AWG conductor. The attachment plug cap (NEMA 14-50P) is rated 50 amperes.

A 50-AMPERE CORD SET CONTAINS:

- **3-wire:** two No. 6 AWG, and one No. 8 AWG conductors. The attachment plug cap (NEMA 10-50P) is rated 50 amperes

- **4-wire:** three No. 6 AWG, and one No. 8 AWG conductors. The attachment plug cap (NEMA 14-50P) is rated 50 amperes.

Terminal Identification for Receptacles and Cords

Receptacle and cord terminals are marked:

- "X" and "Y" for the ungrounded conductors.

- "W" for the white grounded conductor. The "W" terminals will generally be whitish or silver (tinned) in color, in accordance with *Section 200-9* and *200-10* of the *NEC.*

- "G" for the equipment grounding conductor. This terminal is green-colored, hexagon shaped, and is marked "G," "GR," "GRN," or "GRND" in accordance with *Sections 250-126* and *410-58(b)* and *(d)* of the *NEC.* The grounding blade on the plug cap must be longer than the other blades so that its connection is made before the ungrounded conductors make connection in accordance with *Section 410-58(d)* of the *NEC.*

CONNECTIONS AND SPLICES

There are a number of Code related issues relating to the proper making up of electrical connections and splices. The following discusses these concerns.

Screw Terminals

Wiring devices (switches and receptacles) usually have screw terminals under which the conductors are terminated. For the best connection, bend a "hook" on the wire. Then place the wire under the screw, so that as the screw is tightened in the clockwise direction, the hook will tend to turn tighter. Putting the wire under the screw in the opposite direction will tend to allow the wire to be pushed out from under the screw terminal as the screw is tightened (Figure 4–38). A screw terminal is designed for one conductor. Do not connect more than one conductor under these screws.

Push-In Terminals

Some wiring devices have back-wiring holes in addition to the screw terminals. Figure 4–18 (page 61) is an exploded illustration of a typical grounding-type receptacle, showing the back-wiring holes. You have the choice of using the screw terminals or the

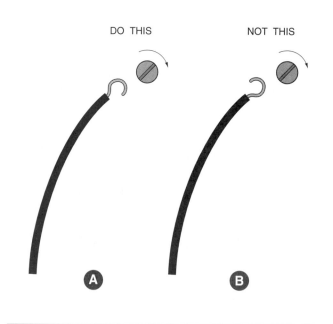

DO THIS   NOT THIS

A   B

Figure 4–38 The best way to attach a conductor to a screw terminal is illustrated in (A) because the loop on the wire will pull tighter as the screw terminal is turned tighter (clockwise). Placing the wire under the screw terminal as shown in (B) is not recommended because the loop on the wire will tend to unwind and come out from under the terminal as the screw terminal is turned tighter (clockwise).

back-wiring holes. The insulation is stripped from the conductor about ½ inch, then inserted into the holes or under the screw. The screw is then tightened.

Attaching Wires to a Receptacle

The inexperienced person, seeing two holes and two screw terminals on each side of a receptacle, quite often will use all of the back-wiring holes and the screw terminals, using these terminals to make up the necessary splices.

On a receptacle, this could mean that as many as four wires would be connected to the white terminal and four wires to the "hot" terminal. Laying the wires back into the box neatly and without damaging the insulation on the conductors is very difficult, if not impossible. Replacing a receptacle connected in this manner is extremely difficult. Although using the wiring device as a splicing means is permitted, there is a better way to do it. Figure 4–39 is an actual photo showing how the conductor splicing has been done independent of the receptacle. Pigtails (short lengths of wire) are included in the splice, and these pigtails in turn connect to the receptacle. If the receptacle ever has to be replaced, it can be disconnected without opening any of the splices.

Correct Size of Pigtails

Because the short pigtail is actually part of the branch circuit, it **must** be the same size as the branch-circuit conductors. For example, the required small appliance branch circuits in kitchens and dining rooms are rated 20 amperes and are wired with No. 12 AWG copper conductors. Therefore, the pigtails also must be No. 12 AWG copper conductors.

Limitations for Push-In Terminals

"Push-in" terminals on receptacles are no longer permitted for use with No. 12 AWG conductors. They are "listed" by Underwriters Laboratories for use **only** with No. 14 AWG solid copper conductors or with No. 14 AWG solid copper-clad conductors. By design, the holes are only large enough to take one No. 14 AWG solid conductor.

Do not use "push-in" terminals with stranded conductors. Do not use "push-in" terminals with aluminum conductors.

If the receptacle also has screw terminals, these terminals can be used for terminating a No. 12 AWG

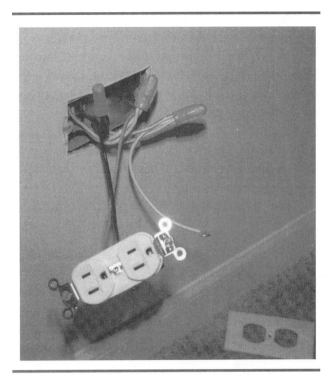

Figure 4–39 Photo of a receptacle connected to the branch-circuit wiring with short pigtails. One outlet is switched and one outlet is on continuously. Six conductors enter the box. A 4–inch square outlet box is trimmed with a single-gang raised plaster cover. All of the splices are independent of the receptacle, and can be neatly folded back into the box. It is then easy to lay the three conductors that are connected to the receptacle back into the box and fasten the receptacle to the box. Since the pigtails are really an extension of the branch circuit, they must have the same current-carrying capacity (ampacity) as the branch-circuit ampere rating.

conductor. "Push-in" terminals for No. 12 AWG solid copper and No. 12 AWG solid copper-clad aluminum conductors are still acceptable on switches.

Requirements for Electrical Connections and Terminations

The *National Electrical Code®* covers electrical connections in *Section 110-14.* You can also find easy-to-understand information regarding electrical connections in the UL White Book.

When splicing wires or connecting a wire to a switch, fixture, circuit breaker, panelboard, meter socket, or other electrical equipment, usually some type of solderless wire connector is used.

Wire connectors are known in the trade by such names as *screw terminal, pressure terminal connec-*

tor, wire connector, Wing nut, Wire-nut, Scotchlok, split-bolt connector, pressure cable connector, solderless lug, soldering lug, solder lug, and others. Solder-type lugs and connectors are rarely used today. In fact, connections that depend on solder are not permitted for connecting service-entrance conductors to service equipment, *Section 230-81.*

The labor costs and time spent make the use of solder-type connections prohibitive. Solderless con-

nectors, designed to establish connections by means of mechanical pressure, are quite common. Examples of some types of wire connectors, and their uses, are shown in Figure 4–40.

As with the terminals on wiring devices (switches and receptacles), wire connectors must be marked *AL* when they are to be used with aluminum conductors. This marking is found on the connector itself, or appears on or in the shipping carton.

CRIMP CONNECTORS USED TO SPLICE AND TERMINATE NO. 20 AWG TO 500 KCMILS ALUMINUM-TO-ALUMINUM, ALUMINUM-TO-COPPER, OR COPPER-TO-COPPER.

PROPERLY CRIMP THEN TAPE

CONNECTORS USED TO CONNECT WIRES TOGETHER ON COMBINATIONS OF NO. 18 AWG THROUGH NO. 6 AWG. THEY ARE TWIST-ON, SOLDERLESS, AND TAPELESS.

*WIRE-NUT® and WING-NUT® are registered trademarks of IDEAL INDUSTRIES, INC. Scotchlok® is a registered trademark of MINNESOTA MINING.

WIRE CONNECTORS VARIOUSLY KNOWN AS WIRE-NUT,® WING NUT,® AND SCOTCHLOK.®

CONNECTORS USED TO CONNECT WIRES TOGETHER IN COMBINATIONS OF NO. 16, NO. 14, AND NO. 12 AWG. THEY ARE CRIMPED ON WITH A SPECIAL TOOL, THEN COVERED WITH A SNAP-ON INSULATING CAP OR WRAPPED WITH INSULATING TAPE.

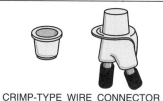

CRIMP-TYPE WIRE CONNECTOR AND INSULATING CAP

SOLDERLESS CONNECTORS ARE AVAILABLE IN SIZES NO. 14 AWG THROUGH 500 MCM. THEY ARE USED FOR ONE SOLID OR ONE STRANDED CONDUCTOR ONLY, UNLESS OTHERWISE NOTED ON THE CONNECTOR OR ON ITS SHIPPING CARTON. THE SCREW MAY BE OF THE STANDARD SCREWDRIVER SLOT TYPE, OR IT MAY BE FOR USE WITH AN ALLEN WRENCH OR SOCKET WRENCH.

SOLDERLESS CONNECTORS

COMPRESSION CONNECTORS ARE USED FOR NO. 8 AWG THROUGH 1000 MCM. THE WIRE IS INSERTED INTO THE END OF THE CONNECTOR, THEN CRIMPED ON WITH A SPECIAL COMPRESSION TOOL.

COMPRESSION CONNECTOR

SPLIT-BOLT CONNECTORS ARE USED FOR CONNECTING TWO CONDUCTORS TOGETHER, OR FOR TAPPING ONE CONDUCTOR TO ANOTHER. THEY ARE AVAILABLE IN SIZES NO. 10 AWG THROUGH 1000 MCM. THEY ARE USED FOR TWO SOLID AND/OR TWO STRANDED CONDUCTORS ONLY, UNLESS OTHERWISE NOTED ON THE CONNECTOR OR ON ITS SHIPPING CARTON.

SPLIT-BOLT CONNECTOR

Figure 4–40 Various types of solderless wire connectors.

Connectors marked *AL/CU* are suitable for use with aluminum, copper, or copper-clad aluminum conductors. This marking is found on the connector itself, or it appears on or in the shipping carton. Connectors not marked AL or AL/CU are for use with copper conductors only.

Unless marked on or in the shipping carton, or on the connector itself, conductors made of copper, aluminum, or copper-clad aluminum shall not be used in combination in the same connector. When combinations of conductors are permitted, the connector will be identified for the purpose and for the conditions of where they may be used. The conditions usually are limited to dry locations only. There are some "twist-on" wire connectors that are "listed" for use with combination of copper and aluminum conductors.

Where moisture or dampness are expected, twist-on connectors are available that are filled with a silicone sealant so that after the splice is completed, the conductors will be completely encapsulated with silicone. This keeps the moisture away from the conductors, thus preventing corrosion of the conductors and the connector's internal metal spring.

Wire connectors and other splicing devices are not permitted to be used underground unless they are "listed" for such use.

Figure 4–41 is a chart showing how wiring devices and wire connectors are marked relative to their permitted use with copper, copper-clad, and aluminum conductors.

Always read the label on the box or package to be sure that you are using these solderless connectors in a proper manner.

ALUMINUM CONDUCTORS

When using aluminum conductors, most problems will show up at the termination. That is why it is so important to check the markings on terminals, lugs, and connectors to make sure of the type of conductors that are permitted to be used with that particular connector. Trouble-free connections for aluminum conductors require the use of the proper connector.

Some common problems associated with aluminum conductors when not properly connected can be summarized as follows:

TYPE OF DEVICE	MARKING ON TERMINAL OR CONDUCTOR	CONNECTOR PERMITTED
15- or 20-ampere receptacles and switches	CO/ALR	aluminum, copper, copper-clad aluminum
15- and 20-ampere receptacles and switches	NONE	copper, copper-clad aluminum
30-ampere and greater receptacles and switches	AL/CU	aluminum, copper, copper-clad aluminum
30-ampere and greater receptacles and switches	NONE	copper only
Screwless pressure terminal connectors of the push-in type	NONE	copper or copper-clad aluminum
Wire connectors	AL	aluminum
Wire connectors	AL/CU	aluminum, copper, copper-clad aluminum
Wire connectors	CC	copper-clad aluminum only
Wire connectors	CC/CU	copper or copper-clad aluminum
Wire connectors	CU	copper only
Any of the above devices	COPPER OR CU ONLY	copper only

Figure 4–41 Terminal and connector identification markings showing the acceptable types of conductors permitted to be connected to a specific type of terminal or connector. This data is found in UL Standards.

- A corrosive action is set up when dissimilar wires come in contact with one another when moisture is present.

- The surface of aluminum oxidizes as soon as it is exposed to air. If this oxidized surface is not broken through, a poor connection results. When installing aluminum conductors, particularly in large sizes, an inhibitor is brushed onto the aluminum conductor, then the conductor is scraped with a stiff brush where the connection is to be made. The process of scraping the conductor breaks through the oxidation, and the inhibitor keeps the air from coming into contact with the conductor. Thus, further oxidation is prevented. Aluminum connectors of the compression-type usually have an inhibitor paste already factory installed inside of the connector.

- Aluminum wire expands and contracts to a greater degree than does copper wire, lug, or connector for a given load. Technically, aluminum and copper have a different coefficient of expansion. When heavily loaded, the aluminum conductor will expand. If the connector does not expand at the same rate, the aluminum conductor will *cold flow* or be squeezed ever so slightly out of the connector. When the load is removed, the aluminum conductors contract more than the copper wire, lug, or connector. This action is repeated over and over again, until finally, we have a poor connection unable to carry the current that the connection is trying to carry. Tremendous heat can build up. Arcing can occur. Crimp connectors for aluminum conductors are usually longer than those for comparable copper conductors, thus resulting in greater contact surface of the conductor in the connector.

Trouble-free connections for aluminum or copper-clad aluminum conductors require terminals, lugs, and connectors that are suitable for the type of conductors being installed.

Terminals on receptacles and switches must be suitable for the conductors being attached.

The *National Electrical Code®* discusses all of the above in *Sections 110-14, 380-14(c),* and *410-56(b).*

REVIEW QUESTIONS

1. Switches used for controlling lighting are referred to in the *National Electrical Code®* and in the UL Standards as:

   _____

2. General-use snap switches rated "AC Only" can be used on resistive loads not to exceed _____ of the switches' rating.

   a) 50%

   b) 80%

   c) 100%

3. General-use snap switches rated "AC Only" can be used on tungsten-filament lamp loads not to exceed _____ of the switches' rating.

   a) 50%

   b) 80%

   c) 100%

4. General-use snap switches rated "AC Only" can be used on motor loads not to exceed _____ of the switches' rating.

   a) 50%

   b) 80%

   c) 100%

5. General-use snap switches rated "AC/DC" can be used on resistive loads not to exceed _____ of the switches' rating.

   a) 50%

   b) 80%

   c) 100%

6. General-use snap switches rated "AC/DC" can be used on tungsten-filament lamp loads not to exceed _____ of the switches' rating.

    a) 50%

    b) 80%

    c) 100%

7. General-use snap switches rated "AC Only" can be used on motor loads not to exceed _____ of the switches' rating.

    a) 50%

    b) 80%

    c) 100%

8. Does the *National Electrical Code®* permit switches to be installed in wet locations such as in tub and shower spaces? What section of the Code applies to this question?

    a) Yes

    b) No

9. Dimmers are connected in _____ with a lighting load.

    a) series

    b) parallel

10. Electronic dimmers are very popular to control lighting levels. The most common electronic dimmer used in residential applications is rated 600 watts, 120 volts. Fluorescent lighting fixtures are installed above and on both sides of a mirror in a bathroom. May a standard electronic dimmer be used to dim the fluorescent lamps? Explain.

    a) Yes

    b) No

_____

11. Should a standard electronic dimmer switch be used to control the speed of a ceiling paddle fan? Explain.

    a) Yes

    b) No

_____

12. Loose wall receptacles are a problem. This usually is the result of the wall box set back, and not flush with the plaster, dry wall, or wood paneling. The plaster ears on the receptacle may not result in a permanent tight support for the receptacle. To ensure tight securing of the receptacle, what else could be done?

_____

_____

13. Is it always necessary to connect the bare equipment grounding conductor to the green hexagon-shaped equipment grounding screw on a receptacle? Explain.

    a) Yes

    b) No

_____

_____

14. How are most conventional duplex receptacles converted into a split-circuit receptacle?

_____

15. Ground-fault circuit-interrupter receptacles are easily identified because they have two "buttons" on them. How are these "buttons" marked?_____

_____

16. The *National Electrical Code®* requires that switches and receptacles located outdoors be mounted in weatherproof enclosures. This is easily accomplished by using suitable spring-loaded covers that will close over the switch or receptacle, thereby keeping water out. The *NEC®* requires that if a cord is to be permanently plugged into a receptacle, then the weatherproof cover must be such that it will be weatherproof with the plug inserted. This requirement is found in *Section* _____ of the *NEC.®*

17. In a flower/shrub/rock garden, you are installing a weatherproof receptacle outlet into which outdoor decorative lighting will be plugged. You will run ½-inch conduit underground, and bring the conduit up and out of the ground into one of the bottom ½-inch threaded openings. Does this ½-inch conduit adequately support the enclosure (box)? What section of the Code covers this type of installation?

    a) Yes   b) No

_____

18. What hazard might present itself if an electric baseboard heater is installed below a wall receptacle? Give a *NEC®* reference that discusses this issue._____

_____

_____

19. You are going to install a through-the-wall room air conditioner. In addition to the installation instructions furnished with the unit, where in the *National Electrical Code®* would you look for more information? _____

20. On a new installation, you are to install a 50-ampere, 250-volt receptacle for an electric range. *Section 250-140* of the *National Electrical Code®* prohibits grounding the frame of an electric range to the grounded neutral conductor of the branch circuit. It will be necessary to install a _____ receptacle. What is the NEMA designation for this receptacle?

    a) 3-wire
    b) 4-wire

21. Match the following statements with the correct letter.

    1. ___ A 30-ampere, 3-wire receptacle   a) For the equipment grounding conductor
    2. ___ A 50-ampere, 4-wire receptacle   b) For the white neutral conductor
    3. ___ The "L"-shaped blade on a   c) For the "hot" ungrounded conductors
       30-ampere plug cap
                                                   d) NEMA 10-30R
    4. ___ The "L" shaped slot on a
       50-ampere receptacle   e) NEMA 14-50R

    5. ___ A 50-ampere, 4-wire plug cap   f) NEMA 14-50P
    6. ___ Terminals "X" and "Y"
    7. ___ Terminal "W"
    8. ___ Terminal "G"

22. In a wall box, short 6-inch "pigtails" are spliced to the No. 12 AWG conductors of a 20-ampere small appliance branch circuit. These pigtails will be connected to the terminal of the duplex receptacle. What size conductors must be used for these pigtails?

_____

23. The marking on a duplex receptacle shows the letters *CO/ALR*. This means that the terminals are suitable for use with:

   a) copper conductors only.

   b) copper or aluminum conductors.

24. The marking on a duplex receptacle provides no reference to the type of conductors permitted to be connected to the terminals of the receptacle. This means that the terminals are suitable for use with:

   a) copper conductors only.

   b) copper or aluminum conductors.

Switch Control of Lighting Circuits, Bonding and Grounding of Receptacles

OBJECTIVES

After studying this unit, you will be able to:

- **Learn about the required length of conductors to be left at switch and outlet boxes for ease in making up the connections of the wiring devices and lighting fixtures.**

- **Understand how conductors are identified by different colored insulation.**

- **Grasp the importance for proper color coding according to the *National Electrical Code*.**

- **Learn about the different types of switches used for residential wiring.**

- **Learn how to make connections for switch control of lighting circuits from one or more than one location.**

- **Know the proper way to make connections to a lampholder.**

- **Learn how to bond the grounding terminals of receptacles to the box.**

INTRODUCTION

Lights and appliances work properly when circuits are hooked up properly. This unit provides a variety of wiring diagrams for connecting switches and receptacles in conformance to the *National Electrical Code*. Because the color of the insulation on the conductors in a cable is determined by industry standards, the electrician must know how to make connections using established color coding practices.

REQUIRED LENGTH OF CONDUCTORS IN ELECTRICAL BOXES

This unit contains many different wiring diagrams for wiring a typical residence.

One of the things that bothers electricians is working with conductors that are not long enough.

Making up connections, terminations, and splices properly is difficult enough, let alone having to work with conductors that are too short.

Section 300-14 sets forth the requirements for how much conductor length must be left at outlet boxes, device boxes, junction boxes, and at switch points for splicing and for the connections at fixtures and wiring devices. These measurements are shown in Figure 5–1. Conductors that are too short are difficult to work with. Leaving too much wire length will usually result in having to crowd, push, and jam the wires into the box. The insulation on the conductors could be damaged, resulting in a short circuit or ground fault, which is not good. Be practical. During the "rough-in" stage of wiring, allow 8 to 10 inches, then cut to proper length when installing receptacles, switches, and lighting fixtures.

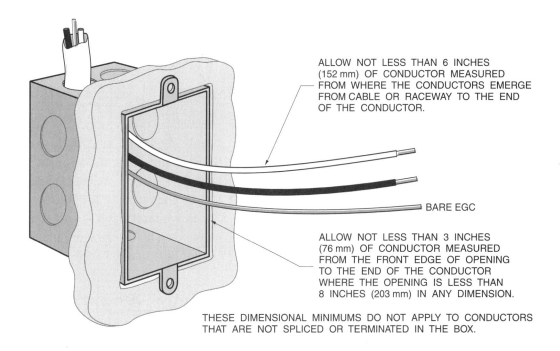

ALLOW NOT LESS THAN 6 INCHES (152 mm) OF CONDUCTOR MEASURED FROM WHERE THE CONDUCTORS EMERGE FROM CABLE OR RACEWAY TO THE END OF THE CONDUCTOR.

BARE EGC

ALLOW NOT LESS THAN 3 INCHES (76 mm) OF CONDUCTOR MEASURED FROM THE FRONT EDGE OF OPENING TO THE END OF THE CONDUCTOR WHERE THE OPENING IS LESS THAN 8 INCHES (203 mm) IN ANY DIMENSION.

THESE DIMENSIONAL MINIMUMS DO NOT APPLY TO CONDUCTORS THAT ARE NOT SPLICED OR TERMINATED IN THE BOX.

Figure 5–1   Illustration of how much wire must be left at each outlet, junction, or switch point for splices and connections of fixtures and wiring devices in accordance with *Section 300-14*.

CONDUCTOR IDENTIFICATION (COLOR CODING)

Before making any connections to wiring devices, one must become familiar with the different ways in which conductors are identified. The *National Electrical Code*® references to the identification of conductors and terminals is found in *Sections 200-6, 200-7, 200-9, 200-10, 200-11, 210-4(d),* and *210-5.*

Do not confuse the term *grounded* conductor with the term *equipment grounding* conductor. They have different meanings.

A *grounded* conductor is *"A system or circuit conductor that is intentionally grounded."*

An *equipment grounding* conductor is *"The conductor used to connect the noncurrent-carrying metal parts of equipment, raceways, and other enclosures to the system grounded conductor, the grounding electrode conductor, or both, at the service equipment or at the source of a separately derived system."*

In residential wiring, the white neutral conductor is always the grounded conductor in conformance with *Section 250-26(2)*. The grounding of the

neutral conductor is established at the transformer secondary and again at the main service where the neutral is grounded to the building's incoming metal water piping and ground rod(s). This is covered in Unit 15. Touching a properly grounded conductor and a grounded surface at the same time **will not** give you an electrical shock because they are both at the same potential. For grounded conductors in sizes No. 6 AWG and smaller, the insulation must always be white or natural gray in color, or have three continuous white stripes, *Section 200-6(a)*. For grounded conductors in sizes larger than No. 6 AWG, the insulation may be white or gray or have three continuous white stripes for the entire length of the conductor, or identified by wrapping the conductor with white tape or painting it white at its terminations. This taping or painting must be done at the time of installation. These requirements are found in *Section 200-6(b),* and is needed *at each location where the conductor is visible and accessible."*

Section 200-7 states that white or natural gray-colored covering (insulation) on a conductor or with three continuous white stripes, or white or natural gray color marking at a termination be used only for the grounded conductor. But there are a few important exceptions that must be used when it comes to wiring houses with cable.

Section 200-7(c)(1) permits the white conductor in cable to be used as an ungrounded conductor where permanently re-identified to indicate its use, by painting, colored taping, or other effective means at its termination, and at each location where the conductor is visible and accessible. An example of this would be a two-conductor nonmetallic-sheathed cable for the hookup of a 240-volt, single-phase motor.

Section 200-7(c)(2) permits the white conductor in cable to be used as a switch "loop" for single-pole, three-way, or four-way switches. As seen in many of the wiring diagrams in this unit, the white conductor of the cable is spliced to the "hot" conductor in order for the switch return conductor not to be white.

To re-identify the white wire as an ungrounded conductor, electricians will put a small piece of black plastic tape or will slide a short length of heat shrink sleeve over the white conductor. This accomplishes the re-identification required by the Code. This is illustrated in Figure 5–2.

The ungrounded phase conductors of a circuit must be identified by **any** color other than green or green with one or more yellow strips. *Sections 250-119* and *310-12* reserves the color green for equipment grounding conductors. *Sections 250-118(1)* and *250-119* permit the equipment grounding conductor to be bare, insulated, or covered.

The ungrounded conductors generally are called the "hot" conductors. An electrical shock is felt if the grounded conductor and the "hot" ungrounded phase conductor. Also, a shock is felt when the "hot" ungrounded phase conductor and a grounded surface such as a metal water pipe are touched at the same time. An electrical shock can be fatal. This is covered in Unit 7.

There is an exception to the requirement that the insulation on a grounded conductor be white or natural gray. *Section 230-41, Exception* permits bare conductors to be used for the neutral conductor for services. Anywhere on the load side of the main service, the grounded conductor must be identified with white or natural gray insulation, by three continuous white stripes, or identified with white tape or paint as discussed above.

Neutral Conductors

In a multiwire circuit, the grounded conductor is also referred to as a neutral conductor. Although most electricians refer to all white or natural gray conductors as the neutral conductor, this is technically incorrect. Technically, there must be a multiwire circuit or system before there can be a neutral conductor. A grounded white or natural gray conductor is a true neutral conductor when the branch circuit has two or more ungrounded conductors "having a potential difference between them, and a grounded conductor having equal potential difference between it and each ungrounded conductor." In a two-wire branch circuit, the white grounded conductor is not really a neutral. But if calling the grounded white or natural gray conductor a neutral keeps you from making mistakes when discussing electrical circuits, then keep on doing it. Figure 5–3 compares the grounded neutral conductor of a three-wire circuit to the grounded conductor of a two-wire circuit.

Color Coding (Cable Wiring)

Nonmetallic-sheathed cable (Romex) and armored cable (BX) are color-coded as follows:

- Two-wire: one black (the "hot" ungrounded phase conductor)

  one white (the grounded "identified" conductor)

  one bare or green insulated (the equipment grounding conductor)

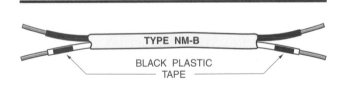

TYPE NM-B

BLACK PLASTIC TAPE

Figure 5–2   The white conductor of a cable may be wrapped with a piece of black tape so that anyone working on the circuit will know that this wire is *not* a grounded conductor. Use red tape if it is desired to mark the conductor red. Colored heat-shrink tubing can also be used.

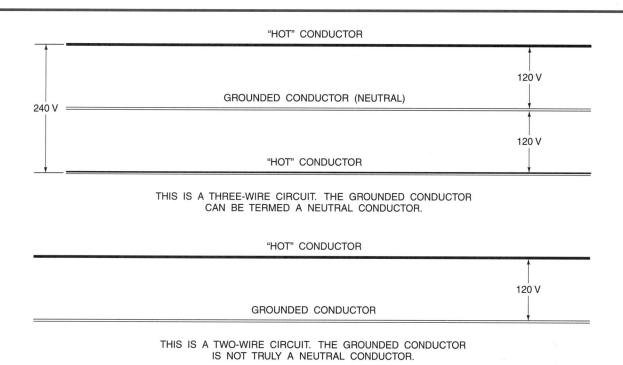

"HOT" CONDUCTOR

GROUNDED CONDUCTOR (NEUTRAL)

"HOT" CONDUCTOR

240 V

120 V

120 V

THIS IS A THREE-WIRE CIRCUIT. THE GROUNDED CONDUCTOR
CAN BE TERMED A NEUTRAL CONDUCTOR.

"HOT" CONDUCTOR

GROUNDED CONDUCTOR

120 V

THIS IS A TWO-WIRE CIRCUIT. THE GROUNDED CONDUCTOR
IS NOT TRULY A NEUTRAL CONDUCTOR.

Figure 5–3 **The top circuit shows a true neutral grounded conductor in a three-wire multiwire branch circuit. The lower circuit shows a grounded conductor in a two-wire branch circuit.**

- Three-wire: one black (the "hot" ungrounded phase conductor)

  one white (the grounded "identified" conductor)

  one red (the other "hot" ungrounded phase conductor)

  one bare or green insulated (the equipment grounding conductor)

- Four-wire: one black (a "hot" ungrounded phase conductor)

  one white (the grounded "identified" conductor)

  one red (a "hot" ungrounded phase conductor)

  one blue (a "hot" ungrounded phase conductor)

  one bare or green insulated (the equipment grounding conductor)

In the case of armored cable (BX), the bonding wire (bonding strip) that lies just under the armor is **not** an equipment grounding conductor. This bonding strip is in direct contact with the metal armor for the entire length of the cable. The bonding strip **does not** have to be terminated. The bonding strip is usually folded back over the cable armor where the cable is terminated in a connector. The bonding strip ensures that the impedance of the ground path is maintained to a certain value as required by the Underwriters Laboratories standards. A low impedance return ground path is necessary so that should a ground fault occur, the cable and the bonding strip **together** would have the ability to safely carry the ground-fault current back to the source, where the branch circuit breaker or fuse would be able to clear the fault.

This is a requirement of *Section 250-2,* and is discussed in Unit 12.

Section 250-118(9) permits the cable armor to serve as an equipment grounding conductor.

In the wiring diagrams that follow, you will see very quickly see that because the conductor insulation colors in cable are "fixed" as discussed above, the Code gives you the freedom to splice a conductor of one color to a conductor of another color, just so the final switch return (switch leg) to the lampholder is **not** white. In some of the wiring diagrams, you will find a white conductor spliced to a black conductor. You might want to establish red and white for the travelers (sometimes referred to as dummies). But even this is not always possible to do. Having to splice different colored wires together applies to cable assemblies, and does not apply when the wiring method is conduit, because there is an unlimited choice of conductor insulation colors to use for the travelers and switch returns.

Color Coding (Conduit Wiring)

When the wiring method is conduit, any color may be used for the "hot" phase conductors except:

Green or green with a yellow strip: reserved for use as the *equipment grounding conductor* only.

White or Gray: reserved for use as the *grounded* identified conductor.

For the ungrounded "hot" phase conductors, black and red are usually the first choice. Then you could use yellow for travelers, and blue for switch legs. The choice is yours. Establish some sort of color coding that will be followed throughout the entire wiring project.

Changing Colors When Conductors Are in a Raceway

Occasionally, it may be necessary to change the actual color of the conductors insulation. The following table provides guidelines for doing this.

TO CHANGE FROM THIS	TO THIS	DO THIS
red, black, blue, etc.	an *equipment grounding* conductor	**The Basic Rule:** No reidentification permitted for conductors No. 6 AWG and smaller. An equipment grounding conductor must be bare, covered, or insulated. If covered or insulated, the outer finish must be green or green with yellow stripes. The color must run the entire length of the conductor. **Conductors larger than No. 6 AWG:** At time of installation, at both ends and at every point where the conductor is accessible, do any of the following, For the entire exposed length: • strip off the covering or insulation • paint the covering or insulation green • use green tape, adhesive labels, or green heat shrink tubing. See *Sections 210-5(b), 250-119, 310-12(b)*.
red, black, blue, etc.	a *grounded* conductor	**Conductors No. 6 AWG or smaller:** For the conductor's entire length, the insulation must be white, natural gray, or have three continuous white stripes on other than green insulation. Reidentification in the field is not permitted when the wiring method is a raceway. **Conductors larger than No. 6 AWG:** For the conductors entire length, the insulation must be white, natural gray, or have three continuous white stripes on other than green insulation. At terminations, reidentification may be white paint, white tape, or white heat shrink tubing. This marking must encircle the conductor or insulation. Do this marking at time of installation. See *Sections 200-2, 200-6(a)* and *(b), 200-7(a)(b)* and *(c), 210-5(a), 310-12(a)*.
White, natural gray	an *ungrounded* (hot) conductor	Insulated conductors that are white, natural gray, or have three continuous white stripes are to be used as grounded conductors only. There is no provision in the *NEC*® for reidentifying white or natural gray conductors for use as ungrounded conductors in raceways. The only exception to use white or natural gray insulated conductors as ungrounded conductors is for cable wiring. See *Sections 200-7(a)(b)* and *(c), and 310-12(c)*.

TYPES OF SWITCHES

As previously mentioned, wall switches in homes are available in many styles. There are toggle switches (toggle is another name for the handle), slider switches, rocker switches, push switches, rotating knob switches. silent switches, quiet switches, pilot light switches, time-delay switches, and mercury switches. There probably are others. But thrown all together in one pot, they come under the category of *snap switches* in both the *National Electrical Code*® and in the Underwriters Laboratories standard.

Single-Pole Switch

A single-pole switch is used when a light, group of lights, receptacle, or other load is to be controlled from one switching point. A single-pole switch has two brass-colored terminals. The handle (toggle) is marked ON/OFF. A single-pole switch is connected in series with the ungrounded "hot" conductor. Figure 5–4 shows the symbol, the internal operation,

and the connection of a single-pole switch into a circuit.

Three-Way Switch

Three-way switches are used to control a light, group of lights, receptacle, or other load from two locations. The name "three-way" is somewhat misleading. What this really means is that the switch has three terminals. One terminal is the *common terminal* to which the feed is connected. The other two terminals are called the *traveler terminals.* The two conductors connected to the traveler terminals are called the *travelers. The travelers* run between the *traveler terminals* on the two three-way switches. The common terminal is a darker color than the two brass-colored traveler terminals. Internally, the switching mechanism alternately switches between the common terminal and either one of the traveler terminals. A three-way switch is technically a single-pole, double-throw switch. A three-way switch does not have an ON/OFF position. Figure 5–5

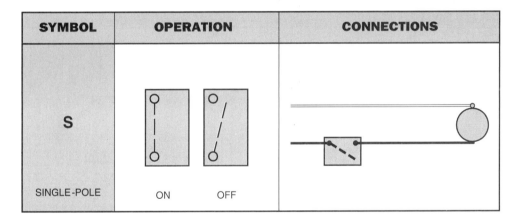

Figure 5–4 This illustration shows the symbol, the internal operation, and the connection of a single-pole switch into a circuit.

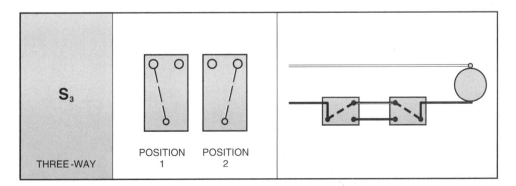

Figure 5–5 This illustration shows the symbol, the internal operation, and the connection of two three-way switches into a circuit.

shows the symbol, the internal operation, and the connection of two three-way switches into a circuit.

Four-Way Switch

Four-way switches are used to control a light, group of lights, receptacle, or other load from three or more locations. The name "four-way" is somewhat misleading. What this really means is that the switch has four terminals. All four terminals are brass colored. The terminals are arranged in pairs. A four-way switch does not have an ON/OFF position. Internally, the switching mechanism alternately switches *straight through* or *criss-cross* between the pairs of terminals. Four-way switches are always connected into the travelers that run between two three-way switches. One set of travelers coming from one of the three-way switches is connected to one pair of terminals on the four-way switch. The other set of travelers coming from the other three-way switch is connected to the other pair of terminals on the four-way switch. Figure 5–6 shows the symbol, the internal operation, and the connection of a four-way switch and two three-way switches into a circuit.

A four-way switch can be mistaken for a double-pole switch because both have four brass-colored terminals. The difference is that a four-way switch handle has no ON/OFF markings. A double-pole switch toggle is marked ON/OFF.

Double-Pole Switch

Double-pole toggle switches are not too common in residential applications. A double-pole switch is really two single-pole switches combined into one switch. The toggle is marked ON/OFF. A double-pole switch has four brass-colored terminals. One pair is for the line-side connections, the other pair for the load connections. A double-pole switch can be used to control a 240-volt load such as an air conditioner, fan, or pump. They are also used when two separate circuits must be controlled by one switch. A four-way switch can easily be mistaken for a double-pole switch because both have four brass-colored terminals. The difference is that a double-pole switch is marked ON/OFF. A four-way switch has no ON/OFF markings.

Figure 5–7 shows the symbol, the internal oper-

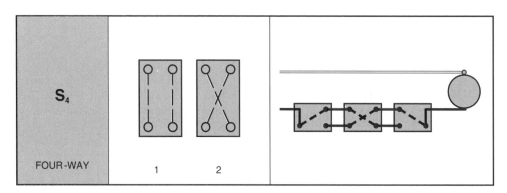

Figure 5–6   **This illustration shows the symbol, the internal operation, and the connection of a four-way switch and two three-way switches into a circuit.**

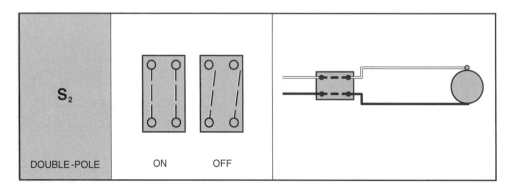

Figure 5–7   **This illustration shows the symbol, the internal operation, and the connection of a double-pole switch into a circuit.**

ation, and the connection of a double-pole switch into a circuit.

Connecting a Lampholder

For safety reasons as required by the *National Electrical Code,*® in all of the following wiring diagrams, the white conductor of a fixture is connected to the white grounded conductor of the cable. When a lampholder with screw terminals is used, the white grounded conductor must be connected to the lampholder's white (silver) terminal, *Section 200-10(c)*. The white conductor lead of a fixture, and the white (silver) terminal of a lampholder are always connected to the outer shell of a lampholder. The "hot" conductor is always the center contact of a screw shell lampholder. In this manner, as a lamp is removed, touching the shell of the lamp and "ground" will not result in an electrical shock. With the lamp removed and the power on, one would have to reach deep down inside of the lampholder to touch the "hot" center contact (Figure 5–8).

TYPICAL WIRING DIAGRAMS FOR HOUSE WIRING

Some of the following wiring diagrams contain receptacles. These receptacles may or may not require ground-fault circuit-interrupter (GFCI) protection, depending on the location. GFCI protection is not discussed in this unit as the subject is discussed in detail in Unit 7. As you read the following text, follow the description of the cable runs and connections as you look at the wiring diagrams.

Switch Control From One Location— Feed At Switch

Figure 5–9 shows a single-pole switch controlling a light from one switching point. The 120-volt, two-wire supply feeds into the switch location. With this configuration, the white conductors of both cables are spliced together in the switch box. The single-pole switch is connected in series with the black conductors of each cable. At the ceiling outlet box, the black and white conductors of the two-wire cable connect to the corresponding white and black conductors of the fixture, or directly to the corresponding silver and brass-colored terminals of a lampholder.

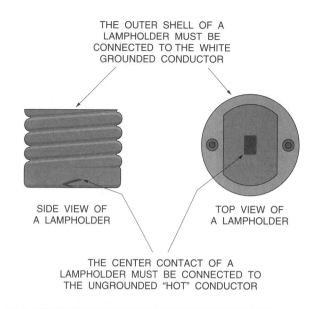

Figure 5–8   A sketch of a typical screw shell lampholder. The white grounded conductor must always be connected to the terminal that connects to the outer shell of the lampholder, *Section 200-10(c)*. The ungrounded "hot" conductor must always connect to the terminal that connects to the center contact of the lampholder.

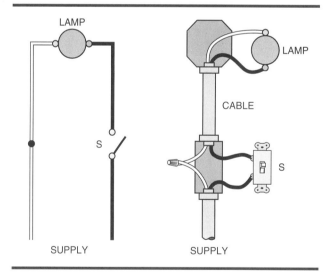

Figure 5–9   Single-pole switch in circuit with feed at switch.

Switch Control From One Location— Feed At Fixture

Figure 5–10 shows the 120-volt, two-wire supply feeding into the outlet box where the fixture will be installed. The fixture is controlled by a separately located single-pole switch. The two-wire cable between the ceiling outlet box and the switch box

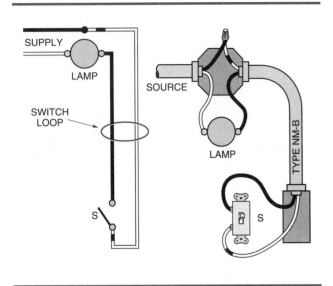

Figure 5–10   Single-pole switch in circuit with feed at fixture.

serves as the switch loop. The white conductor of the supply cable connects to the white conductor of the fixture, or to the silver terminal of a lampholder. The white conductor of the two-wire switch loop cable is connected to the black supply conductor in the ceiling outlet box. The return conductor from the switch to the outlet box is the black conductor,

which is connected to the black wire of the fixture, or directly to the brass terminal of a lampholder. The use of a white wire in a single-pole switch loop is permitted in *Section 200-7(c)(2)*. When so used, the white conductor must be re-identified in the outlet box and in the switch box. For re-identification, colored plastic tape is most often used. For this diagram, black tape would be suitable.

Switch Control From One Location— Receptacle Is Not Switched

Figure 5–11 shows the 120-volt, two-wire supply coming into the switch box where a single-pole switch is installed. A three-wire cable is then run from the switch to the ceiling outlet box where a fixture will be installed. This three-wire cable carries the 120-volt circuit and the switch leg to the outlet box. From the outlet box, a two-wire cable is run to a wall device box where a receptacle will be installed. The receptacle is on all of the time, and is not controlled by the switch. At the switch location, the white conductors of both cables are spliced "straight through" and the black conductors of both cables are spliced together with a short pigtail to feed the switch. The red conductor in the three-wire cable serves as the switch leg. At the fixture loca-

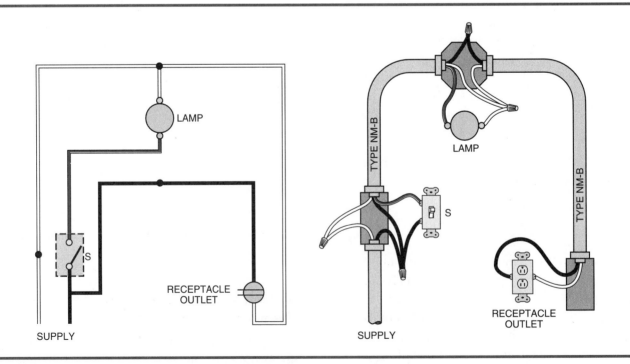

Figure 5–11   Ceiling fixture controlled by a single-pole switch. The feed comes into the box at the switch location. The receptacle is not controlled by the switch, and is on at all times.

tion, the red switch leg conductor connects to the black wire of the fixture or to the brass-colored terminal of a lampholder. The white conductors of the cables and the fixture are spliced together.

Switch Control From Two Locations— Feed Into The First Switch Box

Multiple switch control is generally needed for the control of lighting in stairways and for rooms or other areas that have more than one entry.

Section 210-70(a)(2) requires wall-switch control of interior stairway lighting. This section also requires that if the stairway has six or more steps between floor levels, then switch control for the stairway lighting must be provided at both levels. This rule applies to a stairway between the first and second floor, a stairway to a basement, and a permanent stairway to an attic. Folding fold-down ladders to a scuttle hole providing access to an attic are not considered to be a permanent stairway. See Unit 6 for additional information regarding the requirements for switched lighting outlets.

Figure 5–12 shows a ceiling fixture controlled by two three-way switches. The 120-volt, two-wire supply feeds into the first switch box. The black conductor of the two-wire supply is connected to the first three-way switch. A three-wire cable is run between the two three-way switches. The white conductor is spliced straight through both switch boxes. From the second three-way switch, a two-wire cable is run to the ceiling outlet box where the fixture will be located. The black conductor of this two-wire cable is connected to the black conductor of the fixture, or directly to the brass-colored terminal of a lampholder. The white conductor of this two-wire cable is connected to the white wire of the fixture, or is connected directly to the silver-colored terminal of a lampholder. The black and red conductors in the three-wire cable serve as the travelers. You could re-identify the black conductor with red tape.

Switch Control From Two Locations— Feed at Fixture

Figure 5–13 shows two three-way switches controlling a ceiling fixture. The 120-volt, two-wire supply comes into the ceiling outlet box where the fixture will be installed. A two-wire cable is run from the ceiling outlet box to the first three-way

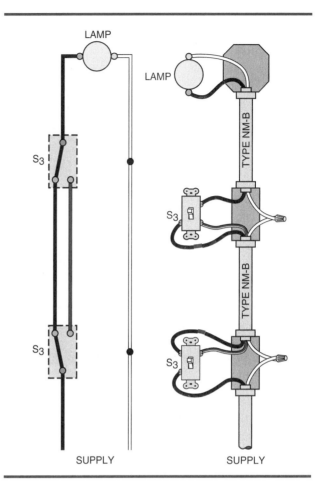

Figure 5–12 Two three-way switches controlling a ceiling fixture. The feed is at the first switch with the fixture fed from the second switch. The black and red conductors in the three-wire cable are used as the travelers. This configuration enables the white conductor of all three cables to be spliced straight through to the fixture.

switch. A three-wire cable is run between the two three-way switches. In the outlet box (the fixture location), the white grounded conductor of the supply is connected to the white wire of the fixture, or to the silver-colored terminal of a lampholder. The black switch return in the two-wire cable that runs between the outlet box and the three-way switch is connected to the black wire of the fixture, or to the brass-colored terminal of a lampholder. The black wire of the two-wire supply cable is connected to the white wire of the two-wire cable that runs between the outlet box and the first three-way switch. At the first three-way switch, the white conductor of the two-wire cable is spliced to the black conductor of the three-wire cable that runs between the two switches. At the second three-way switch, the black

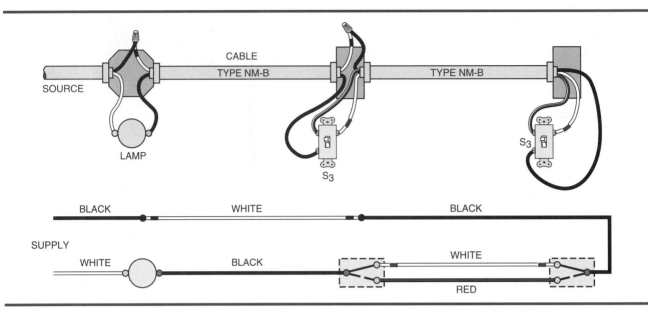

Figure 5–13 Two three-way switches controlling a ceiling fixture. The feed is at the light. The red and white conductors of the three-wire cable are used as the travelers between the three-way switches. The white conductor of the supply cable connects to the white wire of the fixture.

wire connects to the dark-colored terminal (the common terminal) of the three-way switch. The red and white conductors in the three-wire cable are used as travelers between the three-way switches.

Note in the wiring diagram that the white conductor, where used as an ungrounded conductor in the outlet box and in the first device box, has been re-identified with black plastic tape. To re-identify the white conductor where used as one of the

"travelers," red plastic tape could be used, consistent with keeping the "travelers" red in color.

Switch Control From Two Locations— Feed at Fixture—Alternate Connection

Figure 5–14 shows two three-way switches controlling a ceiling fixture. The 120-volt two-wire supply comes into the ceiling outlet box where the

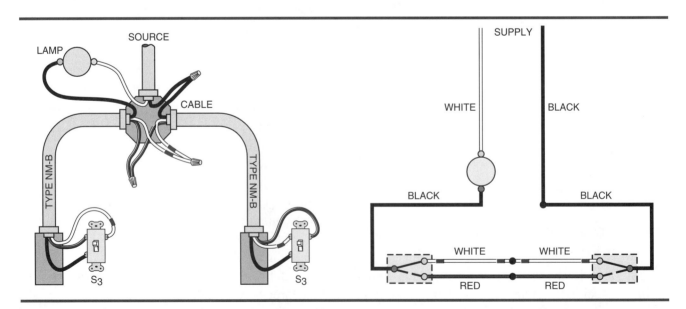

Figure 5–14 Two three-way switches controlling a ceiling fixture. The feed is at the light. The red and white conductors of the three-wire cables are used as the travelers between the three-way switches. The white grounded conductor of the supply connects to the white wire of the fixture. The black ungrounded switch return from one of the three-way switches connects to the black wire of the fixture.

fixture will be installed. From the ceiling outlet box, a three-wire cable is run to each of the switch boxes. The red and white conductors of the three-wire cables are used as the travelers between the three-way switches. The white grounded conductor of the supply connects to the white wire of the fixture or to the silver-colored terminal of a lampholder. The black ungrounded switch return from one of the three-way switches connects to the black wire of the fixture or to the brass-colored terminal of a lampholder. This connection makes it possible for the black conductor to be connected to the dark-colored (common) terminal of the three-way switches.

Note in the wiring diagram that the white conductor, where used as one of the "travelers" in both switch boxes and where running through the outlet box, has been re-identified with plastic tape. For this marking, red plastic tape could be used, consistent with keeping the "travelers" red in color.

Switch Control From Three or More Locations

Figure 5–15 shows a circuit with switch control from three locations. Two three-way switches and one four-way switch are needed, with the four-way switch connected between the three-way switches. The 120-volt, two-wire supply enters in the first switch box. The white grounded conductor in all of the cables is connected straight through to the white wire of the fixture or to the silver-colored terminal on a lampholder. The black switch return from the second three-way switch connects to the black wire of the fixture or to the brass-colored terminal of a lampholder. The red and black conductors in the three-wire cables serve as the travelers.

The black conductors in the three-wire cables could be re-identified with red plastic tape, consistent with keeping the "travelers" red in color.

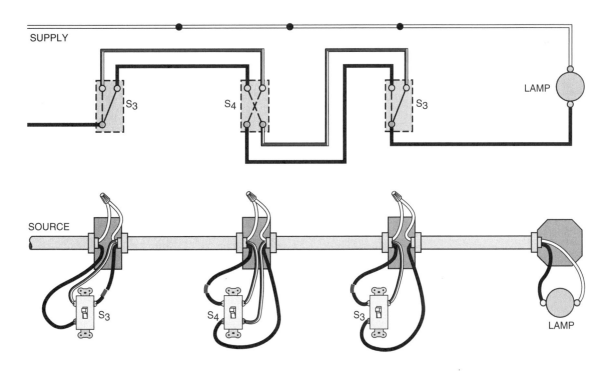

Figure 5–15 A circuit with switch control from three locations. The circuit requires two three-way switches and one four-way switch.

Double-Pole Switch

Figure 5–16 shows a double-pole switch controlling a 240-volt load such as an electric baseboard heater. This switch will disconnect both of the "hot" conductors when the switch is turned off. Using conduit as the wiring method, any number of conductor insulation colors may be selected. Using a two-wire cable, it will be necessary to remark the white wire in the cable so as not to confuse it with a grounded conductor. This is required by *Section 200-7(c)(1)*. This was illustrated in Figure 5–2.

Switches with Pilot Light That Glows When Switch is in OFF Position

Figure 5–17 shows how switches with pilot lights in their handles are connected. Referred to as

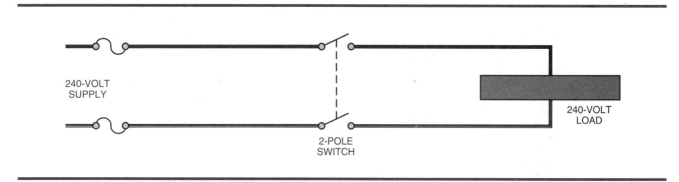

Figure 5–16 A double-pole switch controlling a 240-volt load. Both of the "hot" phase conductors are disconnected when the switch is in the OFF position.

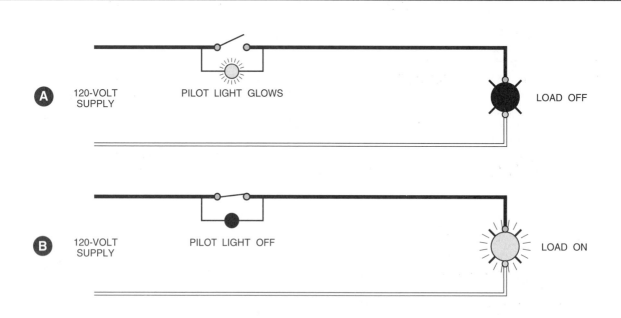

Figure 5–17 These wiring diagrams show a single-pole switch that has a neon pilot light in the handle. This type of switch is sometimes referred to as a locator switch. When the switch is in the OFF position (A), the neon lamp in the handle of the switch glows, and the load is off. When the switch is in the ON position (B), the neon lamp goes out, and the load comes on. This type of switch is used where you want to make it easy to locate the switch in the dark.

lighted toggle switches, they are great when you want to be able to locate the switch in the dark. Because the neon lamp has a very high resistance, it "sees" line voltage (120 volts) when the switch is in the OFF position (A), while the connected load "sees" zero volt. Actually, a very, very small voltage might be read across the load, but for all practical purposes the voltage can be considered to be zero. Because the neon lamp and the load are in "series" when the switch is in the OFF position, the circuit follows the rule that "in a series circuit voltage divides proportionately to the resistance."

When the switch is in the ON position (B), the neon lamp is shunted out and will have zero volt across it. The neon lamp will not glow. The connected load will have full voltage supplied to it, and will turn on.

This type of switch might be referred to as a *locator switch* because the lamp glows when the switch is off, and does not glow when the load is on.

Single-pole Switch With True Pilot Light

Figure 5–18 illustrates a true pilot light switch used when the pilot lamp is to indicate that the load is on. The pilot light is in the toggle. This type of switch has three terminals: (1) the "hot" feed, (2) the switch leg, and (3) the grounded white circuit conductor needed for the pilot lamp. This connection also can be made using interchangeable wiring devices of the types shown in Figure 4–2.

Other Switching Arrangements

The following diagrams are "line" diagrams of other switching arrangements. Cable runs are not shown. As can be seen by the choice of insulation colors, these diagrams would indicate that the wiring method is a raceway, where conductors are pulled into the conduit.

Switch Control From Two Locations— Receptacle On At All Times

Figure 5–19 shows a switching arrangement that could be used between two buildings. Consider a typical example such as a house and a detached garage. The lighting outlet (fixture) and the receptacle are both located in the garage. The 120-volt circuit in the house feeds into the first three-way switch. Four conductors are run between the house and the garage to the second three-way switch location in the garage. The two three-way switches control the light—the receptacle is on at all times. The white conductor is connected to the white wire (or terminal) of the fixture and to the white terminal (wide slot) or the receptacle. The black wire connects to the brass-colored terminal (narrow slot) of the receptacle. The blue or yellow conductors serve as the travelers between the two three-way switches. Red has been selected as the switch leg.

This circuit could also be wired with underground direct burial conductors, in which case the identification of the conductors would have to be various colored tapes.

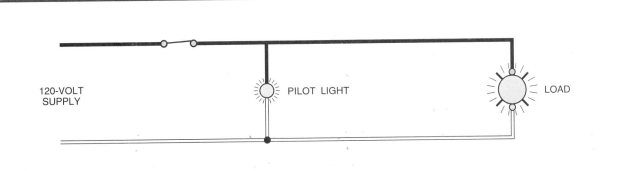

Figure 5–18 This wiring diagram illustrates a true pilot light switch. The pilot lamp is an integral part of the switch. When the switch is in the ON position, both the pilot lamp and the connected load come on. When the switch is turned off, both pilot lamp and connected load turn off.

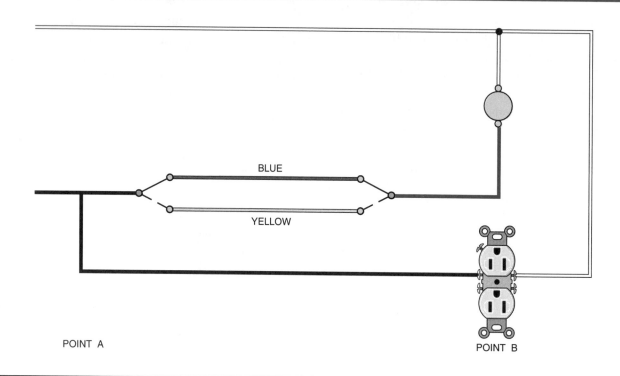

BLUE

YELLOW

POINT A

POINT B

Figure 5–19 A lighting outlet controlled by two three-way switches. The receptacle is "live" at all times. Four conductors are run between Point A and Point B. A typical example of this might be between a house and a detached garage or other remote building. This is a conduit installation where you have the option to choose the color of the insulation on the conductors.

Switch Control From Two Locations— Receptacle On At All Times

Figure 5–20 is similar to Figure 5–19. The difference is that a second lighting outlet has been added. This switching arrangement could be used between two buildings such as a house and a detached garage. One lighting outlet (fixture) is located on the house. The second lighting outlet (fixture) is located on or in the garage. The receptacle is located in the garage. The 120-volt circuit in the house feeds into the first three-way switch. Five conductors are run between the house and the garage to the second three-way switch location in the garage. The two three-way switches control the lights—the receptacle is on at all times. The white conductor is connected to the white wire (or terminal) of the fixtures and to the white terminal (wide slot) of the receptacle. The black wire connects to the brass-colored terminal (narrow slot) of the receptacle. The blue or yellow conductors serve as the travelers between the two three-way switches. Red has been selected as the switch leg.

Dimmers

Dimming of lighting is accomplished by using a dimmer switch. Today, most dimmers for residential use are electronic. They are available in rotary, toggle, rocker, and slider types in both single-pole (to control the light from one location) and three-way (to control the light from more than one location). Figure 5–21 (A) shows the connections for single-location control, and (B) for two-location control. If control is needed from more than two locations, then a four-way switch needs to be connected between the three-way switches.

Dimming Fluorescent Lighting

With one exception, you cannot dim an ordinary fluorescent lighting fixture! It takes a special dimming ballast and a special fluorescent dimmer switch. Figure 5–22 shows the hookup of one type of "slider" fluorescent dimmer switch. The marking on the label of the ballast will indicate that it is a dimmer ballast. The marking on the dimmer switch will indicate that it is a fluorescent lamp dimmer.

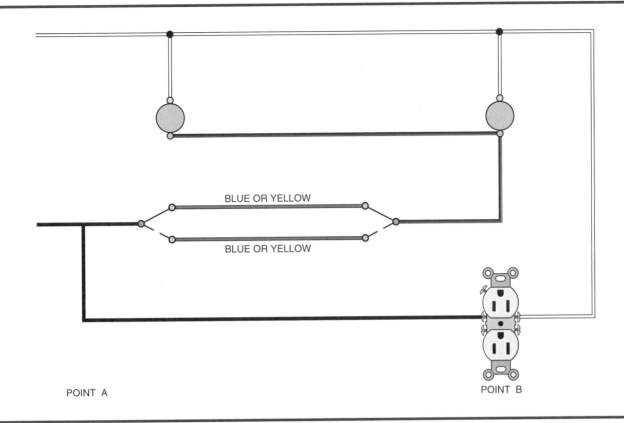

POINT A   POINT B

Figure 5–20   Two lighting outlets controlled by two three-way switches. The receptacle is "live" at all times. Five conductors are run between Point A and Point B. An example of this hookup might be between a house and detached garage, where one lighting outlet (fixture) is on the house, and the other lighting outlet (fixture) is on or in the garage. This is a conduit installation where you have the option to choose the color of the insulation on the conductors.

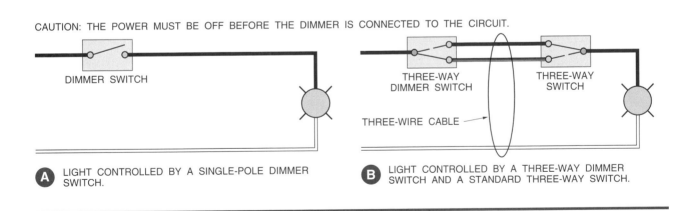

Figure 5–21   Use of single-pole and three-way electronic solid-state dimmer switches for the control of lighting.

Note the number of wires connected to the switch. Adjustment of light output varies from 20% to 100% of the lamp.

At least one manufacturer has a compact fluo-rescent lamp that can be dimmed by a standard dim-mer switch. The lamp has integral built-in electronic circuitry that compensates for the variable output of the dimmer.

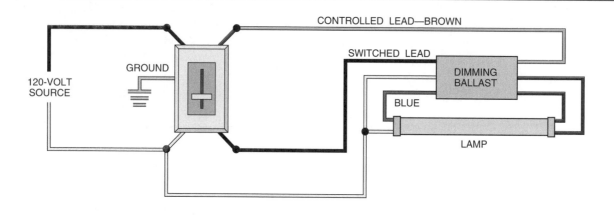

Figure 5–22 An electronic "slider" dimmer switch used to control fluorescent lighting. This requires a special dimmer switch and a special rapid start fluorescent lamp dimming ballast.

UNUSUAL CONNECTIONS

The following wiring diagram is one of the unusual ways to connect switches.

Switch Control From Two Locations— Receptacle On At All Times.

This is a very confusing hookup. **Do not** use this connection when making a new installation. The hookup in Figure 5–23 accomplishes the same thing as Figure 5–20, but with four conductors instead of five. Insofar as the Code is concerned, there are no violations. However, if for some reason there are only four conductors on an existing installation between the two points, and to add a fifth conductor would be impossible, redoing the connections as shown in Figure 5–23 could be considered—**but only as the last resort.**

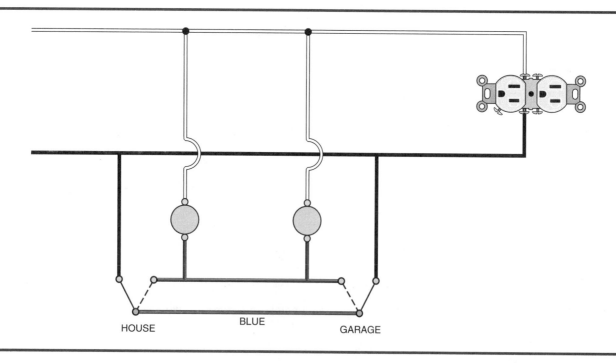

Figure 5–23 An unusual connection! Two lighting outlets controlled by two three-way switches. The receptacle is "hot" at all times. This hookup accomplishes the same thing as Figure 5–20, but with four wires instead of five. This connection does not violate any *NEC*® rules, but is very confusing. Trying to troubleshoot this connection is extremely difficult. Do not use this connection. It is illustrated here only for informational purposes.

CODE VIOLATIONS

There are also some ways to connect switches that are not in conformance to the *NEC.*® The following wiring diagrams illustrate some of these unorthodox connections.

Connection that Reverses the Polarity to a Lampholder

No matter how many books are written about the proper way to connect switches, there will always be someone who is "creative," even though it violates the requirements of the *National Electrical Code.*® Figure 5–24 is just such a violation. Note how the polarity of the lampholder changes each time one of the switches is operated. Never switch the grounded circuit conductor, except for rare instances in industrial wiring.

Sometimes the shell of the lampholder is connected to the grounded conductor, and at other times the shell is connected to the "hot" conductor. Sometimes, both terminals of the lampholder are connected to the grounded neutral conductor (no voltage across the lampholder), and sometimes both terminals of the lampholder are connected to the ungrounded "hot" conductor (no voltage across the lampholder). This is a clear violation of *Section 200-10(c),* and is a very unsafe hookup. This diagram

has been included in this text so should you come across one of these confusing and unsafe connections, you will recognize it as a violation of the Code, and will make the necessary corrections to bring the installation into conformance with the Code.

Trace the circuit for each possible position of the two switches. Here is what happens to the polarity at the screw shell lampholder.

	SHELL	CENTER CONTACT	LAMP
POSITION 1	neutral	"live"	on
POSITION 2	"live"	neutral	on
POSITION 3	neutral	neutral	off
POSITION 4	"live"	"live"	off

Never, never make an installation using this hookup!

Switch Legs That Are Too Small

On occasion, we find electricians installing switch legs using conductors that are smaller than the actual branch-circuit conductors. This is incorrect!

Sometimes there is confusion in explaining "what is branch-circuit wiring, and what is a tap." *Section 210-3* states that the rating of a branch-circuit is based upon the ampere rating of the

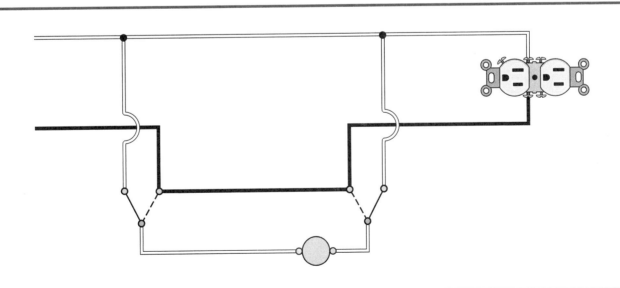

Figure 5–24 This wiring diagram illustrates a hookup that is in violation of *Section 200-10(c)* of the *National Electrical Code.*® With this connection, the polarity on the shell of the lampholder reverses each time a switch is snapped. This could result in a serious electrical shock. Never make a hookup like this! It is shown here for informational purposes only.

branch-circuit overcurrent device (circuit breaker or fuse). *Section 210-20* states that the branch-circuit conductors must be protected by a circuit breaker or fuse that does not exceed the ratings of the branch circuit established by *Section 210-3*. One of the most common examples of a "tap" to a branch circuit in-house wiring are the small fixture wires supplied with a lighting fixture. The fixture conductors might be No. 18 AWG, yet they are spliced to the No. 14 or in some case No. 12 AWG branch-circuit conductors that are protected by 15-ampere and 20-ampere overcurrent devices. *Section 210-19(c), Exceptions No. 1 and 2* permit fixture "taps" to be smaller than the branch-circuit wiring.

A switch leg (loop) is **not** a "tap." It is a part of the branch-circuit wiring. Even though the connected load such as a lighting fixture has a much, much smaller current draw than the branch-circuit ampere rating, these switch legs must also be capable of safely carrying short circuits and ground faults.

Therefore, if the branch-circuit rating is 20 amperes using No. 12 AWG conductors, then switch legs and travelers must also be the same size.

Figure 5–25 clearly illustrates the Code violation.

The same situation occurs where 20-ampere small appliance branch circuits are installed in a home. In Figure 4–39, we showed the recommended practice of making up all of the splices, independent of the receptacle, then using short pigtails to make the connections to the receptacle. The pigtails must also be rated 20 amperes because they are an extension of the 20-ampere branch circuit.

As with most rules, there are exceptions. Sometimes to minimize voltage drop, No. 12 AWG conductors are used for long "home runs" in residential lighting branch circuits, but the branch-circuit breaker is rated 15 amperes. This is permitted by the last sentence of *Section 210-3*, which states that *"where conductors of higher ampacity are used for any reason, the ampere rating or setting of the specified overcurrent device shall determine the circuit rating."* A "home run" is the branch-circuit wiring from the panel to the first outlet or junction box in that circuit. From this point on, the branch-circuit wiring continues with No 14 AWG.

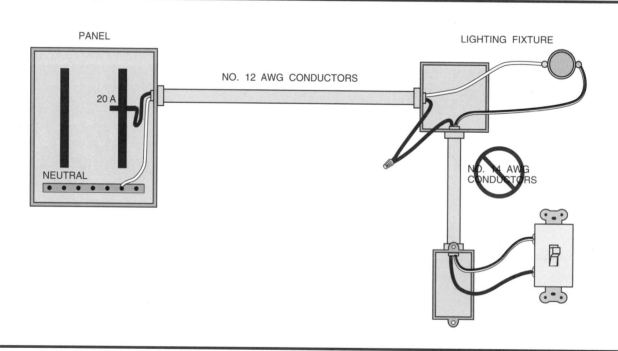

Figure 5–25 This diagram shows a common Code violation of Section *210-3*. The switch loop is made up with 15–ampere No. 14 AWG conductors. The branch circuit is made up of 20-ampere No. 12 AWG. The branch circuit breaker has a rating of 20 amperes. The correct way to make this hookup is to use 20-ampere No. 12 AW conductors for the switch loop.

Connections That Result in Induction Heating

When alternating-current circuits are run through metal raceways or through openings in metal boxes, the circuiting must be arranged to prevent *induction heating* of the metal, *Section 300-20*. When all of the conductors of a circuit are run together, the magnetic flux surrounding each conductor cancels each other. When one conductor only is run in a metal raceway or through an opening in a metal box, induction heating of the metal will result.

Section 300-3(b) of the Code states in part that: *All conductors of the same circuit and, where used, the grounded conductor and all equipment grounding conductors shall be contained within the same raceway, auxiliary gutter, cable tray, trench, cable, or cord . . ."*

Some individuals believe they can save substantial lengths of three-wire cable when hooking up three-way and four-way switches in usual ways. In fact, they may brag about the fact that they can wire a house using no three-wire cable. They use two-wire cable, and make up the connections as shown in Figure 5–26.

Do not make up any switching or branch-circuit arrangement as shown in Figure 5–26. The sole purpose of showing this diagram is to show a violation of the *National Electrical Code.*®

If you are using a metal raceway, run all of the conductors of a given circuit through the same raceway. A nontechnical way to understand this is to ask yourself "in the same metal raceway, is the amount of current going out the same as the amount of current coming back?" If the answer is no, then do not do it. Figure 5–27 illustrates the intent of this Code requirement.

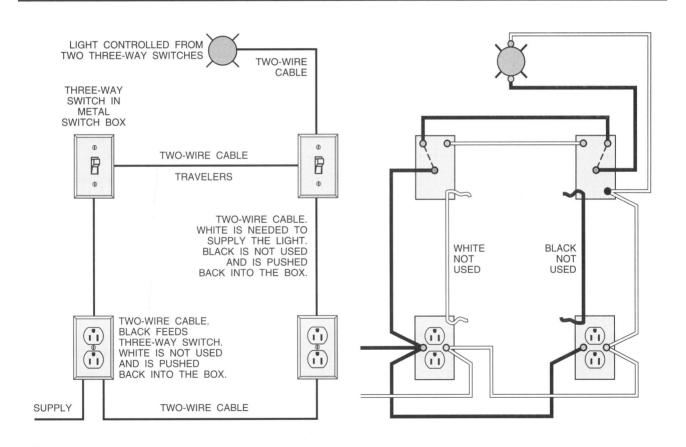

Figure 5–26 Do not make connections like this! This connection does not "Meet Code." It is in violation of *Sections 300-3(b), 300-20,* **and** *380-2(a), National Electrical Code*®. **The diagram is shown here for informational purposes only.**

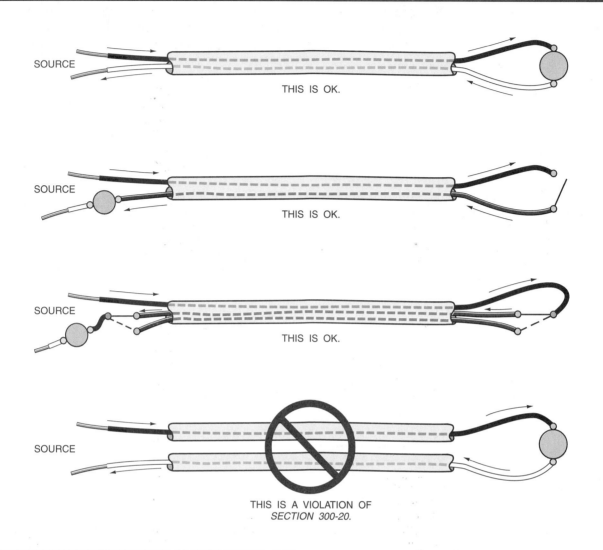

SOURCE — THIS IS OK.

SOURCE — THIS IS OK.

SOURCE — THIS IS OK.

SOURCE — THIS IS A VIOLATION OF *SECTION 300-20.*

Figure 5–27   A few examples of arranging circuitry to avoid induction heating according to *Sections 300-3(b)* and *300-20.* Always ask yourself, "Is the same amount of current flowing in both directions in the metal raceway?" If the answer is "no," there will be induction heating that over time can damage the insulation on the conductors.

BONDING AT RECEPTACLES

Grounding and bonding are covered in Unit 12.

Because we are discussing wiring diagrams and how to make connections when "trimming" out a house, we need to discuss the why and how of bonding of receptacles.

All receptacles used on 15- and 20-ampere branch circuits are required to be of the grounding type, *Section 210-7(a).* These receptacles have a No. 10-32 green hexagon-shaped equipment grounding screw. In Figure 5–28, the equipment grounding screw is hidden from view. This equipment grounding

terminal is connected to the metal yoke and to the horseshoe-shaped grounding slot of the receptacle.

An equipment grounding conductor connects noncurrent-carrying metal parts of equipment, raceways, and other enclosures to the system grounded conductor.

New branch-circuit wiring contains an equipment grounding conductor. This might be the separate equipment grounding conductor of a non-metallic-sheathed cable, the metal armor and bonding strip of BX, or a metal raceway such as electrical metallic tubing or flexible metal conduit. See

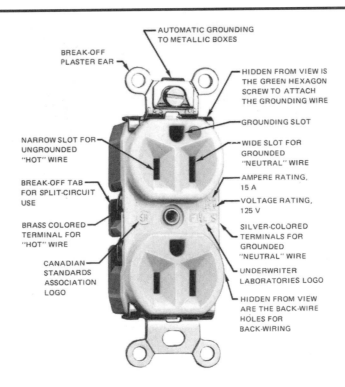

Figure 5–28 Note the metal clip on the metal yoke of the receptacle that provides the required metal-to-metal contact. When used with grounded metal wall boxes, this metal clip holds the No. 6-32 mounting screws in place, and provides continuity between the metal box, and the metal yoke and green hexagon equipment grounding screw on the receptacle. See *Section 250-146(a)* and *(b)*.

Section 250-118 for a list of acceptable equipment grounding conductors.

Section 370-40(d) of the *National Electrical Code®* requires that *a means shall be provided in each metal box for the connection of equipment grounding conductors.* Generally, holes are tapped in the box for No. 10-32 screws. These holes are marked GR, GRN, or GRND, or similar identifications. The screws used for terminating the equipment grounding conductor must be hexagon-shaped, and must be green. Electrical distributors sell green, hexagon-shaped No. 10-32 screws for this purpose.

Section 250-146 requires the connection of a separate equipment grounding conductor to the grounding screw of the grounding-type receptacle. But there are some exceptions.

A bonding jumper is not needed if the receptacle provides metal-to-metal contact between the receptacle yoke and the metal switch box, *Section 250-146(a)*.

The receptacle illustrated in Figure 5–28 has its

No. 6-32 mounting screws held captive by a special clip that provides the required metal-to-metal contact between the receptacle yoke and the metal box, *Section 250-146(b)*.

Figure 5–29 shows the equipment grounding conductor from the nonmetallic-sheathed cable passing under the No. 10-32 green equipment grounding screw in the back of the box, then brought out and connected to the hexagon-shaped green equipment grounding screw on the receptacle. This properly grounds the metal box and also provides the equipment ground for the receptacle.

To make sure that the continuity of the equipment grounding conductor path is ensured, *Section 250-148* of the Code requires that when more than one equipment grounding conductor enters a box, they shall be spliced together with devices "suitable for the use." These splices shall not be soldered. The splicing must be done so that if the receptacle is removed, the continuity of the ground path will not be interrupted.

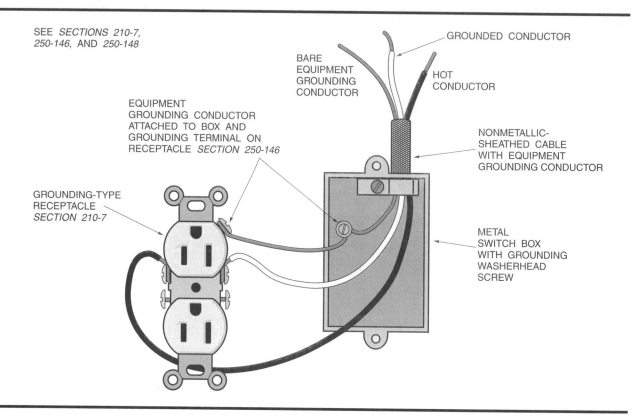

SEE *SECTIONS 210-7, 250-146,* AND *250-148*

GROUNDED CONDUCTOR

BARE EQUIPMENT GROUNDING CONDUCTOR

HOT CONDUCTOR

EQUIPMENT GROUNDING CONDUCTOR ATTACHED TO BOX AND GROUNDING TERMINAL ON RECEPTACLE *SECTION 250-146*

NONMETALLIC- SHEATHED CABLE WITH EQUIPMENT GROUNDING CONDUCTOR

GROUNDING-TYPE RECEPTACLE *SECTION 210-7*

METAL SWITCH BOX WITH GROUNDING WASHERHEAD SCREW

Figure 5–29 A grounding-type receptacle. The branch circuit's equipment grounding conductor is attached to the metal box and to the grounding screw on the receptacle in conformance to *Section 250-146* of the *National Electrical Code.*®

Figure 5–30 shows a crimp-type sleeve that splices two equipment grounding conductors together. One of the equipment grounding conductors is left longer than the other, and is attached to

GROUNDING CLIP

Figure 5–30 One method commonly used to attach equipment grounding conductors to a switch (device) box. The grounding clip slips onto the box, and is held tightly in place because of its spring metal design and shark barb that "digs" into the metal.

the metal switch box by a grounding clip "listed" for that application.

Figure 5–31 illustrates the use of a special grounding-type connector that provides the splicing together of multiple equipment grounding conductors of the nonmetallic-sheathed cables, and also has a conductor coming out of the end of the connector to terminate on the receptacle's grounding screw.

The methods for terminating equipment grounding conductors in metal switch boxes are also used for metal outlet boxes.

Where nonmetallic switch and outlet boxes are used, the equipment grounding conductor(s) can be spliced together using a crimp sleeve as illustrated in Figure 5–30, or spliced together using a special grounding-type twist-on connector as illustrated in Figure 5–31. The equipment grounding conductor must be connected to the equipment grounding terminal of a receptacle.

Metal Faceplates—Grounding

Section 380-9(b) requires that a means be provided to ground metal faceplates.

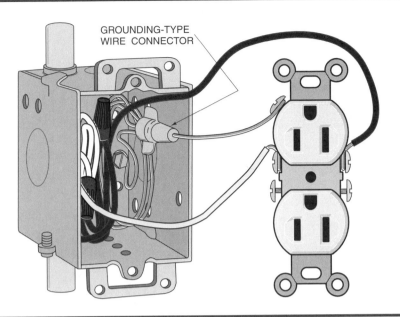

GROUNDING-TYPE WIRE CONNECTOR

Figure 5–31   Another method of connecting the equipment grounding conductor to a metal box and to the grounding terminal of the receptacle using a special grounding-type twist-on wire connector.

When metal faceplates are used with properly grounded metal boxes, the path for the faceplate grounding is accomplished through the No. 6-32 screws that attach the faceplate to the wiring device, then through the No. 6-32 screws that attach the wiring device to the metal box.

When metal faceplates are used with receptacles fastened to nonmetallic boxes, the metal faceplate is adequately grounded through the No. 6-32 screw that attaches the faceplate to the receptacle. This, of course, assumes that the required equipment grounding conductor has been connected to the grounding terminal of the receptacle.

When metal faceplates are used with switches fastened to nonmetallic boxes, the equipment grounding conductor must be connected to the equipment grounding terminal of a switch as illustrated in Figure 4–10.

REVIEW QUESTIONS

1. To provide adequate conductor length to work with for making splices and connecting switches and receptacles, what is the minimum length of conductor from where the conductor emerges from the cable that must be left at outlet boxes and switch boxes?

   a) 3 inches

   b) 6 inches

   c) 10 inches

2. This conductor is usually bare or green, and is used to connect noncurrent-carrying metal parts of equipment to the electrical system's grounding electrode conductor. Which term best describes this definition?

   a) grounded conductor

   b) equipment grounding conductor

3. This conductor is usually white, and is the conductor that is intentionally grounded at the main electrical service. Which term best describes this definition?

    a) grounded conductor

    b) equipment grounding conductor

4. When wiring with nonmetallic-sheathed cable, the insulation color on the conductors is determined by the manufacturer of the cable. This makes it necessary to use the white conductor as a "hot" conductor in many instances, such as for switch legs. What section of the *National Electrical Code®* allows this practice?

    a) *Section 200-7*

    b) *Section 210-8*

    c) *Section 250-122*

5. You are pulling some No. 14 AWG conductors into a raceway. May a green insulated No. 14 AWG conductor be pulled into the raceway for use as a grounded circuit conductor?

    a) Yes

    b) No

    As an ungrounded "hot" conductor?

    a) Yes

    b) No

    What sections of the *NEC®* support your answer?

    _____

6. Match the following statements with the letter that best describes the type of wiring device needed for the following.

    1. ____ Control a lighting fixture from one location.   a) A split-circuit receptacle

    2. ____ Control a lighting fixture from two locations.   b) A single-pole switch

    3. ____ Control a lighting fixture from three locations   c) Two three-way switches

    4. ____ The type of duplex receptacle needed when one   d) One four-way switch
        receptacle is to be controlled by a switch and   and two three-way
        the other receptacle is to be "hot" at all times.   switches.

7. The circuit comes into an outlet box in the ceiling where a lighting fixture will be installed. A two-wire nonmetallic-sheathed cable is run from the outlet box to the wall switch location that will control the lighting fixture. This means that in the ceiling outlet box, there are two white wires and two black wires. Which of the following statements best describes how the wires in the outlet box should be connected?

    a) Splice the black wire of the feed (the incoming circuit) to the black wire of the switch loop cable. This results in a white switch leg return conductor.

    b) Splice the black wire of the feed (the incoming circuit) to the white wire of the switch loop cable. This results in a black switch leg return conductor.

    c) Splice the two white wires in the outlet box together. This results in a black switch leg return conductor.

8. A split-level home has seven steps leading from one level to another. Because of the small number of steps, the electrician decided to install one switch at the top of the stairs to control the lighting fixture that will provide lighting to the stairway. The electrical inspector cited the electrician for a Code violation. To support the citation, the electrical inspector referenced *Section* _____ of the *National Electrical Code.*

9. A dimmer switch simply replaces a standard switch. Dimmer switches are connected in _____ with the lighting fixture.

   a) series
   b) parallel

10. A standard dimmer switch may be used to control fluorescent lighting.

   a) True
   b) False

11. The receptacles in a kitchen are supplied by 20-ampere small appliance branch circuits as required by the *National Electrical Code.* The nonmetallic-sheathed cables contain No. 12 AWG copper conductors. The cables are run from box to box. Each box has two white conductors and two black conductors. Which of the following statements is correct?

   a) Use No. 12 AWG copper conductors for the short pigtails that connect to the terminals on the receptacles.
   b) Use No. 14 AWG copper conductors for the short pigtails that connect to the terminals on the receptacles.

12. To prevent induction heating, what does *Section 300-3(b)* require? _____
   _____
   _____

13. The *NEC*® in *Section 210-7(a)* requires that receptacles be of the grounding type. What methods are used to connect the equipment ground to the receptacle?
   _____
   _____
   _____

14. How are metal face plates grounded?
   _____
   _____
   _____

Branch Circuit Calculations, Required Receptacle and Lighting Locations

OBJECTIVES

After studying this unit, you will be able to:

- Understand the definition of a branch circuit, a feeder, and ampacity.
- Understand the four different types of branch circuits used for house wiring.
- Understand the maximum permissible loads for branch circuits.
- Learn how to calculate electrical loads based on the square foot area of a home.
- Become familiar with conductor sizes and the overcurrent protection for branch circuits.
- Learn where lighting outlets and receptacle outlets must be installed.
- Learn how many receptacles and lighting outlets may be connected to a branch circuit.

INTRODUCTION

At the main panel in a home, the main power is divided into many smaller pieces called branch circuits. General purpose branch circuits supply lighting outlets and receptacles. Appliance branch circuits supply the receptacles in kitchens, dining rooms, and laundry area. Individual branch circuits supply specific appliances. This unit covers the fundamentals for determining the number and rating of the required branch circuits.

It is standard practice in house wiring for the electrician to plan and design the circuitry. The floor plans may show where outlets, switches, and special purpose outlets are to be located, but the plans do not show the actual wiring layout. The electrician must be familiar with all of the *National Electrical Code®* requirements regarding branch-circuit wiring.

And remember, the *National Electrical Code®* is *minimum*. Homes are always running short of places to plug in cords. So as we discuss minimum code requirements, it is highly recommended to install more receptacles and branch circuits than required by the *National Electrical Code®*. Try to think of locating wall receptacles so that extension cords and "cube taps" will not be needed. Consider installing "quadplex" instead of "duplex" receptacles in those rooms where there will be a great number of "things" that need to be plugged in, such as computers, printers, stereos, VCRs, CD players, television sets, CATV cable boxes, radios, fax machines, answering machines, typewriters, adding machines, lamps, etc. You will never have too many receptacles. You will never have too many branch circuits.

Article 210 of the *National Electrical Code®* covers most of the general requirements for branch circuits.

BRANCH CIRCUIT

The *National Electrical Code®* defines a branch circuit as **Branch Circuit:** *"The circuit conductors between the final overcurrent device protecting the circuit and the outlet(s)."*

See Figure 6–1. In homes, the wiring to receptacle outlets, lighting outlets, electric ranges, electric dryers, and other fixed-in-place appliances are all examples of a branch circuit. The cords on portable appliances are not considered to be part of the branch circuit.

TYPES OF BRANCH CIRCUITS

The *National Electrical Code®* breaks down branch circuits into four specific types. These are:

- **Appliance Branch Circuit:** *"A branch circuit supplying energy to one or more outlets to which appliances are to be connected, and that has no permanently connected lighting fixtures not a part of an appliance."*

  These are the 20-ampere small appliance branch circuits discussed later in this unit.

- **General Purpose Branch Circuit:** *"A branch circuit that supplies a number of outlets for lighting and appliances."*

In a home, these are the circuits that feed lighting fixtures and receptacles that are not connected to the 20-ampere required small appliance branch circuits.

- **Individual Branch Circuit:** *"A branch circuit that supplies only one utilization equipment."*

  Some examples of an individual branch circuit are separate circuits supplying a single appliance such as an air conditioner unit, a dishwasher, a food waste disposer, an electric range, and electric clothes dryer. An individual branch circuit can supply a single receptacle. If the circuit supplies a duplex receptacle, then by definition, it is not an individual branch circuit.

- **Multiwire Branch Circuit:** *"A branch circuit consisting of two or more ungrounded conductors having a potential difference between them, and a grounded conductor having equal potential difference between it and each ungrounded conductor of the circuit and that is connected to the neutral or grounded conductor of the system."*

  Multiwire branch circuits can be used effectively in house wiring for certain applications. Multiwire branch circuits were discussed in Unit 2.

FEEDER

Quite often, the terms *branch circuit* and *feeder* are not used properly. We have already defined a branch circuit.

- **Feeder:** *"All circuit conductors between the service equipment, the source of a separately derived system, or other power supply source and the final branch circuit overcurrent device."*

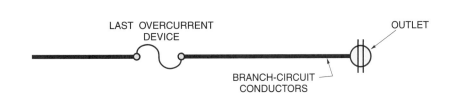

Figure 6–1 A branch circuit is that part if the wiring that runs from the final overcurrent device (usually in the panel) to the outlet. The ampere rating of the overcurrent device determines the rating of the branch circuit—not the ampere rating of the conductor.

An example of a feeder would be the conductors between the main service panel and a subpanel. It is quite common in larger homes to locate one or more subpanels some distance from the main service panel where there might be a heavy concentration of load.

AMPACITY

Let us define a special word that is unique and quite common in the electrical trade.

Ampacity: "The current in amperes that a conductor can carry continuously under the conditions of use without exceeding its temperature rating."

The ampacity of branch-circuit conductors must not be less than the maximum load to be served, *Section 210-19(a)*. Figure 6–2 illustrates four 5-ampere loads that total 20 amperes, requiring conductors that have a minimum 20-ampere rating.

Feeder conductors must have an ampacity not less than the computed load, *Section 215-2*.

Where the branch circuit supplies receptacle outlets for cord- and plug-connected portable loads, the conductor's ampacity must not be less than the ampere rating of the overcurrent device protecting that conductor, *Section 210-19(b)*. This concept is illustrated in Figure 6–3.

BRANCH-CIRCUIT RATING

The rating of a branch circuit is determined by the ampere rating of the overcurrent device protecting the branch circuit, *Section 210-3*. Figure 6–4 shows conductors that have three different ampere ratings, yet the branch-circuit overcurrent device in all cases is 15 amperes. Because it is the rating of the over-

current device (OCD) that determines the rating of a branch circuit, all three of the branch circuits in Figure 6–4 are considered to be 15-ampere branch circuits.

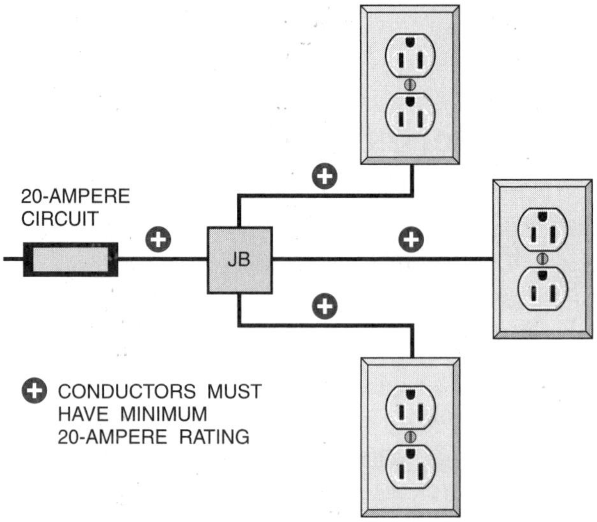

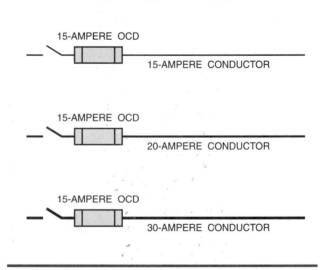

Figure 6–3   **All conductors in this 20-ampere branch circuit must have an ampacity of not less than the 20-ampere rating of the branch circuit because the circuit supplies more than one receptacle, *Section 210-19(a)* and *Table 210-24*.**

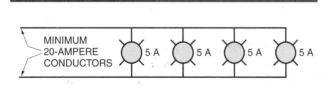

Figure 6–2   **Branch-circuit conductors shall have an ampacity not less than the maximum load to be served, *Section 210-19(a)*. See *Section 210-23* for permissible loads.**

Figure 6–4   **All three of these branch circuits are rated as 15–ampere circuits, even though larger conductors were used for some reason, such as solving a voltage drop problem. The rating of the overcurrent device (OCD) determines the rating of a branch circuit, *Section 210-3*.**

A single receptacle connected to a circuit must have a rating not less than the rating of the branch circuit, *Section 210-21(b)(1)*.

Branch circuits rated 15 amperes supplying two or more receptacles shall not contain receptacles rated over 15 amperes. *See Table 210-21(b)(3)*.

Branch circuits rated 20 amperes supplying two or more receptacles may have 15- or 20-ampere-rated receptacles. See *Table 210-21(b)(3)*.

Exceptions

There are some exceptions to the basic rule that the ampere rating of the overcurrent device must match the ampacity of the conductor. Some of these exceptions are quite common in residential wiring, such as:

- Motor branch circuits: The conductors are based on 125% of the motor's full-load current draw, *Section 430-22(a)*. The motor's branch-circuit short-circuit and ground-fault protection is sized at 175% to 400% of a motor's full-load current, depending on the type of overcurrent protective device used, *Section 430-52* and *Table 430-152*. The motor's and branch-circuit overload protection is sized at 125% of the motor's full-load current draw, *Section 430-32(a)*. The overload protection for the motor will also protect the branch-circuit conductor from becoming overloaded.

- Small lighting fixture tap wires are generally smaller than the branch-circuit wires, *Table 210-24*.

- Where the ampacity of the conductor does not match a standard rating of a fuse or circuit breaker, the next higher standard size overcurrent device may be used, provided the overcurrent device does not exceed 800 amperes. This permission is found in *Section 240-3(b)*. For example, from *Table 310-16*, we find that a No. 6 TW 60°C copper conductor has an ampacity of 55 amperes. It would be all right to protect this conductor with a 60-ampere fuse or circuit breaker, which is a standard ampere rating as listed in *Section 240-6(a)*. This exception is not permitted for circuits that supply more than one receptacle because too many "plug-in" loads could overload the circuit, *Section 210-19(b)*.

PERMISSIBLE LOADS ON BRANCH CIRCUITS

Section 210-23 covers the permitted loading for a typical branch circuit.

- The load shall not exceed the branch circuit rating.

- The branch circuit must be rated 15, 20, 30, 40, or 50 amperes when serving two or more outlets.

- An individual branch circuit may supply any size load.

- 15- and 20-ampere general purpose branch circuits:

    a. Can supply lighting, other equipment, or both.

    b. Cord- and plug-connected equipment shall not exceed 80% of the branch-circuit rating.

    c. Equipment fastened in place shall not exceed 50% of the branch-circuit rating if the branch circuit also supplies lighting, other cord- and plug-connected equipment, or both.

- The two 20-ampere small appliance branch circuits required by *Section 210-11(c)(1)* must not supply any other loads, *Section 210-52(b)(2)*.

- The separate 20-ampere branch circuit for laundry equipment must not supply any other loads, *Section 210-11(c)(2)*.

- The separate 20-ampere branch circuit for bathrooms required by *Section 210-11(c)(3)* must not supply any other loads.

- 30-ampere branch circuits may supply equipment such as electric dryers, cooktops, ovens, etc. If cord- and plug-connected, these appliances shall not exceed 80% of the branch-circuit rating. See *Section 210-23(b)*.

- 40- and 50-ampere branch circuits may supply cooking appliances that are fastened in place, such as an electric range.

- Branch circuits rated over 50 amperes may be needed to connect large loads, such as electric furnaces, heat pumps, and double-ovens.

CONTINUOUS LOADS

In commercial and industrial wiring, it is rather easy to determine what loads will be on continuously. But in homes, so many of the lighting fixtures and receptacles are not going to be on continuously. There is so much diversity that for these loads, it is hard to define them as continuous. We need to refer to the *National Electrical Code®* to look up the definition of a continuous load.

> *Continuous Load: "A load where the maximum current is expected to continue for three hours or more."*

An electric furnace, a heat pump, an air conditioner, snow melting cables, and other specific loads could very well be on for three hours or more. For these types of loads, we find in *Sections 210-19* and *210-20(a)* that the branch-circuit rating and the ampacity rating of the conductors shall not be less than 125% of the continuous load. These sections also tell us that the overcurrent protective device shall not be loaded to more than 80% of the OCD rating. In *Section 384-16(d),* we find the same requirement, except the wording is slightly different. *Section 384-16(d)* states that the load shall not exceed 80% of the OCD ampere rating.

Example: Find the minimum branch circuit required for a 40-ampere electric furnace.

Answer: $40 \times 1.25 = 50$ amperes.

Refer to Unit 13 for specific code requirements for appliances where the 125% and 80% loading factors are discussed in detail.

VOLTAGE FOR CODE CALCULATIONS

For uniformity, all calculations for house wiring are based on 120 volts and 240 volts. This is mentioned in *Section 220-2*. In Appendix D of the *National Electrical Code,®* we find a similar statement that for uniform application of *Articles 210, 215,* and *220,* nominal voltages of 120 and 240 volts shall be used. The word *nominal* means **in name only, not a fact**. In calculations then, we use 120 volts even though the actual voltage might be 110, 115, or 117 volts, and we use 240 volts even though the actual voltage might be 220 or 230 volts.

CALCULATING FLOOR AREA

The starting point to determine the number of general purpose branch circuits is to calculate the square footage of occupied floor area. Some areas in a home such as unfinished basements, open porches, garages, or other unfinished or unused spaces that are not adaptable for future use need not be included in the "volt-amperes per square foot" calculations, *Section 220-3(a)*. Yet many patios, terraces, open porches can be used as recreation and entertainment areas, in which case you should include these areas in the square footage calculation. It is somewhat of a judgment call. If in doubt, include these areas. The results will be more "plus" as opposed to being on the conservative side, and not having a large enough service, and not having enough branch circuits. It is always a good idea to provide some extra "head room" when it comes to electrical calculations.

Let us use a typical home with a full finished basement (Figure 6–5). The outside dimensions of the home is 56×36 feet.

Step 1: 56 feet $\times$ 36 feet = 2,016 square feet
Step 2: Deduct areas that need not be included.

195 square feet (part of an attached garage)
121 square feet (front porch)
  84 square feet (more of the front porch)
400 square feet

Therefore: 2,016 – 400 = 1,616 square feet

As stated above, we will also include the same square footage for the finished basement.

First Floor	1,616 ft²
Basement	1,616 ft²
Total	3,232 ft²

DETERMINING THE MINIMUM NUMBER OF LIGHTING BRANCH CIRCUITS

The term "lighting branch circuit" is a general purpose branch circuit used to supply lighting, receptacle outlets, or a combination of lighting and receptacle outlets.

Table 220-3(a) shows the minimum load required for dwelling units is 3 volt-amperes (VA) per square foot of occupied area.

To determine the minimum number of 15-ampere, 120-volt branch circuits:

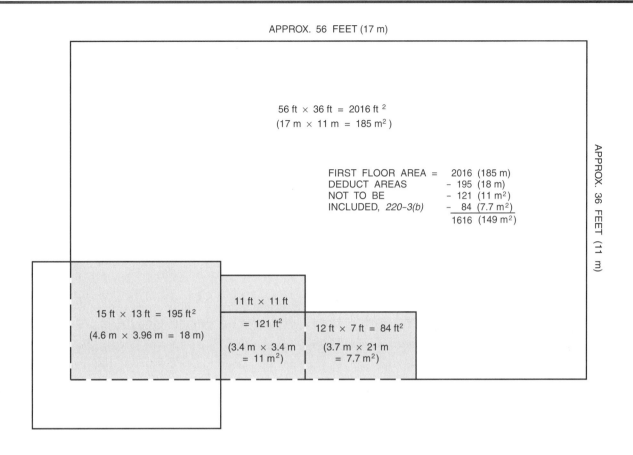

APPROX. 56 FEET (17 m)

56 ft × 36 ft = 2016 ft ^{2}
(17 m × 11 m = 185 m^2)

FIRST FLOOR AREA =	2016	(185 m)
DEDUCT AREAS	- 195	(18 m)
NOT TO BE	- 121	(11 m^2)
INCLUDED, *220–3(b)*	- 84	(7.7 m^2)
	1616	(149 m^2)

APPROX. 36 FEET (11 m)

15 ft × 13 ft = 195 ft^2

(4.6 m × 3.96 m = 18 m)

11 ft × 11 ft

= 121 ft^2

(3.4 m × 3.4 m
= 11 m^2)

12 ft × 7 ft = 84 ft^2

(3.7 m × 21 m
= 7.7 m^2)

Figure 6–5   Determining the square foot area of a house. This example shows the outside dimensions of the house and the dimensions for areas that need not be included. In our example, the full basement is finished, and is included in the calculation for determining the total square foot area.

Step 1:   total volt-amperes =
square feet × 3 VA per square foot

Step 2: $amperes = \dfrac{total\ volt - amperes}{120\ volts}$

Step 3:   Minimum number of 15-ampere
circuits required
$= \dfrac{amperes}{15}$

This equates to a minimum of one 15-ampere branch circuit for every 600 square feet.

When 20-ampere, 120-volt lighting branch circuits are used in residential installations, the same procedure as indicated in Steps 1, 2, and 3 above would result in one 20-ampere branch circuit for every 800 square feet.

In our example, we calculated the total square foot area to be 3,232 ft². At 3 VA per square foot, we have:

Step 1: $3,232 \times 3 = 9,696$ volt-amperes

Step 2: The total amperage is:

$Amperes = \dfrac{volts - amperes}{volts} = \dfrac{9,696}{120} = 80.8$ amperes

Step 3: The minimum number of 15-ampere lighting circuits required is:

$\dfrac{80.8}{15} = 5.39$ circuits (round up to six circuits)

We also could have calculated the minimum number of 15-ampere lighting circuits by using the factor of 600 square feet per 15-ampere circuit.

Doing this, we have:

$\dfrac{3,232}{600} = 5.39$ circuits (round up to six circuits)

In our example, the minimum number of lighting 15-ampere branch circuits is six.

When we talk about lighting branch circuits in homes, all general purpose receptacles and lighting are included in the 3 VA per square foot load factor. This is in accordance with *Section 220-3(b)(10)*. Where resistive loads are involved, watts and volt-amperes are the same.

TRACK LIGHTING

No additional load need be added for track lighting in homes for the purpose of calculating branch circuits, feeders, and service-entrance equipment. Track lighting in homes is included in the general lighting load of 3 VA per square foot. This is confirmed in *Section 220-3(b)(10)*.

As required by *Section 410-101(b)*, the connected load on a lighting track shall not exceed the rating of the track. Also, be sure that a lighting track is supplied by a branch circuit having a rating not more than that of the track.

DETERMINING THE MINIMUM NUMBER OF SMALL APPLIANCE BRANCH CIRCUITS

Small high wattage appliances such as toasters, microwave ovens, coffee pots, hot plates, and similar appliances draw quite a bit of current. They need to be plugged into receptacles that are connected to 20-ampere "heavy-duty" branch circuits. They should not be plugged into 15-ampere general purpose lighting branch circuits as this could lead to overloading.

In a home, at least two 20-ampere small appliance branch circuits are required by *Section 210-11(c)(1)*. In accordance with *Section 210-52(b)(1)*, these 20-ampere small appliance branch circuits supply the receptacles in kitchens, pantries, dining rooms, breakfast rooms, and similar rooms.

Although the *NEC®* requires a **minimum** of two 20-ampere small appliance branch circuits, installing more than two 20-ampere small appliance branch circuits is recommended. Two 20-ampere branch circuits are just not enough when one considers how many small appliances are used today.

Section 210-52(b)(1), Exception No. 1 allows a switched receptacle to be supplied by a 15-ampere general lighting circuit rather than connecting it to one of the 20-ampere small appliance branch circuits. This switched receptacle would be in addition to the required small appliance receptacle outlets. Figure 6–6 is an example of how this Code rule applies.

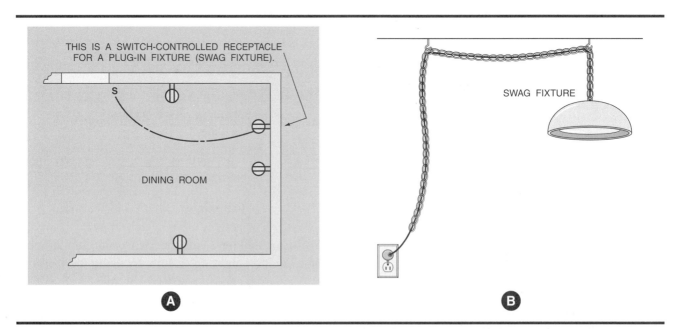

Figure 6–6 *Section 210-52(b)(1), Exception No. 1* allows a switch-controlled receptacle outlet to be installed in rooms where the receptacle outlets normally would be supplied by 20-ampere small appliance circuits (breakfast rooms, dining rooms, etc.). This switched receptacle must be in addition to the required small appliance receptacle outlets, and would be connected to one of the 15–ampere general lighting branch circuits. (A) is a circuit layout and (B) is a pictorial view of the intent of *Section 210-52(b)(1), Exception No. 1.*

Section 210-52(b)(1) allows the receptacle for a refrigerator in a kitchen to be connected to either one of the two 20-ampere small appliance branch circuits. But recognizing that the refrigerator load takes up some of the 20-ampere small appliance branch circuit's capacity, *Section 210-52(b)(1), Exception No. 2* states that the receptacle outlet for refrigeration equipment may be supplied by a separate 15- or 20-ampere branch circuit rather than connecting it to one of the two required 20-ampere small appliance circuits. Because this separate branch circuit diverts the refrigerator load from the receptacles that serve the countertop areas, it is not considered to be a small appliance branch circuit. When you study Unit 16, you will learn that the Code does not require that an additional 1,500 volt-amperes be included in the load calculations for the separate refrigerator 15- or 20-ampere branch circuit. However, there is nothing in the Code that prohibits adding another 1,500 volt-amperes to the load calculations for this additional branch circuit, as

this would result in more capacity in the electrical system.

Section 210-52(b)(2) states that these 20-ampere small appliance branch circuits must not supply other outlets. Exceptions to this are clock receptacles, and receptacles that are installed to plug in the cords of gas-fired ranges, ovens, or counter-mounted cooking units in order to power up their timers and ignition systems (Figure 6–7).

Do not connect food waste disposers and dishwashers to the required minimum of two 20-ampere small appliance branch circuits. These circuits are intended solely for countertop appliances. Food waste disposers and dishwashers must be supplied by their own separate branch circuits.

Section 210-52(b)(3) requires that not less than two 20-ampere small appliance branch circuits must supply the receptacle outlets serving the countertop areas in kitchens.

Section 210-11(c)(2) requires that at least one additional 20-ampere branch circuit must be provided to supply the laundry receptacle outlet(s). A receptacle for laundry equipment is required by *Section 210-52(f)*.

LIGHTING AND SMALL APPLIANCE BRANCH CIRCUITS BASICS

The following text discusses the minimum conductor size and maximum overcurrent protection requirements for 15- and 20-ampere branch circuits.

15-Ampere Branch Circuits

Fifteen-ampere branch circuits are wired with No. 14 AWG copper conductors. The maximum overcurrent protection rating is 15 amperes. Refer to *Table 310-16* and *Section 240-3(d)*. Some electricians prefer using No. 12 AWG conductors throughout the house to keep voltage drop to a minimum, but protect the lighting branch circuits with 15-ampere overcurrent protection. This is a matter of choice, and not the *NEC.*

20-Ampere Branch Circuits

Twenty-ampere branch circuits are wired with No. 12 AWG copper conductors. The maximum overcurrent protection rating is 20 amperes. Refer to *Table 310-16* and *Section 240-3(d)*. The use of

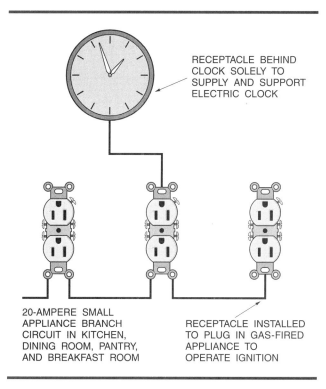

RECEPTACLE BEHIND CLOCK SOLELY TO SUPPLY AND SUPPORT ELECTRIC CLOCK

20-AMPERE SMALL APPLIANCE BRANCH CIRCUIT IN KITCHEN, DINING ROOM, PANTRY, AND BREAKFAST ROOM

RECEPTACLE INSTALLED TO PLUG IN GAS-FIRED APPLIANCE TO OPERATE IGNITION

Figure 6–7   A receptacle installed solely to supply and support an electrical clock, or a receptacle that supplies power for clocks, clock timers, and electric ignition for gas ranges, ovens, and cooktops may be connected to a 20-ampere small appliance branch circuit, or to a general purpose lighting branch circuit, *Section 210-52(b)(2), Exceptions.*

No. 12 AWG wire will minimize the voltage drop in the circuit, improves appliance performance, and lessens the danger of overloading the circuits.

LOAD CALCULATIONS FOR SMALL APPLIANCE BRANCH CIRCUITS

Section 220-16(a) and *(b)* state that an additional load of not less than 1,500 watts shall be included for each 20-ampere, two-wire small appliance circuit when calculating feeders and services. This is covered in detail in Unit 16.

SUMMARY OF WHERE RECEPTACLES MUST BE INSTALLED

The *NEC®* sets forth very specific rules for receptacles that must be installed in homes. The following text presents these requirements.

Habitable Rooms (General Purpose Receptacle Outlets 125-volt, single-phase, 15- or 20-ampere)

- Branch circuits supplying wall receptacle outlets for plug-in lighting loads are generally 15-ampere circuits, but are permitted by the Code to be 20-ampere circuits.

- Wall receptacle outlets must be placed so that no point along the floor line is more than 6 feet (1.83 m) measured horizontally from an outlet. Fixed room dividers and railings are considered to be walls for the purpose of this requirement, *Section 210-52(a)(1)* and *(2)*.

- Nonsliding fixed glass panels on exterior walls are considered to be wall space and are to be figured in when applying the rule that no point along the floor line is more than 6 feet (1.83 m) from a receptacle. Sliding panels on exterior walls are not considered to be a wall, *Section 210-52(a)(2)*.

- Receptacle outlets located in the floor more than 18 inches (457 mm) from the wall are not to be counted as meeting the required number of wall receptacle outlets, *Section 210-52(a)(3)*. An example of this would be a floor receptacle installed under a dining room table for plugging in warming trays, coffee pots, or similar appliances.

- Any wall space 2 feet (610 mm) or more in width must have a receptacle outlet, *Section 210-52(a)(2)*. Figure 6–8 shows the preceding requirements.

Hallways

Hallways 10 feet or longer in homes must have at least one receptacle outlet, *Section 210-52(h)*. See Figure 6–9.

Do not get caught up in a play on words. A home may have a *front hall,* a *front entry* a *vestibule,* or a *foyer.* These are hallways insofar as the *National Electrical Code®* is concerned, and require at least one receptacle outlet in conformance to *Section 210-52(h)*.

Crawl Spaces and Attics

In crawl spaces and in attics, a receptacle outlet must be installed in an accessible location within 25 feet of heating, air-conditioning, or refrigeration equipment. This receptacle is for plugging in trouble lights and portable tools when servicing the equipment. Therefore, do *not* connect this receptacle outlet to the load side of the equipment disconnecting means, *Section 210-63*.

Roof Tops

- A receptacle is not required on the rooftop of one- and two-family dwellings, *Section 210-63, Exception.*

- For buildings containing more than two-family dwellings (for example, a four-flat multifamily building), a GFCI-protected receptacle outlet per *Section 210-8(b)(2)* must be installed on the roof if there is heating, air-conditioning, or refrigeration equipment that requires servicing on the roof. This receptacle outlet must be located in an accessible location within 25 feet of the equipment on the same level as the equipment, *Section 210-63*.

Basements

- Unfinished basements or unfinished portions of basements with areas not "lived in," such as work areas and storage areas, are not subject to the "6-foot rule" for wall receptacle spacing.

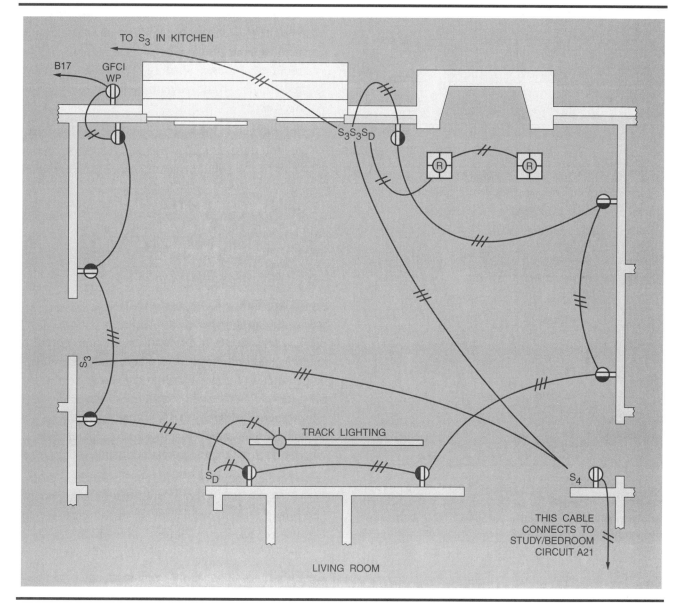

Figure 6–8 The scale of this electrical plan is ¼" = 1 foot. No point along the floor line is more than 6 feet (1.83 m) from a receptacle. The fixed glass panel is considered to be wall space. The sliding glass door is not considered to be wall space. A receptacle is required in the short wall space that is 2 feet wide or greater in the lower right-hand corner of the floor plan.

These areas are not required to have as many receptacles as habitable rooms.

• In unfinished basement areas that are not intended to be "lived in" such as storage areas or workshop areas, all 125-volt, single-phase, 15- or 20-ampere receptacles must have GFCI protection, *Section 210-8(a)(5)*. This includes receptacles that are an integral part of porcelain or plastic pull-chain or keyless lampholders. See Unit 7 for detailed coverage of GFCI protection.

• In finished basements or finished portions of basements, the required number, spacing, and location of receptacles must be in conformance to *Section 210-52(a)* as discussed above under General Purpose Receptacle Outlets.

• At least one receptacle outlet must be installed in addition to the required laundry outlet, *Section 210-52(g)*. The required laundry outlet may be located in a location other than in the basement. A receptacle is required for the laundry area, *Section 210-52(f)*.

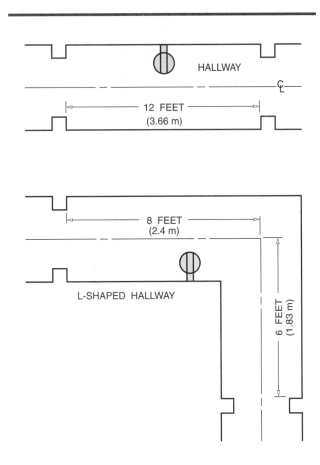

HALLWAY

12 FEET
(3.66 m)

C_L

8 FEET
(2.4 m)

L-SHAPED HALLWAY

6 FEET
(1.83 m)

Figure 6–9   Hallways 10 feet or longer must have at least one receptacle outlet, *Section 210-52(h)*. Hallway length is determined using the centerline measurements of the hall. The receptacle may be installed anywhere in the hall.

- Connect to 15- or 20-ampere, 120-volt circuit.

- GFCI protection is *not* required for receptacles that are not readily accessible, or for a single receptacle or duplex receptacle for two appliances that are not easily moved, are cord- and plug-connected, and are located in a dedicated space, *Section 210-8(a)(5), Exceptions No. 1* and *2*. Examples of this include the receptacle for laundry equipment, a sump pump, a water softener, a freezer, or a refrigerator. These appliances are not readily moved from one location to another. Refer back to Figure 3–8 for a clear understanding of the meaning of a single receptacle and a multiple receptacle.

- In finished basements, GFCI protection for receptacle outlets is not required unless the receptacles are located within 6 feet (1.83 m) of

the outer edge of a wet-bar sink and are intended to serve the countertop areas around the wet-bar sink, *Section 210-8(a)(7)*. The required number, spacing, and location of the receptacles would have to be in conformance to *Section 210-52(a)*.

- If a kitchen is located in a basement, GFCI protection must be provided for all receptacles. See Unit 7 for detailed coverage of GFCI protection.

Refer to *Sections 210-8(a), 210-52(f)* and *210-52(g)* of the Code.

Kitchens and Countertops

Small Appliance Receptacle Outlets (125-volt, single-phase, 20-ampere)

To discourage the use of extension cords in kitchens, the *NEC*® specifies where receptacles must be installed in kitchens for the convenience of plugging in small electrical appliances. There are also requirements relating to the ampere rating of the branch circuits supplying these receptacles, and to where GFCI receptacles must be installed.

- For the receptacle outlets in the kitchen, dining areas, breakfast room, or pantry, at least two 20-ampere small appliance circuits must supply these receptacle outlets. *See Sections 210-11(c)(1), 210-52, and 220-16(a)* of the Code.

- In kitchens and dining rooms, receptacle outlets must be installed above all countertops 12 inches (305 mm) or wider (Figure 6–10).

- Branch circuits supplying electrical outlets in kitchens, dining rooms, and pantries where appliances will be plugged in are required to be 20 amperes.

- Receptacle outlets that serve countertops must be supplied by at least two 20-ampere small appliance circuits. These two 20-ampere circuits are required to supply the other receptacle outlets in the kitchen, dining room, pantry, or breakfast room. Of course, more than the required minimum of two 20-ampere small appliance branch circuits can be installed.

- Receptacles are not permitted to be installed "face-up" on countertops.

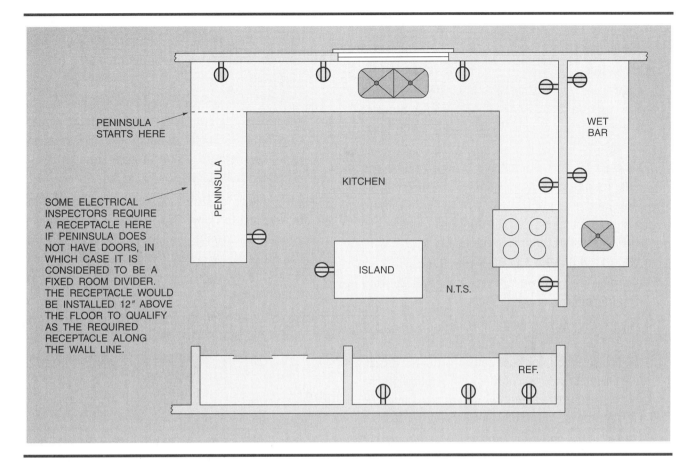

Figure 6–10 **Floor plan showing placement of receptacles that serve countertop surfaces in kitchens and wet-bar areas in homes, in conformance to *Sections 210-8(a)(6)* and *(7)* and *210-52(c)* of the Code. Drawing is not to scale. See text for detailed explanation.**

- Receptacle outlets that serve countertop surfaces in kitchens must be GFCI protected. This includes receptacles installed along the wall that serve the countertops, and on peninsulas and freestanding islands. See Unit 7 for detailed coverage of GFCI protection.

- These receptacles *shall not* be installed more than 18 inches (458 mm) above the countertop. When special permission is given by the authority having jurisdiction (usually the electrical inspector), receptacles that serve countertops may be mounted not more than 12 inches (305 mm) below the countertop. You cannot use this special permission when the countertop extends more than 6 inches (153 mm) beyond the base cabinet. This permission can only be given:

    1. if the receptacles must be located so that the physically impaired can reach them, or

    2. in the case of islands and peninsulas, where

the construction makes it impractical to mount the receptacles above the countertop.

- A receptacle located behind a refrigerator, or under the sink serving a food waste disposer and/or a dishwasher does not have to be GFCI protected because these receptacles do not serve countertop surfaces. See Unit 7 for detailed coverage of GFCI protection.

- Receptacle outlets installed within 6 feet (1.83 in) from the outside edge of a wet-bar sink must be GFCI protected.

- Receptacle outlets in kitchen and dining areas located above countertops shall be positioned so that no point along the wall line is more than 24 inches (610 mm), measured horizontally, from a receptacle outlet in that space. This results in the receptacle being installed not more than 4 feet apart. In Figure 6–9, the wall line measurement to the right of the sink starts at the edge of the sink, to the upper right-hand corner of the counter, then to the edge of the range. See Unit 7 for detailed coverage of GFCI protection.

Example: A section of countertop measures 5' 3" along the wall line. The minimum number of receptacles is: 5' 3" ÷ 4 = 1+ (install two receptacles).

Example: The countertop length from the edge of a sink to a range space measures 11' 2" measured along the wall line. The minimum number of receptacles for this space is: 11' 2" ÷ 4 = 2 + (install three receptacles).

Note: Dividing by 4 assures that there will be a receptacle outlet for every 4 linear feet or fraction thereof of counter length.

- Receptacle outlets behind refrigerators or appliances fastened in place must not be thought of as meeting the previous requirements for receptacle outlets above countertops, because these outlets would be impossible to use once the refrigerator or appliance is fastened in place.

- See *Sections 210-8(a), 210-11(c)(2), 210-50(c), 210-52(f)*, and *220-16(b)*.

Peninsulas and Islands

Peninsulas and freestanding islands that have a long dimension of 24 inches (619 mm) or greater, and a short dimension of 12 inches (305 mm) or greater must have at least one receptacle outlet. Measure the peninsula length from the edge where the peninsula connects to the wall base cabinet unit. If the peninsula or island contains a counter-mounted range that would create a space 12 inches (305 mm) or greater on both sides of the appliance, then a receptacle must be installed in each space because this arrangement is considered to be two spaces. Note in Figure 6–9 that one receptacle is indicated on the peninsula and one receptacle on the island. The receptacle outlet(s) must be GFCI protected.

See *Sections 210-8(a)(6) and (7), 210-11(c), 210-50(c), 210-52(c)(1) and (2),* and *220-16(a).*

Laundry

- At least one receptacle outlet is required within 6 feet (1.83 m) of the intended location of the washing machine.

- This branch circuit must be a 20-ampere circuit.

- This branch circuit must not serve any other outlets, unless the other outlet is for laundry purposes.

- This receptacle *need not* be GFCI protected because acceptable levels of leakage current of some clothes washing machines, as determined by Underwriters Laboratories, can cause a GFCI to nuisance trip.

- This outlet and circuit are in addition to the basement receptacle outlet requirements already discussed, and must be in addition to the required two small appliance circuits that serve the receptacle outlets in the kitchen and dining areas.

- See *Sections 210-8(a), 210-11(c)(2), 210-50(c), 210-52(f)*, and *220-16(b)*.

Bathrooms

The Code defines a bathroom as:

- **Bathroom:** *"A bathroom is an area with a basin and one or more of the following: a toilet, a tub, or a shower."*

- At least one receptacle must be installed within 36 inches (914 mm) from the outside edge of a basin. This receptacle shall be located on a wall that is adjacent to the basin. In a two basin bathroom, two receptacles are generally required. However, if you can arrange to install the receptacle in the middle of the wall space between the two basins, one receptacle could serve both basins and would meet the intent of the Code.

- Do not install receptacles face up in a bathroom countertop.

- A receptacle that is an integral part of a fixture or medicine cabinet does *not* qualify for this required outlet. However, if there is a receptacle on the cabinet or fixture, it must be GFCI protected. If providing GFCI protection for a receptacle in a cabinet or fixture presents a problem, it is best to leave the receptacle on the cabinet or fixture unconnected or removed altogether.

- Connect the receptacle(s) to at least one separate 20-ampere branch-circuit. This is because of the use of high wattage appliances such as electric hair dryers used in bathrooms. This separate circuit is permitted to supply other receptacles in the bathroom, and receptacles in other bathrooms. No other outlets are permitted to be connected to this circuit. However, if the circuit supplies only one bathroom, then other fastened in place equipment that does not exceed 10 amperes (50% of the 20-ampere branch circuit) such as an exhaust fan or a hydromassage tub, is permitted to be connected to the separate 20-ampere branch circuit. Of course, if the instructions furnished with the hydromassage tub requires a separate branch-circuit, then those instructions must be followed.

Although a minimum of one separate 20-ampere branch-circuit is required by the Code, more than one branch-circuit may be provided. For service and feeder load calculations, no additional load needs to be included for this separate circuit(s). Calculations are covered in Unit 16.

• The receptacle(s) must be GFCI protected. Refer to Unit 7 for details.

• See *Sections 210-8(a)(1), 210-11(c)(3), 210-23(a), 210-52(d)* and *220-16.*

Receptacles and Switches in Bathtub and Shower Spaces. Because of the obvious hazards associated with water and electricity, receptacles are not permitted to be installed in bathtub and shower spaces, *Section 410-57(c).*

Switches are not permitted in the wet locations in and around tub and shower locations, *Section 380-4.* The only exception to this is for switches that are part of a "listed" tub or shower assembly such as a hot tub or hydromassage tub.

Outdoors

• At least one receptacle outlet must be installed at grade level in the front and another in the back of all one-family dwellings and each unit of a two-family dwelling. These receptacles must be accessible and not more than 6½ feet (1.98 m) above grade level (Figure 6–11).

• A receptacle on a porch or deck is acceptable, provided there is easy access to the receptacle from grade level. A screened-in porch or deck, or a porch or deck that requires access by climbing up a number of steps would not meet the intent of the Code. This is a judgment call.

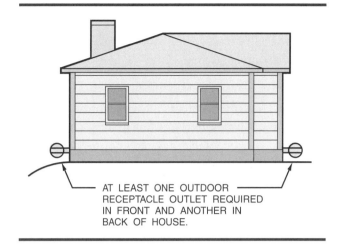

AT LEAST ONE OUTDOOR RECEPTACLE OUTLET REQUIRED IN FRONT AND ANOTHER IN BACK OF HOUSE.

Figure 6–11 For residential wiring, at least one receptacle outlet must be installed in the front and another in the back of the house. These receptacles must be GFCI protected, and must not be installed more than 6½ feet above grade. See text for details of *Sections 210-8(a)(3), 210-52(e)* **and** *426-28.*

One or two steps from grade level onto the porch is O.K. More steps might make it difficult to access the receptacle.

• Connect to 15- or 20-ampere, 120-volt circuit.

• All outdoor receptacles must be GFCI protected. An exception is that personnel GFCI protection is not required for outdoor receptacle(s) installed for snow and ice-melting equipment (i.e., heating cables in rain gutters) that is not readily accessible, and is supplied by a separate branch circuit that has ground-fault protection for equipment. See Unit 7 for detailed coverage of GFCI protection.

• See *Section 426-28* for ground-fault protection *of equipment.* GFPE protection must be provided for branch circuits that supply "fixed" outdoor de-icing and snow melting equipment. An example of "fixed" outdoor de-icing and snow melting equipment is Type MI cable that is buried in the concrete of a driveway or sidewalk. These systems are not plugged in, but instead, are permanently wired. Ground-fault protection devices for equipment (GFPE) trip at approximately 30 milliamperes, whereas GFCI protection for personnel trips in the range of 4 to 6 milliamperes. GFPE is provided by the manufacturer of the equipment.

• See *Sections 210-8(a)(3), 210-52(e)* and *426-28.*

Garages

- At least one receptacle outlet must be provided for an attached garage.

- Connect to 15- or 20-ampere circuit.

- Must be GFCI protected. An exception to this is for receptacles that are not readily accessible, such as a receptacle mounted on the ceiling for plugging in a garage door opener. Also exempt from the GFCI requirement are single receptacles or duplex receptacles intended for cord- and plug-connected appliances that are not easily moved, and are located in dedicated spaces. Examples are a freezer, refrigerator, or central vacuum equipment. See Unit 7 for detailed coverage of GFCI protection.

- If the garage is detached, the Code does not require electric power, but if electric power is provided, then at least one GFCI-protected receptacle outlet must be installed.

- See *Sections 210-8(a)(2)* and *210-52(g)*.

Receptacles in Other Locations

All of the required receptacles for general use, small appliances, countertops, bathrooms, outdoors, laundry, basements, garages, and hallways discussed above are *in addition* to any receptacles that are part of lighting fixtures or appliances, are located inside of cabinets, or are located more than 5½ feet (1.68 m) above the floor, *Section 210-52,* first paragraph.

GFCI Exemption Recap

Under certain conditions, GFCI protection is not required for receptacle outlets in garages, in crawl spaces at or below ground level, or in unfinished basements if the receptacles are:

- not readily accessible.

- single-type receptacles or duplex-type receptacles located in dedicated spaces to be used for cord- and plug-connected appliances that are not easily moved.

Examples of these would be overhead door openers, refrigerators, freezers, water softeners, central vacuum equipment, or sump pumps. These would be the receptacles that are dedicated to specific plug-in appliances. The homeowner would normally not unplug the appliance to plug in an extension cord. These receptacles are exempt from the GFCI requirement because they are not intended to be used for plugging in extension cords for portable electric tools or portable appliances. These GFCI exempt receptacles are **in addition** to the receptacles required by *Section 210-52(g)* for garages and basements. *See Section 210-8(a).*

Specific Loads

- Appliances such as electric ranges and ovens, large attic fans, air conditioners, electric furnaces, electric baseboard heaters, heat pumps, electric water heaters, and similar large loads are connected to their own separate circuits. They are not connected to the branch circuits discussed in this unit. See Index for specific appliances.

- Central heating equipment such as a gas furnace must be supplied by an individual branch circuit, *Section 422-12.* An electrical furnace would obviously require a separate branch circuit because of its load requirements, but even a gas furnace requires a separate branch circuit so that in the event of problems on other branch circuits in the home, the gas furnace would still be operative.

SUMMARY OF WHERE LIGHTING OUTLETS MUST BE INSTALLED

The *National Electrical Code®* has minimum requirements for providing lighting for homes. In conformance to *Section 210-70,* lighting outlets in dwellings must be installed as follows:

Switched Lighting Outlets (Figure 6–12)

At least one wall switch-controlled lighting outlet shall be installed:

- in every habitable room. A habitable room is defined as "space in a structure for living, sleeping, eating, or cooking. Bathrooms, toilets, closets, halls, storage or utility space, and similar areas are not considered habitable space."

- in bathrooms.

- in hallways.

- in "interior" stairways. Where the permanent stairway has six or more steps (risers), switch

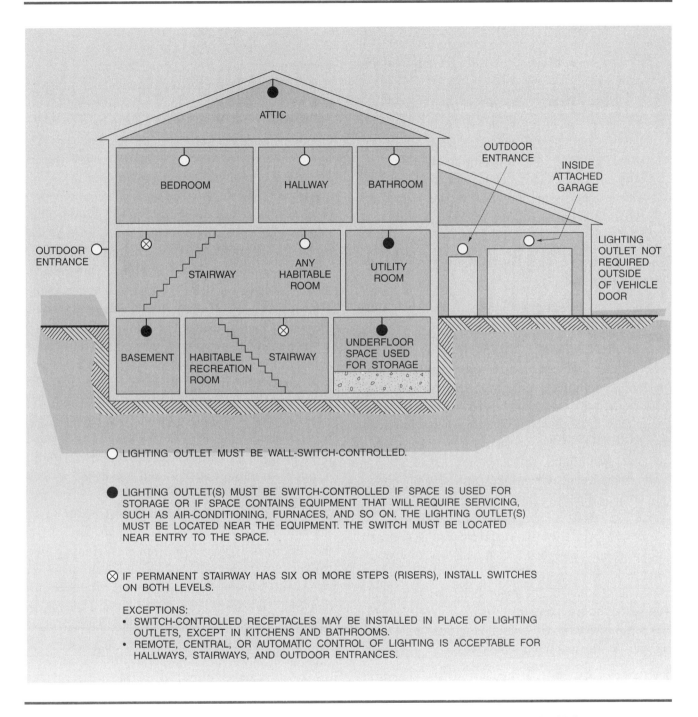

○ LIGHTING OUTLET MUST BE WALL-SWITCH-CONTROLLED.

● LIGHTING OUTLET(S) MUST BE SWITCH-CONTROLLED IF SPACE IS USED FOR STORAGE OR IF SPACE CONTAINS EQUIPMENT THAT WILL REQUIRE SERVICING, SUCH AS AIR-CONDITIONING, FURNACES, AND SO ON. THE LIGHTING OUTLET(S) MUST BE LOCATED NEAR THE EQUIPMENT. THE SWITCH MUST BE LOCATED NEAR ENTRY TO THE SPACE.

⊗ IF PERMANENT STAIRWAY HAS SIX OR MORE STEPS (RISERS), INSTALL SWITCHES ON BOTH LEVELS.

EXCEPTIONS:
• SWITCH-CONTROLLED RECEPTACLES MAY BE INSTALLED IN PLACE OF LIGHTING OUTLETS, EXCEPT IN KITCHENS AND BATHROOMS.
• REMOTE, CENTRAL, OR AUTOMATIC CONTROL OF LIGHTING IS ACCEPTABLE FOR HALLWAYS, STAIRWAYS, AND OUTDOOR ENTRANCES.

Figure 6–12 Lighting outlet requirements in a typical home, *Section 210-70(a).*

control for the lighting outlet(s) that light up the stairway must be provided at each level. This includes permanent stairways to basements and attics, but does not include fold-down storable ladders commonly installed for access to attics.

• in attached garages.

• in detached garages if they have electric power.

• in attics, underfloor spaces, utility rooms, and basements where these areas are used for storage, or if these areas contain equipment that will require servicing (furnace, air-conditioning, heat pumps, etc.). This lighting outlet must be near the equipment and must be switch controlled near the point of entry. Pull-chain lampholders are all right for this application.

- at the exterior side of outdoor entrances or exits. A sliding glass door **is** considered to be an outdoor entrance. A vehicle door in an attached garage **is not** considered to be an outdoor entrance.

- in clothes closets. The present edition of the *National Electrical Code®* does not require lighting outlet(s) in clothes closets. However, some building codes do require lighting outlet(s) in clothes closets. Be sure to check this out in your community. See *Section 410-8* for the installation of lighting in clothes closets.

Exceptions to the Switching Requirement

There are exceptions to the basic switching requirements in *Section 210-70(a)(1)*.

Exception No. 1: This exception recognizes the fact that in most residences, a considerable amount of general lighting is accomplished by plugging table and floor lamps into wall receptacle outlets. Other than in kitchens and bathrooms, switch-controlled wall receptacles are permitted in lieu of ceiling and/or wall lighting outlets.

Switched wall receptacles must be supplied by one of the general purpose lighting branch circuits, *Section 210-52(b)(1), Exception No. 1.*

Where switched wall receptacles are installed in dining rooms or breakfast rooms, the switched receptacles are **in addition** to and are **not** to be connected to any of the required 20-ampere small appliance branch circuits, *Section 210-52(b)(1), Exception No. 1.* An example of this is "swag" lighting as illustrated in Figure 6–6.

Exception No. 2: Occupancy sensors (motion detectors) may also be used **in addition** to wall switch control. When an occupancy sensor that has a manual override is installed where a wall switch would normally be located, a separate wall switch is not required.

Another exception is found in *Section 210-70(a)(2)*. A separate wall switch is not required for hallways, stairways, or outdoor entrances where automatic timers, photocells, central or remote control systems have been installed for the control of these lighting outlets. For example, a post light controlled by a photocell does not require a separate switch.

COMMON AREAS IN TWO-FAMILY OR MULTIFAMILY DWELLINGS

Be careful when wiring a two-family or multifamily dwelling. *Section 210-25* of the *National Electrical Code®* contains specific requirements for circuits that serve common areas in two-family or multifamily dwellings. For areas such as a common entrance, common basement, or common laundry area, the branch circuits for lighting, central fire/smoke/security alarms, communication systems, or any other electrical requirements that serve the common area *shall not* be served by circuits from any individual dwelling unit's power service. Common areas will require separate circuits, associated panels, and metering equipment. The reasoning for this requirement is that power could be lost in one occupant's electrical system, resulting in no lights or alarms for the other tenant(s) using the common shared area.

For example, a two-family dwelling would probably have two separate stairways to the basement. But if only one common stairway is provided, then lighting would have to be provided and served from a branch circuit originating from each tenant's electrical system so that each tenant would have control of the lighting. For a common entrance, two lighting fixtures would be installed and connected to each tenant's electrical system.

Likewise, fire/smoke/security systems in a two-family dwelling should be installed and connected to branch circuits originating from each tenant's electrical panel. This becomes an architectural design issue. This is not generally a problem in two-family dwellings, but can become a problem in dwellings having more than two tenants. Rather than having to install a third watthour meter and associated electrical equipment on a two-family dwelling, the design of the structure can be such that there are no common area. (Figure 6–13).

CORRECT NUMBER OF RECEPTACLE OUTLETS AND LIGHTING OUTLETS

Up to this point in this unit, branch-circuit calculation requirements and location requirements have been discussed. Now it is time to learn about just how many receptacle outlets and lighting outlets can be supplied by one branch circuit.

The *National Electrical Code®* does not specify the maximum number of receptacle outlets or light-

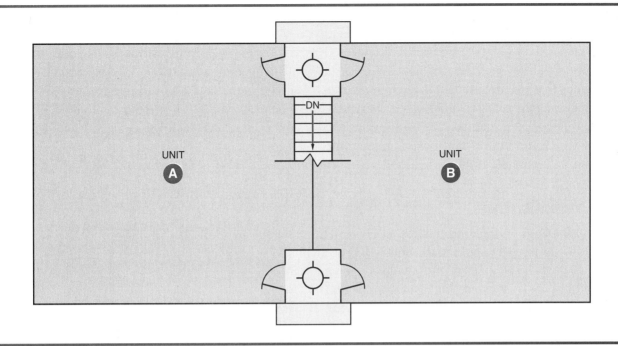

Figure 6–13 This outline of a two-family dwelling shows a common rear entry, a common stairway to the basement, and a common front entry. Note that each entry has one lighting fixture. These common area lighting fixtures are not permitted to be connected to either of the tenant's electrical systems in the individual dwelling units, *Section 210-25*. The choices are (1) architect/designer must redo the layout of the building to eliminate the common entrances; (2) provide two lighting fixtures in each entry, with each fixture connected to the respective individual tenant's electrical circuit; or (3) install a third meter and panel to supply the electrical loads in the common areas.

ing outlets that may be connected to one 120-volt lighting or small appliance branch circuit in a residence. It may seem ridiculous and illogical that ten, twenty, or more receptacle outlets and lighting outlets can be connected to one branch circuit and not be in violation of the Code. Consider the fact, however, that having many "convenience" receptacles is "safer" because many receptacle outlets will virtually eliminate the use of extension cords, one of the highest reported causes of electrical fires. Rarely, if ever, would all receptacle outlets and lighting outlets be fully loaded at the same time. There is much diversity of load in residential occupancies. Check with your local electrical inspector to see if there are requirements regarding the maximum number of outlets permitted on one circuit.

The general-use receptacle outlets in homes are considered to be general lighting. Their load is included in the general lighting load calculation, and no additional load calculations are necessary, *Table 220-3(a)*, asterisk (*) footnote and *Section 220-3(b)(10)*.

CIRCUIT LOADING GUIDELINES

A good rule to follow in residential wiring (in fact, any wiring) is **never load the circuit to more than 80% of its rating.** A 15-ampere, 120-volt branch circuit would be calculated as:

$$15 \times 0.80 = 12 \text{ amperes}$$
$$\text{or}$$
$$12 \text{ amperes} \times 120 \text{ volts} = 1,440 \text{ volt-amperes}$$

Although 20-ampere lighting circuits are generally not installed in residences, the maximum allowable load in volt-amperes for such a circuit is:

$$20 \times 0.80 = 16 \text{ amperes}$$
$$\text{or}$$
$$16 \text{ amperes} \times 120 \text{ volts} = 1,920 \text{ volt-amperes}$$

You can use the preceding guideline to **estimate** residential lighting loads, for both permanently installed lighting fixtures and general-use receptacles.

Certain fixtures, such as recessed lights and fluorescent lights, are marked with their maximum lamp wattage and ballast current (for fluorescent fixtures only). Other fixtures, however, are not marked. No one knows the exact load that will be connected to the receptacle outlets. Also, unknown is the size of the lamps that will be installed in the lighting fixtures (other than the recessed and fluorescent types). At best, it becomes a judgment call—a guess. The room in which the outlets are located gives some indication as to possible loads.

LOAD ESTIMATIONS

The lamp loads in the lighting fixtures can be estimated by assuming the lamp wattages that will probably be needed in each fixture to provide adequate lighting for the area involved. Wattage recommendations to attain adequate lighting levels are found in the American Lighting Association publications and in various manufacturers' publications usually available at electrical distributors, lighting show rooms, and home centers.

ESTIMATING NUMBER OF OUTLETS

One method of determining the number of lighting and receptacle outlets to be included on one circuit is to assign a value of 1 to 1½ amperes to each outlet to a total of 15 amperes. Thus, a total of 10 to 15 outlets could be connected to a 15-ampere branch circuit, 10 being the recommended number, and 15 being the maximum number. Working in your favor is the big diversity in a home.

All outlets will not be required to deliver 1 to 1½ amperes. For example, closet lights, night lights, and clocks will use only a small portion of the allowable current. A 60-watt closet light will draw less than one ampere.

$$\text{Amperes} = \frac{\text{watts}}{\text{volts}} = \frac{60}{120} = 0.5 \text{ amperes}$$

If low-wattage fixtures are connected to a circuit, it is quite possible that 15 or more lighting outlets (many times referred to as "openings") could be connected to the circuit without a problem. On the other hand, if the load consists of high-wattage lamps, the number of outlets would be less.

After installing (roughing-in) all of the ceiling boxes, wall boxes, and recessed fixtures, it is time to think about laying out the circuit runs and to decide on which receptacles and lighting fixtures will be

connected to each circuit, limiting the number to 10 or 15, or less.

A quick way to **estimate** the total number of 15-ampere lighting branch circuits desired for a new house is to count all of the receptacles and fixtures. For example, let us say the total count of lighting outlets and receptacle outlets is 80.

$$\frac{80}{10} = 8 \text{ (eight 15-ampere lighting branch circuits)}$$

If the circuit consists of low-wattage loads, then:

$$\frac{80}{15} = 5.3 \text{ (six 15-ampere lighting branch circuits)}$$

CONNECTING THE CORRECT NUMBER OF RECEPTACLE OUTLETS TO THE 20-AMPERE SMALL APPLIANCE BRANCH CIRCUITS

The 20-ampere small appliance circuits in the kitchen, laundry area, and other areas supply heavy concentrations of plug-in appliances. Keep the required laundry circuit and equipment separate from the required 20-ampere small appliance branch circuits. Put the hair dryer on the required separate bathroom branch circuit.

The 10 to 15 "openings" (1 to 1½ amperes per outlet) would not be applicable. Here are some examples:

- Toaster: 1,000 watts, 8.33 amperes
- Microwave oven: 1,000 watts, 8.33 amperes
- Coffee maker: 900 watts, 7.5 amperes
- Toaster/oven: 1,400 watts, 11.66 amperes
- Hot-shot water heater: 1,450 watts, 12 amperes
- Flat-iron: 1,200 watts, 10 amperes
- Hair dryer: 1,250–1,500 watts, 10.4–12.5 amperes

Therefore, one or two duplex receptacles per 20-ampere small appliance branch circuit will usually be adequate. Many electricians will install split-circuit duplex receptacles above the kitchen counters, and feed each split-circuit duplex receptacle with two 20-ampere branch circuits. This is usually done with a 120/240-volt, 3-wire, multiwire branch circuit, discussed in Unit 2. Another advantage of a three-wire, multiwire branch circuit is that as they are properly connected up in the panel, the loads will also be balanced across the two "hot" phases.

It can be seen that the *NEC®* minimum requirement of two 20-ampere small appliance branch circuits in a residence is really not adequate. Install more 20-ampere branch circuits than required by the Code. You will not be sorry.

CORRECT NUMBER OF BRANCH CIRCUITS NEEDED FOR OTHER LOADS

Do not play games by trying to outguess what 120-volt loads that might or might not be plugged in at the same time. Always be on the conservative side. The following are recommendations and/or required by the *National Electrical Code.®*

Install separate branch circuits for:

- Refrigerator—15- or 20-ampere branch circuit permitted by *Section 210-52(b)(1), Exception No. 2.*

- Food-waste disposer—15- or 20-ampere branch circuit

- Dishwasher—15- or 20-ampere branch circuit

- Trash compactor—15- or 20-ampere branch circuit

- Sump pump—15- or 20-ampere branch circuit

- Clothes washing machine—Minimum of one receptacle required by *Section 210-52(f)* and minimum of one 20-ampere laundry branch circuit required by *Section 210-11(c)(2).*

- Bathroom receptacle(s)—Minimum of one receptacle required by *Section 210-52(d)*; at least one separate 20-ampere branch circuit required by *Section 210-11(c)(3)* for one or more bathrooms.

- Outdoor receptacles—15- or 20-ampere branch circuit

- Computer, printer, scanner—consider a separate 15- or 20-ampere branch circuit and install a transient voltage surge suppresser (TVSS) receptacle

- Copy machine—15- or 20-ampere branch circuit

- Security lighting branch circuits

- Security systems.

Refer to Unit 7 for text on which of the above receptacles must be GFCI protected.

Divide the Loads Evenly

Although this is common sense, it is addressed in the Code. *Section 210-11(b)* makes it mandatory to divide loads evenly among the various circuits. The obvious reason is to not experience overload conditions on some circuits, and underload conditions on other circuits. As the 120-volt branch circuit loads come together and are connected at the main panel or subpanel, the panels will also have balanced loads. The 240-volt loads will automatically be balanced.

Example: After calculating and estimating all of the 120-volt loads in a new residence, the result of 140 amperes is derived. Connecting all of the branch circuit "home runs" should result in as close a balance as possible, like 70 amperes on each phase. Without paying close attention to the "home run" branch-circuit hookups, loads could end up being too much out of balance, such as 100 amperes on one phase, and 40 amperes on the other phase.

CIRCUIT DESIGN

The *National Electrical Code®* does not dictate the actual circuit layout (circuit design) for general lighting branch circuits. In commercial and industrial work, the electrical engineer/architect does the circuit design, showing on the plans and on the panel schedules which receptacles and lighting fixtures are connected to what circuit. In house wiring, it is up to the electrician to lay out the circuits, but within the parameters of the *National Electrical Code.®* There are many possible combinations or groupings of outlets.

In addition to limiting the number of outlets per branch circuit, give some thought to bringing more than one circuit into a room. Should one circuit have a problem, the second circuit would still continue to supply power to the other outlets in that room (Figure 6–14).

Another idea that can save material and labor is to mount receptacles back-to-back where possible (Figure 6–15). Of course, you will need to be careful when installing back-to-back boxes when the wall partitions are fire-rated, such as between a garage and the inside of the house. This is discussed in Unit 10.

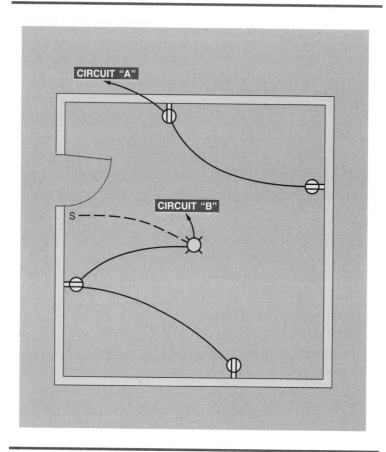

Figure 6–14 Wiring layout showing one room that is fed by two different circuits. If one of the circuits goes out, the other circuit will still provide electricity to the room.

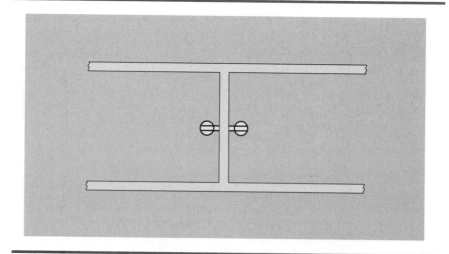

Figure 6–15 Receptacle outlets connected back-to-back. This can reduce the cost of the installation because of the short distance between the outlets.

DRAWING A WIRING DIAGRAM OF A TYPICAL CIRCUIT

The electrician must take information from the building plans and convert it to a workable wiring layout, always in conformance to the *National Electrical Code.* Generally, the best way to do this is to first make a cable layout as shown in Figure 6–16.

Next, prepare an actual wiring diagram showing all of the conductors, wiring devices, and all connections in the circuit as shown in Figure 6–17. Colored pencils or colored marking pens work nicely for this. A white pencil or white marker on white paper is impossible to see, so you might want to use a gray pencil or marking pen to represent the white conductors.

Complete information regarding switch and receptacle connections is covered in Unit 5.

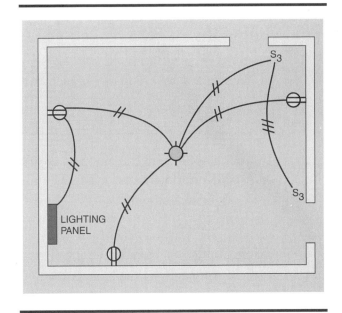

Figure 6–16 Typical cable layout.

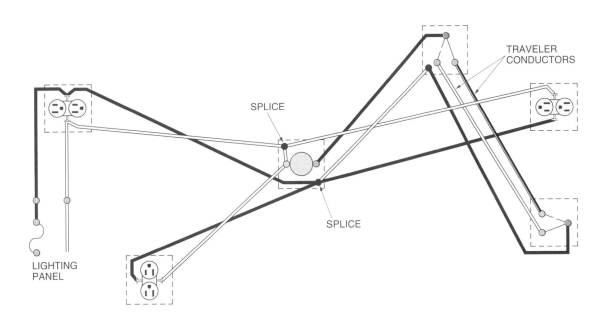

Figure 6–17 Wiring diagram of the circuit shown in Figure 6–16.

REVIEW QUESTIONS

1. What is the definition of a branch circuit?

   _____

   _____

2. Conductors rated 100 amperes are run from the main distribution panel to a second panelboard (load center). These conductors are _____ conductors.
   a) branch circuit
   b) feeder

3. The rating of a branch circuit is determined by the:
   a) ampacity of the branch circuit conductor.
   b) the ampere rating of the overcurrent device.
   c) the ampere rating of the wiring devices (receptacles) connected to the branch circuit.

4. The duplex receptacles on the 20 ampere small appliance branch circuits in the kitchen:
   a) must be rated 20 amperes.
   b) may be rated 15 amperes or 20 amperes.
   c) must be rated 15 amperes.

   What section(s) and table(s) of the *National Electrical Code®* support your answer?
   Sections _____ and *Tables* _____

5. For a typical electric motor, *Section 430-22(a)* of the *NEC®* states that the conductors must be sized at a minimum of _____ percent of the motor's full-load current draw.
   a) 100
   b) 110
   c) 125

6. Where the ampacity of a conductor does not match a standard rating of a fuse or circuitbreaker, it is _____ to use the next larger size fuse or circuit breaker.
   a) all right
   b) not all right

7. An outdoor receptacle is located back-to-back to one of the receptacles in the dining room. May the outdoor receptacle be connected to the same branch circuit as the dining room receptacle?
   a) Yes
   b) No

   What section(s) of the Code support your answer?

   _____

8. Loads that can be expected to be on for periods of three hours or more are considered continuous loads. As such, the branch-circuit rating and the ampacity of the conductors must be sized at _____ % of the load.

   a) 100

   b) 125

   c) 150

   What would be an example of a continuous load that you are familiar with?

   _____

9. For uniformity in making electrical code calculations for house wiring, what voltage values must be used?

   a) 110 volts

   b) 115 volts

   c) 120 volts

   d) 220 volts

   e) 230 volts

   f) 240 volts

10. A house has 2,400 square feet of usable floor space. What is the minimum number of 15-ampere, 120-volt lighting branch circuits required?

11. No. _____ AWG copper conductors are normally used for 15-ampere lighting branch-circuits in homes?

    a) No. 14

    b) No. 12

    c) No. 10

12. No. _____ AWG copper conductors are normally used for 20-ampere small appliance receptacle branch circuits in homes.

    a) No. 14

    b) No. 12

    c) No. 10

13. Receptacles must be located so that no point along the floor line is more than _____ feet from a receptacle.

    a) 6 feet

    b) 9 feet

    c) 12 feet

14. In a hallway, 10 feet or longer, in a home, at least _____ receptacle(s) must be installed.

    a) one

    b) two

    c) three

15. As you rough in the wiring of a home, you find a number of narrow wall spaces between closets and doorways, and between bedroom doors and bathroom doors. To save money, the homeowner tells you "you don't have to put a receptacle there. The space is too small to place furniture there." Because you know the Code, what section of the Code do you show the homeowner?

    _____

16. *Section* _____ of the *NEC*® states that in dwelling units, at least one receptacle outlet shall be installed for the laundry.

17. What section in the *NEC*® requires that a separate 20-ampere circuit be provided for the laundry receptacle outlet(s), and that this circuit shall not supply any other loads?

    _____

18. You are wiring a new home. There is a crawl space under the house. Under what condition is it necessary to install a receptacle in this crawl space? The *NEC*® code reference is *Section* _____, which states that _____

    _____

    _____

    If the crawl space is at or below grade, the receptacle must be _____ protected.

19. You are wiring a new home. At least one receptacle must be installed outdoors at grade level _____. This requirement is found in *Section* _____ of the *NEC*® These receptacles must be _____ protected in accordance with *Section* _____.

20. The number of receptacles required above countertops around the kitchen work area depends on how many feet of countertop there is. These receptacles shall be positioned so that no point along the wall line of the countertop is more than _____ inches from a receptacle.

    a) 18

    b) 24

    c) 36

21. Above countertops, a receptacle must be installed at each wall counter space wider than12 inches.

    a) True

    b) False

22. The receptacle located behind the refrigerator in a kitchen:

    a) must be connected to one of the 20-ampere small appliance branch circuits.

        1. True

        2. False

    b) may be connected to one of the 20-ampere small appliance branch circuits.
       1. True
       2. False
    c) must be connected to a separate 20-ampere small appliance branch circuit.
       1. True
       2. False
    d) must be connected to a separate 15- or 20-ampere small appliance branch circuit.
       1. True
       2. False
    e) may be connected to a separate 15- or 20-ampere small appliance branch circuit.
       1. True
       2. False
    f) must be GFCI protected.
       1. True
       2. False

23. Does the *NEC*® permit the dishwasher and/or a food waste disposer to be connected to one of the required 20-ampere small appliance branch circuits that supply the receptacles above the counter top?
    a) Yes
    b) No

The *NEC*® code reference is *Section* _____.

24. In kitchens, each peninsula and free-standing island must have a minimum of _____ receptacles.
    a) one
    b) two
    c) three

Fill in the blank and select the correct answer. This (these) receptacles must be _____ protected, and must be supplied by one of the required _____ small appliance branch circuits.
    a) 15-ampere
    b) 20-ampere

25. The receptacle(s) in a bathroom shall be installed on a wall that is adjacent to the basins, and must be supplied by a _____ branch circuit.
    a) separate 15-ampere
    b) separate 20-ampere
    These requirements are found in *Sections*_____.

26. *Section 210-70* sets forth the requirements of where lighting outlets must be installed in dwellings. Many times the question arises as to whether or not a lighting outlet is required in an attic or in a crawl space. In your own words, explain how you would decide whether or not to install a lighting outlet in an attic or in a crawl space.

_____

_____

_____

_____

27. The *National Electrical Code®* does not have a limitation on the number of lighting outlets and general purpose receptacle outlets permitted to be connected to a branch circuit for residential wiring. Common sense prevails. In your community, are there any guidelines or restrictions as to the maximum number of lighting outlets and general-purpose receptacle outlets permitted to be connected to a branch circuit? What are they?

_____

_____

28. The *National Electrical Code®* does not specify a maximum number of receptacles-permitted to be connected to a 20-ampere small appliance branch circuit. What is your experience relative to this issue? How many duplex receptacles do you connect to one 20-ampere small appliance branch circuit? _____

Ground-Fault Circuit Interrupters and Other Types of Personnel Protection

OBJECTIVES

After studying this unit, you will be able to:

- **Learn about the hazards of electricity.**
- **Understand how ground-fault circuit interrupters (GFCIs) operate.**
- **Learn how to connect a GFCI into a branch circuit.**
- **Know where GFCIs are required in residences.**
- **Become familiar with the types of receptacles to use when replacing an existing receptacle.**
- **Become aware that GFCI protection must be provided for temporary wiring.**

INTRODUCTION

Water and electricity do not mix! Electrical shocks can be fatal. Working in and around sinks in kitchens and bathrooms, standing or lying on concrete floors in basements and garages, and working outdoors with electrical extension cords and tools all are scenarios that set the stage for the possibility of receiving a lethal electrical shock. This unit introduces a unique device called a ground-fault circuit interrupter that minimizes the likelihood of receiving a deadly shock. What are they? Where must they be installed? This unit addresses these questions.

The *National Electrical Code®* has many requirements relating to ground-fault circuit interrupters. In the electrical trade these are referred to as GFCIs, and are installed for personnel protection against possible injury or death from an electrical shock. There can be no compromise where human life is involved! The Code recognizes this by requiring that ground-fault circuit interrupters be installed

in specific locations in every home. Let us begin by discussing electrical shock hazards, how injury or death is minimized by the use of GFCIs, then continue by explaining what GFCIs do, what GFCIs do not do, and where they must be installed.

ELECTRICAL SHOCK HAZARDS

Ground-fault circuit interrupters (GFCIs) are required to be installed in all locations where there is a potential for electrical shocks. We will pinpoint these areas later.

Over the years, many lives have been lost due to "ground faults." A ground fault might result in an electrical shock, a burn, or electrocution. Even if the electrical shock does not cause injury or death, the reaction from the shock might knock the person off of a ladder, or jump into moving equipment. These hazards exist whenever a person comes in contact with a "hot" wire, or touches an appliance or equipment where the "hot" conductor has come in contact with the metal frame of the appliance or equipment. This condition may be due to the breakdown of the insulation within the appliance because of wear and tear, overloading, defective construction, or misuse of the equipment.

Touching **only** a "hot" conductor by itself is not a hazard. Note that birds can sit on a high voltage line with no danger because they are touching only one wire. There is not a complete circuit present. The real danger arises when a person touches a **"hot" conductor** (or "hot" appliance or "hot" equipment) **and a "grounded" object at the same time,** thus completing the circuit. Examples of grounded objects are metal water pipes, metal ducts, metal downspouts and gutters, aluminum siding, metal framing members (joists and studs), faucets, grounded metal lighting fixtures, earth, concrete in contact with the earth, water, or other grounded surfaces. Appliances that have a three-wire plug and cord are grounded when they are plugged into a grounding-type receptacle. When a grounding-type receptacle is properly connected, the horseshoe shaped slot of a receptacle is connected to the *equipment grounding conductor,* the wide slot is connected to the *grounded conductor,* and the narrow slot is connected to the *"hot" live ungrounded conductor.*

SEVERITY OF AN ELECTRICAL SHOCK

The severity of a shock, a burn, or possible electrocution that a person might receive depends upon:

- The path of current through the body. The path might be finger to finger, hand to hand, hand to foot, foot to foot, hand touching a "hot" wire while the other hand is holding a pair of pliers, electric drill or other appliance, touching a faucet or some other grounded metal object, touching a "hot" wire with hand in water, or standing in water or on damp ground. A path of current through the heart is extremely dangerous.

- The length of time the current flows through the body. The length of time that a person is exposed to the electrical shock also has an effect on the severity of the electrical shock.

- The amount of current that flows through the body. The amount of current flowing through

the body depends upon internal resistance, body (surface of skin) resistance at points of contact, moisture, and voltage. The flow of current produces heat, heat burns, and the body is burnt when current flows through it. Experts say that the resistance of the human body varies from a few hundred ohms to many thousand ohms, but generally agree on values of 800 to 1,000 ohms as typical.

It is the amount of current flowing through the body that kills. This current is measured in milliamperes (thousandths of an ampere)—not in full amperes. Because it is voltage that "pushes" current through a given circuit, 120 volts can be just as deadly as 240 volts.

Figure 7–1 is a chart showing the effect of an electric shock for different values of electrical current and time on human beings. Children are more sensitive to electric shock than adults. The current

EFFECT OF ELECTRIC SHOCK		
	Current in milliamperes @ 60 hertz	
	Men	Women
• cannot be felt	0.4	0.3
• a little tingling—mild sensation	1.1	0.7
• shock—not painful—can still let go	1.8	1.2
• shock—painful—can still let go	9.0	6.0
• shock—painful—just about to point where you can't let go—called "threshold"—you may be thrown clear	16.0	10.5
• shock—painful—severe—can't let go—muscles immobilize—breathing stops	23.0	15.0
• ventricular fibrillation (usually fatal) length of time . . . 0.03 sec. length of time . . . 3.0 sec.	1000 100	1000 100
• heart stops for the time current is flowing—heart may start again if time of current flow is short	4 amperes	4 amperes
• burning of skin—generally not fatal unless heart or other vital organs burned	5 or more amperes	5 or more amperes
• cardiac arrest, severe burns, and probable death	10,000	10,000

Figure 7–1 Chart showing the effect of different values of electrical current and time on human beings. Children are more sensitive to electric shock than adults.

values in Figure 7–1 are shown in milliamperes, and are not great enough to trip an ordinary 15- or 20-ampere circuit breaker or fuse. To sense and de-energize a circuit for such extremely low values of ground-fault current, something else is needed . A ground-fault circuit interrupter (GFCI) answers this need.

GROUND-FAULT CIRCUIT INTERRUPTERS

Figure 7–2 is a time/current curve showing the tripping characteristics of a typical Class "A" GFCI device for different amounts of current that a normal healthy adult can withstand for a certain length of time. For example, follow the 6 mA (6/1000 of an ampere) line vertically to the crosshatched typical time/current curve. Note that the GFCI will shut off the power between approximately 0.035 second and 0.1 second. One electrical cycle is $\frac{1}{60}$ of a second (0.0167 second). To make a comparison of how fast a GFCI device operates, it is interesting to note that an air bag in an automobile inflates in approximately $\frac{1}{20}$ of a second (0.05 second).

Types of Ground-Fault Circuit Interrupters

For house wiring, ground-fault circuit interrupters are available in both circuit receptacle types and breaker types as illustrated in Figure 7–3.

Classes of Ground-Fault Circuit Interrupters

There are two classes of ground-fault circuit interrupters used in homes: Class "A" and Class "B."

Class "A" GFCI Devices. Class "A" GFCI devices are designed to trip when a ground fault in the range of 4 to 6 milliamperes (4/1000 to 6/1000 of an ampere) or more occurs.

Class "B" GFCI Devices. Class "B" GFCI devices are pretty much obsolete. They were designed to trip when a ground fault exceeds 20 milliamperes (20/1000 of an ampere). One manufacturer lists single-pole 120-volt and two-pole 120/240-volt 15-, 20-, and 30-ampere ground-fault circuit breakers with a trip setting of 30 milliamperes. These may be used on fixed outdoor deicing, snow melting, pipe

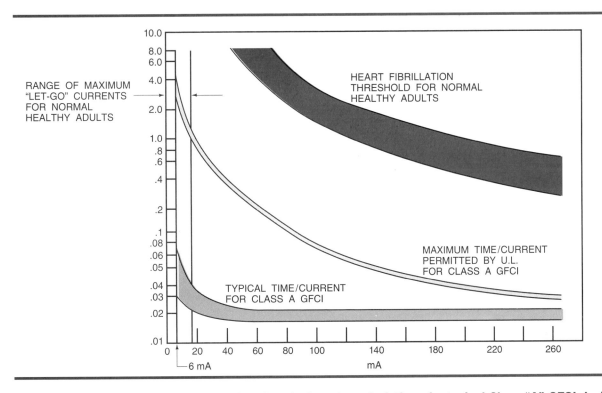

Figure 7–2 Time/current curve showing the tripping characteristics of a typical Class "A" GFCI device. Follow the 6 mA line vertically to the crosshatched typical time/current curve. Note that the GFCI will open between approximately 0.035 second and 0.1 second. One electrical cycle is $\frac{1}{60}$ of a second (0.0167 second). An air bag in an automobile inflates in approximately $\frac{1}{20}$ of a second (0.05 second).

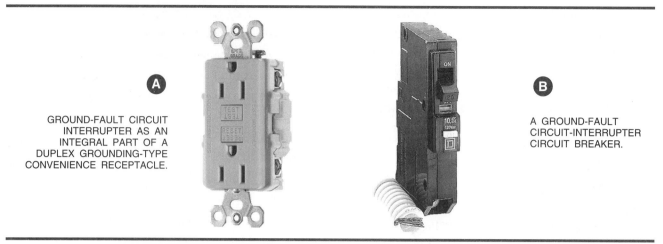

Figure 7–3 Two common types of ground-fault circuit interrupters (GFCI) used in homes: a duplex GFCI receptacle (*Courtesy Pass & Seymour Legrand*); and a GFCI circuit breaker. (*Courtesy Square D Company*)

line and vessel heating as covered in *Articles 426* and *427* of the *NEC.* The NEMA Standard 280 and the UL White Book state that these would also be used for swimming pool underwater lighting installed **before** the adoption of the 1965 edition of the *National Electrical Code.* For these older swimming pool installations, a Class "A" GFCI is too sensitive and would nuisance trip.

OPERATION OF GROUND-FAULT CIRCUIT INTERRUPTERS

Figure 7–4 is a pictorial sketch showing the basic operating principles of a ground-fault circuit interrupter.

Figure 7–5 shows the internal components and connections of a ground-fault circuit interrupter.

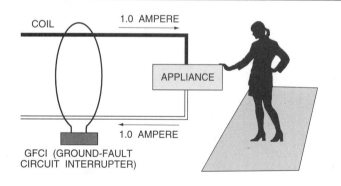

NO CURRENT IS INDUCED IN THE COIL BECAUSE BOTH WIRES ARE CARRYING THE SAME CURRENT. THE GROUND-FAULT CIRCUIT INTERRUPTER DOES NOT TRIP THE CIRCUIT OFF.

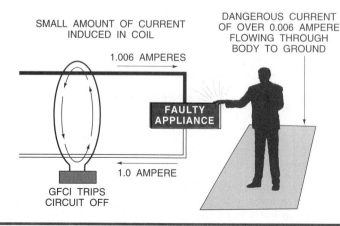

A SMALL AMOUNT OF CURRENT IS INDUCED IN THE COIL BECAUSE OF THE UNBALANCE OF CURRENT IN THE CONDUCTORS. THIS CURRENT DIFFERENCE IS AMPLIFIED SUFFICIENTLY BY THE GROUND-FAULT INTERRUPTER TO CAUSE IT TO TRIP THE CIRCUIT OFF BEFORE THE PERSON TOUCHING THE FAULTY APPLIANCE IS INJURED OR KILLED. **NOTE: CURRENT VALUES ABOVE 6 MILLIAMPERES ARE CONSIDERED DANGEROUS.** GROUND-FAULT CIRCUIT INTERRUPTERS MUST SENSE AND OPERATE WHEN THE GROUND CURRENT EXCEEDS 6 MILLIAMPERES.

Figure 7–4 Basic principle of how a ground-fault circuit interrupter operates.

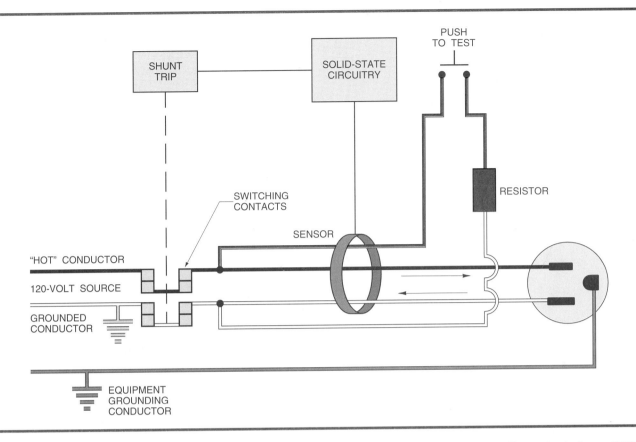

Figure 7–5 Ground-fault circuit interrurpter internal components and connections. Receptacle-type GFCIs switch both the "hot" and grounded conductors. When the test button is pushed, the path of the test current can be traced (1) from the "hot" conductor through the test button's electrical contacts; (2) through the resistor that limits the ground-fault test current; (3) through the sensor; (4) back around (bypasses) the sensor; then (5) back to the opposite circuit conductor. This is how the "unbalance" is created and monitored by the solid-state circuitry that signals the GFCI's contacts to open. Under normal load conditions, the current in both conductors passing through the sensor is the same. No unbalanced currents are present. As soon as the GFCI senses an unbalanced condition, it goes into action—fast.

Receptacle-type GFCIs switch both the "hot" and grounded conductors. A circuit breaker GFCI trips only the "hot" conductor. When the test button is pushed, the path of the test current can be traced:

1. from the "hot" conductor through the test buttons electrical contacts

2. through the resistor that limits the ground-fault test current

3. through the sensor

4. back around (bypasses) the sensor

5. then back to the opposite circuit conductor

This is how the "unbalance" is created and monitored by the solid-state circuitry that signals the GFCIs contacts to open. Under normal load conditions, the current in both conductors passing through the sensor is the same. No unbalanced currents are present. As soon as the GFCI senses an unbalanced condition, it goes into action—fast.

CONNECTING GROUND-FAULT CIRCUIT INTERRUPTERS INTO A CIRCUIT

A GFCI device is a life-saving device, but it must be properly connected in order for it to work correctly.

There are a number of ways in which a GFCI can be connected in a circuit to provide ground-fault protection for personnel.

Some GFCI receptacles are the "feedthrough-type," Figure 7–6. This means that one GFCI receptacle can be connected to provide ground-fault protection for the entire circuit downstream of that receptacle. Connected in this manner, a ground fault anywhere downstream of the GFCI receptacle will shut off the GFCI receptacle and everything down-

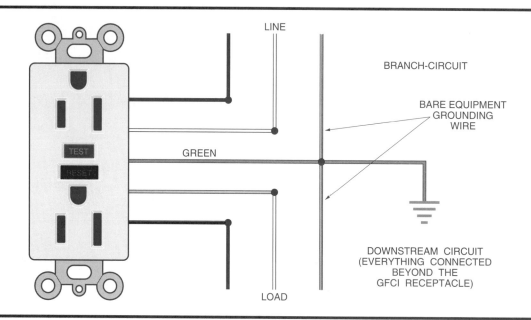

Figure 7–6   This illustration shows the electrical connections for a typical feedthrough ground-fault circuit-interrupter receptacle.

stream of the GFCI receptacle. It is very difficult to locate a ground-fault problem in this circuit arrangement. The decision to use a number of GFCI receptacles rather than trying to protect many downstream receptacles through one GFCI receptacle becomes one of economy and practicality. The same situation is present when using a GFCI circuit breaker in a panel. The entire branch circuit is ground-fault circuit interrupter protected, and is subject to the same total power outage in the branch circuit as using one GFCI receptacle as discussed above. There are no hard and fast rules for this. Although GFCI receptacles are more expensive than conventional receptacles, the real issue is that of **safety.** Use common sense when laying out a branch circuit!

Figure 7–6 shows the electrical connections for a feedthrough ground-fault circuit-interrupter receptacle. Note that the line side terminals (leads) are

black and white. The load side terminals (leads) are gray and red. The green lead is the equipment ground and is common to both line and load.

Figure 7–7 shows a GFCI receptacle connected at the end of a branch circuit. Should the GFCI receptacle sense a ground fault, only the GFCI receptacle shuts off. The rest of the branch circuit remains energized.

Figure 7–8 shows a feedthrough GFCI receptacle as the first receptacle in the circuit. The remaining receptacles are the conventional grounding-type. When the feedthrough GFCI trips, the feedthrough GFCI receptacle and all of the remaining receptacles are de-energized. All of the wiring downstream of the GFCI receptacle is ground-fault protected.

Figure 7–9 shows a feedthrough GFCI receptacle connected midway in the branch circuit. When the feedthrough GFCI receptacle is called upon to

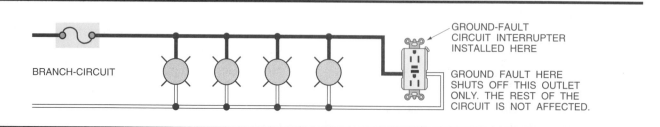

Figure 7–7   Diagram showing a GFCI receptacle at the end of a branch circuit. Should a ground fault occur in anything plugged into this receptacle, only the GFCI receptacle will shut off. The remaining upstream portion of the circuit will remain energized.

trip, the feedthrough GFCI receptacle and all downstream receptacles are de-energized. All of the circuit downstream of the GFCI receptacle is ground-fault protected. If the feedthrough feature of the GFCI receptacle is not used, then that GFCI receptacle would "stand alone," and would be the only device to trip off should a ground fault occur on anything plugged into the GFCI receptacle.

Figure 7–10 shows a GFCI-type circuit breaker in a panel. The entire branch circuit is ground-fault

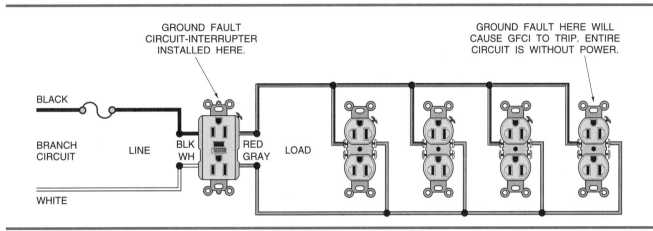

Figure 7–8   This diagram shows a feedthrough ground-fault circuit-interrupter receptacle protecting all of the other downstream receptacles. A ground fault anywhere in the circuit will trip the GFCI feedthrough receptacle. This also will shut off the power to all of the downstream receptacles.

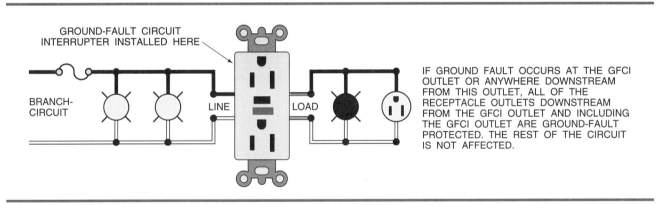

Figure 7–9   This diagram shows a feedthrough ground-fault circuit-interrupter receptacle connected mid-way into the branch circuit so that in the event of a ground fault on the load side of the GFCI receptacle, only part of the circuit is affected. All of the circuit upstream from the GFCI receptacle remain energized.

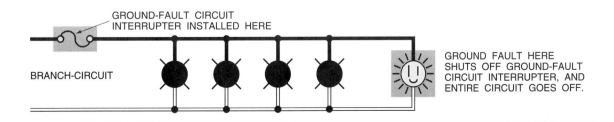

Figure 7–10   This diagram shows a GFCI circuit breaker in a panel protecting the entire branch circuit. A ground fault anywhere in the circuit will cause the GFCI circuit breaker to trip.

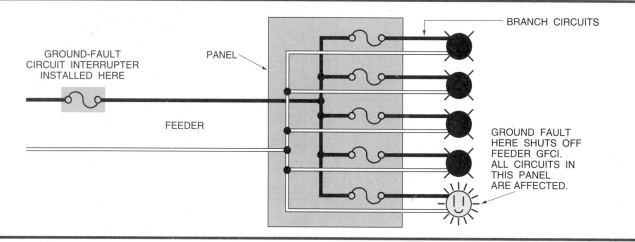

Figure 7–11 This diagram shows a ground-fault circuit interrupter connected ahead of the feeder that supplies the panel. The feeder, panel, and everything connected to the panel is ground-fault protected. Should a ground fault occur on the feeder, in the panel, or on any circuit fed from the panel, the feeder GFCI would trip, resulting in a total loss of power to the panel. Although the *National Electrical Code*® in *Section 215–9* recognizes this as a means of obtaining ground-fault protection, it is rarely, if ever, used in residential wiring because of the possibility of a total "blackout."

protected. A ground fault anywhere in the circuit will cause the GFCI circuit breaker to trip off, de-energizing the entire circuit.

Figure 7–11 shows a ground-fault circuit interrupter connected ahead of the feeder that supplies the panel. One manufacturer catalogs a 60-ampere, two-pole GFCI circuit breaker that could be used for this application. It would provide ground-fault protection for a small panel or load center containing 15- and/or 20-ampere branch circuits. Should a ground fault occur on any branch circuit fed from the panel, the feeder GFCI would trip off, resulting in a total loss of power to the panel. The *National Electrical Code*® in *Section 215-9* recognizes this as a means of obtaining ground-fault protection. It is rarely, if ever, used in residential wiring.

WHAT A GFCI DOES

- A GFCI **does** work only on grounded electrical systems. The wiring in a home is a grounded electrical system.

- A GFCI **does** operate only when it senses line-to-ground current.

- A GFCI **does** monitor the current balance between the ungrounded "hot" conductor and the grounded "neutral" conductor. When the

current flowing through the "hot" conductor is 4 to 6 milliamperes greater than the current flowing in the "return" grounded conductor, the GFCI senses this unbalance and trips (opens) the circuit off. This unbalanced condition indicates that part of the current flowing in the circuit is being diverted to some path other than the normal return path along the grounded return conductor. If the "other" path is through a human body as illustrated in Figure 7–4, the outcome could be fatal.

- A GFCI **does** operate on a two-wire circuit even though an equipment grounding conductor is not included with the circuit conductors. This is true because a GFCI senses any unbalance or current that might occur in the two circuit conductors.

Many old homes still contain knob-and-tube wiring. In its day, knob-and-tube wiring was adequate, but it did not have an equipment grounding conductor run with the branch-circuit conductors. There was also a period of time prior to, during, and after World War II when houses were wired with nonmetallic-sheathed cable that did not have an equipment grounding conductor. Today, typical house wiring methods are nonmetallic-sheathed cable that contains a separate equipment grounding conductor, armored cable where the metal armor and the internal bonding strip together provide the equipment ground. In the case of metal raceways such as EMT, the equipment grounding conductor is the metal raceway. *Section 250-118* shows the many types of acceptable equipment grounding conductors.

WHAT A GFCI DOES NOT DO

* A GFCI **does not** protect a person from an electrical shock. The person will feel a shock during the time it takes the GFCI device to trip when it senses a ground fault.

* A GFCI **does not** protect against electrical shock when a person touches both circuit conductors at the same time (two "hot" wires, or one "hot" wire and one grounded neutral wire) because the current flowing in both conductors is the same. There is no unbalance of current for the GFCI to sense and trip.

* A GFCI **does not** limit the magnitude of ground-fault current. It **does** limit the time it takes to sense and trip off. Figure 7–2 shows the time in seconds that it takes a GFCI to open.

* A GFCI **does not** sense short-circuits that occur between the "hot" conductor and the grounded "neutral" conductor. It is up to the branch-circuit fuse or circuit breaker to provide this type of protection.

* A GFCI **does not** sense short circuits that occur between two "hot" conductors. It is up to the branch-circuit fuse or circuit breaker to provide this type of protection.

* A GFCI **does not** sense "arcing" faults that occur between conductors or at loose connections. An example of this is when an extension cord becomes frayed and the conductors break open. The heat generated by the arc can result in a fire easily, yet the amount of current flowing is less than the rating of the branch-circuit fuse or circuit breaker. Remember, a GFCI will operate **only** when it senses ground-fault current. To protect against arcing faults, replace frayed or damaged extension cords. Make sure all electrical connections are tight.

* A GFCI **does not** provide overload protection for a branch circuit.

* A GFCI **does not** detect ground faults on the secondary side of an isolation transformer (separate primary winding and separate secondary winding) when the GFCI is protecting the branch circuit that supplies the primary of the isolation transformer. Isolation transformers are used in certain applications for swimming pool underwater lighting fixtures.

SOME DO'S AND SOME DON'TS

* **Do** use wall boxes that are large enough. GFCI receptacles take up very much room. They are much larger than standard receptacles. Consider using 4-inch square boxes with suitable raised plaster rings where GFCI receptacles will be installed. The same holds true for dimmers, which are also much larger than conventional toggle switches. Unit 10 discusses how to calculate "box fill."

* **Do not** use metal staples. **Do** use nonmetallic staples. Although not a code requirement, this is the recommendation of some electricians and some electrical inspectors as a means of reducing nuisance tripping of GFCIs.

* **Do not** ground a neutral conductor (the white wire) anywhere in the home except at the service equipment. An exception to this is where a second building such as a garage, tool shed, or some other outbuilding is fed from the main house service. This is covered in Unit 12. Grounding the neutral beyond the service equipment sets up a parallel return path whereby the neutral conductor and the ground return path would both be carrying current back to the source. Each parallel path would carry current inversely proportional to the path's impedance. The return path could be metal raceways, water pipes, metal sheet metal ducts, etc. There is no way to predict how the return current will divide. GFCI devices could be inoperative or otherwise not operate properly because they would only be sensing a portion of the ground-fault current.

* **Do not** connect the neutral of one circuit to the neutral of another circuit. GFCI devices could be inoperative or could nuisance trip.

- **Do not** install extremely long runs of branch circuit wiring on the load side of GFCI devices. The GFCI might nuisance trip. Check the manufacturer's instructions to see if the maximum length of branch-circuit wiring is mentioned.

- **Do not** reverse the LINE and LOAD connections on a feedthrough GFCI. This would result in the receptacle still being "live" even though the GFCI mechanism has tripped. UL requires that a "safety yellow" adhesive label must cover the load terminals, or be wrapped around the load wire leads. The wording on the label will read something like this: *"Attention. The load terminals under this label are for feeding additional receptacles. Miswiring can leave this outlet without ground fault protection. Read instructions prior to wiring."*

It is easy to tell if a GFCI receptacle has the line and load leads reversed. Push the test button. The GFCI mechanism will trip. If the GFCI receptacle is still "hot," it has been wired incorrectly. Some GFCI receptacles have an indicator light that comes on when you push the test button. This indicates that the line and load connections have been properly connected. If the indicator light does not come on, the line and load connections are reversed.

INSTALLING GFCIs IN HOME

The *National Electrical Code®* in *Section 210-8(a)* requires that all 125-volt, 15- and 20-ampere receptacle outlets in certain areas in homes be GFCI protected using either GFCI receptacles or GFCI circuit breakers; 250-volt electric dryer and electric range receptacles do not come under the GFCI requirement. These areas are:

Bathrooms

The Code, in *Definitions,* defines a bathroom as *"An area including a basin with one or more of the following: a toilet, a tub, or a shower."* See Figure 7–12.

Garages

GFCI-protected receptacle outlets must be installed in garages (attached or detached) **and** at grade levels in unfinished outbuildings such as a tool shed or work shed. Easily reached porcelain or plas-

tic lamp holders that have a plug on them also must be GFCI protected. There are two exceptions to this requirement.

Exceptions. GFCI protection is **not** required for receptacles that are not readily accessible, such as a receptacle installed on the ceiling for an overhead garage door opener. The Code, in *Definitions,* defines "readily accessible" as being able to be reached without the use of ladders or chairs. However, it is common practice to install easy-to-plug in "reel lights" on the ceiling of a garage to use for working on an automobile and similar projects. Many of these reel lights have a receptacle on them. This sets the stage for an electric shock hazard, the same as if the cord had been plugged into a wall outlet, which of course is required to be GFCI protected. So give careful consideration to installing a duplex GFCI receptacle in the ceiling. One receptacle will serve the overhead door operator, and the other will serve the "reel light." There can be no compromise to the safety issue.

GFCI protection is **not** required for a single receptacle, or a duplex receptacle for two appliances that is:

1. installed for cord-and plug-connected appliances, and

2. where the appliance is in a dedicated space, and

3. where the appliance is not easily moved from one location to another, such as a freezer or a refrigerator. *Section 210-52(b)(1), Exception No. 2* permits a separate 15- or 20-ampere branch circuit to supply the refrigerator receptacle.

Outdoors

All outdoor receptacles *anywhere* on the property. Even receptacles installed on a balcony, or installed high up under an eave or overhang of a house for plugging in Christmas lighting requires GFCI protection.

There is one exception.

GFCI protection is *not* required for outdoor receptacles that are supplied from a dedicated branch circuit for snow-melting and deicing equipment, and are not readily accessible.

The branch circuit supplying this type of equipment must have Class "B" ground-fault protection

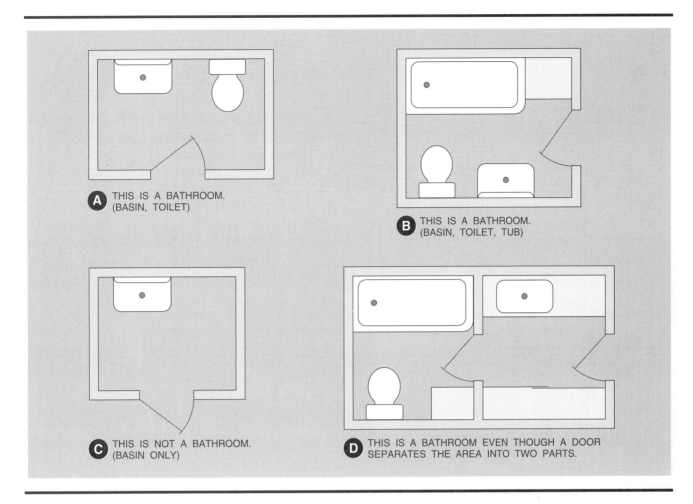

A THIS IS A BATHROOM.
(BASIN, TOILET)

B THIS IS A BATHROOM.
(BASIN, TOILET, TUB)

C THIS IS NOT A BATHROOM.
(BASIN ONLY)

D THIS IS A BATHROOM EVEN THOUGH A DOOR
SEPARATES THE AREA INTO TWO PARTS.

Figure 7–12 These floor plans help clarify what a bathroom is as defined by the *National Electrical Code®* in *Article 100*. The Code defines a bathroom as *"An area including a basin with one or two of the following: a toilet, a tub, or a shower."*

for equipment as required in *Section 426-28*. Class "B" ground-fault protection devices *for equipment* trip on ground-faults that exceed 20 milliamperes. Class "A" ground-fault circuit-interrupter (GFCI) protection devices *for personnel* trip on ground-faults in the range of 4 to 6 milliamperes. Snow melting and deicing cables that lay in a rain gutter or down spout could cause a fire should the insulation of the cable fail and allow current to flow to ground (via the metal gutter or down spout). The required ground-fault protection for *equipment* will sense this leakage current, and will trip off.

In Crawl Spaces That Are At or Below Grade.

Porcelain or plastic lamp holders that have a plug-in receptacle on them also must be GFCI protected.

In Unfinished Basements.

Porcelain or plastic lamp holders that have a plug on them also must be GFCI protected. There are two exceptions:

Exceptions. GFCI protection is **not** required for receptacles that are not readily accessible. The Code defines "readily accessible" as being able to be reached without the use of ladders or chairs.

GFCI protection is **not** required for a single-receptacle, or a duplex receptacle for two appliances that is:

1. installed for cord-and plug-connected appliances,

2. where the appliance is in a dedicated space, and

3. where the appliance is not easily moved from one location to another, such as a freezer, re-frigerator, water softener, or gas-fired electric

controlled water heater. *Section 210-52(b)(1), Exception No. 2* permits a separate 15- or 20-ampere branch circuit to supply the refrigerator receptacle.

Although a sump pump occupies a dedicated space, it is probably not a good idea to plug a sump pump into a GFCI receptacle because of the possibility of the GFCI tripping off, unknown to the homeowner, resulting in a flooded basement. Because a single receptacle is exempt from the GFCI requirement, it is better to install a single receptacle for the sump pump.

Also, plan on installing a single receptacle for the laundry receptacle that is required by *Section 210-52(f)*. The laundry receptacle must be supplied by a separate 20-ampere branch circuit,

and must not supply any other loads, *Section 210-11(c)(2)*.

Some individuals try to get around the GFCI rule in their basement claiming their basement is "finished" as a habitable room. If that is true, then the Code requirements for the spacing of receptacle, lighting outlets, and the switch control of the lighting outlets must be followed.

Kitchens

All receptacles that serve countertop surfaces, including receptacles on islands and peninsulas must be GFCI protected. Receptacles under the sink for a food waste disposer, behind the refrigerator, behind a clock, or a receptacle inside of an upper cabinet for plugging in a microwave oven do not have to be GFCI protected (Figure 7–13).

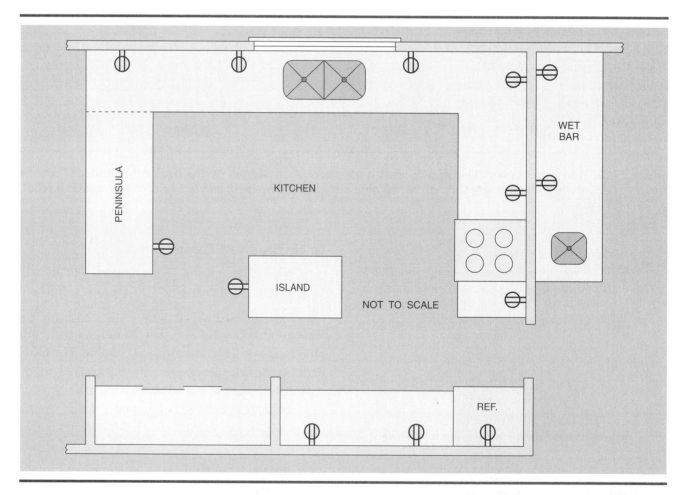

Figure 7–13 In kitchens, **all** 125–volt, single-phase, 15– and 20-ampere receptacles that serve countertop surfaces (along walls, on islands, on peninsulas) must be GFCI protected. The key words are "that serve countertop surfaces." Where wet-bar sinks are installed, **all** 125–volt, single-phase, 15– and 20-ampere receptacles within 6 feet (1.83 m) from the outer edge of the wet-bar sink that serve the countertops must be GFCI protected if the receptacles are. The refrigerator receptacle, clock outlet receptacles, receptacles located under the sink or inside of the cabinets are not required to be GFCI protected because they do not serve countertop surfaces. Refer to *Section 210-8(a)(6)* and *(7)*. This drawing is for illustrative purposes, and is not to scale.

Areas Near Wet-Bar Sinks

All receptacles that serve the countertop within 6 feet (1.83 m) of the edge of a wet-bar sink must be GFCI protected (Figure 7–13).

Swimming Pool Equipment and Other Items

GFCI protection is required for certain swimming pool equipment, spas, hot tubs, fountains, and specific receptacles located near these items. This topic is covered in *Article 680* of the *NEC.®*

Residential Underground Branch-Circuit Wiring

There is a special rule for residential underground branch circuit wiring if the underground wiring is GFCI protected. *Table 300-5* of the *NEC®* indicates that for residential underground 120-volt or less branch circuits rated 20 amperes or less, the underground wiring need only be buried 12 inches below grade when the circuit is GFCI protected. Type UF cable must be buried at least 24 inches below grade if the circuit is not GFCI protected. If

the underground circuit is installed in rigid or intermediate metal conduit, a depth of not less than 6 inches is required. Refer to Unit 9 for more detailed information regarding underground wiring.

TESTING AND RECORDING OF TEST DATA FOR GFCI RECEPTACLES

Referring to Figure 7–3(A), note that GFCI receptacles have "T" (test) and "R"(reset) buttons. Pushing the test button places a small ground fault of the circuit as explained earlier in this unit, Figure 7–5. When connected properly and operating properly, the GFCI receptacle will trip to the OFF position. Pushing the reset button will restore power. GFCI circuit breakers work the same way.

GFCI receptacles and circuit breakers come with detailed installation and testing instructions. Monthly testing is recommended to ensure that the GFCI mechanism will operate properly should a person using that receptacle be subjected to an electrical ground-fault shock. Figure 7–14 is a chart that homeowners can use to record monthly GFCI testing.

OCCUPANT'S TEST RECORD

TO TEST, depress the "TEST" button, the "RESET" button should extend. Should the "RESET" button not extend, the GFCI will not protect against electrical shock. Call qualified electrician.

TO RESET, depress the "RESET" button firmly into the GFCI unit until an audible click is heard. If reset properly, the "RESET" button will be flush with the surface of the "TEST" button.

This label should be retained and placed in a conspicuous location to remind the occupants that for maximum protection against electrical shock, each GFCI should be tested monthly.

Year	Jan	Feb	Mar	Apr	May	Jun	Jul	Aug	Sep	Oct	Nov	Dec

Figure 7–14 Chart for recording GFCI testing dates and results.

GFCI receptacles and GFCI circuit breakers have a test button to make sure that the GFCI mechanism operates properly. For a branch circuit that is protected by a GFCI-type circuit breaker, a simple tester can be used that deliberately puts a small magnitude ground fault on the branch circuit. The tester is plugged into any receptacle on the circuit. If the GFCI circuit breaker mechanism is working properly, this tester will cause the GFCI circuit breaker to trip off. This type of tester also can be used to check the circuit for open ground, reversed polarity of the "hot" and grounded neutral conductor, reversed polarity of the "hot" and grounding conductor, open "hot" conductor, and open grounded neutral conductor. Small indicating lamps indicate the type of fault present. Figure 7–15 shows one of these testers.

REPLACING EXISTING RECEPTACLES

The *National Electrical Code®* is very specific about the types of receptacles permitted to be used as a replacement for an existing receptacle. These Code rules are found in *Section 210-7(d)* and *250-130(c)*.

Figure 7–15 A tester that plugs into a receptacle that can check for the proper operation of a GFCI circuit breaker. It also can check the branch circuit for an open ground, reversed polarity of the "hot" and grounded neutral conductor, reversed polarity of the "hot" and grounding conductor, open "hot" conductor, and open grounded neutral conductor. Small indicating lamps indicate the type of fault present. (*Photo courtesy of Ideal Industries, Inc., Sycamore, IL*)

Replacing a Two-Wire or a Grounding-Type Receptacle Where the *NEC®* Now Requires GFCI Protection.

Of utmost importance is *Section 210-7(d)(2)*, a retroactive ruling requiring that wherever an existing receptacle is replaced in **any** location where the present Code requires GFCI protection (bathrooms, kitchens, near wet bars, outdoors, unfinished basements, garages, crawl spaces, etc.), the replacement receptacle must be GFCI protected. This could be a GFCI receptacle, or the branch circuit could be protected by a GFCI circuit breaker.

In older homes that have "knob-and-tube" wiring or old style nonmetallic-sheathed cable wiring where an equipment grounding means was not provided, a replacement GFCI receptacle will still operate. This was discussed earlier in this unit. Refer back to Figures 7–4 and 7–5.

Replacing Old Two-Wire Receptacles Where Grounding Means Exists or Where a Grounding Means is Added

Refer to Figure 7–16 and *Section 210-7(d)(1)*.

Old style two-wire nongrounding-type receptacles (A) must be replaced with:

• grounding-type receptacles (B), or

• GFCI-type receptacles (C).

Replacing Old Two-Wire Nongrounding-Type Receptacles Where Grounding Means Does Not Exist

Refer to Figure 7–17 and *Section 210-7(d)(3)*.

Old style two-wire nongrounding-type receptacles (A) may be replaced with any of the following:

• Nongrounding-type receptacles (A).

• GFCI-type receptacles (C). These GFCI receptacles must be marked *"No Equipment Ground."* It is not necessary to connect the GFCI's grounding screw to any grounding means. It can be left "unconnected." The GFCI mechanism will still operate as discussed earlier in this unit.

Do not connect an equipment grounding conductor from the green hexagon grounding screw on the GFCI receptacle to any other downstream receptacles fed from the replace-

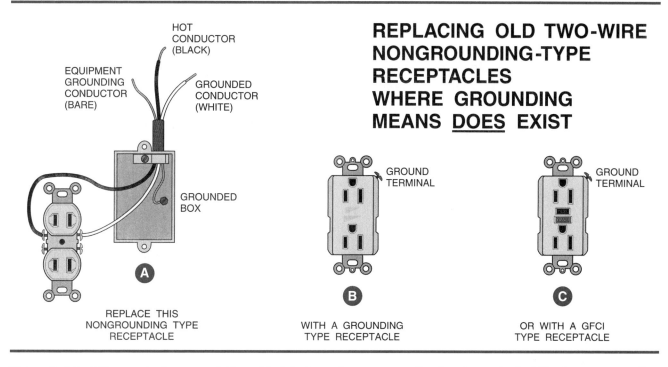

REPLACING OLD TWO-WIRE NONGROUNDING-TYPE RECEPTACLES WHERE GROUNDING MEANS <u>DOES</u> EXIST

HOT CONDUCTOR (BLACK)

EQUIPMENT GROUNDING CONDUCTOR (BARE)

GROUNDED CONDUCTOR (WHITE)

GROUNDED BOX

A

REPLACE THIS NONGROUNDING TYPE RECEPTACLE

GROUND TERMINAL

B

WITH A GROUNDING TYPE RECEPTACLE

GROUND TERMINAL

C

OR WITH A GFCI TYPE RECEPTACLE

Figure 7–16   When replacing an existing old style two-wire nongrounding-type receptacle (A) where a *grounding means exists or is added,* the replacement receptacle must be a grounding-type receptacle (B) or a GFCI receptacle (C). See *Section 210-7(d)(1).* Acceptable grounding methods are specified in *Section 250-118,* such as a separate equipment grounding conductor, metal raceway, or cable armor.

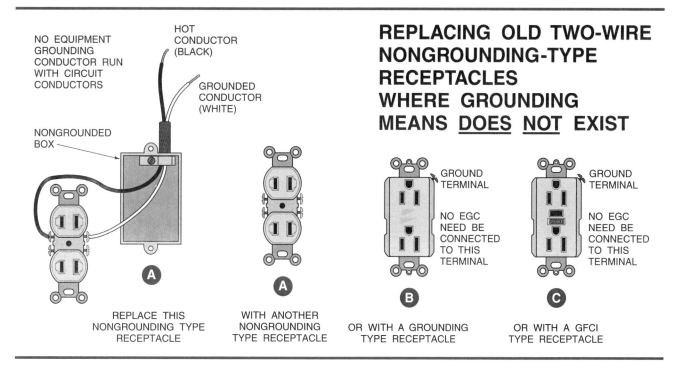

REPLACING OLD TWO-WIRE NONGROUNDING-TYPE RECEPTACLES WHERE GROUNDING MEANS <u>DOES</u> <u>NOT</u> EXIST

NO EQUIPMENT GROUNDING CONDUCTOR RUN WITH CIRCUIT CONDUCTORS

HOT CONDUCTOR (BLACK)

GROUNDED CONDUCTOR (WHITE)

NONGROUNDED BOX

A

REPLACE THIS NONGROUNDING TYPE RECEPTACLE

A

WITH ANOTHER NONGROUNDING TYPE RECEPTACLE

GROUND TERMINAL

NO EGC NEED BE CONNECTED TO THIS TERMINAL

B

OR WITH A GROUNDING TYPE RECEPTACLE

GROUND TERMINAL

NO EGC NEED BE CONNECTED TO THIS TERMINAL

C

OR WITH A GFCI TYPE RECEPTACLE

Figure 7–17   When replacing an existing old style two-wire nongrounding-type receptacle (A) *where a grounding means does not exist,* the replacement receptacle must either be a nongrounding-type receptacle (A), a grounding-type receptacle (B) if supplied through another GFCI receptacle, a grounding-type receptacle (B) if a separate equipment grounding conductor is run from the receptacle to any accessible point on the grounding electrode system, *Section 250-130(c),* or a GFCI-type receptacle (C). Most electricians would make the replacement with a GFCI-type receptacle (C) rather than running an equipment grounding conductor all the way back to the first 5 feet of metal water pipe as it enters the home. See *Section 210-7(d)(3).*

ment GFCI receptacle. To do so could lead someone who is working on the receptacle to think that the grounding terminal is actually properly grounded, when in fact it would not be grounded. We now have a false sense of security, a real shock hazard (Figure 7–18).

• Grounding-type receptacles (B) that are supplied through another GFCI. These GFCI receptacles must be marked *"GFCI Protected"* and must also be marked *"No Equipment Ground."* Repeating what was mentioned above, do not connect an equipment grounding conductor between any of the grounding-type receptacles (Figure 7–18).

• Grounding-type receptacles (B) when equipment grounding is added as illustrated in Figure 7–19 in accordance with *Sections 210-7(d)(1)* and *250-130(c)*. Note that an equipment grounding conductor is run from the green hexagon grounding screw on the replacement

receptacle to a listed ground clamp on the "grounding electrode system." A "grounding electrode system" is the metal underground water pipe in direct contact with at least 10 feet (3.05 m) of the earth, a supplemental ground rod, the grounded metal frame of a building, a concrete-encased electrode, or a ground ring. To form a "grounding electrode system," all of these items must be bonded together. Attaching an equipment grounding conductor anywhere beyond the first 5 feet of the interior metal water piping after it enters the building is not considered to be an adequate ground, because the metal piping might contain sections of plastic (PVC) piping. Generally, when replacing a nongrounding-type receptacle, it is more cost effective to replace it with a GFCI-type receptacle rather than trying to run an equipment grounding conductor all the way back to the first 5 feet of the interior metal water piping as it enters the building. Refer to *Section 250-50*.

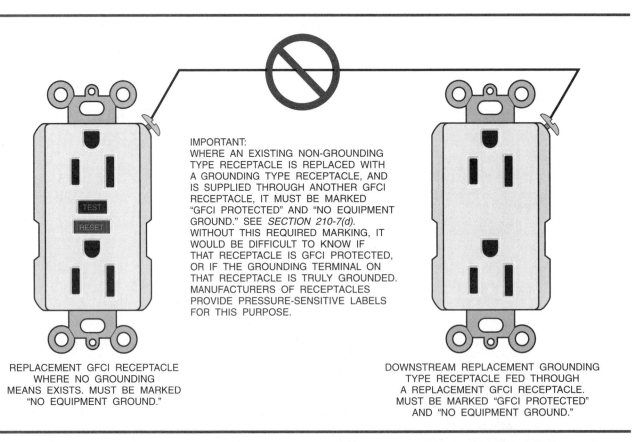

IMPORTANT:
WHERE AN EXISTING NON-GROUNDING TYPE RECEPTACLE IS REPLACED WITH A GROUNDING TYPE RECEPTACLE, AND IS SUPPLIED THROUGH ANOTHER GFCI RECEPTACLE, IT MUST BE MARKED "GFCI PROTECTED" AND "NO EQUIPMENT GROUND." SEE *SECTION 210-7(d)*. WITHOUT THIS REQUIRED MARKING, IT WOULD BE DIFFICULT TO KNOW IF THAT RECEPTACLE IS GFCI PROTECTED, OR IF THE GROUNDING TERMINAL ON THAT RECEPTACLE IS TRULY GROUNDED. MANUFACTURERS OF RECEPTACLES PROVIDE PRESSURE-SENSITIVE LABELS FOR THIS PURPOSE.

REPLACEMENT GFCI RECEPTACLE WHERE NO GROUNDING MEANS EXISTS. MUST BE MARKED "NO EQUIPMENT GROUND."

DOWNSTREAM REPLACEMENT GROUNDING TYPE RECEPTACLE FED THROUGH A REPLACEMENT GFCI RECEPTACLE. MUST BE MARKED "GFCI PROTECTED" AND "NO EQUIPMENT GROUND."

Figure 7–18 Where a grounding means *does not exist*, certain markings are required. Labels are usually furnished by the manufacturer, and are found in the box with the GFCI receptacle. Do not connect an equipment grounding conductor between these receptacles.

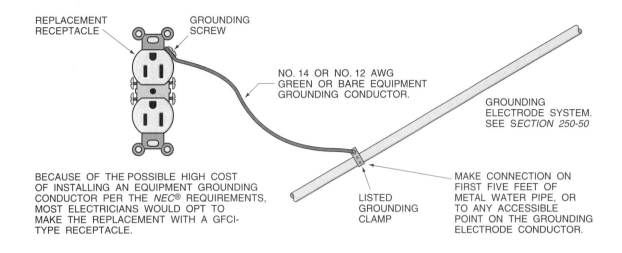

REPLACEMENT
RECEPTACLE

GROUNDING
SCREW

NO. 14 OR NO. 12 AWG
GREEN OR BARE EQUIPMENT
GROUNDING CONDUCTOR.

GROUNDING
ELECTRODE SYSTEM.
SEE *SECTION 250-50*

BECAUSE OF THE POSSIBLE HIGH COST
OF INSTALLING AN EQUIPMENT GROUNDING
CONDUCTOR PER THE *NEC®* REQUIREMENTS,
MOST ELECTRICIANS WOULD OPT TO
MAKE THE REPLACEMENT WITH A GFCI-
TYPE RECEPTACLE.

LISTED
GROUNDING
CLAMP

MAKE CONNECTION ON
FIRST FIVE FEET OF
METAL WATER PIPE, OR
TO ANY ACCESSIBLE
POINT ON THE GROUNDING
ELECTRODE CONDUCTOR.

Figure 7–19   **In existing installations where there is no equipment ground, as in the case of old knob-and-tubing wiring, *Section 210-7(d)(1)* permits replacing a nongrounding-type receptacle with a grounding-type receptacle provided the grounding screw on the receptacle is properly grounded. An acceptable way is to run an equipment grounding conductor from this screw to any accessible point on the grounding electrode system, or to any accessible point on the grounding electrode conductor. See *Section 250-130(c)*.**

Adding a New Receptacle Where The Wiring Does Not Have a Grounding Means

Adding a new receptacle in an old house that has knob-and-tube wiring or old nonmetallic-sheathed cable that does not contain an equipment grounding conductor is different than **replacing** an existing receptacle because new wiring is involved. If the newly added receptacle is connected to newly installed wiring, all of the *National Electrical Code®* requirements must be met. However, if the newly added receptacle is wired by extending the branch circuit from another receptacle, the added receptacle must be:

- a grounding-type, in which case an equipment grounding conductor must be run back to the "grounding electrode system" discussed previously, or
- a GFCI-type receptacle marked "No Equipment Ground."

Caution!

Even though the *NEC®* allows replacing existing receptacles with nongrounding-type receptacles, remember that personal computers require a proper equipment ground connection. Every manufacturer of personal computers, printers, scanners, and other electronic equipment state in their installation instructions that the equipment be plugged into a three-wire grounding-type, 120-volt, 60-hertz receptacle. These manufacturers recommend further that the equipment be plugged into a surge suppressor outlet. One manufacturer states that "Lack of a proper ground connection will eliminate common-mode protection and cause a small leakage potential on the case." The personal computer may work if plugged into a two-wire nongrounding-type receptacle, but it also could be damaged by voltage transients. Voltage transients, called **surges** or **spikes,** can cause abnormal current to flow through the sensitive electronic components. This can stress, degrade, and/or destroy components, can cause loss of memory in the equipment, or "lock up" the microprocessor.

So if you are replacing an existing two-wire receptacle, give serious thought to installing new branch-circuit wiring complete with an equipment grounding conductor, then install a three-wire grounding-type receptacle. If the purpose of the receptacle is to plug in a computer, consider installing a transient voltage surge suppressor (TVSS) receptacle or use a TVSS plug-in strip. These are discussed in *Electrical Wiring– Residential*.

TEMPORARY WIRING: GROUND-FAULT PROTECTION FOR PERSONNEL

Temporary wiring has always been a problem.

Because of the nature of temporary wiring, there is a continual presence of shock hazard that can lead to serious personal injury or death through electrocution. Workers are standing in water, standing or kneeling on damp or wet ground, touching metal ducts or pipes, or are in contact with steel framing or concrete that is in direct contact with the earth. Electric cords and cables are lying on the ground, being stepped on, draped over ducts and pipes, and are subject to severe mechanical abuse. All of these conditions spell "Danger."

The following *"Ground-fault protection for personnel"* Code requirements are found in *Article 305* of the *National Electrical Code®* and apply to all temporary power used *"during the period of construction, remodeling, maintenance, repair, or demolition of buildings, structures, equipment or similar activities."* Read these words carefully! The words leave few, if any, situations where GFCI protection would not be required.

Section 305-6(a) of the Code requires that **all** 125-volt, single-phase, 15-, 20-, and 30-ampere receptacle outlets that are not part of the permanent wiring of the building and that will be used by workers on the construction site must be GFCI protected. If the receptacles that are part of the actual permanent wiring are used for temporary power, they must be GFCI protected—either with a GFCI receptacle or with a GFCI circuit breaker.

Some contractors use "listed" manufactured power outlets to provide temporary power on construction sites (Figure 7–20). These meet the requirements of *Section 305-6(a)*.

Many electricians and other construction site workers carry their own "listed" ground-fault circuit-interrupter cord sets to be sure they are protected. These are acceptable in accordance with *Section 305-6(a)*. These portable GFCI devices are easy to carry around. They have "open neutral" protection should the neutral in the circuit supplying the GFCI open for whatever reason. Portable GFCI devices are available with manual reset, which is advantageous should a power outage occur or if the GFCI is unplugged, so that equipment (drills, saws, etc.) would not start up again when the power comes

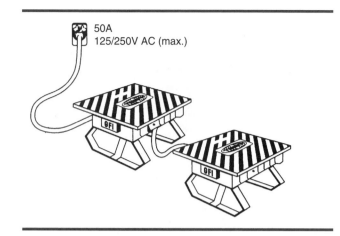

50A
125/250V AC (max.)

Figure 7–20 A portable power outlet commonly used on construction sites that provide ground-fault protection for workers. This particular unit is connected with a 50-ampere 125/250-volt power cord. Each of the six 20-ampere receptacles in the unit are connected to individual GFCI circuit breakers in the unit that provide both ground-fault and overload protection for anything plugged into one of the receptacles. (*Courtesy Hubbell Incorporated, Wiring Device Division*)

back on, causing injury to anyone using equipment when the power is restored. Portable GFCI devices are also available with automatic reset, which is advantageous should a power outage occur or if the GFCI is unplugged when lighting, portable signs, engine heaters, or water pumps are the connected load. Two types of portable GFCI devices are shown in Figure 7–21.

Exempt from the GFCI protection requirement are small two-wire, single-phase portable or vehicle-mounted generator's rated not over 5 kW when the generators windings and conductors are insulated from the frame of the generator and other grounded surfaces, *Section 305-6(a), Exception No. 1.*

For receptacles **other** than the 125-volt, single-phase, 15-, 20-, and 30-ampere receptacles, *Section 305-6(b)* permits a written *Assured Equipment Grounding Conductor Program* that must be continuously enforced on the construction site. A designated person must keep a written log, ensuring that all electrical equipment are properly installed and maintained according to the applicable requirements of *Section 210-7(c), 250-114, 250-138, and 305-4(d).* This option has proved to be difficult to enforce. Many "authorities having jurisdiction" still require GFCI protection as stated in *Section 305-6(a).*

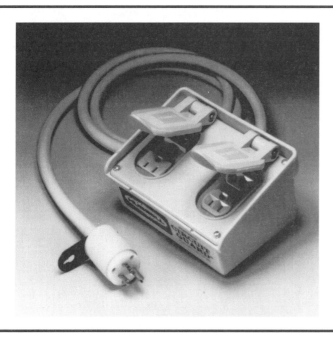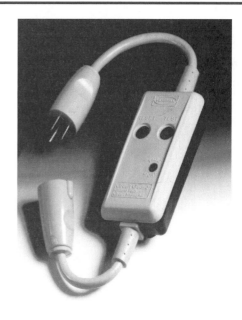

Figure 7–21 Plug- and cord-connected "portable" GFCI devices are easy to carry around. They can be used anywhere when working with electrical tools and extension cords. These devices operate independently and are not dependent upon whether or not the branch circuit is GFCI protected. These portable GFCI in-line cords are relatively inexpensive, and should always be used when using portable electric tools. (*Courtesy Hubbell Incorporated, Wiring Device Division*)

ADDITIONAL KINDS OF PERSONNEL PROTECTION

In addition to *ground-fault circuit interrupters,* electricians should be aware that there are other types of personnel protection in use today.

Because most older homes do not have GFCI protection, industry standards are slowly but surely requiring manufacturers to provide "people protection" as an integral part of specific appliances.

Appliances such as hair dryers, hair stylers, heated air combs, curling irons, and similar personal grooming appliances might have an *Immersion Detection Circuit Interrupter (IDCI)* in the attachment plug that will shut off the appliance if dropped into the water.

There also are *Appliance Leakage Circuit Interrupters (ALCI), Equipment Leakage Circuit Interrupters (ELCI), Leakage Monitoring Receptacles (LMR), Transient Voltage Surge Suppressors (TVSS),* and *Isolated Ground Receptacles.*

Although these devices are not required by the *National Electrical Code®* they might be required by an Underwriters Laboratories standard, or they might be installed to protect electronic equipment against line surges or "noise."

These devices and how they are used in homes are discussed in the *Electrical Wiring—Residential* text.

Arc-Fault Circuit Interrupters (AFCI)

Arc-fault circuit interrupters are the new kid on the street.

Electrical arcing is one of the leading causes of electrical fires in homes. The heat of an arc is extremely high, enough to ignite most combustible material. Arcing faults can be line-to-ground, line-to-neutral, or line-to-line. When a conductor that is carrying current breaks, or when a loose connection occurs, it is a "series" arcing fault because the arcing is traveling in the intended path. A broken or frayed cord is a good example of this.

An *Arc-Fault Circuit Interrupter (AFCI)* is designed to sense the rapid fluctuations of current flow typical of an arcing condition. It will open the circuit when arcing occurs. The sensing range of an AFCI is 5 amperes up through the expected instantaneous tripping current value of a typical circuit breaker. However, the design of the electronic circuitry is such that it will not nuisance trip because of the normal momentary arc that occurs when a switching device is turned on or off.

Effective January 1, 2002, AFCI protection must be provided for all 125-volt, single-phase, 15- and 20-ampere receptacles in dwelling bedrooms. This protection can be provided with AFCI circuit breakers or AFCI receptacles. This requirement is found in *Section 210-12, NEC.*

REVIEW QUESTIONS

1. Electricity can kill! The *National Electrical Code®* has recognized this hazard for years by requiring _____ in specific locations in a dwelling.

2. Match the following conductors with their proper receptacle connection slot.

   1. _____ Wide slot   a. Equipment grounding conductor

   2. _____ Narrow slot   b. Grounded conductor

   3. _____ Horseshoe-shaped slot   c. Ungrounded "hot" conductor

3. The severity of an electrical shock depends upon three things. Name them.

   _____

   _____

   _____

4. Most GFCI receptacles and GFCI circuit breakers used in residential applications are referred to as Class A GFCIs, and will trip off when the device senses a ground fault in the range of _____ milliamperes. The term *mil* means 1/1,000.

   a) 2 to 4

   b) 4 to 6

   c) 6 to 8

5. In the spaces provided, mark T or F (True or False) for the following statements.

   a) A GFCI will operate on line-to-ground faults.   _____

   b) A GFCI will operate on line-to-line short circuits.   _____

   c) A GFCI will operate when you touch the "hot" wire and the
      grounded (neutral) circuit conductor   _____

   d) It is absolutely necessary when connecting "feedthrough"
      GFCIs to make sure that the leads marked *line* are con-
      nected to the circuit, and that the leads marked *load* are
      connected to the downstream GFCIs   _____

   e) You will receive an electrical shock when you touch a
      "hot" conductor and a grounded surface when the circuit
      is protected by a GFCI   _____

6. Check each of the following locations of a dwelling unit in which the Code specifically requires GFCI receptacles, or receptacles that are GFCI protected by a GFCI circuit breaker.

    a) ____ Bedroom   g) ____ Bathrooms

    b) ____ Garage   h) ____ Outdoors

    c) ____ Unfinished basements   i) ____ Crawl spaces

    d) ____ Family room   j) ____ Living room

    e) ____ Kitchen   k) ____ Wet bar

    f) ____ Laundry room   l) ____ Dining room

7. You are replacing an old receptacle in a bathroom. The receptacle is a two-wire receptacle. The house was built long before the advent of GFCI devices. You are required to replace the defective receptacle. In the spaces provided, mark T or F (True or False) for the following statements.

    a) Replace the receptacle with the same type as the existing
       receptacle—a two-wire type   _____

    b) Replace the receptacle with a duplex grounding type receptacle   _____

    c) Replace the receptacle with a GFCI-type receptacle   _____

8. You are replacing a worn out receptacle in the living room of an old house. The receptacle is a two-wire receptacle. You note that the wiring is armored cable (BX). When you test from "hot" to the metal wall box, your voltmeter tester reads 120 volts. In the spaces provided, mark T or F (True or False) for the type of receptacle to be installed as the replacement.

    a) A grounding-type receptacle   _____

    b) A GFCI-type receptacle   _____

    c) A two-wire nongrounding-type receptacle—
       the same type as being replaced   _____

9: You are replacing a worn-out receptacle in the living room of an old house. The receptacle is a two-wire receptacle. You note that the wiring is knob-and-tube, and that there is no equipment ground in the wall box behind the defective receptacle. In the spaces provided, mark T or F (True or False) for the type of receptacle to be installed as the replacement.

    a) A grounding-type receptacle   _____

    b) A GFCI-type receptacle   _____

    c) A two-wire nongrounding-type receptacle—
       the same type as being replaced   _____

10. You are adding a new receptacle in a bedroom of an old house that is wired with old nonmetallic-sheathed cable of the type that does not contain an equipment grounding conductor. To connect the new receptacle, you will extend the branch circuit from one of the other receptacles in the room. The new receptacle must be:

    a) A standard grounding- type receptacle. Nothing else need be done.

    b A standard grounding-type receptacle. An equipment grounding conductor must also be run from this receptacle back to the "grounding electrode system," which would be all the way back to the first 5 feet of the interior metal water piping where it enters the house.

    c) A GFCI receptacle. Nothing else need be done.

11. All 125-volt, single-phase, 15- and 20-ampere receptacles installed for temporary power on construction sites must be of the:
    a) nongrounding type.
    b) grounding type.
    c) GFCI type.

12. A receptacle is installed in the ceiling of a garage. Its sole purpose is to plug in the overhead garage door operator. Is this receptacle required to be GFCI protected?
    a) Yes
    b) No

13. A dedicated space in a garage is to be used for a large freezer. In the spaces provided, mark T or F (True or False) for the type of receptacle to be installed.
    a) A standard duplex grounding-type receptacle installed low and out of reach behind the freezer _____
    b) A standard duplex grounding-type receptacle installed 4 feet above the garage floor so the receptacle can be used for other things _____
    c) A GFCI receptacle _____
    d) A single grounding-type installed low and out of reach behind the freezer _____
    e) A single grounding type installed 4 feet above the garage floor above the freezer _____

Conductors

OBJECTIVES

After studying this unit, you will be able to:

- Learn about the different types of conductors used for residential wiring.
- Discover how conductors are sized.
- Understand that different types of insulation on a conductor are permitted for specific uses and applications.
- Know how to find the allowable ampacity of a given conductor, using *National Electrical Code®* tables.
- Know how to determine the allowable ampacity for conductors located in extreme heat, and when there are more than three current-carrying conductors in the same raceway or cable.
- Understand how to determine conductor ampacity based on the temperature rating of the equipment and the termination.

  Know the *National Electrical Code®* rules for flexible cords.
- Understand the color coding of conductor insulation.

INTRODUCTION

Conductors in cables and raceways are the arteries and veins of a wiring system. Conductors carry electrical power from one point to another. This unit discusses the different types of conductors generally used for house wiring, their uses, and their limitations.

CONDUCTORS— GENERAL DISCUSSION

Article 310 of the *National Electrical Code®* is devoted to conductors used for general wiring. To learn about other wiring such as for telephones, cable television, low-voltage wiring for chimes and thermostats, and flexible cords, refer to the Index.

The *National Electrical Code®* defines different kinds of conductors.

Bare Conductor: A conductor having no covering or electrical insulation. Example: A bare No. 8 grounding electrode conductor.

Covered Conductor: A conductor encased within material of composition and thickness that is not recognized by this Code as electrical insulation. Example: An equipment grounding conductor in an armored cable that is wrapped with paper.

Insulated Conductor: A conductor encased within material of composition and thickness that is recognized by this Code as electrical insulation. Example: A Type THHN conductor.

When we talk about conductors, it is important to understand that conductors by themselves are not considered to be a wiring method. Single conductors that are listed in *Table 310-13* (shown in part later) must be installed as part of a recognized wiring method, *Section 300-3(a)*. For example, you are not permitted to use a Type THHN conductor by itself as a branch-circuit conductor. It must be in a raceway or cable.

In the electrical trade, the terms *wire* and *conductor* are used interchangeably. There is no problem in doing this. Do not get hung up on this issue. Everyone will know that *pulling in the wires*, or *pulling in the conductors* mean one and the same thing.

Conductors may be solid; for example, a No. 12 AWG Type THHN conductor.

Conductors may be stranded; for example, a No. 8 Type THHN conductor that is made up of seven individual solid conductors. *Section 310-3* of the Code requires that conductors No. 8 and larger must be stranded when they are installed in raceways, like electrical metallic tubing.

A *cable* can be a single conductor or can be made up of multiple conductors. When referring to a single conductor cable, the term is generally used for large sizes. An example of a single conductor cable is a 500 kcmil Type XHHW. An example of a multiple conductor cable is Type NMC nonmetallic-sheathed cable.

Small cables are called *cords*.

TYPES OF CONDUCTORS

Conductors used in house wiring are rated for 600 volts. Some common exceptions are the low-voltage conductors used for hooking up chimes and thermostats, and the small 300-volt fixture wires that are an integral part of the lighting fixture. The cables used for telephone, cable television, computers, and similar applications also have a low-voltage rating. This is discussed in Unit 14.

Conductors are made of copper, aluminum, and copper-clad aluminum. The most commonly used conductors for residential wiring are copper, although you will find some aluminum or copper-clad aluminum conductors used in the larger sizes, such as for services.

CONDUCTOR SIZE

Conductors used in electrical installations are sized according to the American Wire Gauge (AWG) Standard. The wire diameter in the AWG standard is expressed as a whole number. AWG sizes vary from fine, hairlike wire used in clocks and small transformers, to large conductors used for branch circuits, feeders, and services. Although actual AWG sizes include **all** numbers from No. 40 through No. 1, you will find in *Table 310-16* and *Table 8, Chapter 9* of the *NEC®* that many of the AWG numbers are skipped, as they are not used for the manufacturing of typical building wire.

Size designation for conductors is *retrogressive.* This means that the higher the AWG number, the smaller the wire. For example, a No. 14 AWG conductor is smaller than a No. 10 AWG conductor. This retrogressive numbering is true through No. 1 AWG. After No. 1 AWG, the numbering system changes somewhat. The wire sizes continue to get progressively larger, and the numbering is 1/0, 2/0, 3/0, and 4/0. After 4/0, wire sizes are designated with their actual circular mil area, starting with 250,000 circular mils and ending with 2,000,000 circular mils.

The diameter of a wire can be measured in inches or mils. A *mil* is defined as one-thousandth of an inch (0.001).

The cross-sectional area of a wire can be measured in square inches (inches squared) or circular mils (mils squared).

The letter *k* designates 1,000 in the metric system.

Large conductors such as 500,000 circular mils are generally expressed as 500 kcmils. The term kcmils means "thousand circular mils." This is much easier to express in both written and verbal terms.

Older texts used the term MCM, which also meant "thousand circular mils." The first letter *M* refers to the Roman numeral that represents 1,000. Thus, the term 500 MCM means the same as 500 kcmils. Roman numerals are no longer used in the electrical industry for expressing conductor sizes.

Conductor stranding, diameters, circular mil areas, square inch areas, and resistance values for copper and aluminum conductors are found in *Table 8, Chapter 9* of the *NEC®.*

Approximate Conductor Size Relationship

Although the size designations for conductors might seem a bit confusing, there are two mathematical relationships between sizes. These relationships provide us with an easy way to approximate the cross-sectional areas and resistance values of conductors, which is the information needed to do voltage drop calculations.

Try to fix in your mind that No. 10 AWG conductor has a cross-sectional area of 10,380 circular mils, and that it has a resistance of 1.2 ohms per 1,000 feet. Then, note the following size relationships that conductors have to one another.

Relationship No. 1. For all wire sizes No. 40 through No. 4/0, every third size doubles or halves in circular mil area and resistance for a given length. The results are approximate but very acceptable. For example:

- A No. 1 AWG is twice as large as a No. 4 AWG and has one-half the resistance for a given length.

- A 1/0 is one-half the size of a 4/0 and has twice the resistance for a given length.

Relationship No. 2. For all wire sizes No. 40 through No. 4/0, every consecutive size is approximately 1.26 larger or smaller than the preceding size, and has 1.26 more or less resistance for a given length than the preceding size. We use the 1.26 factor by dividing or multiplying, depending upon whether we are increasing or decreasing. The results are approximate, but very acceptable. For example:

- A No. 3 AWG is approximately 1.26 (26%) larger than a No. 4 AWG and has approximately 79% the resistance.

$$1 \div 1.26 = 0.79 \ (79\%).$$

- A No. 2 AWG is approximately 1.26 smaller than a No. 1 AWG and has approximately 126% greater resistance for a given length.

Example: What is the approximate cross-sectional area (CSA) and resistance of 1,000 feet of No. 6 AWG copper conductor? Refer to *Table 8, Chapter 9, NEC.* Note how close your results are when compared to the actual values.

Answer: A No. 10 AWG has a CMA of 10, 380 and an approximate resistance of 1.2 ohms per 1,000 feet. A No. 7 (remember the "every third size" rule) would be twice as large as a No. 10 AWG, which would calculate to be:

$$10,380 \times 2 = 20,760 \text{ circular mils}$$

A No. 6 is 1.26 larger (remember the "consecutive size" rule) than a No. 7.

$$20,760 \times 1.26 = 26,158 \text{ circular mils}$$

A No. 7 would have an approximate resistance of:

$$1.2 \div 2 = 0.6 \text{ ohm}$$

A No. 6 has 79% (1 ÷ 1.26 = 0.79) of the resistance of a No. 7.

$$0.6 \times 0.79 = 0.0474 \text{ ohm}$$

This quick and easy "Rule of Thumb" is useful when making voltage drop calculations without having the wire tables readily available.

Figure 8–1 is *Table 310-16* of the *National Electrical Code.* It shows the allowable ampacities of insulated conductors, 0 through 2,000 Volts, 60° (140°F) through 90°C (194°F), not more than three current-carrying conductors in a raceway or cable, based on an ambient temperature of 30°C (86°F). The table lists conductors in sizes from No. 18 AWG through 4/0. Some tables will show 1/0 as "0," 2/0 as "00," 3/0 as "000," and 4/0 as "0000." Conductor sizes larger than 4/0 are given in circular mil area. *Table 8* in *Chapter 9* of the *NEC* shows conductor details such as sizes, area in circular mils, number of strands, diameters, and resistance values.

When installed in raceways, conductors No. 8 AWG and larger must be stranded, *Section 310-3.* Smaller-size conductors may be stranded. For example, if the wiring for an electric clothes dryer is installed in electrical metallic tubing (EMT), stranded No. 10 AWG conductors would pull in and handle easier than solid No. 10 AWG conductors.

Branch-circuit wiring with nonmetallic-sheathed cable and armored cable in sizes Nos. 14, 12, and 10 AWG is always done with solid conductors. These cables have stranded conductors in sizes No. 8 and larger for flexibility.

The minimum conductor size permitted in house wiring is No. 14 AWG as required by *Section 210-19(d).* The exceptions to *Section 210-19(d)* list such

Table 310-16. Allowable Ampacities of Insulated Conductors Rated 0 through 2000 Volts, 60°C through 90°C (140°F through 194°F) Not More than Three Current-Carrying Conductors in Raceway, Cable, or Earth (Directly Buried), Based on Ambient Temperature of 30°C (86°F)

Size AWG or kcmil	Temperature Rating of Conductor (See Table 310-13)						Size AWG or kcmil
	60°C (140°F)	75°C (167°F)	90°C (194°F)	60°C (140°F)	75°C (167°F)	90°C (194°F)	
	Types TW, UF	Types FEPW, RH, RHW, THHW, THW, THWN, XHHW, USE, ZW	Types TBS, SA, SIS, FEP, FEPB, MI, RHH, RHW-2, THHN, THHW, THW-2, THWN-2, USE-2, XHH, XHHW, XHHW-2, ZW-2	Types TW, UF	Types RH, RHW, THHW, THW, THWN, XHHW, USE	Types TBS, SA, SIS, THHN, THHW, THW-2, THWN-2, RHH, RHW-2, USE-2, XHH, XHHW, XHHW-2, ZW-2	
	COPPER			ALUMINUM OR COPPER-CLAD ALUMINUM			
18	—	—	14	—	—	—	—
16	—	—	18	—	—	—	
14*	20	20	25	—	—	—	
12*	25	25	30	20	20	25	12*
10*	30	35	40	25	30	35	10*
8	40	50	55	30	40	45	8
6	55	65	75	40	50	60	6
4	70	85	95	55	65	75	4
3	85	100	110	65	75	85	3
2	95	115	130	75	90	100	2
1	110	130	150	85	100	115	1
1/0	125	150	170	100	120	135	1/0
2/0	145	175	195	115	135	150	2/0
3/0	165	200	225	130	155	175	3/0
4/0	195	230	260	150	180	205	4/0
250	215	255	290	170	205	230	250
300	240	285	320	190	230	255	300
350	260	310	350	210	250	280	350
400	280	335	380	225	270	305	400
500	320	380	430	260	310	350	500
600	355	420	475	285	340	385	600
700	385	460	520	310	375	420	700
750	400	475	535	320	385	435	750
800	410	490	555	330	395	450	800
900	435	520	585	355	425	480	900
1000	455	545	615	375	445	500	1000
1250	495	590	665	405	485	545	1250
1500	520	625	705	435	520	585	1500
1750	545	650	735	455	545	615	1750
2000	560	665	750	470	560	630	2000

CORRECTION FACTORS							
Ambient Temp. (°C)	For ambient temperatures other than 30°C (86°F), multiply the allowable ampacities shown above by the appropriate factor shown below.						Ambient Temp. (°F)
21–25	1.08	1.05	1.04	1.08	1.05	1.04	70–77
26–30	1.00	1.00	1.00	1.00	1.00	1.00	78–86
31–35	0.91	0.94	0.96	0.91	0.94	0.96	87–95
36–40	0.82	0.88	0.91	0.82	0.88	0.91	96–104
41–45	0.71	0.82	0.87	0.71	0.82	0.87	105–113
46–50	0.58	0.75	0.82	0.58	0.75	0.82	114–122
51–55	0.41	0.67	0.76	0.41	0.67	0.76	123–131
56–60	—	0.58	0.71	—	0.58	0.71	132–140
61–70	—	0.33	0.58	—	0.33	0.58	141–158
71–80	—	—	0.41	—	—	0.41	159–176

*See Section 240-3.

Figure 8–1 *Table 310-16* from the *National Electrical Code.* (*Reprinted with permission from NFPA 70-1999*)

items as "taps," fixture wires, and cords. Low-voltage wiring comes under the provisions of *Article 725* and is discussed in Unit 14.

Figure 8–2 shows the sizes of conductors used for some common applications in house wiring.

CONDUCTOR INSULATION AND TEMPERATURE RATINGS

The *National Electrical Code®* requires that all wires used in electrical installations be insulated, *Section 310-2(a)*. Exceptions to this rule are clearly indicated in the Code, such as the permission to use a bare neutral conductor for services. This is discussed in Unit 9.

Table 310-13 is a rather complete listing of conductor names, letter designation, maximum operating temperature, permitted uses, insulation type, size availability, and the material used (if any) for the outer covering.

Most conductors used for wiring houses are

CONDUCTOR APPLICATIONS CHART

CONDUCTOR SIZE	OVERCURRENT PROTECTION	TYPICAL APPLICATIONS (CHECK WATTAGE AND/OR AMPERE RATING OF LOAD TO SELECT THE CORRECT SIZE CONDUCTORS BASED ON *TABLE 310-16*)
No. 20 AWG	Class 2 Transformers provide overcurrent protection.	Recommended by some manufacturers of low-voltage control.
No. 18 AWG	7 amperes See *Table 430-72(b)* and *Section 752-3*. Class 2 circuit transformers provide overcurrent protection.	Low voltage control circuits such as thermostat wiring and chime wiring. Telephone wiring usually is No. 20 or 22 AWG. Also used for the LV wiring in "Advanced Home Systems" often times referred to as "Home Automation."
No. 16 AWG	10 amperes See *Table 430-72(b)* and *Section 752-3*. Class 2 circuit transformers provide overcurrent protection.	Same applications as above. Good for long runs to minimize voltage drop.
No. 14 AWG	15 amperes	Typical lighting branch circuits
No. 12 AWG	20 amperes	Small appliance branch circuits for the receptacles in kitchens and dining rooms. Also laundry receptacles and work shop receptacles. Often used as the "home run" for lighting branch circuits. Some water heaters.
No. 10 AWG	30 amperes	Most clothes dryers, built-in ovens, cooktops, central air conditioners, some water heaters, heat pumps.
No. 8 AWG	40 amperes	Ranges, ovens, heat pumps, some large clothes dryers, large central air conditioners, heat pumps.
No. 6 AWG	50 amperes	Electric furnaces, heat pumps.
No. 4 AWG	70 amperes	Electric furnaces, feeders to sub-panels.
No. 3 AWG and larger	100 amperes	Main service entrance conductors, feeders to sub-panels, electric furnaces

Figure 8–2   This chart shows typical applications for different size conductors.

insulated with thermoplastic material. These conductors are marked with the letter "T" on the insulation.

Conductors that are suitable for use in wet locations are marked with the letter "W." Examples of a wet location are conductors installed in underground raceways, and service-entrance conductors that are exposed to the weather such as the "drip loop" where the conductors come out of a service head for overhead services.

Conductors that have a thin nylon outer jacket are marked with the letter "N."

Without question, copper conductors with Type THHN insulation are the most popular and most commonly used. Type THHN conductors have a small diameter and corresponding small cross-sectional area. This makes Type THHN easy to handle, allows more conductors of a given size in a given size raceway, and can be installed where high ambient temperatures are encountered such as in attics, buried in insulation, and connecting lighting fixtures where the label on the lighting fixture might require 90°C supply conductors.

The temperature rating of the conductors found in nonmetallic-sheathed cable is 90°C. The temperature rating of the conductors found in armored cable varies, depending on the marking of the armored cable. This is covered later on in this chapter. For both nonmetallic-sheathed cable and armored cable, the allowable ampacity of the conductors is that of the 60°C column of *Table 310-16.* This requirement is stated in *Section 336-26* for nonmetallic-sheathed cable, and in Section 333-20 for armored cable.

The basic temperature rating for conductors is 60°C. A conductor that has one letter "H" marked on the insulation has a 75°C temperature rating. A conductor that has two letter "Hs" marked on the insulation has a 90°C temperature rating. Each "H" designation is equal to 15°C

When these letters are combined and marked on the conductor insulation, it is very easy to look in *Table 310-13* to find out the permitted uses and limitations for the conductor.

Some conductors have been "listed" for more than one application. For example, A type THHN has a 90°C temperature rating when used in a dry location, but if also marked THWN, this conductor would have a 75°C temperature rating when used in a wet location. A Type THHW has a 75°C rating when used in a wet location, and has a 90°C rating when used in a dry location. When in doubt about a conductor's application, check the markings on the insulation, then refer to *Table 310-13* for specifics.

Figure 8–3 is a table that contains some of the more commonly used conductors. Refer to *Table 310-13* for other types.

WET, DAMP, AND DRY LOCATIONS

The definitions are found in *Article 100* of the *National Electrical Code.*

- Wet Location: Installations underground or in concrete slabs or masonry in direct contact with the earth, and locations subject to saturation

TRADE NAME	TYPE LETTER	MAXIMUM OPERATING TEMP.	APPLICATION PROVISIONS	INSULATION	AWG	OUTER COVERING
Heat resistant thermoplastic	THHN	90°C (194°F)	Dry and damp locations	Flame retardant, heat resistantant thermoplastic	14 through 1,000 kcmil	Nylon jacket or equivalent
Moisture and heat resistant thermoplastic	THHW	75°C (167°F) 90°C (194°F)	Wet location Dry location	Flame retardant, moisture and heat resistantant thermoplastic	14 through 1,000 kcmil	None
Moisture and heat resistant thermoplastic	THWN Note: If marked THWN-2, OK for 90°C (194°F) in dry or wet locations	75°C (167°F)	Dry and wet locations	Flame retardant, moisture and heat resistantant thermoplastic	14 through 1,000 kcmil	Nylon jacket or equivalent
Moisture and heat resistant thermoplastic	THW Note: If marked THW-2, OK for 90°C (194°F) in dry or wet locations	75°C (167°F)	Dry and wet locations	Flame retardant, moisture and heat resistantant thermoplastic	14 through 2,000 kcmil	None
Underground feeder and branch circuit single conductor or multi-conductor cable. See *Article 339*	UF	60°C (140°F) 75°C (167°F)	See *Article 339* See *Article 339*	Moisture resistant Moisture resistant	14 through 4/0	Integral with insulation

Figure 8–3   Typical conductors used for residential wiring.

with water or other liquids, such as vehicle washing areas, and locations exposed to weather and unprotected.

- Damp Location: Partially protected locations under canopies, marquees, roofed open porches, and like locations, and interior locations subject to moderate degrees of moisture, such as some basements, some barns, and some cold-storage warehouses.

- Dry Location: A location not normally subject to dampness or wetness. A location classified as dry may be temporarily subject to dampness or wetness, as in the case of a building under construction.

AMPACITY

The term *ampacity* is unique to the electrical industry. The term was derived by combining the words *ampere* and *capacity*. It means "the current in amperes a conductor can carry continuously under the condition of use without exceeding its temperature rating."

This value depends upon the conductor's cross-sectional area, the type of insulation around the conductor, and how the conductor is used.

Section 310-10 states that "*No conductors shall be used in such a manner that its operating temperature will exceed that designated for the type of insulated conductor involved.*" What factors affect a conductor's operating temperature? The major contributing factors are:

- the size of the conductor

- the heat generated by the current flowing through the conductor

- the added heat from adjacent current-carrying conductors

- the effectiveness of how this heat can be dissipated

- the ambient (surrounding) temperature.

For common types of conductors in raceways or cables, the ampacity values are found in *Table 310-16,* shown in Figure 8–1. *Article 310* contains other tables for other types of conductors.

Table 310-16 applies when there are not more than three current-carrying conductors in one raceway or cable, and the temperature does not exceed 86°F.

Conductors must have an ampacity not less than the maximum load that they supply. This was discussed in Unit 6. There are some exceptions to this basic rule, such as for electric ranges, ovens, and cooktops, electric clothes dryers, motors, and electric water heaters. This is discussed in Unit 13.

When a bare or covered conductor is used with insulated conductors, the bare or covered conductor's allowable ampacity is the same as if the bare conductor were insulated with the same insulation as the other conductors in the raceway or cable. This rule is found in *Section 310-15(b)(3)*. This simply means that if you are looking up the allowable ampacity of a conductor in the 75°C insulated conductor column in *Table 310-16,* then you would also look up the allowable ampacity of the bare conductor in the same column.

OVERCURRENT PROTECTION

The basic rule for conductors is that the overcurrent protection must not exceed the ampacity of the conductor, *Section 240-3*. There are exceptions to this basic rule, such as the branch-circuit overcurrent protection for a motor. This is covered in Unit 17.

When the ampere rating of the fuse or circuit breaker does not match the ampacity of the conductor, it is all right to install the next standard size fuse or circuit breaker, *Section 240-3(b)*.

Unless a specific section of the *National Electrical Code®* allows a larger size overcurrent device, such as for a motor branch circuit, *Section 240-3(d)* restricts the overcurrent protection for certain small conductor sizes. This is important. Figure 8–4 shows the maximum size overcurrent protection for copper conductors in sizes No. 14, 12, and 10 AWG.

ALUMINUM CONDUCTORS

Aluminum conductors are rarely used in residential wiring, but are used in certain applications in commercial and industrial installations. All nonmetallic-sheathed cable, armored cable, and individual conductors used for house wiring are made up with copper conductors. Large size conductors for service entrances might be aluminum, but the popular choice is copper.

The conductivity of aluminum conductor is not as great as that of a copper conductor for a given

COPPER CONDUCTOR SIZE	MAXIMUM AMPERE RATING OF THE CIRCUIT BREAKER
No. 14 AWG	15-amperes
No. 12 AWG	20-amperes
No. 10 AWG	30-amperes

Figure 8–4 This table shows the maximum ampere rating of a fuse or circuit breaker for the protection of copper conductors sizes No. 14, 12, and 10 AWG. See *Section 240-3(d)*.

NUMBER OF CURRENT-CARRYING CONDUCTORS	PERCENT OF VALUES IN *TABLE 310-16*. IF HIGH AMBIENT TEMPERATURE IS PRESENT, ALSO APPLY CORRECTION FACTORS FOUND AT BOTTOM OF *TABLE 310-16*
4 through 6	80
7 through 9	70
10 through 20	50
21 through 30	45
31 through 40	40
41 and above	35

Figure 8–5 Derating factors necessary to determine conductor ampacity when there are more than three current-carrying conductors in a raceway or cable. See *Section 310-15(b)(2)*.

size. For example, checking Figure 8–1 (*Table 310-16*), it is found that a No. 8 Type THHN copper conductor has an allowable ampacity of 55 amperes, whereas a No. 8 Type THHN aluminum or copper-clad aluminum wire has an ampacity of only 45 amperes.

When aluminum conductors are used, the terminations must be "listed" for use with aluminum conductors. The markings would be *AL* or *AL/CU*. If you cannot find a marking on the termination, the termination is for use with copper conductors only.

The resistance for a given size and length of an aluminum conductor is considerably higher than for the same size and length of a copper conductor. For long runs, this has a dramatic impact on voltage drop, because the basic formula for calculating voltage drop involves the resistance of the circuit.

Voltage Drop (E_d) = Amperes (I) × Resistance (R)

ADJUSTMENT FACTORS

As stated previously, the allowable ampacity values for conductors shown in *Table 310-16* is based on not more than three current-carrying conductors in a raceway or cable, and an operating temperature of not more than 30C (86F). Should either or both of these two basic factors change, the ampacity of the conductor will change.

When a raceway or cable contains four or more current-carrying conductors, the allowable ampacity for a given size conductor must be reduced. Figure 8–5 shows the adjusting factors that must be applied.

According to *Section 310-15(b)(4)*, a neutral conductor of a multiwire circuit that carries the unbalanced currents does not have to be counted in

determining current-carrying capacities. Figure 8–6 shows the amount of current flowing in each of the conductors of a 120/240-volt, three-wire, single-

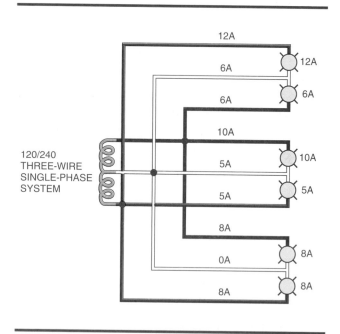

Figure 8–6 A neutral conductor carrying the unbalanced currents from the other conductors in a multiwire circuit need not be counted when determining the ampacity of the conductor, according to *Section 310-15(b)(4)*. In the three multiwire circuits shown, the actual number of conductors is nine, but only six are considered current-carrying conductors for applying the derating factors of Figure 8–5.

phase system. This supports the permission that it is not necessary to count the neutral of a multiwire circuit as a current-carrying conductor. Unit 2 covers the theory of multiwire circuits.

You do not have to count grounding or bonding conductors as current-carrying conductors, *Section 310-15(b)(5)*. You do have to count grounding or bonding conductors when determining conductor fill in a raceway. Raceway conductor fill is discussed in Unit 9.

Adjusting factors are not required for short conduit runs of 24 inches or less, referred to in the Code as nipples. This is found in *Section 310-15(b)(2)(a), Exception No. 3*. An example of where this might apply is in an upscale home that has complex remote control panels and electrical distribution panels. To connect these panels together, one or more conduit nipples may be used (Figure 8–7).

Example: Four No. 12 AWG copper THW conductors are installed in a single raceway to supply two separate 240-volt, 20-ampere branch circuits for snow melting cables in the sidewalk and driveway. There is no neutral conductor. All four conductors must be considered to be current-carrying. The equipment (circuit breaker and panel) is marked as being "listed" and rated for 75°C.

Answer: From the 75°C column of *Table 310-16* we find that the allowable ampacity of No. 12 AWG Type THW copper conductors is 25 amperes. From the table in Figure 8–5 we find that we need to apply a factor of 0.80:

$$25 \times 0.80 = 20 \text{ amperes}$$

Section 240-3(d) states that the maximum overcurrent protective device for No. 12 AWG copper

conductors is 20 amperes. The ampacity of the conductor after making the adjustment and the OCD are perfectly matched.

When nonmetallic-sheathed cables are run together in a "bunch" for distances of more than 24 inches, the adjusting factors in Figure 8–5 must be applied. This requirement is found in *Section 310-15(b)(2)(a)*. The Code recognizes the fact that when cables are stacked or bundled together without space between them, there will be a heat buildup because the heat will not dissipate properly. It is best not to run nonmetallic-sheathed cables in a "bunch" as illustrated in Figure 8–8 because derating will be necessary.

Example: Four branch-circuit "home runs" made up of four No. 12/2 w/ground nonmetallic-sheathed cables are run together through a single set of holes that are bored in-line for a distance of 25 feet. What is the ampacity of the conductors in the NMC cables?

Answer: The current-carrying conductor count is eight. The four equipment grounding conductors need not be counted as current-carrying conductors according to *Section 310-15(b)(5)*. The conductors in the NMC cable are rated 90°C per *Section 336-30(b)*. We can use the 90°C ampacity for derating purposes per *Section 110-14(c)*. The allowable ampacity of 90°C No.12 conductors is 30 amperes per *Table 310-16*. The adjustment factor (the penalty for running four NMC cables with eight current-carrying conductors together) from *Table 310-15(b)(2)(a)* is 70%. The final ampacity of the conductors is:

$$30 \times 0.70 = 21 \text{ amperes}$$

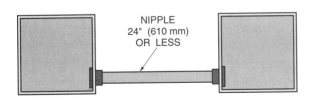

NIPPLE
24" (610 mm)
OR LESS

Figure 8–7 Derating of a conductor's allowable ampacity is not required for a conduit nipple that does not exceed 24 inches. See *Section 310-15(b)(2), Exception No. 3*.

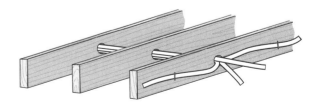

Figure 8–8 When cables are "bundled" together for distances of more than 24 inches, their ampacities must be reduced (derated) according to the percentages found in *Section 310-15(b)(2)(a)*.

CORRECTION FACTORS

When conductors are installed in locations where the temperature is higher than 30°C (86°F), the correction factors noted below *Table 310-16* must be applied.

Example: Two No. 8 Type TW copper conductors are run inside of EMT along the side of a house to connect an air-conditioning unit. The conductors will be subjected to the rays of the sun, and the temperature is expected to reach 120F. What is the ampacity of these conductors after applying correction factors?

Answer: From *Table 310-16,* find the allowable ampacity of No. 8 Type TW copper conductors, 40 amperes. Here are the steps.

1. Find No. 8 in the first column marked "Size."

2. Find Type TW copper conductors in the second column marked "60°C (140°F)."

3. Steps 1 and 2 intersect at 40 amperes.

4. Next, look at the correction factors found at the bottom of *Table 310-16.* In the far right-hand column, find 120°F. This falls in the range of 114–122°F.

5. Follow the 114–122°F row to the left until you reach the second column. This is the "60°C (140°F)" column.

6. Steps 5 and 6 intersect at the correction factor of 0.58.

$$40 \times 0.58 = 23.2 \text{ amperes}$$

In the above example, the new ampacity of these No. 8 Type TW conductors is 23.2 amperes. A 25-ampere OCPD could be used because *Section 240-3* permits the next larger standard size.

Example: Two No. 8 Type THHN copper conductors are run in EMT along the side of a house to connect an air-conditioning unit. The conductors will be subjected to the rays of the sun, and the temperature is expected to reach 120F. What is the ampacity of these conductors after applying correction factors?

Answer: From *Table 310-16,* find the allowable ampacity of No. 8 Type THHN copper conductors, 55 amperes. Next, apply the correction factor for 120°F found at the bottom of *Table 310-16.*

$$40 \times 0.82 = 32.8 \text{ amperes}$$

Applying Adjustment Factors and Correction Factors

Adjustment factors and *correction factors* must both be applied when there are four or more current-carrying conductors in the same raceway or cable, and the conductors are subject to high ambient temperature conditions.

Example: Four No. 10 THHN conductors supply two 2-wire, 240-volt, 30-ampere branch circuits. The conductors are installed in the same EMT. No neutral conductors are involved. All four conductors must be considered current-carrying. There is concern that because the conductors are in an EMT that runs through the attic where the ambient temperature might reach at least 110°F. What is the reduced ampacity of these No. 10 THHN conductors? The equipment (circuit breaker and panel) is marked as being "listed" and rated for 75°C.

Answer: From the 90°C column of *Table 310-16,* we find that the allowable ampacity of No. 10 AWG Type THHN copper conductors is 40 amperes. We apply an adjustment factor of 0.80 because there are more than three current-carrying conductors in the raceway, and a correction factor of 0.87 because of the high ambient temperature.

Section 110-14(c) allows us to use conductors with temperature ratings higher than specified for terminations when adjusting, or correcting, or both. *Section 110-14(c)* also allows us to use the ampacity of the higher rated conductor, provided the equipment is rated for use with the higher temperature conductors. In the above example, the equipment is rated 75°C.

$$40 \times 0.80 \times 0.87 = 27.84 \text{ amperes}$$

Section 240-3(d) states that the maximum overcurrent protective device for No. 10 AWG copper conductors is 30 amperes. The ampacity of the conductor after making the adjustment is 27.84 amperes.

From Figure 8–1 (*Table 310-16*), the ampacity of a No. 10 AWG 75°C copper conductor is 35 amperes. After applying the derating factors, we ended up with 27.84 amperes for the 90°C THHN conductor. *Section 240-3(b)* permits the use of the next standard OCD, which in this case is a 30-ampere fuse or circuit breaker.

TERMINATIONS

The Weakest Link of The Chain!

A conductor has two ends! Selecting a conductor ampacity based solely upon the values found in *Table 310-16* is a very common mistake that can prove costly.

Maximum temperature ratings are found in:

Conductors—UL 83 and *Tables 310-13, 310-16*

Receptacles—UL 498

Snap Switches—UL 20

Panelboards—UL 67

Circuit breakers—UL 489

Disconnect switches—UL 98

Wire splicing devices (connectors)—UL 486

It is apparent that in any given circuit, we have an assortment of electrical components that have different maximum temperature ratings. After sorting this all out, the allowable ampacity of a conductor is based upon the temperature limitations of the lowest temperature rating of any component of the circuit.

Most branch circuits rated 15- and 20-amperes have receptacles and snap switches that have a maximum temperature rating of 60°C. Therefore, no matter what the temperature rating of the panelboard or conductors might be (i.e. 90°C for THHN), the conductor ampacity must be determined based on the 60°C column of *Table 310-16* because the "weakest link" of the chain are the receptacles and snap switches.

Section 110-14(c) and UL standards require that unless otherwise marked, the ampacity of conductors connected to electrical equipment shall be based upon the ampacity of 60°C conductors where the circuit is rated 100 amperes or less, and the conductor sizes are No. 14 through No. 1 AWG. It is acceptable to install conductors having a higher temperature rating, such as 90°C THHN, but you must use the 60°C ampacity values.

For circuits rated over 100 amperes, or marked for use with conductors larger than No. 1 AWG, the conductors installed must have a minimum 75°C rating. The ampacity of the conductors is based upon the 75°C values. It is acceptable to install conductors having a higher temperature rating, such as 90°C THHN, but you must use the 75°C ampacity values.

Conductors having a temperature rating higher than the rating of the terminations may be used. When using high temperature insulated conductors where adjustment factors are to be applied, such as *derating* for more than three current-carrying conductors in a raceway or cable, or *correcting* where high temperatures are encountered, the final results of these adjustments (in amperes) must comply with the requirements of *Section 110-14(c)*.

Individual conductors for most premise wiring are rated 60°C, 75°C, or 90°C. In the case of armored cable and nonmetallic-sheathed cable, we must use the 60°C ampacity column of *Table 310-16*. This requirement is found in *Sections 333-20 and 336-25*.

Snap switches and 15- and 20-ampere receptacles are rated for 60°C.

Some 30-ampere receptacles are rated for 60°C and some are rated for 75°C. Larger 50-ampere range receptacles are rated for 75°C.

Residential panelboards are generally marked 75°C. The temperature rating of the panelboard has been established as an assembly with circuit breakers in place. Don't use the circuit breakers temperature marking by itself.

Circuit breakers and switches are generally marked "75°C only" or "60°/75°C."

It is interesting to note that in *Section 240-3(d)* there is somewhat of a built-in code compliance limitation for applying small branch circuit conductors. Here we find that the maximum overcurrent protection for No. 14 is 15 amperes, for No. 12 is 20 amperes, and for No. 10 is 30 amperes.

From *Table 310-16* we find that:

No. 14 copper conductors allowable ampacity is:
20 amperes in the 60°C column
20 amperes in the 75°C column
25 amperes in the 90°C column

No. 12 copper conductors allowable ampacity is:
25 amperes in the 60°C column
25 amperes in the 75°C column
30 amperes in the 90°C column

No. 10 copper conductors allowable ampacity is:
30 amperes in the 60°C column
35 amperes in the 75°C column
40 amperes in the 90°C column

But no matter what allowable ampacity the conductor has, the maximum size overcurrent protection for No. 14, No. 12, and No. 10 as stated in *Section 240-3(d)* would keep these conductors from becoming overloaded.

In summary, we cannot arbitrarily check *Table 310-16* to find the allowable ampacity of a given conductor. Conductor ampacity is not a simple "stand alone" issue. We must also consider the temperature limitations of the equipment, such as panelboards, receptacles, snap switches, etc. The lowest maximum temperature rated device in an electrical system is the "weakest link," and it is the "weakest link" that we must base our ultimate conductor ampacity decision on.

Figure 8–9 is a chart for matching conductors and terminations.

COLOR CODING OF CONDUCTORS

When wiring with nonmetallic-sheathed cable (Romex) or armored cable (BX), the colors of the insulation on the conductors are established.

The conductors in nonmetallic-sheathed cable (Romex) are color coded as follows.

Two-wire: one black ("hot" phase conductor)
one white ("grounded" "identified" conductor)
one bare (equipment grounding conductor)

Three-wire: one black ("hot" phase conductor)
one white ("grounded" "identified" conductor)
one red ("hot" phase conductor)
one bare (equipment grounding conductor)

Four-wire: one black ("hot" phase conductor)
one white ("grounded" "identified" conductor)
one red ("hot" phase conductor)
one blue ("hot" phase conductor)
one bare (equipment grounding conductor)

Four-wire nonmetallic sheathed cable and armored cable are not commonly used for house wiring. In fact, it may be impossible to find this type of cable in stock at electrical supply houses, hardware stores, or home centers. It may need to be

TERMINATION MARKING	CONDUCTOR INSULATION TEMPERATURE RATING		
	60°C	75°C	90°C
60°C	OK to use at 60°C ampacity	OK to use at 60°C ampacity	OK to use at 60°C ampacity
75°C	Do not use	OK to use at 75°C ampacity	OK to use at 75°C ampacity
60°C/75°C	OK to use at 60°C ampacity	OK to use at 60°C or 75°C ampacity	OK to use at 60°C or 75°C ampacity
90°C	Do not use	Do not use	OK to use at 90°C ampacity, but only if equipment is rated 90°C. This would be very uncommon in residential applications.

Figure 8–9 This chart shows the insulation temperature ratings for conductors and the temperature markings typically found on electrical terminations.

ordered specially. The cost is very high when compared to two-wire and three-wire cables. It is generally easier and more economical to lay out the wiring using two-conductor and three-conductor cables.

When conductors are installed in a raceway, the electrician is permitted by *Section 210-5* to use any color for the "hot" ungrounded phase conductors except green, green with yellow stripes, white, or natural gray. Figure 8–10 shows the *National Electrical Code®* requirements for color coding of conductors applicable to house wiring.

FLEXIBLE CORDS AND CABLES

Flexible cords and cables are not to be used for premise wiring. They are not an acceptable wiring method as discussed in Unit 9. Flexible cords and cables come under the jurisdiction of *Article 400* in the *National Electrical Code.®*

Table 400-4 is a tabulation of the different types of flexible cords, giving details such as type designation, sizes, number of conductors, insulation type, insulation thickness, braiding, outer covering, and permitted use.

Keeping our coverage focused on residential wiring, we find in *Section 400-7(a)(8)* that flexible

Ungrounded (hot) conductors	Any color except white, natural gray, or green.
	Insulated conductors that are white, natural gray or have three continuous white stripes are to be used as grounded conductors only. There is no provision in the *NEC®* for re-identifying white or natural gray conductors for use as ungrounded conductors in raceways. The only exception to use white or natural gray insulated conductors as ungrounded conductors is for cable wiring.
	See *Sections 200-7(a)(b) & (c)*, and *310-12(c)*
Grounded conductors	**Conductors No. 6 AWG or smaller:** For the conductors entire length, the insulation must be white, natural gray, or have three continuous white stripes on other than green insulation. Re-identification in the field is not permitted when the wiring method is a raceway.
	Conductors larger than No. 6 AWG: For the conductors entire length, the insulation must be white, natural gray, or have three continuous white stripes on other than green insulation.
	At terminations, re-identification may be white paint, white tape, or white heat shrink tubing. This marking must encircle the conductor or insulation. Do this marking at time of installation.
	See *Sections 200-2, 200-6(a) & (b), 200-7(a)(b) & (c), 210-5(a), 310-12 (a)*.
	Fixture wires used as the grounded conductor must be distinguishable from ungrounded fixture wire. A white or natural gray braid or insulation, a tracer, or tinning is acceptable. See *Sections 402-8* and *400-22*. Do not use fixture wires for branch circuit wiring, *Section 402-11.*
Equipment grounding conductors	**Conductors No. 6 AWG or smaller:** No reidentification permitted for conductors No. 6 AWG or smaller. An equipment grounding conductor must be bare, covered or insulated. If covered or insulated, the outer finish must be green or green with yellow stripes. The color must run the entire length of the conductor.
	Conductors larger than No. 6 AWG: At time of installation, at both ends and at every point where the conductor is accessible, do any of the following: for the entire exposed length:
	• strip off the covering or insulation • paint the covering or insulation green • use green tape, adhesive labels, or green heat shrink tubing.
	See *Sections 210-5(b), 250-119, 310-12(b)*

Figure 8–10   A summary of the *National Electrical Code®* color-coding requirements for conductors.

cords and cables are permitted on *"appliances where the fastening means and mechanical connections are specifically designed to permit ready removal for maintenance and repair, and the appliance is intended or identified for flexible cord connection."* The term "identified" means that the appliance would be UL "listed" for a cord connection. If the appliance is not listed for connection with a cord, then it must be connected with a recognized wiring method such as armored cable, flexible metal conduit, etc. For example, a food waste disposer might come with a flexible cord in which case the instructions and marking would indicate that it is all right to make the connection with the cord. The appliance would have the proper type of cord connector to fasten the cord to it.

When a cord is used for the above application, the cord must be equipped with an attachment plug cap so that the power will come from a receptacle. Making the connections directly into an outlet or junction box, sometimes referred to as "hard wiring," is not permitted. This restriction is found in *Section 400-8.*

Flexible cords shall not be used:

• as a substitute for the fixed wiring of a structure.

• where run through holes in walls, ceilings (structural, suspended, dropped), or floors.

• where run through doorways, windows, or similar openings.

• where attached to the building surface.

• where concealed behind building walls, ceilings (structural, suspended, dropped), or floors.

• where installed in raceways.

APPLIANCE HOOK-UPS

Article 422, Part C covers this topic. The rules are much the same as in *Article 400.* Again, we find that flexible cords are permitted for the hookup of specific appliances.

For food waste disposers, a flexible cord is all right to use if:

• the cord is identified for the purpose.

• the cord has a grounding-type attachment plug cap.

• the cord is not shorter than 18 inches and not longer than 36 inches.

• the receptacle is so located that the cord will not be subject to physical damage.

• the receptacle must be accessible.

For built-in dishwashers and trash compactors, a flexible cord is OK to use if:

- the cord is identified for the purpose.
- the cord has a grounding-type attachment plug cap.
- the cord is not shorter than 36 inches and not longer than 48 inches.

- the receptacle is so located that the cord will not be subject to physical damage.
- the receptacle is located in the same space or in the space next to the appliance.
- the receptacle must be accessible.

It is obvious that the *National Electrical Code®* is very restrictive on the use of flexible cords.

REVIEW QUESTIONS

1. Type THHN conductors are probably the most commonly used conductors today. May Type THHN conductors be used by themselves (i.e., not in a raceway or cable)?

   a) Yes

   b) No

   The answer to this question is found in *Section(s)* _____

2. What do the letters AWG mean? _____

3. No. 14 AWG conductors are larger in circular mil area than No. 12 AWG conductors.

   a) True

   b) False

4. A No. 10 AWG conductor has a cross-sectional area of 10,380 circular mils. What is the approximate circular mil area of a No. 6 AWG conductor? (Try to answer this question without referring to *Table 8, Chapter 9, NEC®*

5. In the spaces provided, mark T or F (True or False) for the following statements.

   a) Conductors installed in raceways must be stranded when larger than No. 8.   _____

   b) Conductors installed in raceways must be stranded when No. 8 and larger.   _____

6. What do the letters THHN mean when marked on the insulation of a conductor?

   _____

7. Conductors installed in wet locations such as in underground raceways, and service-entrance conductors that are exposed to the weather must be suitable for _____ locations. The type designation will generally include the letter _____.

8. The insulation on conductors deteriorates when subjected to excessive temperature. What four factors are recognized in the *National Electrical Code*® that have an effect on a conductor's temperature?

   1._____

   2._____

   3._____

   4._____

9. In general, the ampere rating of fuses or circuit breakers must not exceed the ampacity of a conductor. In some instances, the ampere rating of the fuse or circuit breaker does not exactly match the ampacity of the conductor. In the spaces provided, mark T or F (True or False) for the following statements.

   a) Requires that the next smaller ampere rating be used.   _____

   b) Permits the next larger ampere rating to be used.   _____

10. For a given size conductor, which of the following has the *lesser* current-carrying ability?

    a) Aluminum

    b) Copper

11. Terminations that are suitable for use with aluminum conductors will be so marked. In the spaces provided, mark T or F (True or False) for terminal markings indicating that the terminal is suitable for use with an aluminum conductor.

    a) CU   _____

    b) AL   _____

    c) AL/CU   _____

12. You are connecting an electric water heater. The water heater has an upper heating element that will be supplied by one branch circuit (two conductors). The lower heating element will be supplied by another branch circuit (two conductors). The EMT that is installed from the panel to the water heater will, therefore, contain four current-carrying conductors. The penalty for installing four current-carrying conductors in one raceway is that the conductors' ampacity values from *Table 310-16* must be _____ to _____ percent. Choose from the following.

    First blank:

    a) reduced

    b) increased

    Second blank:

    c) 70%

    d) 80%

    e) 90%

13. You are to connect a heat pump. The nameplate on the heat pump is marked "Minimum circuit ampacity 40 amperes." You will install an EMT on top of the flat roof of the home where the outside temperature will reach 115°F in the sun. You will install two Type THHN conductors in the EMT. Using this temperature as the determining factor for selecting the conductors, what is the minimum size Type THHN conductors that are suitable for the load and the anticipated high temperature? The terminations and equipment are suitable for 75°C. Show your calculations.

14. Six current-carrying conductors are installed in a raceway. The temperature is expected to reach 100°F. The connected load is 30 amperes. The terminals and the equipment are marked 75°C. What are the minimum size Type THHN conductors required for this installation?

15. Back at the main electrical panel, the "home runs" from all of the branch circuits are bundled together so as to make a very neat installation above the panel. In the spaces provided, mark T or F (True or False) for the following statements. Refer to *Section 310-15(b)(2)* and *Table 310-15(b)(2)(a)*.
    a) If the "bundling together" is more than 24 inches, the allowable ampacity of the conductors must be reduced as shown in the table. _____
    b) If the "bundling together" is 24 inches or less, the allowable ampacity of the conductors must be reduced as shown in the table. _____
    c) If the "bundling together" is more than 48 inches, the allowable ampacity of the conductors must be reduced as shown in the table. _____

16. One of the most common mistakes electricians make is to apply conductors at their ampacity taken directly from *Table 310-16*. *Section 110-14(c)* provides us with the information needed to properly select conductors based on the terminations and equipment. Remember that a conductor is only as good as its weakest link, which could very well be the terminations. In the spaces provided, mark T or F (True or False) for the following statements.
    a) Equipment rated 100 amperes or less shall be used with conductors that are selected based upon the 60°C ampacity value _____
    b) Equipment rated 100 amperes or less shall be used with conductors that are selected based upon the 75°C ampacity value. _____
    c) Equipment rated 100 amperes or less shall be used with conductors that are selected based upon the 90°C ampacity value. _____

  d) Equipment marked for use with conductors sizes No. 14 through No. 1 shall be used with conductors that are selected based upon the 60°C ampacity value. ———

  e) Equipment marked for use with conductors sizes No. 14 through No. 1 shall be used with conductors that are selected based upon the 75°C ampacity value. ———

  f) Equipment marked for use with conductors sizes No. 14 through No. 1 shall be used with conductors that are selected based upon the 90°C ampacity value. ———

  g) Equipment rated over 100 amperes shall be used with conductors that are selected based upon the 60°C ampacity value. ———

  h) Equipment rated over 100 amperes shall be used with conductors that are selected based upon the 75°C ampacity value. ———

  i) Equipment rated over 100 amperes shall be used with conductors that are selected based upon the 90°C ampacity value. ———

17. Many appliances such as electric ranges and electric clothes dryers are permitted to be connected with flexible plug and cord sets. Flexible cords are not a substitute for regular wiring. Flexible cords are not considered to be a "wiring method." Appliances must be marked that they are suitable for connection using a flexible cord. Permitted uses, restrictions, and other code requirements for flexible cords are found in *Articles* _____ and _____.

Wiring Methods

OBJECTIVES

After studying this unit, you will be able to:

- **Understand the different wiring methods commonly used for house wiring.**
- **Know how to install various wiring methods.**
- **Be able to discuss minimum depths for underground wiring.**
- **Be able to discuss minimum clearances above ground for overhead wiring.**
- **Have a better understanding for old knob-and-tube wiring.**
- **Know how to calculate the minimum size raceway for different sizes and types of conductors in one raceway.**

INTRODUCTION

One of the key ingredients for wiring houses is to select the type of wiring method to be used. In all probability, the wiring method will be nonmetallic-sheathed cable (Romex). However, there are other wiring methods. In this unit we discuss some of the wiring methods generally used for residential wiring. We will also discuss the *National Electrical Code®* rules applicable to these wiring methods.

Chapter 3 of the *National Electrical Code®* covers wiring methods and materials. Chapter 3 starts out with *Article 300,* which contains general requirements that pertain to **all** types of wiring installations. Chapter 3 then goes on to discuss each of the different types of wiring methods. The following text covers some of the more commonly accepted wiring methods used for wiring houses.

NONMETALLIC-SHEATHED CABLE

Nonmetallic-sheathed cable is covered in Article 336 of the *National Electrical Code®.* Nonmetallic-sheathed cable is probably the least expensive, easiest to install, and most widely used wiring method for wiring houses than any other wiring method.

Most electricians still refer to nonmetallic-sheathed cable as *Romex,* named years ago after the Rome Wire and Cable Company.

Nonmetallic-sheathed cable is defined as *"a factory assembly of two or more insulated conductors having an outer sheath of moisture-resistant, flame-retardant, nonmetallic material."*

The conductors in nonmetallic-sheathed cable range in size from No. 14 AWG through No. 2 AWG for copper conductors. If the conductors are aluminum or copper-clad aluminum, the size range is No. 12 AWG through No. 2 AWG.

The conductor combinations are:

- Two-wire: one black, one white, and one bare equipment grounding conductor
- Three-wire: one black, one white, one red, and one bare equipment grounding conductor

Nonmetallic-sheathed cable usually has an equipment grounding conductor that is wrapped with paper or fiberglass. The wrapping acts as a "filler" needed in the manufacturing process. The equipment grounding conductor can be insulated with green insulation. The norm is for the equipment grounding conductor to be bare with wrapping

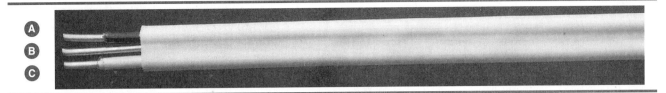

Figure 9–1 A nonmetallic-sheathed cable with (A) a white "grounded" conductor, (B) a bare wrapped "equipment grounding conductor," and (C) a black "ungrounded" (hot) conductor.

around it. Figure 9–1 shows a nonmetallic-sheathed cable with an uninsulated but wrapped equipment grounding conductor.

The equipment grounding conductor is never permitted to be used as a current-carrying conductor. The size of the equipment grounding conductor is found in *Table 250-122.* Note that equipment grounding conductors are the same size as the circuit conductors for 15-, 20-, and 30-ampere ratings.

Underwriters Laboratories lists three types of nonmetallic-sheathed cables:

- Type NM-B has a flame-retardant, moisture-resistant, nonmetallic outer jacket. The conductors have a 90°C (194°F) rating. The ampacity is based on 60°C (140°F).

- Type NMC-B has a flame-retardant, moisture-resistant, fungus-resistant, corrosion resistant, nonmetallic outer jacket. The conductors have a 90°C (194°F) rating. The ampacity is based on 60°C (140°F).

- Type NMS is a special cable assembly that contains power, low-voltage signaling, and communication conductors. This type of cable is used in portions of the wiring for upscale "intelligent" homes such as the Smart House®, as permitted in *Article 780* of the *National Electrical Code.*®

The suffix "B" indicates that the conductors have a 90°C (194°F) rating—really needed because of the extreme heat buildup in attics where the cables are buried in insulation, and subject to the high temperatures encountered with recessed and surface-mounted lighting fixtures. Old style Type NM without the suffix had 60°C (140°F) conductors that just could not stand the heat. Insulation became brittle and broke off. The change to 90°C (194°F) rated conductors solved the problem.

What is nice about the 90°C (194°F) rating is that if it becomes necessary to derate the conduc-

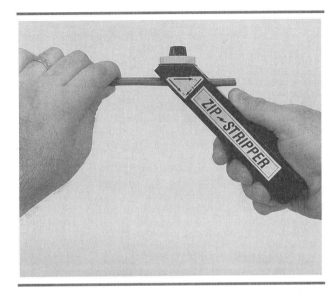

Figure 9–2 One type of stripper that is used to strip the outer jacket from nonmetallic-sheathed cable. (*Courtesy of Seatek Co., Inc.*)

tor's ampacity for high temperatures, the 90° (194°F) ampacity values found in Figure 8–1 (*Table 310-16*) are permitted to be used as the starting point to make the adjustments. After derating, the final derated ampacity must not exceed the ampacity for 60°C (140°F) rated conductors, *Section 336-26.*

The outer jacket of nonmetallic-sheathed cable can be removed by carefully using a knife, or a wire stripper like the one shown in Figure 9–2.

Permitted Uses

Figure 9–3 compares the uses and characteristics of Type NM-B and NMC-B nonmetallic-sheathed cables.

Installation Requirements

Here is a summary of the installation requirements for nonmetallic-sheathed cable as found in *Article 336.* Refer to Figure 9–4 as you read these requirements:

WHERE CAN YOU USE TYPES NM, NMC, AND NMS CABLE?

	TYPE NM	TYPE NMC	TYPE NMS
May be used on circuits of 600 volts or less.	Yes	Yes	Yes
May be run exposed or concealed in dry locations.	Yes	Yes	Yes
May be installed exposed or concealed in damp and moist locations.	No	Yes	No
Has flame-retardant and moisture-resistant outer covering.	Yes	Yes	Yes
Has fungus-resistant and corrosion-resistant outer covering.	No	Yes	No
May be used to wire one- and two-family dwellings, or multifamily dwellings that do not exceed three floors above grade.	Yes	Yes	Yes
May be embedded in masonry, concrete, plaster, adobe, fill.	No	No	No
May be installed or fished in the hollow voids of masonry blocks or tile walls where not exposed to excessive moisture or dampness.	Yes	Yes	Yes
May be installed or fished in the hollow voids of masonry blocks or tile walls where exposed to excessive moisture.	No	Yes	No
In outside walls of masonry block or tile.	No	Yes	No
In inside wall of masonry block or tile.	Yes	Yes	Yes
May be used as service-entrance cable.	No	No	No
Must be protected against physical damage.	Yes	Yes	Yes
May be run in shallow chase of masonry, concrete, or adobe if protected by a steel plate at least 1/16 inch (1.59 mm) thick, then covered with plaster, adobe, or similar finish.	No	Yes	No

Figure 9–3 Summary of uses permitted for Type NM, NMC and NMS. The letter "B" behind these type designations indicates that the conductor insulation is rated 90 degrees C. For example, NM-B, NMC-B, and NMS-B.

1. The cable must be strapped or stapled not more than 12 inches (305 mm) from a box or fitting, *Section 336-18*. Figure 9–5 shows a typical insulated-type staple for securing nonmetallic-sheathed cable. Also illustrated are a one-hole strap for use with large sizes of nonmetallic-sheathed cable and a staple for armored cable (BX). Staple guns that use insulated staples may also be used if available.

2. The "12 inches from a box" stapling requirement can be reduced to 8 inches when using nonmetallic boxes not larger than 2¼" × 4 inches (single-gang device box) where the outer jacket of the nonmetallic-sheathed cable extends at least ½ inch into the box. This is found in *Section 370-17(c), Exception.*

3. More than one nonmetallic-sheathed cable may be run through the same knockout, but only when using nonmetallic boxes, *Section 370-17(c), Exception.*

4. The interval between straps or staples must not exceed 4½ feet (1.37 m), *Section 336-18.*

5. The 4½-foot securing requirement is not needed where nonmetallic-sheathed cable is

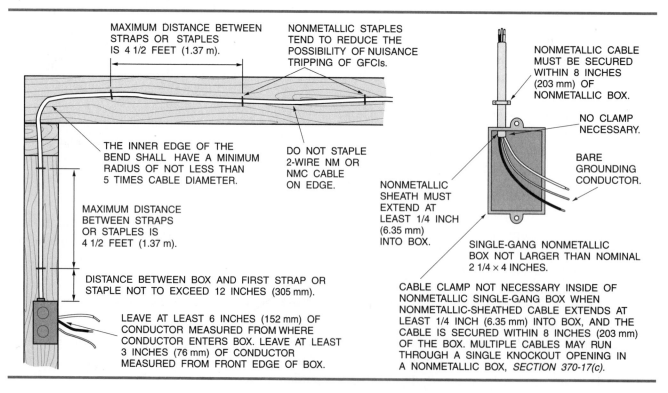

Figure 9–4 Installation requirements for nonmetallic-sheathed cables.

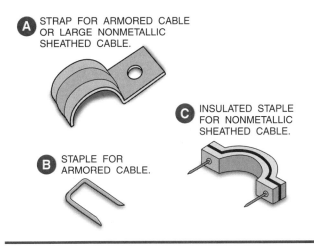

Figure 9–5 (A) One-hole strap used for large nonmetallic-sheathed cables, EMT, and armored cable. (B) Metal staple for nonmetallic-sheathed cable and armored cable. (C) Insulated staple for nonmetallic-sheathed cable and armored cable.

run through holes in framing members such as studs, joists, rafters, *Section 336-18*.

6. The inner edge of a bend shall have a minimum radius not less than 5 times the cable diameter, *Section 336-16*.

7. If passing through a floor where the nonmetallic-sheathed cable might be subject to physical damage, protect the cable with a short length of EMT, rigid metal conduit, Schedule 80 rigid PVC, or other metal pipe. See *Section 336-(b)*. Protect the cable from abrasion on both ends of the protecting conduit (Figure 9–6).

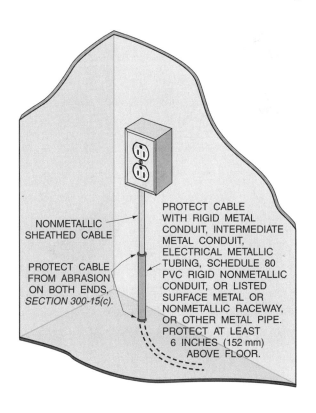

Figure 9–6 Installation requirements for protecting nonmetallic-sheathed cable where passing through a floor.

8. Do not staple flat nonmetallic-sheathed cables on edge, *Section 336-18*. See Figure 9-7.

9. When installing nonmetallic-sheathed cables in unfinished basements, refer to *Section 336-6(c)*. Figure 9–8 illustrates the requirements.

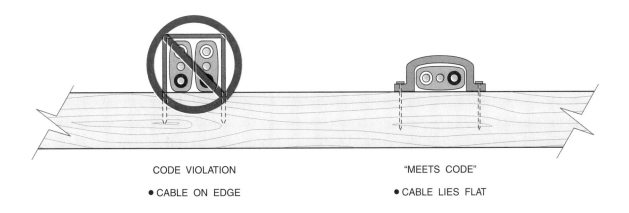

CODE VIOLATION
• CABLE ON EDGE

"MEETS CODE"
• CABLE LIES FLAT

Figure 9–7 Do not staple two-wire nonmetallic-sheathed cables on edge, *Section 310-18*.

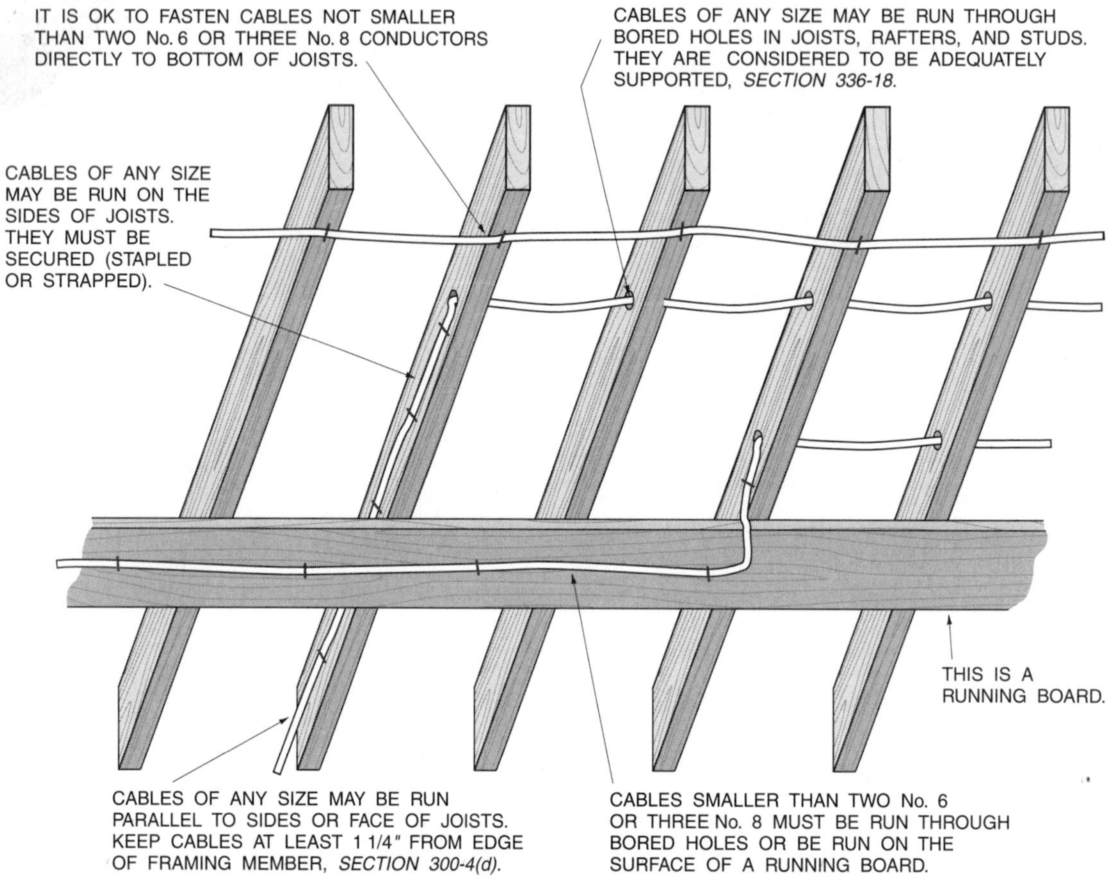

IT IS OK TO FASTEN CABLES NOT SMALLER THAN TWO No. 6 OR THREE No. 8 CONDUCTORS DIRECTLY TO BOTTOM OF JOISTS.

CABLES OF ANY SIZE MAY BE RUN THROUGH BORED HOLES IN JOISTS, RAFTERS, AND STUDS. THEY ARE CONSIDERED TO BE ADEQUATELY SUPPORTED, SECTION 336-18.

CABLES OF ANY SIZE MAY BE RUN ON THE SIDES OF JOISTS. THEY MUST BE SECURED (STAPLED OR STRAPPED).

THIS IS A RUNNING BOARD.

CABLES OF ANY SIZE MAY BE RUN PARALLEL TO SIDES OR FACE OF JOISTS. KEEP CABLES AT LEAST 1 1/4" FROM EDGE OF FRAMING MEMBER, SECTION 300-4(d).

CABLES SMALLER THAN TWO No. 6 OR THREE No. 8 MUST BE RUN THROUGH BORED HOLES OR BE RUN ON THE SURFACE OF A RUNNING BOARD.

Figure 9–8 *Section 336-6(c)* shows the rules for installing nonmetallic-sheathed cable in unfinished basements.

GUARD STRIPS REQUIRED WHEN CABLES ARE RUN ACROSS THE TOP OF JOISTS.

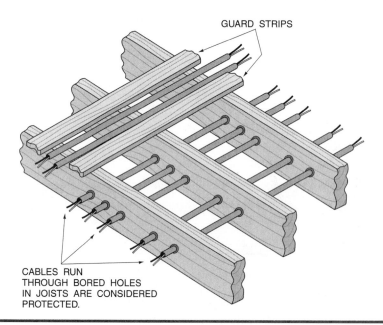

GUARD STRIPS

CABLES RUN THROUGH BORED HOLES IN JOISTS ARE CONSIDERED PROTECTED.

Figure 9–9 Detailed sketch showing guard strips protecting cables in attics. The cables that are run through the holes in the joists are considered adequately protected. See *Sections 336-6(d)* and *333-12*.

10. Because the wiring in accessible attics and accessible roof spaces is subject to physical abuse from storage of boxes, walking across the joists, and possible installation of flooring, protection from physical damage is required. See *Section 336-6(d)*.

Figure 9–9 shows a detail of guard strips installed on the top of floor joists as well as running cables through bored holes. In accessible attics and accessible roof spaces, guard strips are required where the cables (1) are run across the top of floor joists, (2) are run across the face of studs, or (3) are run on rafters within 7 feet (2.13 m) of the floor or floor joist. Guard strips are not required if the cables are (4) run along the sides of rafters, studs, and floor joists. This is illustrated in Figure 9–10(A).

In attics and roof spaces not accessible by permanent stairs or ladders, guard strips are required only within 6 feet (1.83 m) of the nearest edge of the scuttle hole or entrance. This is illustrated in Figure 9–10(B).

Running the cables in the space where the

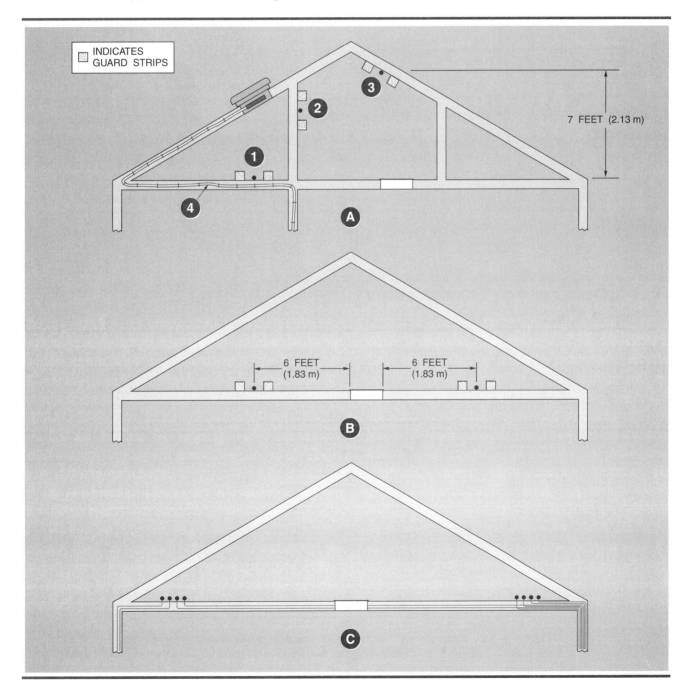

Figure 9–10 Protection of nonmetallic-sheathed cable in accessible attics and roof spaces. See *Sections 336-6(d)* and *333-12*.

ceiling joists and the roof rafters meet as depicted in Figure 9–10(C) provides the protection needed. It is difficult, if not impossible, for a person to crawl into this space or to store cartons in the space because of the low clearance.

Do not "bundle" nonmetallic-sheathed cables together for distances longer than 24-inches, otherwise it may be necessary to derate the conductors ampacity because of possible heat buildup, *Section 310-15(b)(2)*. This was shown in Figure 8–8.

Holes drilled through wood framing members must be drilled so the nearest edge of the bored hole is at least 1¼ inches (31.8 mm) from the face of the framing member. If this distance cannot be attained, then install protective metal plates at least ⅟₁₆ inch (1.59 mm) thick. See *Section 300-4*. This is illustrated in Figure 9–11.

The protective metal plate requirement is also applicable to cables laid in shallow grooves or notches (Figure 9-12).

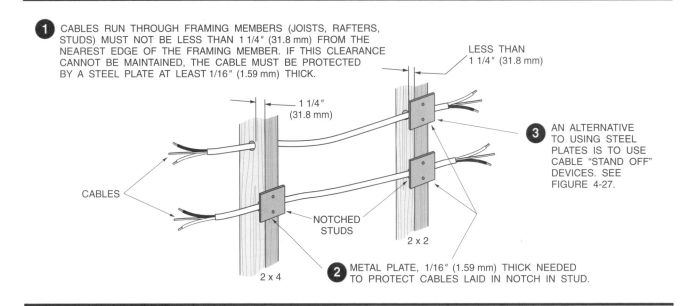

1 CABLES RUN THROUGH FRAMING MEMBERS (JOISTS, RAFTERS, STUDS) MUST NOT BE LESS THAN 1 1/4″ (31.8 mm) FROM THE NEAREST EDGE OF THE FRAMING MEMBER. IF THIS CLEARANCE CANNOT BE MAINTAINED, THE CABLE MUST BE PROTECTED BY A STEEL PLATE AT LEAST 1/16″ (1.59 mm) THICK.

LESS THAN 1 1/4″ (31.8 mm)

1 1/4″ (31.8 mm)

3 AN ALTERNATIVE TO USING STEEL PLATES IS TO USE CABLE "STAND OFF" DEVICES. SEE FIGURE 4-27.

CABLES

NOTCHED STUDS

2 x 2

2 x 4

2 METAL PLATE, 1/16″ (1.59 mm) THICK NEEDED TO PROTECT CABLES LAID IN NOTCH IN STUD.

Figure 9–11 **Methods of protecting nonmetallic-sheathed cables so nails and screws will not damage the cables,** *Section 300-4.*

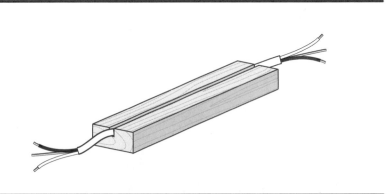

Figure 9–12 **Example of how a solid wood framing member or a styrofoam insulating block might be grooved to receive the nonmetallic-sheathed cable. A ⅟₁₆-inch thick steel plate must cover the groove for the entire length. The alternative is to have the groove deep enough to provide for a 1¼-inch free space for the cable. See** *Section 300-4(e).*

An alternative oftentimes used where the 1¼-inch (31.8 mm) clearance cannot be maintained is to install standoffs of the type shown in Figure 9–13. This keeps the cables away from the stud, and makes it highly unlikely that a nail or screw driven "slightly off course" would penetrate the cables.

Figures 9–14 and 9–15 are examples of wall construction that may require the use of ⅟₁₆" steel

plates or "stand-offs." Figure 9–16 shows wall construction that will not require the use of ⅟₁₆" steel plates or "stand-offs."

Section 300-22 is extremely strict as to what types of wiring methods are permitted in ducts. However, the requirements have been relaxed somewhat in *Section 300-22(c), Exception* to permit Types NM and NMC nonmetallic-sheathed cables to

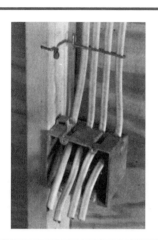

Figure 9–13 Devices that can be used to meet the Code requirement to maintain a 1¼-inch clearance from framing members.

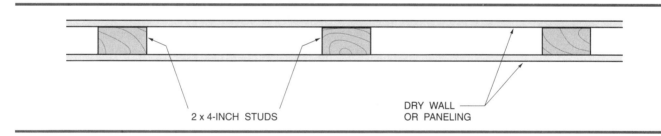

2 x 4-INCH STUDS

DRY WALL OR PANELING

Figure 9–14 When 2 × 4-inch studs are installed "flat," such as might be found in nonbearing partitions, nonmetallic-sheathed cables running parallel to or through the studs must be protected with ⅟₁₆-inch metal plates, or use "standoffs."

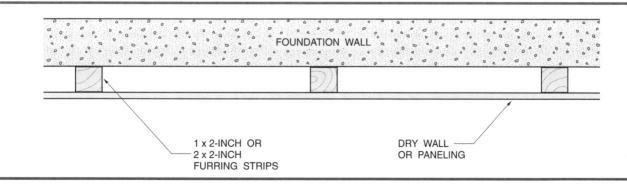

FOUNDATION WALL

1 x 2-INCH OR 2 x 2-INCH FURRING STRIPS

DRY WALL OR PANELING

Figure 9–15 When 1 × 2-inch or 2 × 2-inch furring strips are installed on the surface of a basement wall, the ⅟₁₆–inch metal plate protection or "standoffs" may be required for nonmetallic-sheathed cable installed in the stud spaces, or if the cable is run through the furring strips. See *Section 300-4(a)* and *(d)*.

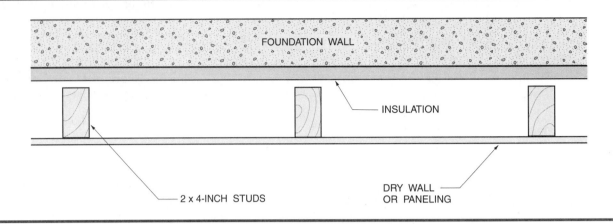

Figure 9–16 Finished basement walls can be constructed in such a way that the ¹⁄₁₆-inch steel plate protection or "standoffs" may not be necessary. Careful planning of the wiring system is important.

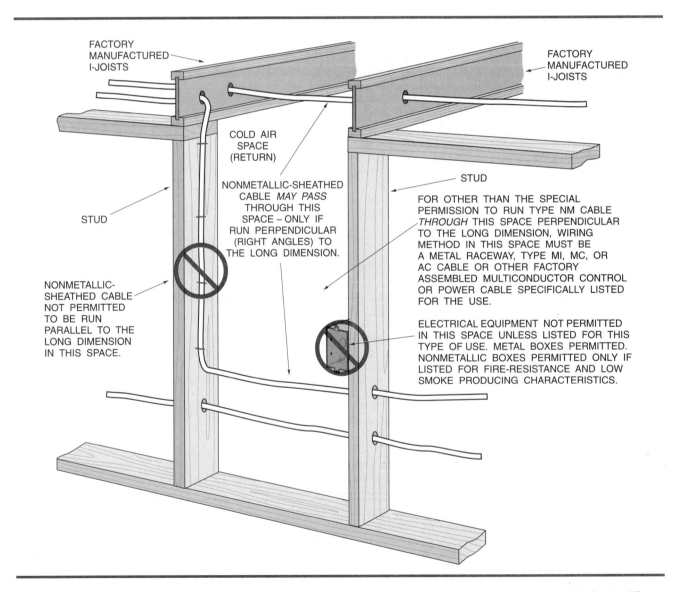

Figure 9–17 Nonmetallic-sheathed cable may pass through cold air return joist and stud spaces in dwellings, but only if the cable is run in right angles to the long dimension of the space, *Section 300-22 (c), Exception.*

pass through such space only if the cables run perpendicular (at a right angle) to the long dimension of joist and stud spaces used as cold air returns. This is illustrated in Figure 9–17.

For metal studs where there is a likelihood that nails or screws could be driven into the cable, the rules for protecting nonmetallic-sheathed cable from damage are the same as for wood construction with one additional requirement. Grommets (bushings) must be installed as shown in Figure 9–18. These grommets are not required when the wiring method is armored cable, EMT, flexible metal conduit, or electrical nonmetallic tubing. Refer to *Section 300-4(b)*.

For concealed wiring in remodel work where cables are "fished" between outlet boxes and other accessible points, *Section 336-18, Exception No. 1* and *Section 300-4(d)* exempt the requirements for supporting the cable and 1⁄16-inch steel plate protection.

MULTISTORY BUILDINGS

Nonmetallic-sheathed cable may be installed in any one- and two-family dwelling regardless of how many floors, *Section 336-4(1)*.

Nonmetallic-sheathed cable is not permitted in multifamily dwellings or other structures that exceed three floors above grade.

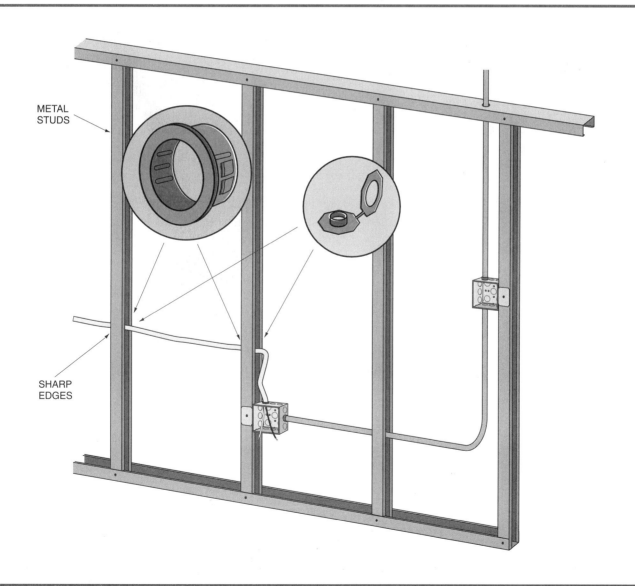

METAL STUDS

SHARP EDGES

Figure 9–18 *Section 300-4(b)* requires protection for nonmetallic-sheathed cables where run through holes in metal framing members.

Section 336-5(a)(1) defines the first floor of a building as *"that floor that has 50% or more of the exterior wall surface area level with or above finished grade."* The *NEC®* allows one additional level for parking, storage, or similar use when such level is not used for human habitation. Walk around the entire building to determine the 50% criteria. Refer to Figure 9–19.

SECURING NM CABLES TO PANEL

Although nonmetallic-sheathed cables are usually secured to a panel, box, or cabinet with a cable clamp or connector, the *Exception to Section 373-5(c)* provides an alternate method for running nonmetallic-sheathed cables into surface-mounted enclosures. This exception is for surface-mounted enclosures only. See Figure 9–19A.

Building Codes

When complying with the *National Electrical Code,®* be sure to also comply with the Uniform Building Code. Building codes are concerned with the weakening of framing members when drilled or notched. Meeting the requirements of the *National Electrical Code,®* can be in conflict with building codes. Here is a recap of some of the key building code requirements that electricians should be aware of.

Studs

- Cuts or notches in exterior or bearing partitions shall not exceed 25 percent of the stud's width.

- Cuts or notches in nonbearing partitions shall not exceed 40 percent of the stud's width.

- Bored or drilled holes in any stud
  — shall not be greater than 40 percent of the stud's width
  — shall not be closer than 5/8 inch to the edge of the stud
  —shall not be located in the same section as a cut or notch

Joists

- Notches on the top or bottom of joists shall not exceed one-sixth the depth of the joist, and shall not be located in the middle third of the span.

- Holes bored in joists shall not be within 2 inches of the top or bottom of the joist, and the diameter of any such hole shall not exceed one-third the depth of the joist.

Engineered Wood Products

Today, we find many homes constructed with engineered wood I-joists and beams. The strength of these factory-manufactured framing members conform to various building codes. It is critical that certain precautions be taken relative to drilling and cutting these products so as not to weaken them. In general:

- Do not drill or notch beams.

- Do not cut, drill, or notch flanges. Flanges are the top and bottom pieces.

- Do not cut holes too close to supports or to other holes.

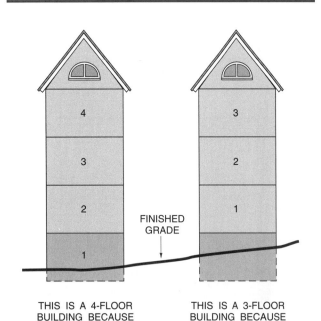

THIS IS A 4-FLOOR
BUILDING BECAUSE
MORE THAN 50% OF
THE FIRST FLOOR IS
ABOVE FINISHED GRADE.

THIS IS A 3-FLOOR
BUILDING BECAUSE
LESS THAN 50% OF
THE FIRST FLOOR IS
ABOVE FINISHED GRADE.

Figure 9–19 Illustration of how *Section 336-5(a)(1)* defines the first floor of a building.

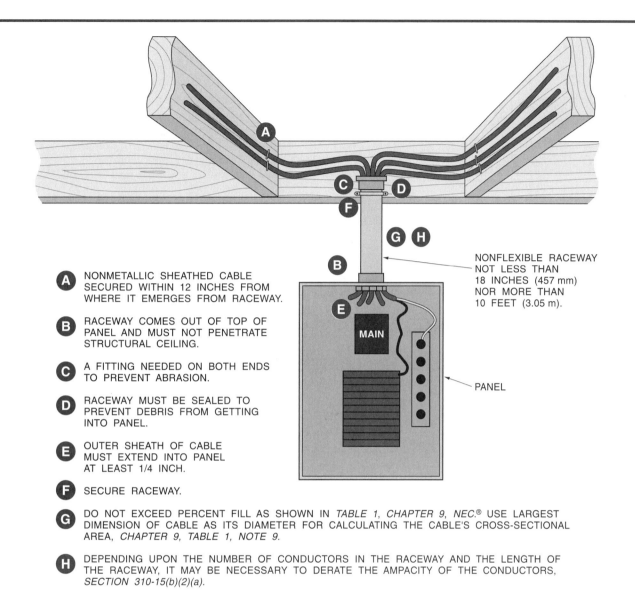

A NONMETALLIC SHEATHED CABLE SECURED WITHIN 12 INCHES FROM WHERE IT EMERGES FROM RACEWAY.

B RACEWAY COMES OUT OF TOP OF PANEL AND MUST NOT PENETRATE STRUCTURAL CEILING.

C A FITTING NEEDED ON BOTH ENDS TO PREVENT ABRASION.

D RACEWAY MUST BE SEALED TO PREVENT DEBRIS FROM GETTING INTO PANEL.

E OUTER SHEATH OF CABLE MUST EXTEND INTO PANEL AT LEAST 1/4 INCH.

F SECURE RACEWAY.

G DO NOT EXCEED PERCENT FILL AS SHOWN IN *TABLE 1, CHAPTER 9, NEC.*® USE LARGEST DIMENSION OF CABLE AS ITS DIAMETER FOR CALCULATING THE CABLE'S CROSS-SECTIONAL AREA, *CHAPTER 9, TABLE 1, NOTE 9.*

H DEPENDING UPON THE NUMBER OF CONDUCTORS IN THE RACEWAY AND THE LENGTH OF THE RACEWAY, IT MAY BE NECESSARY TO DERATE THE AMPACITY OF THE CONDUCTORS, *SECTION 310-15(b)(2)(a).*

NONFLEXIBLE RACEWAY NOT LESS THAN 18 INCHES (457 mm) NOR MORE THAN 10 FEET (3.05 m).

PANEL

MAIN

Figure 9–19A   An alternate method for running nonmetallic-sheathed cables into a surface-mounted panel. This is permitted in *Section 373-5(c), Exception.*

- For multiple holes, the amount of web to be left between holes must be at least twice the diameter of the largest adjacent hole. The web is the piece between the flanges.

- Holes not over 1½" usually can be drilled anywhere in the web.

- Do not hammer on the web except to remove knockout prescored holes.

- Check and follow the manufacturers instructions.

Connectors for Nonmetallic-Sheathed Cable

Outlet boxes and device boxes designed for use with nonmetallic-sheathed cable generally have internal cable clamps supplied with the box. Nonmetallic-sheathed cables can be secured to outlet boxes and device boxes with individual connectors.

Connectors for nonmetallic-sheathed cable come in many varieties. The most common is shown in Figure 9–20. The question continues to be asked: May more than one nonmetallic-sheathed cable be inserted into one connector? Unless the connector is

Figure 9–20 A ½-inch connector for nonmetallic-sheathed cable. Larger connectors are available for use with large sizes of nonmetallic-sheathed cable.

specifically "listed" for use with more than one cable, the rule is *one cable—one connector.*

Plastic nonmetallic-sheathed cable connectors are available that simply snap into a ½-inch knock-out. The nonmetallic-sheathed cable is slipped through the snap-in connector. No screws are needed. The cable is securely held in place by the friction of the flexible plastic tabs on the inside of the snap-in connector, very similar to the tabs that hold cable into a nonmetallic switch box or non-metallic outlet box. They are UL "listed" for one or two cables.

ARMORED CABLE

Armored cable is covered in *Article 333* of the *National Electrical Code.®*

Underwriters Laboratories Inc. tests armored cable per their Standard No. 4. Because UL standards have evolved in numerical order, it is apparent that the armored cable standard is one of the first and oldest standards to be developed. Armored cable first appeared in the 1903 edition of the *NEC.®*

Construction

The *National Electrical Code®* defines *Type AC cable* as a *"fabricated assembly of insulated conductors in a flexible metallic enclosure."* Most electricians still refer to armored cable as BX, a term supposedly derived from an abbreviation of the Bronx in New York City where armored cable was once manufactured. At one time, BX was a trademark owned by the General Electric Company, but over time, the term has become a generic term.

The armor can be steel or aluminum. If aluminum, the words "Aluminum Armor" will appear on the armored cables marking tape **and** on the tag attached to each coil. Figure 9–21 shows the details of how armored cable is constructed.

The conductors in armored cable range in size from No. 14 AWG through No. 1 AWG for copper conductors. If the conductors are aluminum or cop-per-clad aluminum, the size range is No. 12 AWG through No. 1 AWG.

The conductor combinations are:

- Two-wire: one black, one white

- Three-wire: one black, one white, one red

For special situations, armored cable is available with four conductors (black, white, red, blue) and even five conductors (black, white, red, blue, and an additional green insulated equipment grounding conductor). This additional green insulated equipment grounding conductor is used when an isolated equipment ground is required, as might be the case where a significant amount of computer/data processing equipment is connected. These are referred to as nonlinear loads. Nonlinear loads are of little concern in house wiring, but because armored cable is used for commercial installations where an isolated equipment ground might be required, it is mentioned here.

The conductors used in Type AC cable are insulated with thermosetting material. Thermosetting

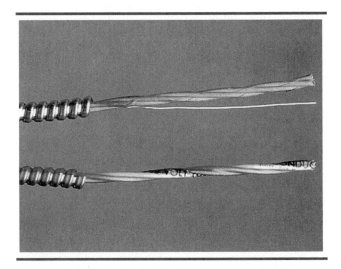

Figure 9–21 Details showing the construction of armored cable. (*Courtesy of AFC Cable Systems, Inc.*)

material **will not** soften, flow, or distort appreciably when subjected to sufficient heat and pressure. Examples are rubber (RH, RHH, RHW) and neoprene (XHH, XHHW) insulation.

The conductors used in Type ACT cable are insulated with thermoplastic material. Thermoplastic material **will** soften, flow, or distort appreciably when subjected to sufficient heat and pressure. Examples are insulation such as THHN, THHW, THW, THWN, and TW.

Adding a suffix further describes armored cable:

- No suffix: The conductors are rated 60°C (140°F).
  Example: Type ACT.

- Suffix "B": Conductors are rated 90°C (194°F). Use ampacity of 60°C (140°F) conductors.
  Example: Type ACTB.

- Suffix "H": Use ampacity of 75°C (167°F) conductors.
  Example: Type ACTH.

- Suffix "HH": Conductors are rated 90°C (194°F).
  Example: Type ACTHH.

Use the ampacity of 90°C (194°F) conductors for derating and temperature correction factors. Use ampacity of 60°C or 75°C conductors for maximum allowable ampacity when the terminations of the equipment are marked 60°C or 75°C. This was covered in Unit 8.

Currently, there is no 600-volt or less electrical equipment marked higher than 75°C. You might want to review *Section 110-14(c)* at this time.

As with nonmetallic-sheathed cable, even though the conductors might have a 90°C (194°F) rating, the ampacity must be based on 60°C (140°F) conductors. Most all armored cable used in residential applications is Type ACTHH.

Bonding Wire in Armored Cable

Section 250-2(d) requires that a ground path be capable of carrying *any* value of fault current that it might be called upon to carry. To ensure that the impedance for a given length of armored cable meets UL standards, an internal bonding strip is used. This bonding strip is in direct contact with the metal armor for the entire length of the cable. The metal armor and the bonding strip are actually in "parallel." In combination, the metal armor and the bonding strip keep the impedance to the required low values. The bonding strip merely needs to be

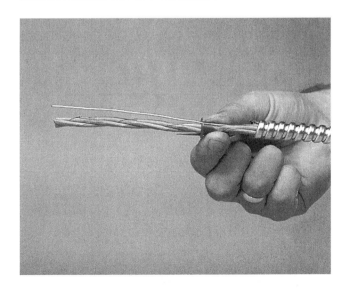 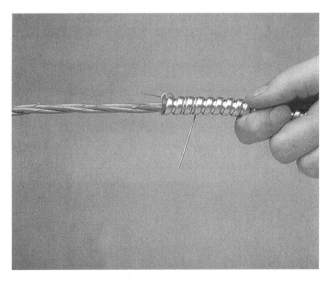

Figure 9–22 These photographs show how to insert the anti-short bushing that keeps the conductors from being cut by the sharp edges of the armor, and what to do with the bonding strip. The bonding strip, when folded back, holds the anti-short bushing in place. (*Courtesy of AFC Cable Systems, Inc.*)

folded back as illustrated in Figure 9–22. The bonding strip must not be used as a "grounded neutral conductor." *It is an internal bonding strip— nothing more!*

This illustration also shows how to insert the "anti-short," a small insulator that keeps the conductors from being cut by the edge of the armored cable.

Connectors for Armored Cable

A connector or fitting is required where armored cables terminate.

Figure 9–23 shows two types of armored cable connectors. One is a straight connector, the other a 90° connector. These connectors can also be used with ⅜-inch flexible metal conduit.

Set screw connectors are not permitted to be used with aluminum armored cable.

Removing the Metal Armor

The armor must be carefully cut using a side cutter pliers, tin snips, a hacksaw (Figure 9–24), or a cable armor cutter, (Figure 9–25).

Permitted Uses and Installation Requirements For Armored Cable

Here are some of the highlights of permitted uses and installation requirements for armored cable. Many of the requirements are the same as for nonmetallic-sheathed cable.

- All right for use on 600 volts or less
- May be used for open and concealed work in dry locations
- May be run through walls and partitions
- May be embedded in the plaster finish on masonry walls, or run through the hollow spaces on such walls if these locations are not considered damp or wet

  Masonry walls in direct contact with the earth are considered to be a wet location. See the definitions of locations (damp, dry, and wet) in *Article 100* of the *NEC.*®

- Must be secured within 12 inches (305 mm) from every outlet box, junction box, cabinet, or fitting, and at intervals not exceeding 4½ feet (1.37 m). Securing is not necessary when the armored cable is "fished" through walls and ceilings.
- The 4½-foot securing requirement is not needed where armored cable is run horizontally through holes in wood or metal framing members.
- The 12-inch and 4½-foot securing requirements are waived when the armored cable not over 6 feet (1.83 m) long is used as a "fixture whip" to connect lighting fixtures or equipment above an accessible ceiling. Where armored cable is exposed, the 12-inch requirement is waived for lengths not over 24 inches (610 mm) used to connect equipment where flexibility is

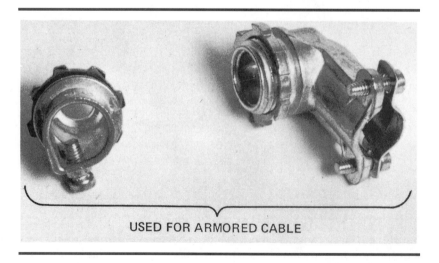

USED FOR ARMORED CABLE

Figure 9–23 Typical connectors for use with armored cable and ⅜-inch flexible metal conduit.

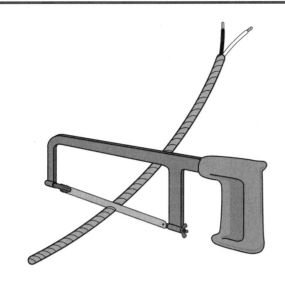

Figure 9–24 Using a hacksaw to cut through a raised convolution of the cable armor at a slight angle, being careful not to cut too deep so as to cut into the insulation on the conductors. A partial cut, then bending the armor, will break the armor off at the cut.

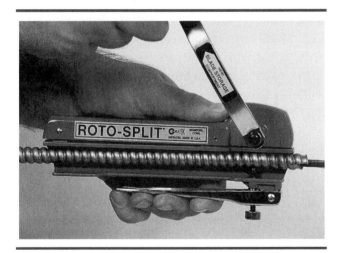

Figure 9–25 A cutter that precisely cuts the outer armor to the proper depth, making it easy to remove the armor with a few turns of the handle. (*Courtesy of Seatek Co., Inc.*)

needed.

- Must not be bent to a radius of less than five times the diameter of the armored cable.

- Must have an approved plastic insulating bushing (anti-short) at the cable ends to protect the conductor insulation.

- May be used where the dwelling or structure

exceeds three floors above grade.

- In attics or roof spaces, running boards or guard strips are required, the same as for nonmetallic-sheathed cables. Refer back to Figures 9–9 and 9–10.

- Where subject to possible damage by nails or screws, ¹⁄₁₆-inch steel plates or "stand-offs" are required, the same as for nonmetallic-sheathed cable. See *Section 300-4.* The protective metal plate requirement is also applicable to cables laid in shallow grooves or notches. Refer back to figures 9–11 and 9–12.

- May be installed in cold air ducts, but only where run at right angles to the long dimension of the space. This is the same requirement as for nonmetallic-sheathed cable. This is shown in Figure 9–17.

- May not be installed underground.

- May not be buried in masonry, concrete, or fill of a building during construction.

- May not be installed where exposed to the weather.

SERVICE-ENTRANCE CABLE

Service-entrance cable is covered in *Article 338* of the *National Electrical Code.*®

Service-entrance cable is defined as *"a single conductor or multi-conductor assembly provided with or without an overall covering, primarily used for services."*

Type SE service-entrance cable is for aboveground use. Where permitted by local electrical codes, SE cable is most often run from the utility's service-point, to the meter base, to the main panel. The outer jacket is suitable where the cable is exposed to the sun. This information is found in the UL White Book. Figure 9–26 shows a Type SE service-entrance cable.

Type USE service-entrance conductors are for underground use. Type USE service-entrance conductors are not permitted for premise wiring, or above-ground wiring except to terminate in service-entrance equipment or metering equipment. Where emerging from the ground to terminate in a meter base, Type USE conductors are protected from physical damage by running the conductors in a raceway,

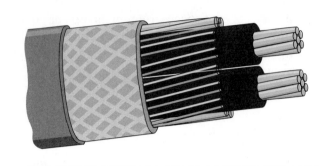

Figure 9–26 A Type SE service-entrance cable. This cable contains two insulated conductors, and one "wrap-around" bare neutral conductor. There is a reinforcement tape plus the final sunlight-resistant outer jacket.

such as from the bottom of a meter base to a point below the ground line as illustrated in Figure 15–5, or from the bottom of a meter pedestal as illustrated in Figure 15–3. See *Section 338-2.* Multiple Type USE conductors are available prebundled in a cable assembly. Figure 9–27 shows a USE conductor.

Some service-entrance cables have a reduced size neutral in which case the cable is so marked. This information is found in *Section 338-5.* The neutral conductor for services and feeders in residential installations is permitted to be smaller than the ungrounded conductors, *Section 310-15(b)(6).*

The UL standards state that where an uninsulated neutral conductor is smaller than the insulated conductors in a service-entrance cable, the current-carrying capacity of the smaller uninsulated neutral

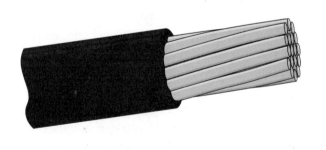

Figure 9–27 A Type USE service-entrance conductor. The insulation is abrasion-, moisture-, heat-, and sunlight-resistant suitable for direct burial. Utility companies generally use Type USE conductors that are prebundled cable assemblies having the proper number of conductors.

conductor is considered to be the same as for the insulated conductor.

Section 338-3 permits Type SE cable for interior wiring to hook up major electrical appliances, such as a range, a wall-mounted oven, a counter-mounted cooktop, or clothes dryer. Type SE cable may also be used as a feeder to a subpanel. For these applications, all of the circuit conductors must be insulated. The equipment grounding conductor in these cables is permitted to be bare.

For many years, there was a cost advantage of using SE cable with two insulated conductors and a bare neutral conductor for the branch circuits supplying ranges and dryers. Previous editions of the *NEC®* permitted the frames of ranges and dryers to be grounded to the neutral conductor. Since the 1996 edition of the *National Electrical Code,®* this is no longer permitted. Now, SE cable and nonmetallic-sheathed cable with insulated grounded and ungrounded conductors are required. The bare equipment grounding conductor in these cables is used to ground the frames of these appliances.

The choice of whether to install SE cable or NMC cable becomes one of cost. Type SE cable with aluminum conductors costs less than Type NMC cable with copper conductors. In some parts of the country, nonmetallic-sheathed cable is used for hooking up these appliances. In other parts of the country, Type SE cable is used.

Section 338-4 requires that when SE cable is used for interior wiring, the installation much comply with the same requirements as nonmetallic-sheathed cable regarding securing and physical protection.

When installing SE cable for interior wiring, the ampacity of SE conductors is found using the 75°C column of *Table 310-16,* assuming the equipment is marked as suitable for use with 75°C conductors. See *Section 110-14(c).*

For 120/240-volt, 3-wire, single-phase dwelling services and feeders, the ampacity of the SE conductors is permitted to be the special values found in *Table 310-15(b)(6).* See Figure 16–1.

If the service-entrance cable has an uninsulated grounded conductor (neutral), a final outer nonmetallic covering, and no conductor in the cable operates over 150 volts to ground as is the case in house wiring, its use is restricted to services or to supply another building on the same property.

To avoid damage to the cable where the cable is bent, the inner edge of the cable shall have a radius of not less than five times the diameter of the cable, *Section 338-6.*

Where service-entrance cable is run from the outside to the inside, passing through a wall, the penetration must be sealed and the cable must be protected using a *sill plate* such as illustrated in Figure 9–28.

ELECTRICAL METALLIC TUBING

Electrical metallic tubing, commonly referred to as EMT is covered in *Article 348* of the *National Electrical Code.*®

In residential wiring, EMT is quite often used where the wiring will be exposed, such as in basements, garages, and as service-entrance raceways. Some communities such as the City of Chicago and the surrounding counties do not permit nonmetallic-sheathed cable or armored cable to be used for house wiring for new construction, in which case EMT is used as the wiring method.

EMT is a thin-walled metal raceway that is not threaded (Figure 9–29).

EMT Fittings

Because EMT is not threaded, threadless fittings such as connectors and couplings are used. Figure 9–30 shows indentor- (crimp-) type fittings. This type of fitting requires the use of a crimping tool (Figure

Figure 9–28   A sill plate used to protect service-entrance cable where it enters a building.

Figure 9–29   Electrical metal tubing, commonly referred to as EMT. (*Courtesy of Allied Tube and Conduit*)

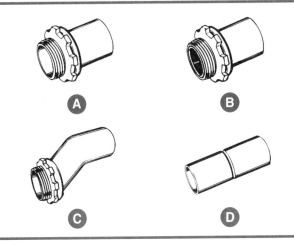

Figure 9–30   (A) Indentor (crimp) type EMT straight connector, (B) an insulated throat straight connector, (C) an offset connector, and (D) a coupling. Offset connectors have just the right amount of offset to change from a flat surface to a knockout in a box without having to bend an offset in the EMT with a bender. (*Courtesy of Hubbell Electrical Products, a Division of Hubbell Incorporated [Delaware]*)

Figure 9–31   A typical indentor (crimping) tool. (*Courtesy of Hubbell Electrical Products, a Division of Hubbell Incorporated [Delaware]*)

9–31). To assure a good tight fit, two crimps, 90 apart are needed. This puts four "indents" into the fitting.

Figure 9–32 shows set screw types fittings that require a screwdriver to tighten the screws. Figure 9–33 shows compression-type fittings that require a wrench or pliers to tighten the end nut. Because of

the compression ring in the end nut that digs into the EMT when the end nuts are properly tightened, compression-type fittings are concretetight and waterproof. You will need to look on the fitting or on the carton to determine if the fitting is suitable for use in poured concrete or exposed to rain. The marking will indicate "raintight" and/or "concretetight."

Bending EMT requires the use of a bender, such as shown in Figure 9–34. When bending EMT, a good bender will not change the internal diameter of the EMT. The *NEC®* in *Table 346-10* shows the minimum radius for bends in EMT. EMT benders available at electrical supply houses conform to the *NEC®* bending requirements. Manufactured 90° and 45° elbows are also available.

For residential use, here are some of the key requirements for EMT:

- May be used for both concealed and exposed work.

- May be buried in concrete or masonry above grade without any supplemental corrosion protection.

- May be buried in concrete or masonry below grade, or underground in contact with the earth only when the EMT has supplemental corrosion protection. Look at the labeling on the EMT to

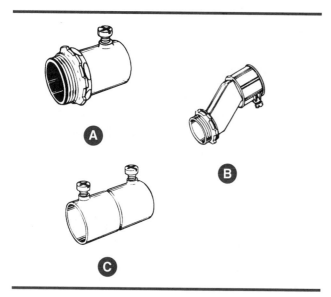

Figure 9–32   Set screw-type EMT fittings. Illustrated are (A) a straight connector, (B) an offset connector, and (C) a coupling. (*Courtesy of Hubbell Electrical Products, a Division of Hubbell Incorporated [Delaware]*)

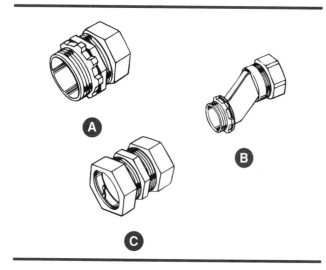

Figure 9–33   Compression-type EMT fittings. Illustrated are (A) a straight connector, (B) an offset connector, and (C) a coupling. (*Courtesy of Hubbell Electrical Products, a Division of Hubbell Incorporated [Delaware]*)

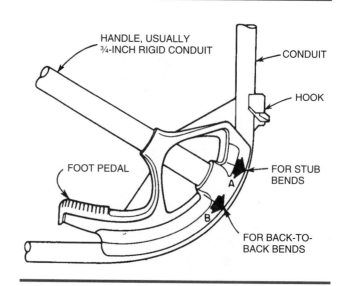

Figure 9–34   A typical EMT bender. Instructions on how to make bends and offsets are furnished with the bender.

see what sort of corrosion protection it has. Here again it is important to know your local electrical code requirements. Some codes do not allow EMT to be used in direct contact with the earth, in concrete slabs or floors poured on the earth, or in exterior concrete walls that are below grade.

- May **not** be installed in cinder concrete or cinder fill where subject to moisture unless protected on all sides by a layer of non-cinder concrete at least 2 inches (50.8 mm) thick.

- In wet locations, secure EMT with straps, screws, and bolts that are corrosion resistant.

- May be installed where it will not be subject to severe physical damage. Normally, residential wiring installations are not considered to be subject to severe physical damage.

- Shall not have more than the equivalent of four 90 bends between pull points. Simply stated, the sum of all bends and offsets must not exceed 360°.

- Must be secured within 3 feet (914 mm) of each outlet box, junction box, device box, cabinet, or conduit body (Figure 9–35).

- Must be secured at least every 10 feet (3.05 m) (see Figure 9–25).

- Ream out the rough edges remaining on the inside of EMT after cutting so conductors will not be damaged.

- Horizontal runs through bored holes in framing members (studs, joists, rafters) are considered

to be adequately supported when the EMT is securely fastened within 3 feet (914 mm) of each outlet box, junction box, device box, cabinet, conduit body.

- When all conductors are the same size and type, the number of conductors permitted in the EMT is found in *Table C1*, in *Appendix C* of the *NEC.*®

- When conductors are different sizes, refer to *Chapter 9* in the *National Electrical Code.*® The number of conductors permitted is determined using the percentage fill shown in *Table 1*, the dimensions of the EMT found in *Table 4*, and the dimensions of the conductors shown in *Tables 5, 5A*, and *8*. Sample calculations are shown later in this unit.

Protecting Conductors

Sections 300-4(f) and *373-6(c)* require that smoothly rounded insulating fittings or equivalent be used where No. 4 AWG or larger conductors enter a raceway, box, cabinet, or other enclosure. Figure 9–36 shows some fittings available to conform to this code requirement.

Where there are smoothly rounded or flared threaded hubs or bosses such as on meter sockets, insulating fittings are not required.

RIGID METAL CONDUIT

Rigid metal conduit finds limited application in residential wiring. *Article 346* of the *National*

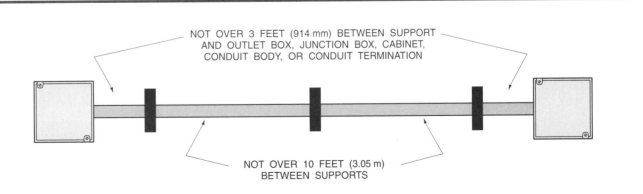

NOT OVER 3 FEET (914 mm) BETWEEN SUPPORT AND OUTLET BOX, JUNCTION BOX, CABINET, CONDUIT BODY, OR CONDUIT TERMINATION

NOT OVER 10 FEET (3.05 m) BETWEEN SUPPORTS

Figure 9–35   This sketch shows the supporting requirements for EMT as required by *Section 348-12.*

Figure 9–36   (A) An EMT connector that has an insulated throat, (B) a bushing that is a combination of metal and insulating material, and (C) a bushing made entirely out of insulating material. (*Courtesy of Hubbell Electrical Products, a Division of Hubbell Incorporated [Delaware]*)

Figure 9–37   Rigid metal conduit. (*Courtesy of Allied Tube and Conduit*)

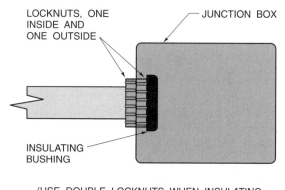

(USE DOUBLE LOCKNUTS WHEN INSULATING BUSHING IS NONMETALLIC)

Figure 9–38   This illustration shows the use of double locknuts plus an insulating bushing as required by *Sections 300-4(f)* and *373-6(c)*.

Electrical Code® covers rigid metal conduit. Figure 9–37 is a photo of galvanized rigid steel conduit.

One of the most common house wiring applications is as a service mast for overhead services where more strength is needed than what EMT or PVC can provide. Mast-type services are covered in Unit 15. Another residential application is where it becomes necessary to run a conduit under a driveway. Depth requirements are found in *Table 300-5* of the *NEC®.*

Rigid metal conduit may be used in all of the same places that EMT is permitted and has the same installation rules.

Rigid metal conduit comes in 10-foot lengths, and can be threaded, *Section 346-8* or can be used with threadless fittings, *Section 346-9.*

When using rigid metal conduit, an insulating bushing cannot be used alone. They are not strong enough to make up a tight connection between the fitting and the box. When the bushing is made of insulating material only, then two locknuts must be used as shown in Figure 9–38.

When all conductors are the same size and type, the number of conductors permitted in rigid metal conduit is found in *Table C8* in *Appendix C* of the *NEC®.*

When conductors are different sizes, refer to *Chapter 9* in the *NEC®.* The number of conductors permitted is determined using the percentage fill as shown in *Table 1,* the dimensions of rigid metal conduit is found in *Table 4,* and the dimensions of the conductors as shown in *Tables 5, 5A,* and *8.* Sample calculations are shown later in this unit.

RIGID NONMETALLIC CONDUIT

Rigid nonmetallic conduit is made of polyvinyl chloride (PVC), and oftentimes referred to as rigid PVC, is covered in *Article 347* of the *National Electrical Code.*

Rigid PVC comes in 10-foot (3.05 m) lengths. It comes in two types:

- Schedule 40 (heavy wall thickness)
- Schedule 80 (extra heavy wall thickness)

The outside diameter is the same for both Schedule 40 and Schedule 80. The inside diameter of Schedule 40 is greater than that of Schedule 80.

Figure 9–39 shows Schedule 40 and Schedule

80 rigid nonmetallic conduit.

Schedule 40 has a thinner wall than Schedule 80 and is used where it is not subject to physical damage. Schedule 80 is stronger and is used where exposed to physical damage.

The fittings used with rigid PVC are attached with a solvent-type cement, much the same as what plumbers use. Plumbing PVC pipe may look the same as electrical PVC—but they are not the same. One is designed to withstand water pressure from the inside, the other is designed to withstand forces from the outside.

Figure 9–40 shows an assortment of fittings and conduit bodies used with rigid PVC nonmetallic conduit.

The applications where rigid PVC are permitted are found in *Article 347.* Here are some of the per-

Figure 9–39   Schedule 40 and Schedule 80 rigid nonmetallic conduit (PVC). (*Courtesy Carlon Electric Products, a Lamson & Sessions Company*)

Figure 9–40   Assorted fittings and conduit bodies used with rigid nonmetallic conduit. These are but a few of the many types available. (*Courtesy Carlon Electric Products, a Lamson & Sessions Company*)

mitted uses and limitations that relate to residential wiring. For residential use, here are some of the key requirements for rigid PVC.

- May be used concealed in walls, ceilings, and floors.

- May be used in wet locations, such as outdoors.

- May be used underground.

- May not be used to support lighting fixtures.

- May be used to support nonmetallic conduit bodies.

- If used to support nonmetallic conduit bodies, the conduit body must not contain wiring devices or support lighting fixtures.

- Is nonmetallic, therefore, an equipment grounding conductor may be needed.

- Shall not have more than the equivalent of four 90° bends between pull points. Simply stated, the sum of all bends and offsets must not exceed 360°.

- Must be supported within 3 feet (914 mm) of

RIGID NONMETALLIC CONDUIT SIZE	MAXIMUM SPACING BETWEEN SUPPORTS
½", ¾", 1"	3 Feet
1¼", 1½", 2"	5 Feet

each outlet box, junction box, device box, conduit body, or other termination.

- Must be supported at intervals as follows:

- Must have expansion fittings on long runs, *Section 347-9.* Instructions furnished with the rigid PVC will provide the necessary details. *Table 347-9(A)* of the *National Electrical Code®* has dimensional requirements for anticipated expansion. Long runs of PVC in direct sunlight can and will pull out of the hubs of boxes, or will pull the boxes off the wall if proper attention is not given to thermal expansion.

- When all conductors are the same size and type, the number of conductors permitted in rigid PVC Schedule 80 is found in *Table C9* in *Appendix C* of the *NEC.®*

- When conductors are different sizes, refer to *Chapter 9* in the *NEC.®* The number of conductors permitted is determined using the percentage fill shown in *Table 1,* the dimensions of

rigid PVC found in *Table 4,* and the dimensions of the conductors as shown in *Tables 5, 5A,* and *8.* Sample calculations are shown later in this unit.

- Conductors emerging from rigid PVC must be protected from abrasion, either by the rounded edges of the box, fitting, or other enclosure.

- *Table 346-10* in the *National Electrical Code®* shows the minimum radius for bends in rigid nonmetallic conduit.

- Bends can be made with a proper bender that electrically heats the PVC to just the right temperature for bending. Manufactured 90°, 45°, and 30° elbows are available.

ELECTRICAL NONMETALLIC TUBING

Electrical nonmetallic tubing, commonly referred to as ENT is covered in *Article 331* of the *National Electrical Code.®* It is a pliable corrugated raceway made of PVC, and required special fittings made specifically for the ENT.

ENT is rarely used in house wiring, but is mentioned here to emphasize that it is a recognized wiring method that could be used.

INSTALLING THE RACEWAY

No matter what type of raceway you are using, install the raceway first—between boxes, fittings, or panels. When the raceway installation is complete, then pull in the conductors. This is a requirement of *Section 300-18* of the *National Electrical Code.®* Figure 9–41 illustrates what *Section 300-18* is saying.

SECURING RACEWAYS AND BOXES

Raceways, boxes, fittings, and cabinets must be securely fastened in place, *Section 300-11(a).* Points "X" are examples of where fastening should take place. Do not hang other raceways, cables, or other nonelectrical equipment from an electrical raceway, *Section 300-11(b).* Point "XX" is an example of a code violation because the shelf is suspended from the electrical raceway. Figure 9–42 illustrates these code requirements.

In *Section 300-11(b)* we find an exception to the basic rule of not allowing any foreign "things" to be suspended or attached to an electrical raceway. That

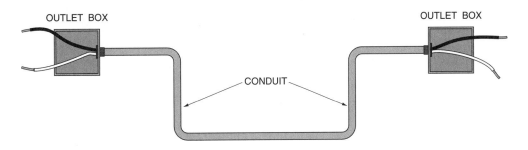

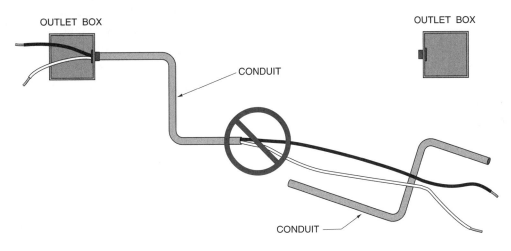

INSTALL CONDUIT COMPLETELY BETWEEN BOXES BEFORE
PULLING IN THE CONDUCTORS, *SECTION 300-18(a)*. NO MORE
THAN THE EQUIVALENT OF FOUR QUARTER BENDS (360° TOTAL)
PERMITTED BETWEEN PULL POINTS, *SECTIONS 331-10*, *345-11*,
346-11, *347-14*, *348-10*, *350-16*, *351-10*, AND *351-30*.

DO NOT PULL CONDUCTORS INTO A PARTIALLY INSTALLED
CONDUIT SYSTEM, AND THEN ATTEMPT TO SLIDE THE
REMAINING SECTION OF RACEWAY OVER THE CONDUCTORS.
THIS IS A VIOLATION OF *SECTION 300-18(a)*.

Figure 9–41 When installing raceways, be sure to install the raceways first as a complete system. After the raceways have been installed, then pull in the conductors, *Section 300-18*.

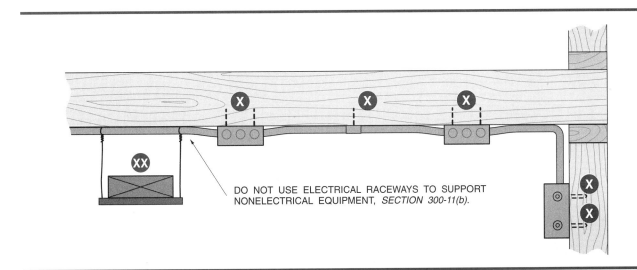

DO NOT USE ELECTRICAL RACEWAYS TO SUPPORT
NONELECTRICAL EQUIPMENT, *SECTION 300-11(b)*.

Figure 9–42 Raceways, boxes, and cabinets must be securely fastened, *Section 300-11(a)*. Do not hang nonelectrical equipment from electrical raceways, *Section 300-11(b)*.

exception refers to the Class 2 control circuit wiring necessary to hook up equipment, such as the thermostat wiring needed to hook up a furnace. To quote the exception, these Class 2 conductors are *"solely for the purpose of connection to the equipment control circuits."*

Figure 9–43 shows the meaning of this exception.

Flexible Connections

The installation of certain equipment requires flexible connections, both to simplify the installation and to stop the transfer of vibrations. In residential wiring, flexible connections are used to hook-up fans, food waster disposers, dishwashers, air conditioners, heat pumps, recessed fixtures, water heaters. and similar equipment.

Figure 9–44 shows some of the common applications where flexible connections are used, for other than cord- and plug-connected appliances. Refer to Unit 8 for detailed data regarding flexible cords and their permitted uses.

There are four types of flexible wiring methods

used for these connections:

• Armored cable, already discussed

• Flexible metal conduit, Figure 9–45(A)

• Flexible liquidtight metal conduit, Figure 9–45(B)

• Flexible liquidtight nonmetallic conduit, Figure 9–45(C)

Flexible Metal Conduit

Article 350 of the *National Electrical Code®* covers the use and installation of flexible metal conduit. In the trade, flexible metal conduit is usually called *Greenfield,* named after Harry Greenfield, who submitted the product for "listing" in 1902. This wiring method is similar to armored cable, except that the conductors are installed by the electrician.

The Code rules for flexible metal conduit in residential wiring applications are:

• Do not use in wet locations unless the conductors are suitable for use in wet locations (TW,

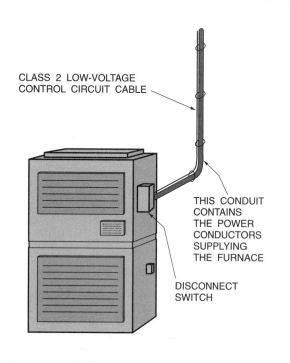

Figure 9–43 The Code in *Section 300-11(b)* allows Class 2 control circuit wiring to be supported by the raceway that contains the power wiring supplying the equipment.

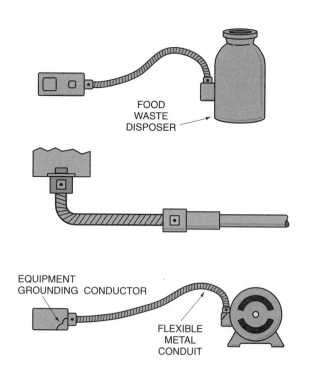

Figure 9–44 Some common uses for flexible connections.

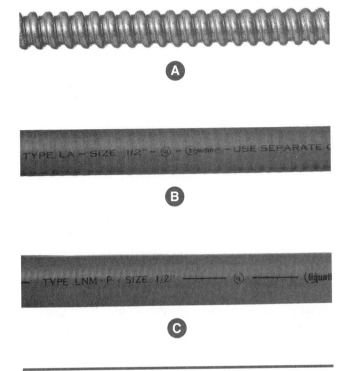

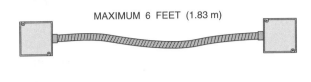

MAXIMUM 6 FEET (1.83 m)

4 FEET (1.22 m) PLUS 2 FEET (.61 m)

Figure 9–46 If a flexible metal conduit is to serve as an equipment ground return path, the total length shall not exceed 6 feet (1.83 m), *Sections 250-118(6)(c), (7)(d), (8)(b),* and *350-14.*

Figure 9–45 (A) Flexible metal conduit, (B) flexible liquidtight metal conduit, and (C) flexible liquidtight nonmetallic conduit. (*Courtesy of Electri-Flexco*)

—the total length of the return ground path through the flex must not exceed 6 feet (1.83 m). See Figure 9–46.

- If the flex is longer than 6 feet (1.83 m), regardless of trade size, the flex is not acceptable as a grounding means. You must install a separate equipment grounding conductor, sized per *Table 250-122, NEC.*

THW, THWN), and the installation is made so water will not enter the enclosure or other raceways to which the flex is connected. Installing the flex so the convolutions shed water instead of collecting and allowing the water to enter the flex might be acceptable to the electrical inspector.

For the flexible connection of outdoor equipment such as air-conditioning equipment, flexible liquidtight metal conduit or flexible liquidtight nonmetallic conduit is the preferred and accepted wiring method.

- Do not bury in concrete.
- Do not bury underground.
- Flexible metal conduit is acceptable as an equipment grounding conductor if:
  —the flex is "listed" for grounding purposes.
  —the fittings are "listed" for grounding purposes.
  —the overcurrent protection (fuse or breaker) does not exceed 20 amperes.

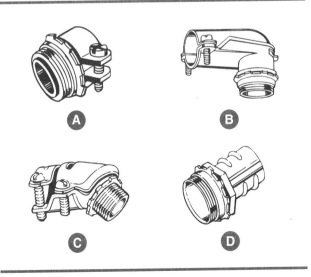

Figure 9–47 Typical connectors for flexible metal conduit. (A) A straight connector, (B) a 90° connector, (C) a 45° connector, and (D) a screw-in connector. (*Courtesy of Hubbell Electrical Products, a Division of Hubbell Incorporated [Delaware]*)

- All fittings must be "listed." Figure 9–47 shows an assortment of fittings used with flexible metal conduit.

- Fittings larger than ¾ inch that have been "listed" as OK for grounding means will be marked "GRND" or some equivalent marking.

- Must be supported every 4½ feet (1.37 m)

- Must be supported within 12 inches (305 mm) of an outlet box, cabinet, fitting, or other termination. The exceptions to this are:
  —Where the flex is "fished,"
  —Where the length is not over 3 feet (914 mm) used for flexibility (like hooking up a food waste disposer), and
  —Where the flex is not over 6 feet (1.83 m) used as a fixture "whip."

- Horizontal runs through holes, and supported by framing members that are not more than 4½ feet (1.37 m) apart. The flex is considered to be adequately supported when the flex is secured within 12 inches (305 mm) of a box, cabinet, fitting, or enclosure.

- Do not have more than four quarter bends (360) between boxes or pull points.

- May be installed exposed or concealed.

- Do not conceal angle-type fittings because of the difficulty that would present itself when pulling in wires.

- Where flexibility is needed, always install a separate equipment grounding conductor inside the flex to ensure a good ground of the equipment connected with the flex.

- When all conductors are the same size and type, the number of conductors permitted in flexible metal conduit is found in *Table C8* in *Appendix C* of the *NEC.*®

- When conductors are different sizes, refer to *Chapter 9* in the *NEC.*® The number of conductors permitted is determined using the percentage fill shown in *Table 1,* the dimensions of flexible metal conduit found in *Table 4,* and the dimensions of the conductors as shown in *Tables 5, 5A,* and *8.* Sample calculations are shown later in this unit.

- ½ inch is the minimum size, except that ⅜-inch trade size is all right in lengths not over 6 feet (1.83 m). Fixture "whips" are a good example of where this rule might apply. See Unit 11 for additional information.

- Conductor fill for ⅜-inch flex is tabulated in *Table 350-12.*

- Be sure to remove rough edges at "cut ends" unless fittings are used that thread internally into the convolutions of the flex.

Liquidtight Flexible Metal Conduit

Article 351, Part A of the *National Electrical Code*® covers the use and installation of liquidtight flexible metal conduit.

Liquidtight flexible metal conduit has a "tighter" fit of its spiral turns as compared to standard flexible metal conduit. It has a thermoplastic outer jacket that makes it liquidtight (watertight). In house wiring applications, liquidtight flexible metal conduit is used to make the flexible connection to a central air-conditioning unit, heat pump, or swimming pool equipment that is located outdoors as shown in Figure 9–48.

Here are some of the uses, limitations, and applications for liquidtight flexible metal conduit:

- May be used for exposed and concealed installations.

- May be buried directly in the ground if so "listed" and marked.

- May not be used where subject to physical abuse.

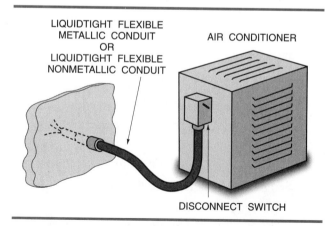

Figure 9–48 Typical use of liquidtight flexible metal conduit.

- May not be used if ambient temperature will exceed the temperature limitation of the non-metallic outer jacket. The "listing" and markings will indicate this.

- Must not be used in sizes smaller than ½ inch except that ⅜ inch is all right for enclosing leads to motors, or as a fixture "whip."

- The flex must be "listed."

- The fittings must be "listed." Figure 9–49 is an assortment of liquidtight flexible metal conduit fittings.

- Must be supported every 4½ feet (1.37 m).

- Must be supported within 12 inches (305 mm) of a box, cabinet, fitting, or enclosure.

- The 4½-foot (1.37 m) and 12-inch (305 mm) securing requirements are waived when the flex is "fished" or when used as a fixture "whip" whose length is not over 6 feet (1.83 m).

- The 12-inch (305 mm) securing requirement is waived for a flexible connection when the length is not over 3 feet (914 mm).

- If flexibility is needed, support within 3 feet (914 mm) of where the flex is terminated.

- Horizontal runs through holes, and supported by framing members that are not more than 4½ feet (1.37 m) apart. The flex is considered to be adequately supported when the flex is secured within 12 inches (305 mm) of a box, cabinet, fitting, or enclosure.

- ⅜ inch and ½-inch trade sizes are acceptable as an equipment grounding means if:
  —the flex is "listed" for grounding purposes.
  —the fittings are "listed" for grounding purposes.
  —the overcurrent protection (fuse or breaker) does not exceed 20 amperes.
  —the total length of the return ground path through the flex must not exceed 6 feet (1.83 m). See Figure 9–46.

- ¾-inch, 1-inch, and 1¼-inch trade sizes are acceptable as an equipment grounding means if:
  —the flex is "listed" for grounding purposes.
  —the fittings are "listed" for grounding purposes.
  —the overcurrent protection (fuse or breaker)

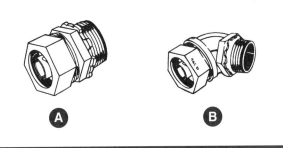

Figure 9–49 (A) A straight connector and (B) a 90° connector for use with liquidtight flexible metal conduit. (*Courtesy of Hubbell Electrical Products, a Division of Hubbell Incorporated [Delaware]*)

does not exceed 60 amperes.
  —the total length of the return ground path through the flex must not exceed 6 feet (1.83 m). See Figure 9–46.

- If the flex is longer than 6 feet (1.83 m), regardless of trade size, the flex is not acceptable as a grounding means. A separate equipment grounding conductor must be installed, sized per *Table 250-122, NEC.*®

- Is not suitable as a grounding means in 1½-inch trade size and larger.

- Do not conceal angle-type fittings because of the difficulty that would present itself when pulling in wires.

- Where flexibility is needed, always install a separate equipment grounding conductor inside the flex to ensure a good ground of the equipment connected with the flex.

- When all conductors are the same size and type, the number of conductors permitted in liquidtight flexible metal conduit is found in *Table C7* in *Appendix C* of the *NEC.*®

- When conductors are different sizes, refer to *Chapter 9* in the *NEC.*® The number of conductors permitted is determined using the percentage fill shown in *Table 1,* the dimensions of the liquidtight metal conduit found in *Table 4,* and the dimensions of the conductors as shown in *Tables 5, 5A,* and *8.* Sample calculations are shown later in this unit.

- Conductor fill for ⅜-inch size, see *Table 350-12.*

When flexible metal conduit and liquidtight

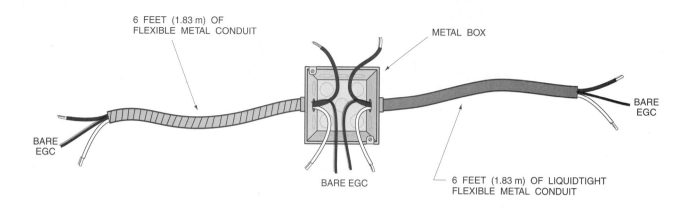

6 FEET (1.83 m) OF
FLEXIBLE METAL CONDUIT

METAL BOX

BARE
EGC

BARE
EGC

BARE EGC

6 FEET (1.83 m) OF LIQUIDTIGHT
FLEXIBLE METAL CONDUIT

Figure 9–50 The combined length of flexible metal conduit and liquidtight flexible metal conduit shall not exceed 6 feet (1.83 m) if they are to serve as the equipment ground path, Section *250-118(6)(c), (7)(d),* and *(8)(b)*. In this illustration, a separate equipment grounding conductor (EGC) is provided.

flexible metal conduit are used in combination, the combined length must not exceed 6 feet (1.83 m) if the flex is to serve as the equipment grounding conductor (Figure 9–50).

Liquidtight Flexible Nonmetallic Conduit

Article 351, Part B of the *National Electric Code®* covers the use and installation of liquidtight flexible nonmetallic conduit.

Figure 9–51 shows liquidtight flexible nonmetallic conduit.

Figure 9–52 is an assortment of fittings for liquidtight flexible nonmetallic conduit.

Liquidtight flexible nonmetallic conduit is available in three types:

- One type has a smooth seamless inner core that is bonded to the outer covering, and has reinforcement layers between the cover and the core.

- Another type has a smooth inner surface. The reinforcement is part of the wall of the flex.

- Another type has a corrugated internal and external surface, and does not have additional reinforcement layers.

As you study the following Code requirements, take note if the rule applies to all three types, or only to a specific type. Unless otherwise mentioned, the rule applies to all three types.

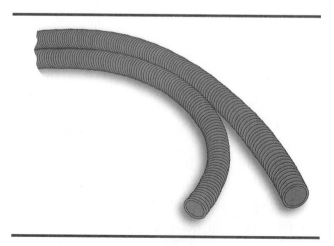

Figure 9–51 Liquidtight flexible nonmetallic conduit.

Figure 9–52 Typical connectors for use with liquidtight flexible nonmetallic conduit. (*Courtesy of Hubbell Electrical Products, a Division of Hubbell Incorporated [Delaware]*)

Liquidtight flexible nonmetallic conduit is flame-resistant, and is similar to liquidtight flexible metal conduit, except that it is made solely of non-metallic material (PVC). Because it is nonmetallic, in most instances an equipment grounding conductor must be installed in addition to the circuit conductors.

The following are some of the important *NEC®* rules pertaining to liquidtight flexible nonmetallic conduit:

- Do not use in direct sunlight unless specifically "listed" and marked for such use.

- May be used in exposed or concealed installations.

- Can become brittle in extreme cold applications.

- May be used outdoors when "listed" and marked as suitable for this application.

- May be buried directly in the earth when "listed" and marked as suitable for this application.

- May not be used where subject to physical abuse.

- May not be used if ambient temperature and heat from the conductors will exceed the temperature limitation of the nonmetallic material.

- Do not have more than four quarter bends (360°) between boxes and pull points.

- Must not be used is sizes smaller than ½ inch except that ⅜ inch is all right for enclosing leads to motors, or as a fixture "whip."

- The flex must be "listed."

- The fittings must be "listed."

- Must be supported every 3 feet (914 mm).

- Must be supported within 12 inches (305 mm) of a box, cabinet, fitting, or enclosure.

- The 3-foot (914 mm) and 12-inch (305 mm) securing requirements are waived when the flex is "fished" or when used as a fixture "whip" whose length is not over 6 feet (1.83 m).

- The 12-inch (305 mm) securing requirement is waived for a flexible connection when the length is not over 3 feet (914 mm).

- If flexibility is needed, support within 3 feet (914 mm) of where the flex is terminated. In this case, the 12-inch (305 mm) securing requirement is waived.

- Horizontal runs through holes, and supported by framing members that are not more than 3 feet (914 mm) apart. The flex is considered to be adequately supported when the flex is secured within 12 inches (305 mm) of a box, cabinet, fitting, or enclosure.

- Do not conceal angle-type fittings.

- If an equipment grounding conductor is needed, it may be run inside or outside of the flex when the flex is not over 6 feet (1.83 m). Run the EGC on the inside for runs longer than 6 feet (1.83 m). The recommended method is to always run an EGC on the inside. EGC are sized according to *Table 250-122*.

- Do not use in lengths longer than 6 feet (1.83 m), unless the flex is of the smooth internal surface construction that has reinforcement built into the walls of the flex, **and** the flex is supported according to the 3-feet (914 mm) and 12-inch (305 mm) requirements mentioned above, or if absolutely necessary for flexibility.

UNDERGROUND WIRING

Residential wiring usually has some underground wiring. Decorative landscape lighting, post lamps, decks, or branch circuits feeding detached garages or tool sheds are examples of where underground wiring might be used. The wiring method could be one of the permitted raceways discussed previously, or it might be in the form of a cable suitable for direct burial. The most common cable for this use is Type UF.

Type UF cable is covered in *Article 339* of the *National Electrical Code.*®

Figure 9-53 shows a Type UF underground cable. Here are some of the key Code rules for Type UF cable when installed in residential applications.

- Is available with copper conductor sizes No. 14 AWG through No. 4/0 AWG and No. 12 AWG through 4/0 AWG with aluminum and copper-clad aluminum conductors.

- May be marked with a suffix "B," which indicates that the conductors are rated 90°C.

- The ampacity of the conductors is found in *Table 310-16,* using the 60°C column no matter what the temperature rating is of the conductors.

- Is marked with the letters UF.

- May be used in direct exposure to the sun if the cable is marked "sunlight resistant."

- May be used as a nonmetallic-sheathed cable when installed according to the rules for non-metallic-sheathed cable, *Article 336.*

- Is flame-retardant.

- Is moisture-, fungus-, and corrosion resistant.

- May be buried directly in the ground.

- May be used for branch-circuit and feeder wiring.

- May be used in interior wiring for wet, dry, or corrosive installations.

- Shall not be used as service-entrance cable.

- Shall not be embedded in concrete, cement, or aggregate.

- When single-conductor conductors, all conductors must be buried in the same trench.

- Do not back-fill a trench with rocks, debris, or similar coarse material that could damage the underground conductors, *Section 300-5(f).*

Submersible Water Pump Cable

Submersible water pump cable is a multiconductor cable made up of two, three, or four single-conductor Type USE, USE-2, or UF cables.

Submersible water pump cable is "tag-marked" for use within the well casing for wiring deep-well water pumps where the cable is not subject to repetitive handling caused by frequent servicing of the pump units. The cable will be marked on its surface "Pump Cable." When made up of Type USE, USE-2, and UF, it is suitable for running down the well casing as well as buried directly in the ground.

Insulated conductors installed in raceways underground must be suitable for use in wet locations, such as Types RHW, TW, THW, THHW, THWN, XHHW. This requirement is found in *Section 310-8* of the *NEC.*®

Cables used in wet locations must be "listed" for use in wet locations, *Section 310-8* of the *NEC.*®

Figure 9–53 A Type UF underground cable. (*Courtesy Southwire Company*)

Depth of Underground Wiring

Section 300-5 of the *National Electrical Code*®
addresses underground installations.

The minimum depends upon what wiring
method is used. *Table 300-5* of the *NEC*® provides
comprehensive tabulation requirements for various
wiring methods. Figure 9–54 is *Table 300-5* from
the *National Electrical Code*.® Note that in this table
the requirements for residential installations are not
quite as severe as for other installations. For example, the basic rule for direct buried cable is that it be

**Table 300-5. Minimum Cover Requirements, 0 to 600 Volts, Nominal, Burial in Inches (Cover
is defined as the shortest distance in inches measured between a point on the top surface of any
direct-buried conductor, cable, conduit, or other raceway and the top surface of finished grade,
concrete, or similar cover.)**

Location of Wiring Method or Circuit	Type of Wiring Method or Circuit				
	Column 1 **Direct Burial Cables** **or Conductors**	**Column 2** **Rigid Metal Conduit** **or Intermediate Metal** **Conduit**	**Column 3** **Nonmetallic Raceways** **Listed for Direct** **Burial Without** **Concrete Encasement** **or Other Approved** **Raceways**	**Column 4** **Residential Branch** **Circuits Rated 120** **Volts or Less with** **GFCI Protection and** **Maximum** **Overcurrent** **Protection of 20** **Amperes**	**Column 5** **Circuits for Control of** **Irrigation and** **Landscape Lighting** **Limited to Not More** **than 30 Volts and** **Installed with Type UF** **or in Other Identified** **Cable or Raceway**
All locations not speci- fied below	24	6	18	12	6
In trench below 2-in. thick concrete or equivalent	18	6	12	6	6
Under a building	0 (in raceway only)	0	0	0 (in raceway only)	0 (in raceway only)
Under minimum of 4- in. thick concrete ex- terior slab with no vehicular traffic and the slab extending not less than 6 in. be- yond the under- ground installation	18	4	4	6 (direct burial) 4 (in raceway)	6 (direct burial) 4 (in raceway)
Under streets, high- ways, roads, alleys, driveways, and park- ing lots	24	24	24	24	24
One- and two-family dwelling driveways and outdoor parking areas, and used only for dwelling-related purposes	18	18	18	12	18
In or under airport run- ways, including ad- jacent areas where trespassing prohibited	18	18	18	18	18

Notes:
1. For SI units, 1 in. = 25.4 mm.

2. Raceways approved for burial only where concrete encased shall require concrete envelope not less
than 2 in. thick.

3. Lesser depths shall be permitted where cables and conductors rise for terminations or splices or where
access is otherwise required.

4. Where one of the wiring method types listed in Columns 1–3 is used for one of the circuit types in
Columns 4 and 5, the shallower depth of burial shall be permitted.

5. Where solid rock prevents compliance with the cover depths specified in this table, the wiring shall be
installed in metal or nonmetallic raceway permitted for direct burial. The raceways shall be covered by
a minimum of 2 in. of concrete extending down to rock.

Reprinted with permission from NFPA 70-1999

Figure 9–54 *Table 300-5* from the *National Electrical Code*® that shows minimum depth requirements for
different wiring methods.

covered by a minimum of 24 inches. However, this minimum depth is reduced to 12 inches for residential installations when the branch circuit is not over 120 volts, is GFCI protected, and the overcurrent protection is not over 20 amperes. This is for residential installations only.

Figure 9–55 is a pictorial view showing the basic depth requirements for underground cables and raceways.

Figure 9–56 is a pictorial view showing the depth requirements for underground cables and raceways under residential driveways and parking areas.

The measurement for the depth requirement is from the top of the raceway or cable to the top of the finished grade, concrete, or other similar cover.

Figure 9–57 shows some of the ways for bringing cables and raceways through a basement wall and upward into the bottom of a post light. Be very careful when penetrating a raceway through a foundation wall. Water will always accumulate in an underground raceway, and in all probability will find its way through the raceway and into the basement. It is better to run the underground raceway out of the ground on the outside of the foundation wall— upward into a conduit body (an LB)—then into the building above grade. This way, water and moisture will not find its way into the basement.

Splices are permitted to be made directly in the ground, but only if the connectors are "listed" by a recognized testing laboratory for such use. Permission to do this is found in *Sections 110-14(b)*

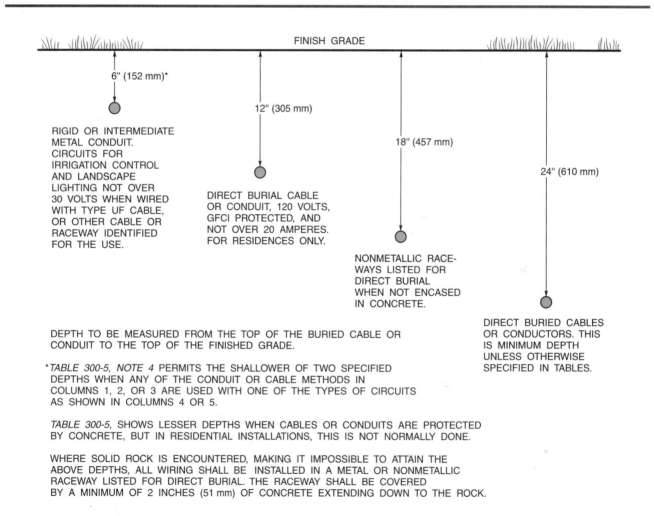

FINISH GRADE

6" (152 mm)*

RIGID OR INTERMEDIATE METAL CONDUIT. CIRCUITS FOR IRRIGATION CONTROL AND LANDSCAPE LIGHTING NOT OVER 30 VOLTS WHEN WIRED WITH TYPE UF CABLE, OR OTHER CABLE OR RACEWAY IDENTIFIED FOR THE USE.

12" (305 mm)

DIRECT BURIAL CABLE OR CONDUIT, 120 VOLTS, GFCI PROTECTED, AND NOT OVER 20 AMPERES. FOR RESIDENCES ONLY.

18" (457 mm)

NONMETALLIC RACE- WAYS LISTED FOR DIRECT BURIAL WHEN NOT ENCASED IN CONCRETE.

24" (610 mm)

DIRECT BURIED CABLES OR CONDUCTORS. THIS IS MINIMUM DEPTH UNLESS OTHERWISE SPECIFIED IN TABLES.

DEPTH TO BE MEASURED FROM THE TOP OF THE BURIED CABLE OR CONDUIT TO THE TOP OF THE FINISHED GRADE.

*TABLE 300-5, NOTE 4 PERMITS THE SHALLOWER OF TWO SPECIFIED DEPTHS WHEN ANY OF THE CONDUIT OR CABLE METHODS IN COLUMNS 1, 2, OR 3 ARE USED WITH ONE OF THE TYPES OF CIRCUITS AS SHOWN IN COLUMNS 4 OR 5.

TABLE 300-5, SHOWS LESSER DEPTHS WHEN CABLES OR CONDUITS ARE PROTECTED BY CONCRETE, BUT IN RESIDENTIAL INSTALLATIONS, THIS IS NOT NORMALLY DONE.

WHERE SOLID ROCK IS ENCOUNTERED, MAKING IT IMPOSSIBLE TO ATTAIN THE ABOVE DEPTHS, ALL WIRING SHALL BE INSTALLED IN A METAL OR NONMETALLIC RACEWAY LISTED FOR DIRECT BURIAL. THE RACEWAY SHALL BE COVERED BY A MINIMUM OF 2 INCHES (51 mm) OF CONCRETE EXTENDING DOWN TO THE ROCK.

Figure 9–55   Minimum depths for cables and raceways installed underground in conformance to *Table 300-5*.

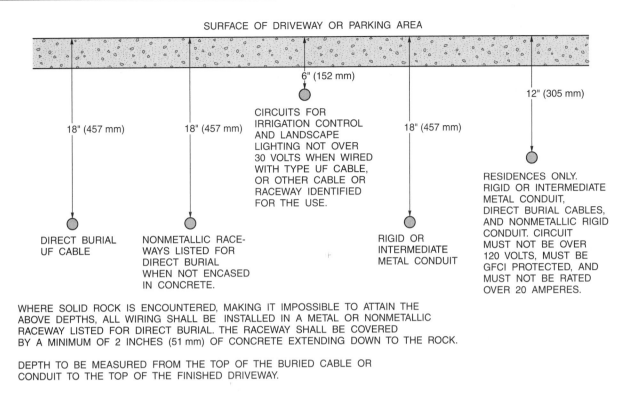

SURFACE OF DRIVEWAY OR PARKING AREA

6" (152 mm)

12" (305 mm)

18" (457 mm)

18" (457 mm)

CIRCUITS FOR
IRRIGATION CONTROL
AND LANDSCAPE
LIGHTING NOT OVER
30 VOLTS WHEN WIRED
WITH TYPE UF CABLE,
OR OTHER CABLE OR
RACEWAY IDENTIFIED
FOR THE USE.

18" (457 mm)

RESIDENCES ONLY.
RIGID OR INTERMEDIATE
METAL CONDUIT,
DIRECT BURIAL CABLES,
AND NONMETALLIC RIGID
CONDUIT. CIRCUIT
MUST NOT BE OVER
120 VOLTS, MUST BE
GFCI PROTECTED, AND
MUST NOT BE RATED
OVER 20 AMPERES.

DIRECT BURIAL
UF CABLE

NONMETALLIC RACE-
WAYS LISTED FOR
DIRECT BURIAL
WHEN NOT ENCASED
IN CONCRETE.

RIGID OR
INTERMEDIATE
METAL CONDUIT

WHERE SOLID ROCK IS ENCOUNTERED, MAKING IT IMPOSSIBLE TO ATTAIN THE
ABOVE DEPTHS, ALL WIRING SHALL BE INSTALLED IN A METAL OR NONMETALLIC
RACEWAY LISTED FOR DIRECT BURIAL. THE RACEWAY SHALL BE COVERED
BY A MINIMUM OF 2 INCHES (51 mm) OF CONCRETE EXTENDING DOWN TO THE ROCK.

DEPTH TO BE MEASURED FROM THE TOP OF THE BURIED CABLE OR
CONDUIT TO THE TOP OF THE FINISHED DRIVEWAY.

Figure 9–56   Depth requirements under residential driveways and parking areas. These depths are permitted for one- and two-family houses.

and *Section 300-5(e)*. However, use extreme care when making underground splices. A poor splice can open, become a shock hazard, and very difficult to find and repair if that becomes necessary. For residential wiring, the need to splice conductors underground is rare. It is best to run the UF cable continuously from box to box, enclosure to enclosure, conduit body to conduit body, and so forth, then make up the splices in these boxes, enclosures, and conduit bodies. Splicing in conduit bodies is permitted only when marked with the conduit body's cubic inch capacity, *Section 370-16(c)*.

Laying Underground Conductors in A Trench

When laying UF multiconductor cables in a trench, the conductors are automatically grouped together. When laying individual conductors in a trench, all of the conductors of the same circuit must

be laid in the same trench, *Section 300-3(b)*. Widely spaced conductors of an alternating circuit can result in excessive voltage drop. To minimize inductive reactance, group the conductors together as shown in Figure 9–58.

Refer to Unit 12 for grounding requirements.

ABOVEGROUND OUTDOOR WIRING

Any wiring done outside must be suitable for use outdoors. We have already discussed various wiring methods common to house wiring. Some are suitable for use outdoors, some are not.

Cables used must be suitable for outdoor use and are usually marked as suitable for use in direct sunlight and must be suitable for use in wet locations.

Conductors installed in raceways outdoors must be suitable for use in wet locations, because these raceways will undoubtedly gather moisture associated with temperature and humidity changes. Types

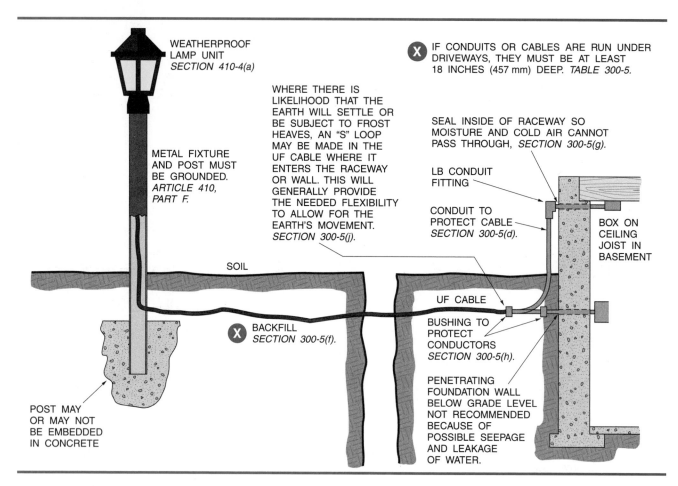

WEATHERPROOF
LAMP UNIT
SECTION 410-4(a)

IF CONDUITS OR CABLES ARE RUN UNDER
DRIVEWAYS, THEY MUST BE AT LEAST
18 INCHES (457 mm) DEEP. *TABLE 300-5.*

WHERE THERE IS
LIKELIHOOD THAT THE
EARTH WILL SETTLE OR
BE SUBJECT TO FROST
HEAVES, AN "S" LOOP
MAY BE MADE IN THE
UF CABLE WHERE IT
ENTERS THE RACEWAY
OR WALL. THIS WILL
GENERALLY PROVIDE
THE NEEDED FLEXIBILITY
TO ALLOW FOR THE
EARTH'S MOVEMENT.
SECTION 300-5(j).

SEAL INSIDE OF RACEWAY SO
MOISTURE AND COLD AIR CANNOT
PASS THROUGH, *SECTION 300-5(g).*

LB CONDUIT
FITTING

METAL FIXTURE
AND POST MUST
BE GROUNDED.
*ARTICLE 410,
PART F.*

CONDUIT TO
PROTECT CABLE
SECTION 300-5(d).

BOX ON
CEILING
JOIST IN
BASEMENT

SOIL

UF CABLE

BACKFILL
SECTION 300-5(f).

BUSHING TO
PROTECT
CONDUCTORS
SECTION 300-5(h).

POST MAY
OR MAY NOT
BE EMBEDDED
IN CONCRETE

PENETRATING
FOUNDATION WALL
BELOW GRADE LEVEL
NOT RECOMMENDED
BECAUSE OF
POSSIBLE SEEPAGE
AND LEAKAGE
OF WATER.

Figure 9–57   Methods of bringing cable and/or raceway through a basement wall and upward into a post light. Penetrating a foundation wall is not a good idea. It is extremely difficult to keep water and moisture out.

DO KEEP WIRES
TOGETHER

DO NOT KEEP
WIRES SEPARATED

Figure 9–58   When laying individual underground conductors in a trench, keep the conductors of the same circuit together to minimize potential voltage drop problems.

RHW, TW, THW, THHW, THWN, XHHW are all right for use in wet locations.

Most outdoor wiring methods such as cables and raceways are secured to the building structure.

Raceways run on the outside of a building must be raintight and must be arranged to drain, *Section 225-22.*

Although underground wiring is the preferred method, there are occasions where overhead wiring is needed. Large conductors run overhead to another building such as a detached garage or storage building could be of a type of assembled cable that contains a messenger wire as well as the insulated conductors. This is the type that electric utilities use for an overhead service-drop to a house. Figure 9–59 shows this type of cable.

Article 321 recognizes messenger-supported wiring methods. In *Section 321-3,* you will find permission to use various types of cables. Whatever type is used, it must be installed according to the conditions described in the specific article referenced for that particular cable. Some types of cable

Figure 9–59 A triplex service-drop cable. Note that two conductors are insulated. The bare messenger wire serves as both the equipment grounding conductor and as the means of supporting the overhead cable between support points.

that could be used for residential outdoor overhead application are:

- Multiconductor service-entrance cable (*Article 338*)

- Multiconductor underground feeder and branch-circuit cable (*Article 339*)

- Other factory-assembled, multiconductor control, signal, or power cables that are identified for the use

Utilities are permitted to use a bare neutral when they supply a service-drop to a house because they come under the regulations of the *National Electrical Safety Code.* The *National Electrical Code®* does not allow bare circuit conductors for residential premise wiring, *Section 310-2(a).* Of course, equipment grounding conductors are per mitted to be bare, covered, or insulated, *Section 250-119.* Therefore, using any of the permitted types of cable for an overhead residential feeder or branch circuit would necessitate that all conductors be insulated. The bare messenger wire serves as the mechanical strength needed for support of the overhead cable, and as the equipment grounding conductors.

Before making an installation with a type of service-drop cable, check it out with your local elec-

Figure 9–60 It is a Code violation to support overhead wiring from vegetation whether live or dead, *Section 225–26.*

trical inspector to see if it is permitted in your area.

Do not support overhead wiring from trees or other vegetation, whether alive or dead, *Section 225-26*. Treated wooden poles or lumber such as those used to build decks is not considered to be vegetation for purposes of *Section 225-26*. Figure 9–60 illustrates a Code violation.

It is all right to support outdoor lighting fixtures and other electrical equipment from a tree, *Section 410-16(h)*. This is shown in Figure 9–61.

CORRECT HEIGHT OF OVERHEAD WIRING

The *NEC®* contains a number of requirements that specify the minimum clearances that conductors have to be above ground, above a roof, and above a swimming pool. *Section 225-18* shows the minimum clearances from ground for different conditions and different voltages. As previously discussed, the electrical system for house wiring are 120/240 volts. This type of system has a voltage to ground of 120 volts. The reason behind maintaining minimum clearances from the ground and above roofs is to make it difficult for someone to reach out and touch the conductor, subjecting that person to possible electrocution.

10 feet (3.05 m)	Above finished grade, sidewalks, or from any platform or projection accessible to pedestrians only. (e.g. a deck or porch)
12 feet (3.66 m)	Over residential property and driveways
18 feet (5.49 m)	Over public streets, alleys, roads,

Figure 9–62 Minimum clearances aboveground for outdoor wiring of a typical residence.

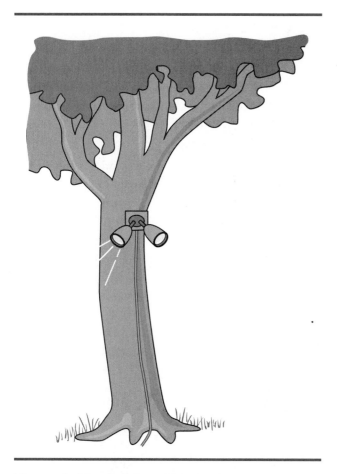

Figure 9–61 *Section 410-16(h)* **allows lighting fixtures to be supported by a tree.**

8 feet (2.44 m)	Minimum vertical clearance above roof. Maintain this clearance for a distance of not less than 3 feet (914 mm) in all directions from edge of roof. The 3 feet (914 mm) does not apply to a final conductor span than is attached to the side of the building.
10 feet (3.05 m)	Minimum vertical clearance above roof if roof can be walked on
3 feet (914 mm)	Minimum vertical clearance above roof that has a 4/12 pitch.
18 inches overhang (457 mm)	Minimum clearance above an _____ portion of roof when not more than 6 feet of the conductor(s) pass over not more than 4 feet of the roof overhang, and terminate in a thru-the-roof mast or other approved support.
3 feet (914 mm)	Minimum vertical, horizontal, and diagonal clearance from television antenna.
3 feet (914 mm)	Minimum clearance from windows that are designed to open, doors, porches, balconies, ladders, stairs, fire escapes, or similar locations.
0 feet	Conductors run above a window do

Figure 9–63 Minimum clearances above roofs for outdoor wiring. Refer to Figures 15–9 and 15–10.

Above Ground

Figure 9–62 shows the clearances from ground for typical residential wiring.

Above a Roof

Figure 9–63 shows minimum clearances above a roof as required by *Section 225-19.*

The above clearances are much the same as for service conductors, discussed and illustrated in Unit 15.

Above Swimming Pools

This is covered in *Article 680* of the *NEC.®*

MULTIOUTLET ASSEMBLIES

Multioutlet assemblies are covered in *Article 353* of the *National Electrical Code.®* In homes, multioutlet assemblies are usually surface mounted. *Section 353-2(b)(1)* permits metal multioutlet assemblies to be recessed into the building finish (e.g. plaster) provided the cover remains exposed, accessible, and removable. Nonmetallic multioutlet assemblies may be recessed into a baseboard.

Metal multioutlet assemblies are permitted to be run through (not within) a dry partition when no outlets are located within the partition, and all exposed portions of the assembly can have the cover removed, *Section 353-3.*

There are a number of locations in a home where multioutlet assemblies can be installed. Examples might be in a workshop, above a desk, under the overhang of a countertop on a peninsula in a kitchen, or behind audio/stereo/television equipment where receptacles are needed for the TV set, the stereo equipment, VCR, and cable box.

A multioutlet assembly consists of prewired units in 3-, 5-, and 6-foot lengths. You can buy multioutlet assemblies with factory-prewired receptacles that are spaced 6, 9, 12, and 18 inches apart. The prewired covers that contain the receptacles snap onto the base channel. Multioutlet assemblies are available in steel, aluminum, and PVC. Multioutlet assemblies can also be made up "on-the-job" with the receptacles placed at whatever spacing is wanted.

Figure 9–64 shows a typical multioutlet assembly.

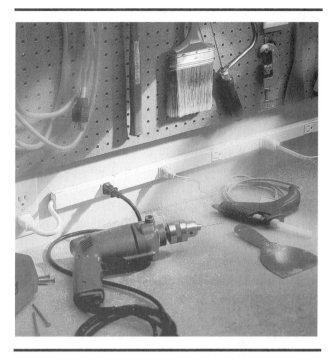

Figure 9–64 A typical factory-wired multioutlet assembly. (*Courtesy The Wiremold Co.*)

Note in Figure 9–64 how the branch-circuit wiring is fed into the back of a multioutlet assembly. Fittings are available to bring the electrical supply in the ends of the multioutlet assembly. Blank end fittings, internal corner couplings, external elbows, flat 90° elbows, tee fittings, straps, and many other fittings are available to simplify the installation.

For residential applications, *Section 220-3(b)(10)* states that *all* general-use receptacle outlets in one, two, and multifamily dwellings are to be considered as outlets for general lighting, and as such have been included in the volt-amperes per square foot calculations. Therefore, no additional load need be added. The number of receptacles on a multioutlet assembly offer convenience—not necessarily more load.

A good example of this is the number of receptacles needed for personal computers. The processing unit, monitor, printer, and scanner all need to be plugged in. The standard duplex receptacle just does not provide enough places to plug equipment into. That is why so many *plug-in strips* are used. And of course, for personal computer equipment, surge suppressor plug-in strips are generally used.

KNOB-AND-TUBE WIRING

Older homes were wired with open, individual, insulated conductors supported by porcelain *knobs* and *tubes*. In its day, when electricity was not used anywhere near as much as today, knob-and-tube wiring was a good wiring method. It served its purpose. This text does not cover knob-and-tube wiring because this wiring method is no longer used. The *National Electrical Code®* still permits knob-and-tube wiring to be used for extensions of existing installations. This usually involves the skill of making taps by soldering and taping the connections. Where knob-and-tube wiring is encountered when doing remodel work, it is probably best to abandon portions of the knob-and-tube wiring, and install all of the new wiring with current acceptable wiring methods.

Many fires have occurred because insulation has been blown in attics of older homes, completely burying the open conductors. The conductors become overheated! *Section 324-4* states in part that *"concealed knob-and-tube wiring shall not be used in hollow spaces of walls, ceilings, and attics where such spaces are insulated by loose, rolled, or foamed-in-place insulating material that envelops the conductors."* Keep a close watch on the companies that blow in insulation.

RACEWAY FILL CALCULATIONS

In Unit 10, you will learn how to calculate the number of conductors permitted in different size outlet and device boxes, so you can select and install the proper size of box for the application.

Just as important as *box fill* is *raceway fill*.

You must be able to select the proper size raceway based on the number, size, and type of conductors to be installed in a particular type of raceway. The necessary information is found in *Chapter 9* and *Appendix C* of the *National Electrical Code.®*

The *National Electrical Code®* has established the following "percent fill" for conduit and tubing to limit the heat that will be developed in the raceway when the conductors are carrying current.

- One conductor: 53%
- Two conductors: 31%
- Three or more conductors: 40%

These percentages are found in *Table 1* of *Chapter 9, National Electrical Code.®*

Raceway Fill When All Conductors Are the Same Size and Type

- Determine how many conductors of one size and type are to be installed in the raceway.
- Refer to *Tables C1* through *C12A* in *Appendix C* of the *NEC®* for the type of raceway, and for the specific size and type of conductors.
- Select the proper size raceway directly from the table.

These tables are easy to use. Here are a few examples of conductor fill for EMT. Look these up in the tables to confirm the accuracy of the examples and to get some practice using the tables.

Examples:

- Not over five No. 10 THHN conductors may be pulled into a ½-inch EMT.
- Not over eight No. 12 THW conductors may be pulled into a ¾-inch EMT.
- Not over three No. 8 THHN conductors may be pulled into a ½-inch EMT.
- Not over six No. 8 THHN conductors may be pulled into a ¾-inch EMT.
- Not over nine No. 8 THHN conductors may be pulled into a 1-inch EMT.

Raceway Fill When Conductors Are Different Sizes and/or Types

- Refer to *Chapter 9, Table 5, 5A,* and *8* of the *NEC.®*
- Find the size and type of conductors used. Determine the cross-sectional area in square inches of the conductors.
- Refer to *Chapter 9, Table 8* for the square inch area of bare conductors.
- Total the square inch areas of all conductors being installed in the raceway.
- Refer to *Chapter 9, Table 4* for the type and size of raceway being used. Actually, *Table 4* consists of twelve separate tables.

- Determine the minimum size raceway based on the total square inch area of the conductors and the allowable square inch area fill for the particular type of raceway used.

- The total area of the conductors must not exceed the allowable percentage fill of the cross-sectional square inch area of the raceway used.

Example 1: Find the minimum size EMT for three No. 6 THW copper conductors and two No. 8 THW copper conductors.

Answer:

1. Find the square inch area of the conductors in *Chapter 9, Table 5* of the *NEC.*

Three No. 6 THW conductors	0.0726 in²
	0.0726 in²
	0.0726 in²
Two No. 8 THW conductors	0.0556 in²
	0.0556 in²
Total cross-sectional area	0.3290 in²

2. In *Chapter 9, Table 4,* look for the EMT table. Look at the fifth column (over two wires, 40% fill) for an EMT that has an area of 0.3290 square inch or more. Note that:

   40% fill for ¾-inch EMT is 0.213 in²

   40% fill for a 1-inch EMT is 0.346 in²

   Therefore, 1-inch EMT is the minimum size for the combination of conductors in this example.

If a separate equipment grounding conductor is to be installed in a raceway, it does take up space and must be counted when making a raceway fill calculation.

Example 2: A feeder to an electric furnace is fed with two No. 3 THHN copper conductors. Because part of the installation will include a short section of flexible metal conduit, a bare equipment grounding conductor will also be needed. This feeder is installed in EMT. The feeder overcurrent protection is 100 amperes. What is the minimum size EMT to be installed?

Answer: According to *Table 250-122,* the minimum size copper equipment grounding conductor for a 100-ampere branch circuit is No. 8 AWG stranded copper conductor. Obtain the cross-sectional area of the insulated conductors from *Table 5* and for the bare EGC conductor from *Table 8, Chapter 9, NEC.*

1. | | |
   |---|---|
   | Two No. 3 THHN conductors | 0.0973 in² |
   | | 0.0973 in² |
   | One No. 8 bare stranded conductor | 0.0170 in² |
   | Total cross-sectional area | 0.2116 in² |

2. In *Chapter 9, Table 4,* check both the EMT table and the flexible metal conduit table. Look at the fifth column (over two wires, 40% fill) for an EMT and flexible metal conduit that has a cross-sectional area of 0.2116 in² or more. We need to check the cross-sectional area for both tables because different types of raceways have different internal dimensions. We find that:

   40% fill for ¾-inch EMT is 0.213 in²

   40% fill for ¾-inch flexible metal conduit is 0.213 in²

   Therefore, install ¾-inch EMT and ¾-inch

Figure 9–65 A variety of common conduit bodies and covers.

flexible metal conduit for the combination of conductors in the example.

CONDUIT BODIES

Sections 370-16(c), 370-23, and *370-29* contain some of the requirements for conduit bodies. Figure 9–65 shows some types of commonly used conduit bodies.

Conduit bodies are used with raceways to provide an easy means to turn corners, change direction, terminate raceways, and to mount switches and receptacles. They provide access to conductors, and provide access for pulling conductors, and space for splicing when permitted.

Conduit bodies:

- Must have a cross-sectional area not less than 2X the cross-sectional area of the largest conduit to which it is attached, Figure 9–66(A).

- Must not contain splices, taps, or devices unless it is marked with its cubic-inch capacity, Figure 9–66(B).

- May contain the same maximum number of conductors permitted for the size raceway attached to it.

- The conductor fill volume must be determined using the conductor volumes found in *Table 370-16(b).*

- Must be supported rigidly and securely.

- That **do not** contain wiring devices or have fixtures attached are considered supported when two or more conduits are threaded wrenchtight into the hubs, but only when the conduits are supported within 3 feet (914 mm) of the conduit body.

- That **do not** contain wiring devices or have fixtures attached are considered supported when the conduit body is not larger than the largest trade size of the rigid metal conduit, intermediate conduit, rigid nonmetallic conduit, or EMT being used with the conduit body.

- That **do** contain wiring devices or have fixtures attached are considered supported when two or

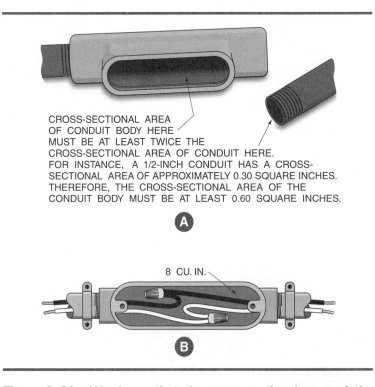

CROSS-SECTIONAL AREA OF CONDUIT BODY HERE MUST BE AT LEAST TWICE THE CROSS-SECTIONAL AREA OF CONDUIT HERE. FOR INSTANCE, A 1/2-INCH CONDUIT HAS A CROSS-SECTIONAL AREA OF APPROXIMATELY 0.30 SQUARE INCHES. THEREFORE, THE CROSS-SECTIONAL AREA OF THE CONDUIT BODY MUST BE AT LEAST 0.60 SQUARE INCHES.

A

8 CU. IN.

B

Figure 9–66 (A) shows that the cross-sectional area of the conduit body must be at least 2X the cross-sectional area of the conduit to which it is attached. (B) shows a conduit body marked with its cubic inch volume.

RIGID METAL, RIGID NONMETALLIC, INTER-
MEDIATE METAL CONDUIT, OR ELECTRICAL
METALLIC TUBING OK TO SUPPORT BOXES
THAT DO NOT HAVE DEVICES OR FIXTURES.
RIGID METAL OR INTERMEDIATE METAL
CONDUIT OK TO SUPPORT BOXES THAT
DO HAVE DEVICES OR FIXTURES.
RIGID NONMETALLIC CONDUIT NOT PERMITTED
TO SUPPORT LIGHTING FIXTURES OR OTHER
EQUIPMENT, *SECTION 347-3(b)*.

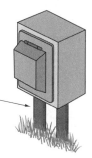

THIS BODY IS CONSIDERED TO
BE ADEQUATELY SUPPORTED.

A WHEN TWO OR MORE CONDUITS ARE TIGHTLY THREADED INTO THE HUBS OF A CONDUIT BODY,
THE CONDUIT BODY IS CONSIDERED TO BE ADEQUATELY SUPPORTED, *SECTION 370-23(e)*. FOR
ENCLOSURES THAT SUPPORT FIXTURES OR CONTAIN WIRING DEVICES, THE ENCLOSURE MUST BE
SUPPORTED WITHIN 18 INCHES (457 mm). CONDUITS COMING OUT OF THE GROUND ARE ACCEPTABLE
AS MEETING THIS SUPPORT REQUIREMENT, *SECTION 370-23(f)*.

B THIS CONDUIT BODY IS NOT ADEQUATELY
SUPPORTED BY THE ONE CONDUIT THREADED
INTO THE HUB. THIS CONDUIT BODY COULD
TWIST VERY EASILY, RESULTING IN DAMAGED
INSULATION ON THE CONDUCTORS AND A
POOR GROUND CONNECTION BETWEEN THE
CONDUIT AND THE CONDUIT BODY. SEE
SECTIONS 370-23(e) AND *(f)*.

THIS BODY IS NOT CONSIDERED
TO BE ADEQUATELY SUPPORTED.

C CONDUCTORS MAY BE SPLICED IN THESE
CONDUIT BODIES ONLY IF THE CONDUIT BODY
IS MARKED WITH ITS CUBIC-INCH CAPACITY
SO THAT THE PERMISSIBLE CONDUCTOR FILL
MAY BE DETERMINED USING THE CONDUCTOR
VOLUME FOUND IN *TABLE 370-16(b)*.

SPLICES MAY BE MADE IN
CONDUIT BODIES WHEN
MARKED WITH THEIR CUBIC-INCH
CAPACITY, *SECTION 370-16(c)(2)*.

Figure 9–67 Illustrated are conduit bodies supported by conduits coming out of the ground. One conduit does not properly support a conduit body.

more conduits are threaded wrenchtight into the hubs, but only when the conduits are supported within 18 inches (457 mm) of the conduit body. Figure 9–67 shows an example of this.

- That **do** contain wiring devices or have fixtures attached are considered supported when the conduit body is not larger than the largest trade size of the rigid metal conduit or intermediate conduit being used with the conduit body.

REVIEW QUESTIONS

1. What wiring method is most commonly used in your community for wiring houses?

    a) For the interior wiring _____

    b) For the service-entrance wiring. _____

    c) For electric ranges, ovens, dryers. _____

2. a) What is the color coding for the conductors in two-wire nonmetallic-sheathed cable?

b) What is the color coding for the conductors in three-wire nonmetallic-sheathed cable?

3. The conductors in nonmetallic-sheathed cable and armored cable are rated 90°C. In the spaces provided, mark T or F (True or False) for the following statements.

a) The ampacity must be based on the 60°C values in *Table 310-16*. _____

b) The ampacity must be based on the 75°C values in *Table 310-16*. _____

c) The ampacity must be based on the 90°C values in *Table 310-16*. _____

4. Nonmetallic-sheathed cable must be properly secured to framing members of building. In the spaces provided, mark T or F (True or False) for the following statements.

a) NMC must be secured within 12 inches of a box. _____

b) NMC must be secured within 18 inches of a box. _____

c) NMC must be secured every 4½ feet. _____

d) NMC must be secured every 36 inches. _____

e) NMC must be secured within 12 inches of a nonmetallic box that does not contain a cable clamp. _____

f) NMC must be secured within 8 inches of a nonmetallic device box that does not contain a cable clamp. _____

g) Two NMC cables may be run through the same knockout in a non-metallic device box. _____

h) NMC (two-wire, flat) must not be stapled on its edge. _____

i) NMC run across the top of joists in accessible attics must be protected with guard strips within 3 feet of a scuttle hole. There are no permanent stairs. _____

j) NMC run across the top of joists in accessible attics must be be protected with guard strips within 6 feet of a scuttle hole. There are no permanent stairs. _____

k) NMC must be protected with metal plates at least ¹⁄₁₆ inch thick if the bored hole for a cable is 1 inch from the face of the stud. _____

l) NMC must be protected with metal plates at least ¹⁄₁₆ inch thick if the bored hole for a cable is 1½ inch from the face of the stud. _____

m) NMC run on the side of a vertical stud must be protected with ¹⁄₁₆ inch metal plates or use "standoffs" if the cable comes within 1 inch of the face of the stud. _____

n) NMC run on the side of a vertical stud must be protected with ¹⁄₁₆ inch metal plates or use "standoffs" if the cable comes within 1½ inch of the face of the stud. _____

o) NMC may be run through the full length of an air duct. _____

p) NMC may be run at a right angle to the long dimension of an air duct. _____

q) NMC run through metal framing members must be protected from abrasion by using grommets (bushings) at each location where the cable runs through metal framing members. _____

r) NMC connectors may contain as many NMC cables as will fit into the connector.   _____

s) NMC may be bent to a radius of 5 times the diameter of the cable.   _____

t) NMC may be bent to a radius of 2½ times the diameter of the cable.   _____

5. Armored cable (BX) is identified by letters. Match the explanations with the correct letters.

   1. _____ Type ACT   a) Conductors are 90°C. All right to use 75°C ampacity if terminals and equipment are suitable for 75°C

   2. _____ Type ACT-HH   b) Conductors are 60°C. Use 60°C ampacity.

   3. _____ Type ACT-B   c) Conductors are 75°C. All right to use 75°C ampacity if terminals and equipment are suitable for 75°C.

   4. _____ Type ACT-H   d) Conductors are 90°C. Use 60°C ampacity.

6. When using armored cable that contains a bonding strip under the armor, which of the following statements is true? Mark T (True) or F (False).

   a) The bonding strip must be connected to the grounding screw inside of a metal box.   _____

   b) The bonding strip may be folded back over the armor. The armored cable is then inserted into a connector or into the hole in a box that contains armored cable clamps.   _____

   c) The bonding strip must be connected to the grounding terminal of a grounding-type receptacle to ensure proper ground.   _____

   d) The bonding strip may be used as a neutral conductor.   _____

7. When using armored cable, what must be inserted into the end of the cable to prevent abrasion to the conductors?

   _____

8. Because armored cable has considerably more mechanical strength than nonmetallic-sheathed cable, it may be installed underground.

   a) True

   b) False

9. In the spaces provided, mark T or F (True or False) for the following statements.

   a) Type SE service-entrance cable is suitable for underground direct burial.   _____

   b) Type USE service-entrance cable is suitable for aboveground use.   _____

   c) Type US service-entrance 3-wire cable with an uninsulated neutral may be used to hook up an electric range.   _____

   d) A reduced size uninsulated neutral in a Type SE service-entrance cable is considered to have the same current-carrying capacity as the insulated ungrounded conductors in that cable.   _____

10. In the spaces provided, mark T or F (True or False) for the following statements.

   a) EMT, when used underground or buried in concrete that is in direct contact with the earth, must have supplemental corrosion protection.   _____

   b) EMT, because of its smooth inside surface, is permitted to have six quarter bends between pull points, such as junction boxes.   _____

   c) EMT shall be secured within 3 feet of an outlet box.   _____

   d) EMT shall be secured at least every 10 feet.   _____

e) EMT running horizontally through framing members is considered to be adequately supported, but in addition, the EMT must still be securely fastened within 3 feet of an outlet box. _____

f) The sum total of all bends and/or offsets between pull points shall not exceed 360 degrees. This is equivalent to a maximum of four quarter bends. _____

g) EMT, when used with conductors No. 4 AWG or larger, must have insulated throat fittings, insulated bushings, or approved separate "slip in" insulators. _____

11. Check which of the following raceways would be best suited to use for a mast type service?

a) _____ electrical metallic tubing (EMT)

b) _____ rigid metal conduit

c) _____ rigid nonmetallic conduit (PVC)

12. Rigid nonmetallic conduit is available in two types. They are Schedule _____ and Schedule _____. Schedule _____ is stronger than Schedule _____.

13. One of the concerns when installing PVC conduit is that it expands and contracts with heat and cold much more than metallic raceways. This is referred to as thermal expansion. Because of this, it may be necessary to use _____ _____ in long runs.

14. Which of the following statements is true? What section of the *National Electrical Code*® supports your answer? *Section* _____.

a) Before pulling conductors into a raceway, the raceway must be completely installed.

b) Because of difficulty in pulling conductors through a raceway, the NEC permits installing a portion of the raceway—pulling the conductors through the partial raceway installation—then slide the remaining section of raceway over the conductors to complete the installation.

15. You have installed EMT throughout a basement. A security system is installed by others. The installer of the security system secures a number of the lightweight, low-voltage cables to the EMT. Do you see anything wrong with this? Explain. What section of the *NEC*® supports your answer?

_____

_____

16. Flexible metal conduit is commonly used to make connections to electrical equipment such as motors, attic fans, food waste disposers, dishwashers, recessed fixtures, and similar equipment where flexibility is needed. In the spaces provided, mark T or F (True or False) for the following statements.

a) Flexible metal conduit size ½ inch is suitable as the equipment grounding means when the overcurrent device does not exceed 20 amperes, and the flex is not over 6 feet long. _____

b) Flexible metal conduit, regardless of size, must have a separate equipment grounding conductor installed in it for lengths longer than 6 feet. _____

c) Flexible metal conduit size ⅜ inch is limited to 6 feet length.

  d) Flexible metal conduit size ⅜ inch is permitted in lengths
   exceeding 6 feet provided an equipment grounding conductor
   is installed in the flex.   _____

  e) Flexible metal conduit size ⅜ inch is permitted to be used as a
   6 foot "fixture whip."   _____

  f) Flexible metal conduit is permitted to hook up an air-conditioner
   located on the outside of the house.   _____

17. When installing liquidtight flexible nonmetallic conduit, an _____
  _____ _____ must be installed.

18. You will be making an underground installation between the house and a tool shed. In
  the spaces provided, mark T or F (True or False) for the following statements.

  a) Use Type NMC cable.   _____

  b) Use Type UF cable.   _____

  c) Use Type ACT cable.   _____

  d) Bury Type THHN directly in the ground, but be sure to protect the
   conductors with a creosoted board ¾ inch thick and 6 inches wide. _____

19. You are installing Type UF underground wiring for a number of 120-volt outdoor
  receptacles, landscape lighting, decorative lighting, and post lights around a dwelling.
  In the spaces provided, mark T or F (True or False) for the following statements.

  a) Bury the cable at least 12 inches. The branch circuit is not GFCI
   protected.   _____

  b) Bury the cable at least 12 inches. The branch circuit is rated
   20 amperes and the branch circuit is GFCI protected.   _____

  c) Bury the cable at least 12 inches. The feeder to the tool shed
   is rated 30 amperes and is GFCI protected.   _____

  d) Bury the cable at least 24 inches. The feeder to the tool shed
   is rated 30 amperes and is *not* GFCI protected.   _____

  e) Bury the cable at least 12 inches under the front driveway. The
   branch circuit is rated 20 amperes and is GFCI protected.   _____

  f) Bury the cable at least 18 inches under the front driveway. The
   branch circuit is rated 20 amperes and is *not* GFCI protected. _____

20. In the spaces provided, mark T or F (True or False) for the following statements. The
  statements relate to residential wiring.

  a) Overhead conductors must be kept a minimum of 10 feet above
   grade where the voltage to ground does not exceed 150 volts.   _____

  b) Overhead conductors in general must be kept a minimum of
   3 feet above a roof where the roof pitch is not less than 4/12.   _____

  c) Overhead conductors must be kept a minimum of 18 inches above
   the overhang of a roof where the conductor length does not exceed
   6 feet where it passes over the roof. The conductors are service
   conductors and are secured to a through-the-roof support, commonly
   referred to as a mast service.   _____

  d) A window is designed to be opened. A minimum clearance of 2 feet
   must be maintained between open conductors and the side edge of
   the window.   _____

e) A window is designed to be opened. A minimum clearance of
    12 inches must be maintained between open conductors and the top
    of the window.   _____

21. You are called upon to install a considerable number of decorative lighting fixtures in
    the trees of the wooded area surrounding a home. In the spaces provided, mark T or F
    (True or False) for the following statements.

    a) To avoid costly digging, you choose to support the overhead wiring
        from the trees, running the overhead conductors from tree to tree.   _____

    b) You choose to run all of the necessary branch-circuit wiring
        underground, then run up the side of the trees wherever a lighting
        fixture will be mounted.   _____

22. List the proper size of electrical metallic tubing (EMT) for the following. Assume that
    it is new work and then THHN conductors are used. Refer to *Chapter 9, Table C1*.

    a) 3 No. 14 _____"   f) 4 No. 12 _____"

    b) 4 No. 14 _____"   g) 5 No. 12 _____"

    c) 5 No. 14 _____"   h) 6 No. 12 _____"

    d) 6 No. 14 _____"   i) 6 No. 10 _____"

    e) 3 No. 12 _____"   j) 4 No. 8 _____"

23. Three electric baseboard heaters rated 230 volts are to be connected to separate branch
    circuits. The instructions require that No. 12 AWG Type THHN conductors be used.
    You decide to install the six conductors in one EMT. What is the minimum size EMT
    required for this installation?

    a) ½ inch

    b) ¾ inch

24. Find the minimum size EMT for four No. 12 AWG Type THHN and four No. 10 AWG
    Type THWN conductors. This is a new installation. Show your work.

25. Find the minimum size EMT for three No. 14 AWG Type THHN and four No. 12
    AWG Type THHN conductors. This is a new installation. Show your work.

26. Find the minimum size flexible metal conduit for two No. 12 AWG Type TW conductors and three No. 8 AWG Type TW. This is a new installation. Show your work.

27. Find the minimum size Schedule 40 PVC for three No. 3/0 AWG Type THWN conductors and one No. 3 AWG bare. This is a new installation. Show your work.

28. You are installing 120-volt outdoor receptacles under shrubbery for the purpose of plugging in decorative outdoor lighting. You are using conduit bodies, referred to as FS boxes for the receptacles. You are running PVC conduit underground, and will stub up into the hubs on the bottom of the conduit bodies. In the spaces provided, mark T or F (True or False) for the following statements.
    a) Run a single PVC conduit into the bottom hub of the conduit body.
       This will provide adequate support for the conduit body. _____
    b) Run two PVC conduits into the bottom hubs of the conduit body.
       This will provide adequate support for the conduit body. _____
    c) Install a grounding-type receptacle in the conduit body. _____
    d) Install a GFCI-type receptacle in the conduit body.
    e) Install the conduit body not more than 24 inches above grade.

Outlet Boxes and Device Boxes

OBJECTIVES

After studying this unit, you will be able to:

- Know that there are basically two types of boxes: metal and nonmetallic.
- Learn the proper application for different types of boxes.
- Know how to install ceiling boxes that will *safely* support a ceiling fan.
- Know that there is a maximum number of wires permitted in a given size box.
- Learn how to determine the proper size box, taking into consideration the number of conductors, the size of the conductors, cable clamps, and wiring devices in a box.
- Know how to make proper use of a junction box that is an integral part of a recessed lighting fixture.
- Understand that there are special outlet and switch boxes designed for remodel (old) work.
- Be aware that certain precautions are necessary when installing boxes back to back to minimize or prevent the spread of fire.

INTRODUCTION

Outlet boxes and switch (device) boxes are main components of a residential electrical system. First, you decide what wiring method you will use (non-metallic-sheathed cable, armored cable, EMT, etc.). Next, you decide whether to install metal outlet and switch boxes, or nonmetallic-type. The wiring method will have a direct bearing on the type of clamps needed in the outlet and switch boxes.

NATIONAL ELECTRICAL CODE® ARTICLES RELATING TO BOXES

Some of the more important requirements for electrical boxes are found in the *National Electrical Code®* in *Articles 300, 370,* and *410.* Where a specific code requirement is involved for a specific application, the code rule might be found in other sections of the Code.

Figure 10–1 shows some of the commonly used electrical boxes in residential wiring.

In conformance to *Section 300-15,* a box, or in some cases, a fitting must be installed at every point in an electrical installation where there are splices, outlets, switches, junctions, or pull points. Figure 10–2 emphasizes this requirement.

In some instances, a fitting could be used instead of a box where a change is made from one wiring method to another. An example of this is illustrated in Figure 10–3.

MATERIAL ELECTRICAL BOXES ARE MADE OF?

Electrical outlet and device boxes for use in house wiring are made of steel and nonmetallic material such as polyvinyl chloride (PVC). *Section 370-3* permits the use of nonmetallic boxes. Conduit

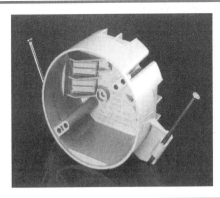

Figure 10–1   Here are some of the types of nonmetallic boxes commonly used in wiring houses. (*Courtesy Carlon Electrical Products, a Lamson & Sessions Company*)

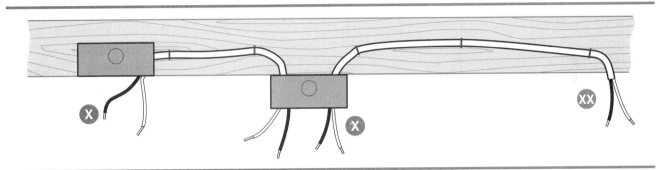

Figure 10–2   *Section 300-15(a)* requires that a box or fitting must be installed wherever there are splices, outlets, switches, or other junction points. Refer to the points marked "X." A Code violation is shown at point "XX."

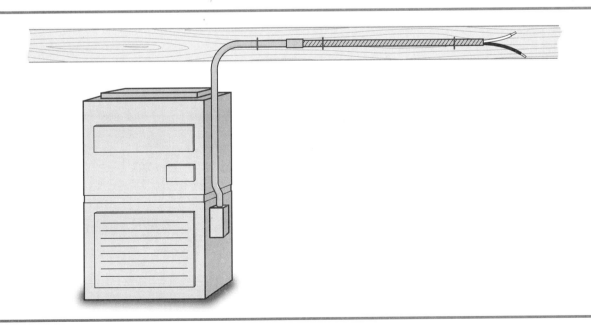

Figure 10–3   *Section 300-15(c)* and *(f)*, permits a transition to be made from one wiring method to another wiring method. In this case, the armor of the Type AC cable is removed, allowing sufficient length of the conductors to be run through the raceway. A fitting must be used at the transition point, and the fitting must be accessible after installation.

bodies are made of malleable iron, die cast zinc, cast aluminum, and PVC. Large pull and junction boxes are made of sheet metal. A more detailed description and installation code requirements for electrical boxes follows in this unit.

TYPES OF BOXES

A visit to an electrical distributor, or to the electrical department of a hardware store or home center will show literally hundreds of types of electrical boxes for use in house wiring. The type of electrical boxes to use depends on how and where they will be used.

There are electrical boxes for new construction and remodel work. Some boxes are mounted with brackets, while others are mounted with nails. Where lighting fixtures will be installed, boxes can be mounted with bar hangers. Where ceiling fans are installed, special attention must be given to the mounting of the outlet box. This is discussed later in this unit. Figure 10–4 shows a metal device box, an outlet box, and outlet box with an adjustable bar hanger.

What's a Gang?

The electrical trade uses many special words unique to the electrical industry. When talking about electrical boxes where wiring devices will be installed, the term "gang" is used. This term is not defined in the *National Electrical Code.*® A gang is one position for a wiring device such as a switch or

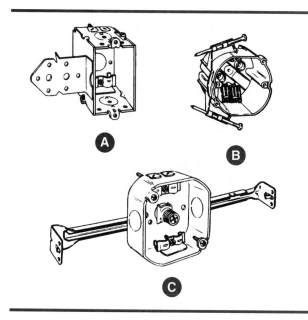

Figure 10–4 (A) A bracket-type device box, (B) a nail-type outlet box, and (C) an outlet box mounted with an adjustable bar hanger. (*Courtesy of Hubbell Electrical Products, a Division of Hubbell Incorporated [Delaware]*)

receptacle. A device (switch) box would be referred to as a "single-gang" box. The term gang is not used for round or octagon boxes, or for square boxes with round plaster rings.

Device boxes with removable sides can be "ganged" together by removing the sides. Figure 10–5 illustrates how three device boxes have been ganged together for three switches. This would be

Figure 10–5 Three sectional device boxes ganged together to accommdate three switches.

referred to as "three-gang," and could be three individual single-gang boxes ganged together, or could be a single three-gang box.

A 4-inch square outlet box trimmed with a 4-inch square single-gang plaster cover provides one position for wiring devices. This would be referred to as single-gang whether there was a single-wiring device on one yoke, or one, two, or three interchangeable wiring devices on one yoke.

A 4-inch square outlet box trimmed with a 4-inch square two-gang plaster cover would be referred to as "two-gang."

The term gang is used in *Table 370-16(a)*.

There are electrical boxes for rework (remodel). These boxes are installed through a cutout hole in the wall or ceiling. A device box having plaster ears can be used for remodel work using a support device mounted through a knockout in the back of the box. When the screw on the support is tightened, the support pulls up tightly behind the plaster or dry wall. Figure 10–6 shows one type of remodel work device box and a device box support.

There are electrical boxes for use with nonmetallic-sheathed cable. Figure 10–7 shows some typical boxes used with nonmetallic-sheathed cable.

There are electrical boxes for use with armored cable (BX). Figure 10–8 shows some typical boxes for use with armored cable.

There are electrical boxes for use with conduit, such as electrical metallic tubing (EMT). These knockouts will accommodate both EMT connectors or cable (NMC and BX) connectors. These boxes might have all knockouts ½ inch, ¾ inch, or combinations of both sizes. Figure 10–9 illustrates some typical boxes for use with conduit or with cable when cable connectors are used.

Figure 10–10 shows some electrical boxes for use with metal framing members such as steel studs. One is for conduit wiring, one for nonmetallic-sheathed cable wiring, and one for armored cable wiring.

Boxes For Use in Concrete, Tile, or Brick

When device boxes are installed in masonry such as concrete, tile, stone, or brick, use boxes that have their ears turned inward. These are called masonry boxes. With the inward turned ears, mortar

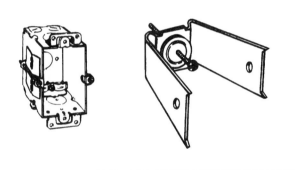

Figure 10–6 A remodel-type device (switch) box. Also shown is a support that is mounted through the back knockout of a device box that has plaster ears. (*Courtesy of Hubbell Electrical Products, a Division of Hubbell Incorporated [Delaware]*)

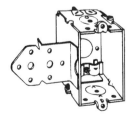   

Figure 10–7 Typical types of boxes for use with nonmetallic-sheathed cable. Shown are a bracket-mounting single-gang device box, a nail-mounting single-gang device box, a two-gang nail-mounting nonmetallic device box, and a bracket-mounting 4–inch octagon outlet box. Note the internal cable clamps that hold the nonmetallic-sheathed cable tightly inside the box. (*Courtesy of Hubbell Electrical Products, a Division of Hubbell Incorporated [Delaware]*)

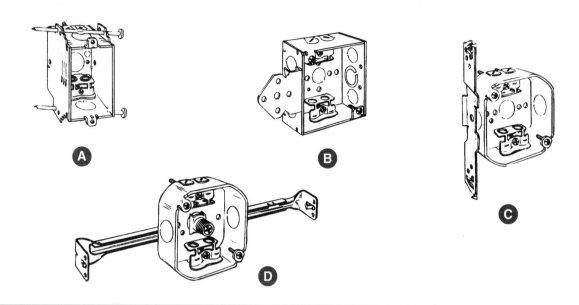

Figure 10–8   Some typical boxes used with armored cable. Shown are (A) a nail-mounting single-gang device box, (B) a bracket-mounting 4-inch square box, (C) a bracket-mounting 4-inch octagon box, and (D) a bar hanger-mounting 4-inch octagon box. Note the internal cable clamps that hold the armored cable tightly inside the box have rounded holes with rounded edges to protect the conductors from abrasion. (*Courtesy of Hubbell Electrical Products, a Division of Hubbell Incorporated [Delaware]*)

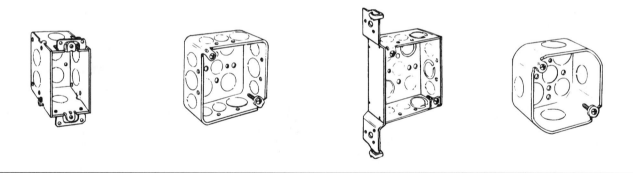

Figure 10–9   Some typical boxes used with electrical metallic tubing. Shown are a single-gang device box, a 4-inch square box that has ½" and ¾" knockouts, a 4-inch square bracket-mounting that has combination ½"–¾" knockouts, and a 4-inch octagon box that has ½" and ¾" knockouts. (*Courtesy of Hubbell Electrical Products, a Division of Hubbell Incorporated [Delaware]*)

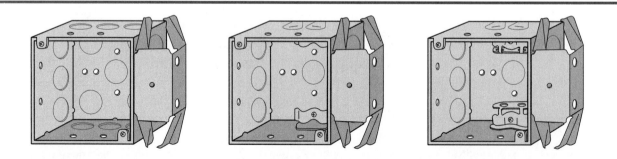

Figure 10–10   Types of boxes that can be used with steel framing construction. Note the side clamps that "snap" directly onto the steel framing members.

will not fill in behind the ears as with standard device boxes having outward turned ears. This makes the job of installing a lighting fixture, switch, or receptacle easy, rather than having to chop out mortar or cement from behind the ears. Masonry boxes fit perfectly in a course of brick when installed in the horizontal position. Because the ears are inward, masonry boxes have more cubic inch space than a similar box that has the ears turned outward. For comparison purposes, in *Table 370-16(a)*, a single-gang 3½" device box has 18 cubic inch capacity, whereas a single-gang 3½" deep masonry device box has 21 cubic inch capacity. Figure 10–11 shows a single-gang and a two-gang masonry box. Note the combination ½-inch and ¾-inch knockouts in these boxes.

Boxes For Use Outdoors

Surface-mounted cast boxes and nonmetallic boxes are available for outdoor use such as on a deck with proper weatherproof covers. Figure 10–12 shows a single-gang box, a switch cover, and a duplex receptacle cover. These covers are spring loaded, and will snap back automatically to keep water from entering the box. This is a requirement of *Sections 370-15(a)* and *410-57*.

Where a lighting fixture is to be mounted, be sure to install boxes that have a "fixture stud" in the box for ease in hanging the lighting fixture, Figure 10–13. Wall-mounted lighting fixtures are permitted to be secured with No. 6/32 screws to device boxes and plaster rings provided they do not weigh more than 6 pounds (2.72 kg) and do not exceed 16 inches (406 mm) in any dimension. This permission is found in *Section 370-27(a)*, and would typically include wall mounted bathroom lighting fixtures and small wall mounted porch bracket lighting fixtures.

Key *National Electrical Code®* Rules

Here is a brief summary of the Code requirements pertaining to boxes. These rules cover most situations found in residential wiring situations.

Section 370-3. Nonmetallic boxes are generally used with nonmetallic-sheathed cable. However, nonmetallic boxes may be used with metal raceways when bonding is provided between raceway entries. This will ensure that proper grounding throughout the system is attained. Although the use of nonmetallic boxes is permitted when using metal raceways, electricians will usually stick with metallic boxes when installing metallic raceways.

Section 370-4. When metallic boxes are installed, they must be properly grounded in accordance with the grounding requirements of *Article*

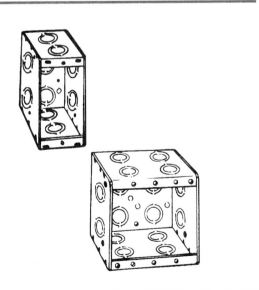

Figure 10–11   A single-gang and a two-gang masonry box that have combination ½" and ¾" knockouts. (*Courtesy of Hubbell Electrical Products, a Division of Hubbell Incorporated [Delaware]*)

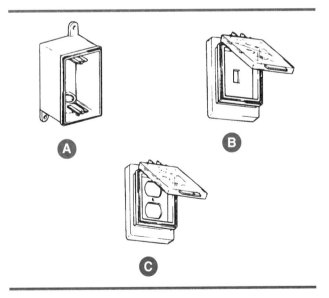

Figure 10–12   (A) A single-gang nonmetallic surface-mounting box for outdoor use. Also shown is (B) a weatherproof switch cover and (C) a weatherproof duplex receptacle cover. (*Courtesy of Hubbell Electrical Products, a Division of Hubbell Incorporated [Delaware]*)

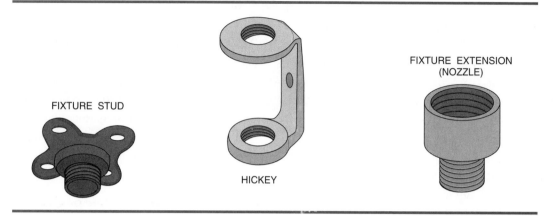

FIXTURE STUD

HICKEY

FIXTURE EXTENSION
(NOZZLE)

Figure 10–13 A fixture stud, a fixture hickey, and a fixture extension (nozzle). These items are used for mounting lighting fixtures to outlet boxes. Bar hangers generally have a fixture stud, so that when the center rear knockout of the box is removed, the fixture stud will protrude into the outlet box.

250. Section 250-148 states that when using metallic boxes, equipment grounding conductors must be attached to the box. Most metal boxes have a No. 10-32 hole tapped in the back of the box marked GND or GRND. This hole is for a No. 10-32 hexagon-shaped green screw to connect the equipment grounding conductor. Grounding clips can be used to connect the equipment grounding conductor to a box as illustrated in Figure 10–14.

Figure 10–15 shows pigtails having No. 14 AWG or No. 12 AWG copper conductors complete with a No. 10-32 hexagon-shaped green grounding

screw. These can be attached to the box, with the other end of the conductor connecting to the grounding terminal of a receptacle.

Figure 10–16 shows three types of nonmetallic boxes.

When nonmetallic boxes are used, *Section 250-148* requires that a means must be provided to

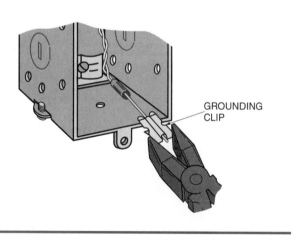

GROUNDING
CLIP

Figure 10–14 This sketch shows a method of attaching a ground clip to a metal device box. Note that two bare equipment grounding conductors have been spliced together with a small crimp-on barrel connector, then one equipment grounding conductor is secured under the metal grounding clip.

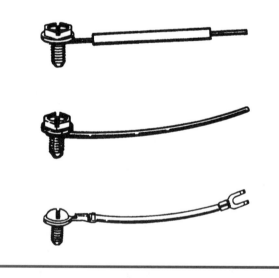

Figure 10–15 Three types of pigtails for use with metal boxes when it is necessary to bring out an equipment grounding conductor to a receptacle. Illustrated are an insulated conductor, a bare conductor, and an insulated conductor with a spade terminal on the end. All three pigtails come with the No. 10–32 grounding screw. (*Courtesy of Hubbell Electrical Products, a Division of Hubbell Incorporated [Delaware]*)

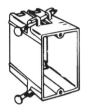

Figure 10–16 An assortment of nonmetallic boxes for use with nonmetallic-sheathed cable. Note that these boxes have integral cable clamps. (*Courtesy of Hubbell Electrical Products, a Division of Hubbell Incorporated [Delaware]*)

connect equipment grounding conductors to the wiring device such as a receptacle. Where a single nonmetallic-sheathed cable is brought into the box, the equipment grounding conductor is merely connected to the grounding terminal of the receptacle. Where multiple nonmetallic-sheathed cables are brought into the box, the equipment grounding conductors must be spliced together, with a pigtail brought out of the splice and connected to the grounding terminal of the receptacle. Grounding is covered in Unit 12.

Section 370-16(c). Conduit bodies, referred to as LB, LL, LR, CC, and so forth, must not contain splices or taps unless they are marked with their cubic inch capacity. When so marked, the number of conductors permitted is based on *Table 370-16(b)*. So called "short radius conduit bodies" covered by *Section 370-5* shall not contain splices or taps. They are really sized to contain the conductors running through the conduit body.

Section 370-15. Boxes and conduit bodies for use in wet or damp locations must be "listed" for such use. They may be cast metal, nonmetallic, or metal that resists rusting. *Section 300-6(a)* states that boxes *"shall be suitably protected against corrosion inside and outside by a coating of approved corrosion-resistant material." Section 300-6(b)* gives permission to install boxes in concrete or in direct contact with the earth, provided the boxes have approved corrosion protection.

Section 370-16. Boxes and conduit bodies shall be of sufficient size to provide free space for all of the enclosed conductors. One of the biggest problems is not using boxes that are large enough for the number of conductors, internal clamps, fixture studs, wiring devices, and so forth, in the box. Crowding the wires into a box leads to short circuits

and ground faults, which in turn can present a shock hazard and fire hazard. A detailed discussion of how to determine the proper size box follows this summary of basic Code requirements.

Section 370-17.

1. Secure raceways and cables to the box.

   • For nonmetallic-sheathed cable and armored cable, boxes with the proper internal cable clamp will secure the cable to the box.

   • For EMT and armored cable used with connectors, the locknut on the connector provides the required securing.

2. When using nonmetallic cable with nonmetallic-sheathed boxes, be sure to extend the cable at least ¼ inch (6.35 mm) into the box.

3. There is an exception to the above rule. When using nonmetallic boxes that are not larger than 2¼ inches by 4 inches, you do not have to secure the cable to the box provided the nonmetallic-sheathed cable is stapled within 8 inches of the box. This rule is for nonmetallic boxes only, not for metallic boxes.

4. When using nonmetallic boxes, it is all right to run more than one nonmetallic-sheathed cable through a single knockout in the box. This is permitted because the box is separately mounted and the cable is separately supported. This rule is for nonmetallic boxes only, not for metallic boxes (Figure 10–17).

Section 370-18. All unused openings (other than the mounting holes for nails and screws) must be closed to prevent dirt and rodents from getting into the box. This also discourages an electrical arc from getting outside of the box in the event of a

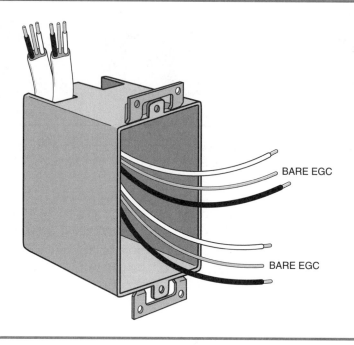

BARE EGC

BARE EGC

Figure 10–17   Multiple nonmetallic-sheathed cables may be run through a single knockout opening in a nonmetallic box, *Section 370-17(c), Exception.*

short circuit or ground fault. *Section 110-12(a)* also requires that unused knockouts must be closed. Figure 10–18 shows knockout closures that can be used to close up unused knockouts. Note that this

KNOCKOUT CLOSURES

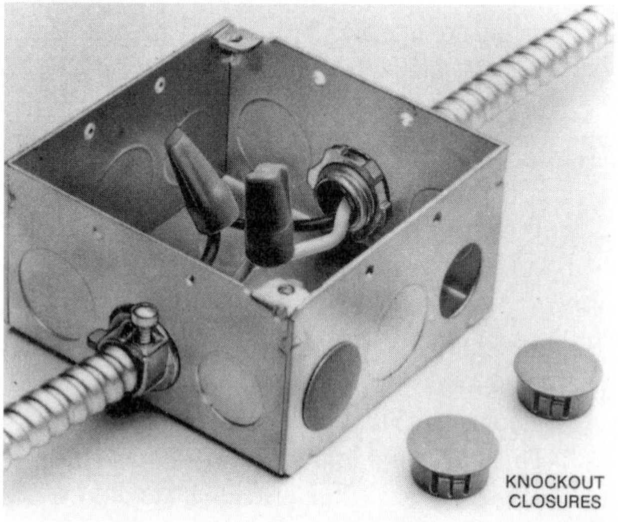

Figure 10–18   Unused openings in boxes must be closed, *Sections 110–12* and *370-18*. Note that the conductors are too short. *Section 300-14* requires that there must be at least 6 inches of free conductor length.

photo shows a Code violation of *Section 300-14* that requires 6 inches of free conductor length. The wires in this photo are too short to work with.

Section 370-19.   The rear and sides must completely enclose devices attached to the box. This section also tells us not to combine the securing of the box and the attachment of the wiring device with the same screw. These rules seem obvious, but the *NEC®* decided to include these statements anyhow.

Section 370-20.   When installing electrical boxes in noncombustible material such as concrete or tile, boxes must not be set back more than 1/4 inch (6.35 mm) from the finished surface. When installing electrical boxes in combustible material such as wood paneling, boxes must be installed so as to be flush with the finished surface (Figure 10–19).

If you "goof," then an item such as an "extension ring" sometimes referred to as a "box extender" could be used to comply with *Section 370-20.* These are available in metal and nonmetallic material (Figure 10–20).

Section 370-21.   To prevent the spread of fire, the space between electrical boxes and drywall, plaster, or plasterboard must be repaired so that

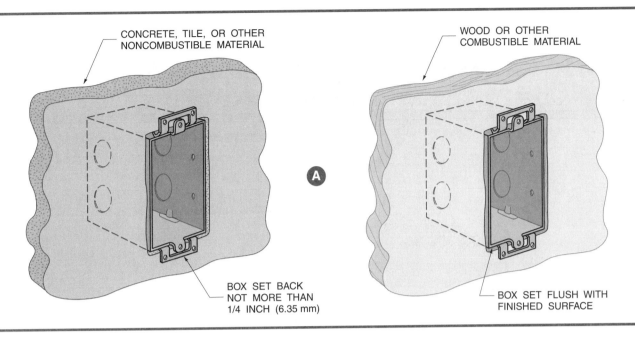

CONCRETE, TILE, OR OTHER NONCOMBUSTIBLE MATERIAL

WOOD OR OTHER COMBUSTIBLE MATERIAL

Ⓐ

BOX SET BACK NOT MORE THAN 1/4 INCH (6.35 mm)

BOX SET FLUSH WITH FINISHED SURFACE

Figure 10–19 Boxes installed in noncombustible material are permitted to be set back not more than ¼ inch from the finished surface. Boxes installed in combustible material must be flush with the finished surface.

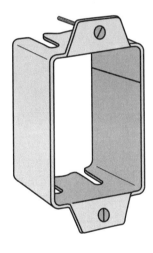

Figure 10–20 A "box extender" that can be used when a wall box is set too far back from the finished surface of the wall.

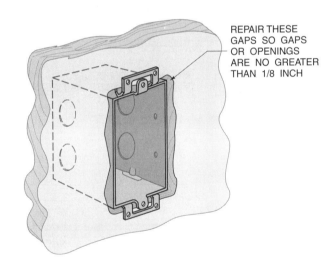

REPAIR THESE GAPS SO GAPS OR OPENINGS ARE NO GREATER THAN 1/8 INCH

Figure 10–21 Close the gap between the box and the wall so that the gap is no larger than ⅛ inch.

there will be no opening or gap greater than ⅛ inch (3.18 mm). Patching plaster is great for this. Drywall installers and drywall finishers will usually do this as they do their work. But the responsibility to see that this is done is on the shoulders of the electrician! (See Figure 10–21.)

Section 370-22. When making an exposed surface extension from a concealed flush-mounted box,

be sure that the mounting to the concealed box is mechanically secure. Do not make the extension as illustrated in Figure 10–22(A) because the connections in the box would not be accessible. The extensions illustrated in Figure 10–22(B) and (C) are all right because the connections in the box are accessible. Note the manner in which the integrity of the equipment grounding conductor is accomplished.

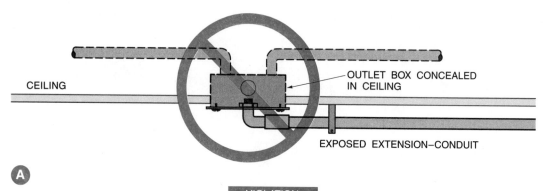

A

VIOLATION. IT IS AGAINST THE CODE
TO MAKE A RIGID EXTENSION FROM A COVER THAT IS ATTACHED TO
A CONCEALED OUTLET BOX, *SECTION 370-22.* IT WOULD BE DIFFICULT TO
GAIN ACCESS TO THE CONNECTIONS INSIDE THE BOX.

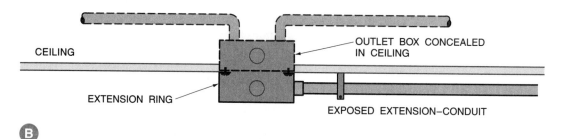

B

THIS MEETS CODE. A BOX OR EXTENSION
RING MUST BE MOUNTED OVER AND MECHANICALLY SECURED TO THE
CONCEALED OUTLET BOX, *SECTION 370-22.*

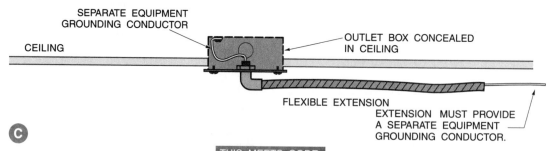

C

THIS MEETS CODE. IT IS PERMITTED TO
MAKE A SURFACE EXTENSION FROM A COVER FASTENED TO A CONCEALED
BOX WHERE THE EXTENSION WIRING METHOD IS FLEXIBLE AND WHERE A
SEPARATE GROUNDING CONDUCTOR IS PROVIDED SO THAT THE GROUNDING
PATH IS NOT DEPENDENT UPON THE SCREWS THAT ARE USED TO FASTEN
THE COVER TO THE BOX.

IT WOULD BE A CODE VIOLATION TO MAKE THE EXTENSION IF A
FLEXIBLE WIRING METHOD WAS USED, BUT HAD NO PROVISIONS FOR
A SEPARATE GROUNDING CONDUCTOR.

Figure 10–22 Making an extension from a flush-mounted box requires that the wiring inside the flush-mounted box be accessible, and that the extension be mechanically secure.

Section 370-23. This section sets forth some obvious and some less obvious requirements for the supporting of electrical boxes. Boxes shall be rigidly and securely mounted. *Section 300-11* also requires that boxes be securely fastened in place

For new work, boxes that have external brackets are most commonly used. For mounting boxes between studs, "Kruse Supports" can be used (Figure 10–23). If wood strips are used, they must have a cross-sectional dimension of not less than 1×2 inches, *Section 370-23(b)(2)*.

For mounting more than one box, Figure 10–24 shows a bracket that spans the distance between two studs. The multiple openings can be used for different purposes, such as line voltage in one box and low voltage (a thermostat) in the other box.

Section 370-24. The shallowest electrical box permitted is ½ inch (12.7 mm). An example of where such a shallow box might be used is where the box is mounted directly below a ceiling joist. These are oftentimes fastened to the bottom of

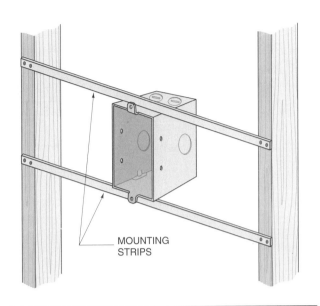

MOUNTING
STRIPS

Figure 10–23 One or more switch (device) boxes can be mounted between studs using metal mounting strips, often referred to as "Kruse Strips." The Kruse Strips slide over the plaster ears on the box.

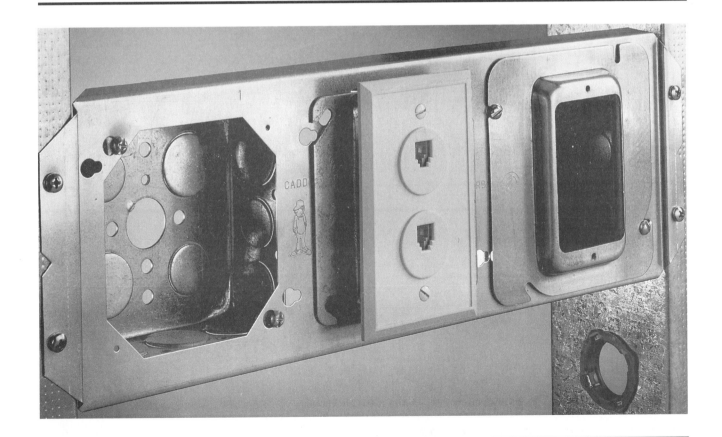

Figure 10–24 A bracket for use on steel framing members. The bracket spans the space between two steel studs. Note that more than one box can be mounted on this bracket. (*Courtesy Erico,® Inc.*)

a ceiling joist and are referred to as "pancake" boxes. Usually, these boxes have a diameter of 4 inches. The number of conductors permitted in this shallow depth box is limited because their cubic inch capacity is only 6 cubic inches. Looking at *Table 310-16(b),* we find the cubic inch volume of:

Two No. 14 AWG conductors is
 $2 \times 2 = 4$ cubic inches

Three No. 14 AWG conductors is
 $3 \times 2 = 6$ cubic inches

Two No. 12 AWG conductors is
 $2 \times 2.25 = 4.5$ cubic inches

Three No. 12 AWG conductors is
 $3 \times 2.25 = 6.75$ cubic inches

Section 370-25(a)* and *(b). This section allows the use of metallic or nonmetallic faceplates with nonmetallic boxes.

Any combustible ceiling or wall finish between an electrical box and a fixture canopy must be covered with noncombustible material. This is easily accomplished because all lighting fixtures come with a piece of fiberglass backing. This keeps the intense heat from the lamps getting to the wires in the box, and to the combustible material of the ceiling or wall. This same requirement is found in *Section 410-13.*

Section 370-27. Boxes installed for the intention of mounting a lighting fixture on them shall have a means of attaching a lighting fixture. Most lighting fixtures can be attached to an outlet box with No. 8/32 screws, or to a fixture stud inside the box.

It is interesting to note that Underwriters Laboratories at the time of the writing of this text has not investigated nor have they "listed" the following for the supporting of lighting fixtures:

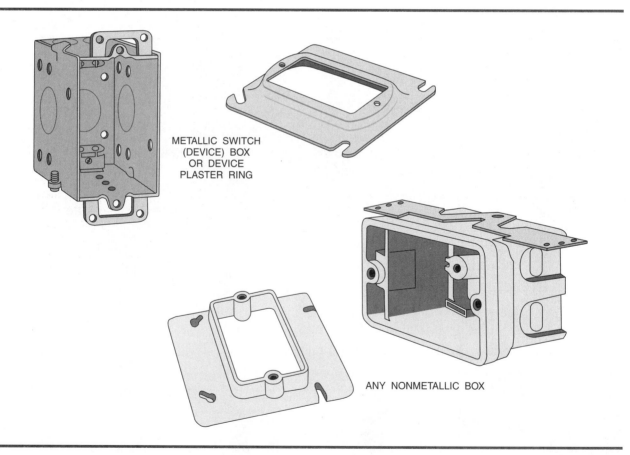

METALLIC SWITCH
(DEVICE) BOX
OR DEVICE
PLASTER RING

ANY NONMETALLIC BOX

Figure 10–25 *Section 370-27(a) Exception* **permits wall-mounted lighting fixtures to be secured with No. 6/32 screws to device boxes and plaster rings provided the fixtures do not weigh more than 6 pounds (2.72 kg) and do not exceed 16 inches (406 mm) in any dimension. These types of boxes and plaster rings are not suitable for supporting ceiling suspended (paddle) fans.**

- Nonmetallic device (switch) boxes
- Nonmetallic device (switch) plaster rings
- Metallic device (switch) boxes
- Metallic device (switch) plaster rings

However, as previously stated, *Section 370-27(a)* permits wall-mounted lighting fixtures to be secured with No. 6/32 screws to device boxes and plaster rings provided the fixtures do not weigh more than 6 pounds (2.72 kg) and do not exceed 16 inches (406 mm) in any dimension.

The boxes and covers illustrated in Figure 10–25 have No. 6-32 holes, and are not suitable for supporting ceiling fans.

Electrical boxes installed in floors must be "listed" for that purpose.

Electrical boxes installed for the support of ceiling fans must be "listed" for the purpose, unless the ceiling fan is supported independently of the electrical box.

Boxes for Supporting Ceiling-Suspended (Paddle) Fans

Let's look at hanging ceiling fans a little closer. Pay careful attention to the marking *on the carton* of metallic and nonmetallic boxes to see if they are "listed" as suitable for the support of lighting fixtures and/or ceiling fans.

The safe supporting of a ceiling-suspended (paddle) fan involves three issues:

—the actual weight of the fan

—the twisting and turning motion when started

—vibration caused by unbalanced fan blades

A typical residential ceiling-suspended (paddle) fan is shown in Figure 10–26.

Most residential-type ceiling-suspended (paddle) fans weigh 35 pounds or less, and may be supported by an outlet box that is clearly marked as being "Acceptable For Fan Support," *Sections 370-27(c)* and *422-18.* This is shown in Figure 10–27.

For ceiling-suspended (paddle) fans that weigh more than 35 pounds, the fan must be supported independent of the box.

The combined weight of the fan and any accessories such as a light kit must be considered when determining compliance with the 35-pound requirement.

Figure 10–26 A typical residential ceiling suspended (paddle) fan. These ceiling fans must be mounted only to boxes that are marked "Acceptable For Fan Support." (*Courtesy NuTone*)

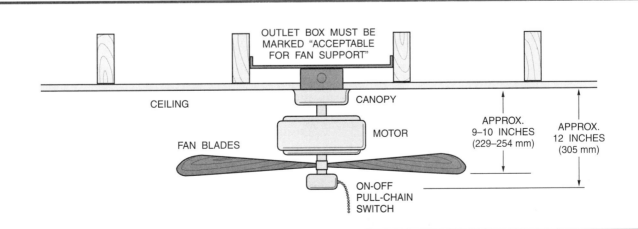

Figure 10–27   View of typical mounting of a ceiling suspended (paddle) fan. Note that the ceiling box must be marked "Acceptable For Fan Support."

Where ceiling-suspended (paddle) fans are likely to be installed, make sure that the outlet box is marked "Acceptable For Fan Support." This would include all habitable rooms such as kitchens, family rooms, dining rooms, parlors, libraries, dens, sun rooms, bedrooms, recreation rooms, some stairways, or similar rooms or areas. Boxes installed close to a wall, or boxes installed specifically for such items as security systems, fire alarms, and smoke detectors are examples of where the outlet box would not be suitable for fan support. Where will the homeowner decide to install a ceiling-suspended (paddle) fan after all of the wiring has been completed? This becomes an important judgment call on the part of the electrician.

Listed assemblies that include a hanger and an outlet box can be installed where a ceiling-suspended (paddle) fan or a particularly heavy lighting fixture is to be installed. Figure 10–28 shows one type of box/hanger assembly that can be installed for new work or for existing installations.

A recent entry into the marketplace is an outlet box that is "listed" for both fixture and fan support for fans of 35 pounds or less, including all accessories. The outlet box has No. 8-32 holes for fixture mounting, and No. 10-32 holes for fan support. The box itself must be securely fastened to framing members of the building structure. Figure 10–29 shows this dual-purpose outlet box.

You must read the markings on the product and on the carton to make sure that the box or cover is suitable for supporting a lighting fixture or ceiling fan. Some electrical inspectors might accept device plaster rings and device boxes where it is known that the lighting fixture is small, such as lightweight wall-mounted lighting fixtures of the type used above or on the sides of medicine cabinets, or small wall-mounted porch bracket lighting fixtures.

Pay careful attention to the marking on metallic and nonmetallic boxes to see if they are "listed" as suitable for the support of ceiling fans. This marking is required to be on the product.

Section 410-16(a). Fixtures that weigh more than 50 pounds must be supported independently from the outlet box.

Section 370-28. This section addresses the required minimum dimensions for pull boxes and junction boxes. This is not a common application in house wiring. The subject matter is covered in detail in *Commercial Wiring.*

Section 370-29. Electrical boxes must be

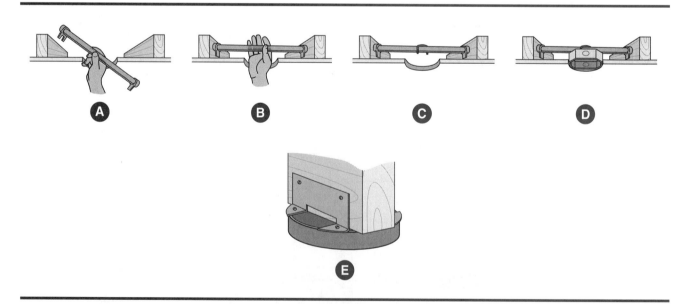

Figure 10–28 (A), (B), (C), and (D) show a "listed" ceiling fan hanger and box assembly that can be used for new work or remodel work. For remodel work, the assembly is installed through a carefully cut hole in the ceiling. The hanger is adjustable for 16–inch and 24–inch joist spacing, but can be cut shorter if necessary. The box shown in (E) is ½ inch deep, and can be used where the fan is supported from the joist, independent of the box, as required by *Section 422–18* for fans that weigh more than 35 pounds. (*Courtesy Reiker Enterprises Inc.*)

Figure 10–29 Picture of a dual-purpose outlet box specifically designed for use with lighting fixtures using No. 8–32 screws and ceiling fan support using No. 10–32 screws. (*Courtesy Reiker Enterprises Inc.*)

accessible without having to remove any part of the building. What good would it do to have an electrical box "buried" somewhere behind the wall? Who would know it is there? What if something in the wiring went wrong? How could it be fixed?

Section 300-22(c). Unless specifically "listed" for the use, this section prohibits equipment such as electrical boxes to be installed in the spaces in ceilings and walls that are used for environmental air. This would include hot air supply and cold air return ducts. Figure 10–30 illustrates this Code rule. Because most electrical outlet and switch (device) boxes do not have this listing, do not install boxes in spaces that serve as hot air supply or cold air returns.

Section 410-12. This section requires that all outlet boxes be covered upon completion of an installation. This requirement is met by using a blank cover, or if the box is covered by a lighting fixture or wiring device, trimmed with a suitable faceplate.

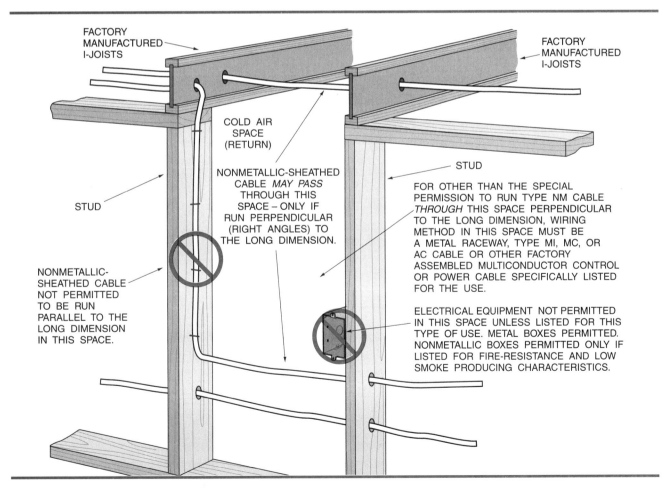

Figure 10–30 Do not install electrical equipment in hot air or cold air ducts.

Remodel (Old Work)

When installing switches, receptacles, and lighting outlets in remodel work where the walls and ceilings (paneling, drywall, sheet rock and plaster, or lath and plaster) are already in place, first make sure you will be able to run cables to where the switches, receptacles, and lighting outlets are to be installed. Then, proceed to cut openings the size of the specific type of box to be installed at each outlet location. Cables are then "fished" through the stud and joist spaces.

Boxes designed for remodel work such as the type shown in Figure 10–31 can be inserted from the front through holes already cut. After the boxes are snapped into place, they become securely locked into place by the brackets.

Another popular method of fastening wall and ceiling boxes in existing walls and ceilings is to use boxes that have plaster ears. Insert the box through the hole in the wall from the front. Then position a metal support into the hole on each side of the box. Be careful not to let the metal supports fall into the hole. Next, bend the metal supports over and into the box. These are referred to as "Madison Holdits" and are shown in Figure 10–32.

Still another type of box for use in remodel work is the type illustrated in Figure 10–33. This box has plaster ears and side screw supports. The plaster ears pull up tight against the wall surface—the metal screw supports pull up tightly behind the wall, firmly holding the box in place. The action is very similar to that of Molly screw anchors.

SPREAD OF FIRE

The first sentence found in the *National Electrical Code®* is *Section 90-1(a)* which states: *"(a) Practical Safeguarding. The purpose of this Code is the practical safeguarding of persons and property from hazards arising from the use of electricity."*

But there are other building codes and standards that are enforced that an electrician must be aware of.

To protect lives and property, fire must be contained and not be allowed to spread. Walls, partitions, and ceilings have a *fire resistance rating* dictated by various building codes. Ratings of fire-resistant materials are expressed in hours. Terms such as "1-hour fire resistance rating" or "2-hour fire resistance rating" are used.

A typical example are the fire resistance-rated walls and ceilings between habitable areas and an

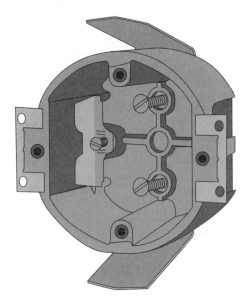
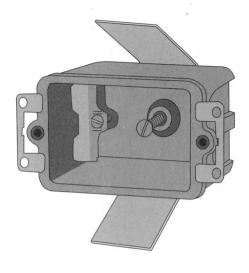

Figure 10–31 Nonmetallic boxes that are used for remodel work so that when they are inserted into an opening cut into a wall or ceiling, they will "snap" into place, adequately supporting the box.

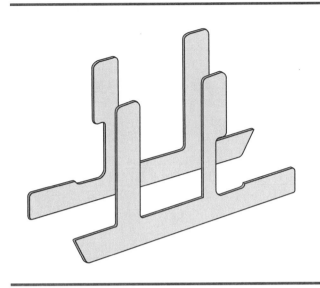

Figure 10–32 Metal supports referred to as "Madison Holdits."

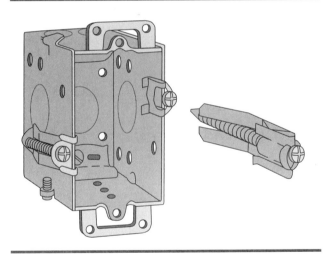

Figure 10–33 These boxes are commonly used for remodel work. The box is installed from the front, then the screws are tightened, securely supporting the box.

attached garage. In multifamily dwellings, fire resistance-rated walls and ceilings are required between occupancies. Installing electrical wall boxes back-to-back, or installed in the same stud or joist space defeats the fire resistance rating of the wall or ceiling.

The NFPA *Standard Methods of Tests of Fire Endurance of Building Construction and Materials,* UL *Standard 263,* and the UL *Fire Resistance Directory* contain wording such as:

- "Listed single and double gang metallic outlet and switch boxes with metallic or nonmetallic cover plates may be used in bearing and non-bearing wood stud and steel stud wall with ratings not exceeding 2 hours."
- "The surface area of individual metallic outlet or switch boxes shall not exceed 16 square inches."
- "The aggregate surface area of the boxes shall exceed 100 square inches per 100 square feet of wall surface."
- "Boxes located on opposite sides of walls or partitions shall be separated by a minimum horizontal distance of 24 inches.'
- "The metallic outlet or switch boxes shall be securely fastened to the studs and the opening in the wallboard facing shall be cut so that the clearance between the box and the wallboard does not exceed ⅛ inch."
- "The boxes shall be installed in compliance with the *National Electrical Code.*®"

One- and two-gang metal boxes are all right to use in fire resistance-rated construction partitions. Some nonmetallic boxes are "listed" for use in fire rated construction. You will need to check the particular manufacturer's "listing" of their nonmetallic boxes.

For both metallic and nonmetallic electric outlet and switch boxes, the **minimum** distance between boxes on opposite sides of a fire-rated wall or partition is 24 inches.

For both metallic and nonmetallic electric outlet and switch boxes, the **maximum** gap between the box and the wall material is ⅛ inch. Of course, as we learned earlier, this is also a requirement found in the *National Electrical Code.*®

Of interest to an electrician is an item "listed" in the UL *Fire Resistance Directory* that is an "*intumescent*" (expands when heated) fire-resistant material that comes in "pads." This moldable putty-like material can be used to wrap electrical boxes that are installed in the same wall cavity. This moldable putty-like material inhibits heat transfer from the fire resistant-rated side of the wall to the nonfire resistant-rated side of the wall. It is pressed onto the metallic electrical outlet or switch box, around electrical raceways or cables that enter the box, and is pressed into the interface where the box, stud, and gypsum wallboard meet. A common size available is 6" × 7" pads, ⅛ inch thick. The ⅛ inch thickness provides a 1-hour fire resistance rating. A 2-hour fire resistance rating is obtained with ¼ inch thickness. When this material is properly installed, the 24-inch separation between boxes is not required, and boxes may be installed in the same wall space, but not back-to-back.

Uniform building codes also require that "fire stopping be provided to cut off all concealed draft openings, both vertical and horizontal, and to form an effective fire barrier between stories, and between

a top story and the roof space." This includes openings around electrical cables, electrical conduits, vents, pipes, ducts, chimneys, and fireplaces at ceiling and floor level. Oxygen supports fire. Fire stopping cuts off or significantly reduces the flow of oxygen. Since most communities have adopted uniform building codes, it is common to find that the insulation contractor does all of the fire stopping around conduits, cables, pipes, and ducts before the drywall installer closes up the walls and ceilings. Figures 10–34, 10–35, 10–36, and 10–37 show typical applications of uniform building code requirements.

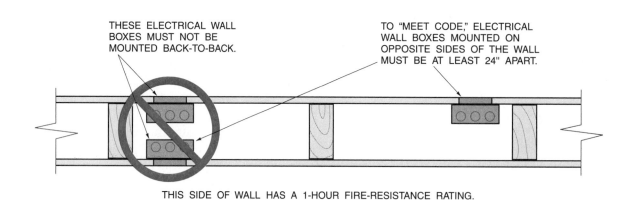

Figure 10–34 In general, building codes do not permit electrical wall boxes to be mounted back-to-back. Electrical boxes must be kept at least 24 inches apart. This is to maintain the integrity of fire resistance rating of the wall.

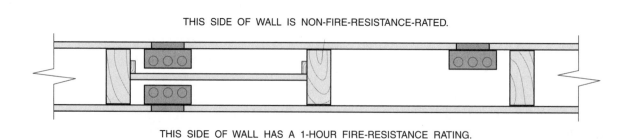

Figure 10–35 Building codes permit boxes to be mounted back-to-back or in the same wall cavity when the fire resistant rating is maintained. In this detail, the gypsum board has been installed the full length (top to bottom, ceiling plate to sole plate) of the wall cavity between the boxes, creating a new fire resistance barrier. Verify if this is acceptable to the building authority in your area.

THIS SIDE OF WALL IS NON-FIRE-RESISTANCE-RATED.

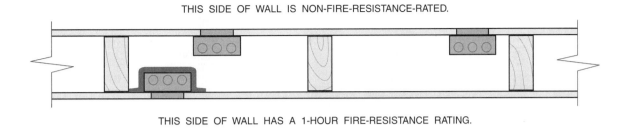

THIS SIDE OF WALL HAS A 1-HOUR FIRE-RESISTANCE RATING.

Figure 10–36   Building codes permit electrical outlet boxes to be mounted in the same stud space when the fire resistance rating is maintained. In this detail, "Wall Opening Protective Material" is packed around the electrical outlet box, thereby maintaining the fire resistance rating of the wall.

Figure 10–37   This illustration shows an electrical outlet box that has been covered with fire resistant "putty-like" material to retain the wall's fire resistance rating per building codes.

CORRECT WIRE LENGTH

Conductors must not be less than 6 inches long measured from where the conductors enter the box to the end of the conductor, and not less than 3 inches from the edge of the box to the end of the conductor, *Section 300-14.* Do not leave too much wire length as this will result in the crowding of the wires into the box and damage to the conductors. This can lead to possible short-circuits and/or ground-faults. Figure 10-38 illustrates the intent of the Code.

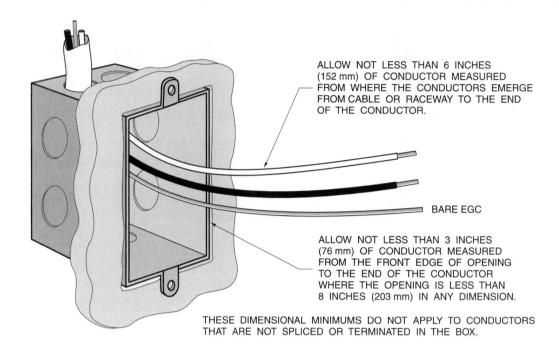

ALLOW NOT LESS THAN 6 INCHES (152 mm) OF CONDUCTOR MEASURED FROM WHERE THE CONDUCTORS EMERGE FROM CABLE OR RACEWAY TO THE END OF THE CONDUCTOR.

BARE EGC

ALLOW NOT LESS THAN 3 INCHES (76 mm) OF CONDUCTOR MEASURED FROM THE FRONT EDGE OF OPENING TO THE END OF THE CONDUCTOR WHERE THE OPENING IS LESS THAN 8 INCHES (203 mm) IN ANY DIMENSION.

THESE DIMENSIONAL MINIMUMS DO NOT APPLY TO CONDUCTORS THAT ARE NOT SPLICED OR TERMINATED IN THE BOX.

Figure 10–38 For ease in splicing conductors, and connecting wiring devices, lighting fixtures, and similar electrical equipment, *Section 300-14* stipulates the minimum length of conductor that must be left at electrical boxes.

NUMBER OF CONDUCTORS IN AN ELECTRICAL BOX

Probably the most common code violation cited in house wiring is "the box not large enough for the number of conductors and devices in the box."

Section 370-16 dictates that outlet boxes, switch boxes, and device boxes must be large enough to provide ample room for the wires in that box, without having to jam or crowd the wires into the box. Jamming too many wires into the box can not only damage the insulation on the wires, but can result in heat buildup in the box that can further damage the insulation on the wires.

The Code in *Table 370-16(a)* specifies the maximum number of conductors allowed in standard metal outlet, device, and junction box.

When all of the conductors are the same size, the proper metal box size can be selected by referring to *Table 370-16(a)*. When conductors are of different sizes, refer to *Table 370-16(b),* and use the cubic inch volume for the particular size of wire used. Because nonmetallic boxes are marked by the manu-

facturer with the cubic inch capacity, refer to *Table 370-16(b)* to determine the maximum number of conductors permitted in a given size box (Figure 10–39).

These tables do not take into consideration the space taken by fixture studs, cable clamps, hickeys, and wiring devices such as switches, receptacles, dimmers, etc. Figure 10–40 provides a summary of the rules necessary to determine box fill.

Figure 10–41 illustrates a cable clamp, a fixture stud on a bar hanger, and a hickey. When calculating box fill, each one of these devices must be counted as one deduction.

To make box selection easier when installing some of the more common types of boxes used for house wiring, refer to the Quik-Chek Box Selection Guide (Figure 10–42).

Counting Fixture Wires and Equipment Grounding Conductors

Where there are four or less fixture wires that are smaller than No. 14 AWG, and/or one equipment

Table 370-16(a). Metal Boxes

Box Dimension in Inches, Trade Size, or Type	Minimum Capacity (in.^{3})	Maximum Number of Conductors*						
		No. 18	No. 16	No. 14	No. 12	No. 10	No. 8	No. 6
4 × 1¼ round or octagonal	12.5	8	7	6	5	5	4	2
4 × 1½ round or octagonal	15.5	10	8	7	6	6	5	3
4 × 2⅛ round or octagonal	21.5	14	12	10	9	8	7	4
4 × 1¼ square	18.0	12	10	9	8	7	6	3
4 × 1½ square	21.0	14	12	10	9	8	7	4
4 × 2⅛ square	30.3	20	17	15	13	12	10	6
4¹¹⁄₁₆ × 1¼ square	25.5	17	14	12	11	10	8	5
4¹¹⁄₁₆ × 1½ square	29.5	19	16	14	13	11	9	5
4¹¹⁄₁₆ × 2⅛ square	42.0	28	24	21	18	16	14	8
3 × 2 × 1½ device	7.5	5	4	3	3	3	2	1
3 × 2 × 2 device	10.0	6	5	5	4	4	3	2
3 × 2 × 2¼ device	10.5	7	6	5	4	4	3	2
3 × 2 × 2½ device	12.5	8	7	6	5	5	4	2
3 × 2 × 2¾ device	14.0	9	8	7	6	5	4	2
3 × 2 × 3½ device	18.0	12	10	9	8	7	6	3
4 × 2⅛ × 1½ device	10.3	6	5	5	4	4	3	2
4 × 2⅛ × 1⅞ device	13.0	8	7	6	5	5	4	2
4 × 2⅛ × 2⅛ device	14.5	9	8	7	6	5	4	2
3¾ × 2 × 2½ masonry box/gang	14.0	9	8	7	6	5	4	2
3¾ × 2 × 3½ masonry box/gang	21.0	14	12	10	9	8	7	4
FS — Minimum internal depth 1¾ single cover/gang	13.5	9	7	6	6	5	4	2
FD — Minimum internal depth 2⅜ single cover/gang	18.0	12	10	9	8	7	6	3
FS — Minimum internal depth 1¾ multiple cover/gang	18.0	12	10	9	8	7	6	3
FD — Minimum internal depth 2⅜ multiple cover/gang	24.0	16	13	12	10	9	8	4

Note: For SI units, 1 in.^{3} = 16.4 cm^3.

*Where no volume allowances are required by Sections 370-16(b)(2) through 370-16(b)(5).

Table 370-16(b). Volume Allowance Required per Conductor

Size of Conductor (AWG)	Free Space Within Box for Each Conductor (in.^{3})
18	1.50
16	1.75
14	2.00
12	2.25
10	2.50
8	3.00
6	5.00

Note: For SI units, 1 in.^{3} = 16.4 cm^3.

Reprinted with permission from NFPA 70-1999

Figure 10–39 Allowable number of conductors in electrical boxes.

- If box contains no fittings, devices, fixture studs, cable clamps, hickeys, switches, receptacles, or equipment grounding conductors ...
  - refer directly to *Tables 370-16(a)* or *(b)*.

- **Clamps.** If box contains one or more internal cable clamps ...
  - add a single-volume based on the largest conductor in the box.

- **Support Fittings.** If box contains one or more fixture studs or hickeys ...
  - add a single-volume for each type based on the largest conductor in the box.

- **Device or Equipment.** If box contains one or more wiring devices on a yoke ...
  - add a double-volume for each yoke based on the largest conductor connected to a device on that yoke.

- **Equipment Grounding Conductors.** If a box contains one or more equipment grounding conductors ...
  - add a single-volume based on the largest equipment grounding conductor in the box.

- **Isolated Equipment Grounding Conductor.** If a box contains one or more additional "isolated" (insulated) equipment grounding conductors as permitted by *Section 250-146(d)* for "noise" reduction ...
  - add a single-volume based on the largest equipment grounding conductor in the box.

- For conductors running through the box without being spliced ...
  - add a single-volume for each conductor that runs through the box.

- For conductors that originated outside of the box and terminate inside the box ...
  - add a single-volume for each conductor that originates outside the box and terminates inside the box.

- If no part of the conductor leaves the box—for example, a "jumper" wire used to connect three wiring devices on one yoke, or pigtails as illustrated in Figure 10-45 ...
  - don't count this (these). No additional volume required.

- For small equipment grounding conductors or not more than four fixture wires smaller than No. 14 that originate from a fixture canopy or similar canopy (like a fan) and terminate in the box ...
  - don't count this (these). No additional volume required.

- For small fittings, such as lock-nuts and bushings ...
  - don't count this (these). No additional volume required.

Figure 10–40 A checklist for things to be considered when determining proper size boxes based on the box fill requirements found in *Article 370*.

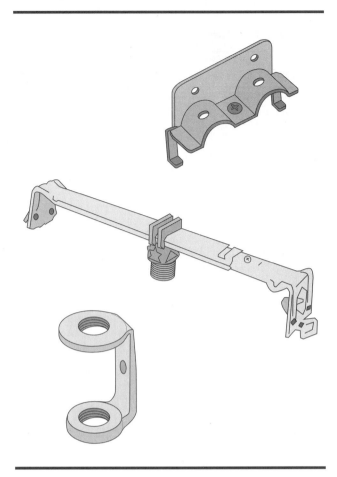

Figure 10–41   This is an illustration of a cable clamp, a hickey, and a fixture stud on a bar hanger.

grounding conductor that originate in a lighting fixture and terminate in an outlet box, they do not have to be counted, *Section 370-16(b)(1), Exception.* (See Figure 10–43.)

Section 410-10 states that the volume of canopies and outlet boxes together to determine must provide adequate space for the conductors. The canopy would have to be marked with its cubic inch capacity if the canopy is to be considered useable space. Canopies are made of thin metal, and do not qualify for additional cubic inch capacity.

Volume of a Raised Cover

Raised plaster rings and extension rings that are marked with their cubic inch volume may be included when determining the overall volume for the combination of a box plus a raised plaster ring or extension ring, *Section 370-16(a)*. Figure 10–44 is an illustration of a plaster ring.

QUIK-CHEK BOX SELECTION GUIDE
FOR METAL BOXES GENERALLY USED FOR RESIDENTIAL WIRING

DEVICE BOXES

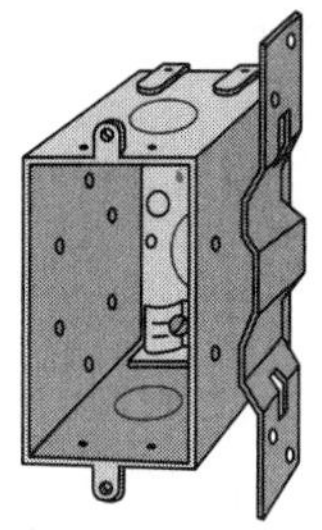

WIRE SIZE	3x2x1½ (7.5 in³)	3x2x2 (10 in³)	3x2x2¼ (10.5 in³)	3x2x2½ (12.5 in³)	3x2x2¾ (14 in³)	3x2x3 (16 in³)	3x2x3½ (18 in³)
#14	3	5	5	6	7	8	9
#12	3	4	4	5	6	7	8

SQUARE BOXES

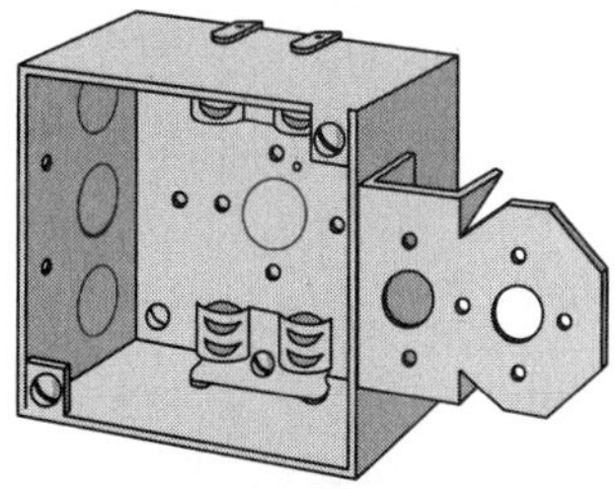

WIRE SIZE	4x4x1½ (21 in³)	4x4x2⅛ (30.3 in³)
#14	10	15
#12	9	13

OCTAGON BOXES

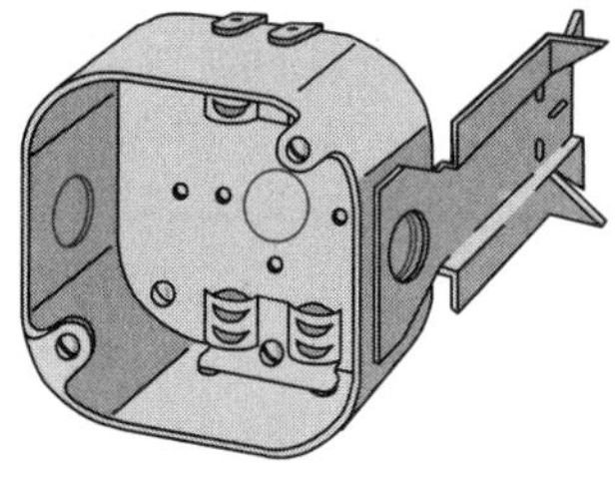

WIRE SIZE	4x1½ (15.5 in³)	4x2⅛ (21.5 in³)
#14	7	10
#12	6	9

HANDY BOXES

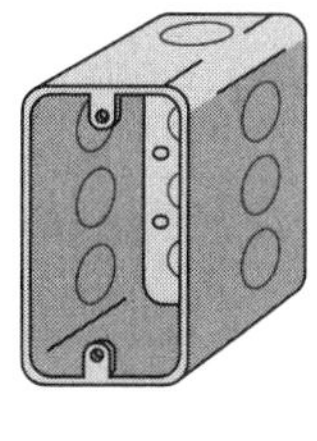

WIRE SIZE	4x2⅛x1½ (10.3 in³)	4x2⅛x1⅞ (13 in³)	4x2x2⅛ (14.5 in³)
#14	5	6	7
#12	4	5	6

RAISED COVERS

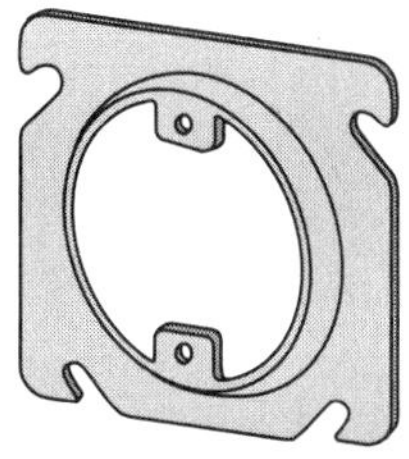

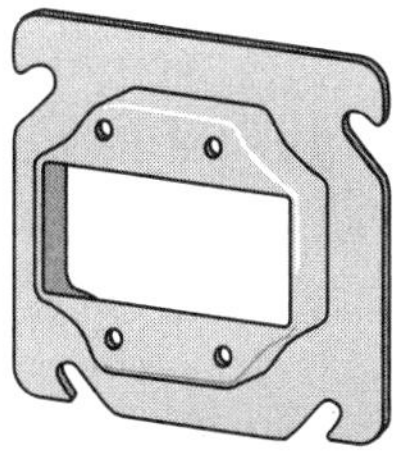

WHERE RAISED COVERS ARE MARKED WITH THEIR VOLUME IN CUBIC INCHES, THAT VOLUME MAY BE ADDED TO THE BOX VOLUME TO DETERMINE MAXIMUM NUMBER OF CONDUCTORS IN THE COMBINED BOX AND RAISED COVER.

NOTE: BE SURE TO MAKE DEDUCTIONS FROM THE ABOVE MAXIMUM NUMBER OF CONDUCTORS PERMITTED FOR WIRING DEVICES, CABLE CLAMPS, FIXTURE STUDS, AND GROUNDING CONDUCTORS. THE CUBIC INCH (IN³) VOLUME IS TAKEN DIRECTLY FROM *TABLE 370-16(a)* OF THE *NEC.*® NONMETALLIC BOXES ARE MARKED WITH THEIR CUBIC INCH CAPACITY.

Figure 10–42   Quik-chek box selector guide for boxes commonly used for residential wiring.

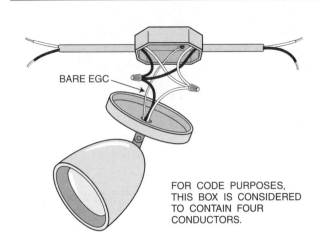

BARE EGC

FOR CODE PURPOSES, THIS BOX IS CONSIDERED TO CONTAIN FOUR CONDUCTORS.

Figure 10–43 Four or less fixture wires that are smaller than No. 14 AWG, and/or one equipment grounding conductor that originate in the lighting fixture and terminate in the outlet box need not be counted when calculating box fill, Section 370-16(a)(1),Exception.

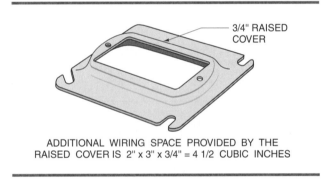

3/4" RAISED COVER

ADDITIONAL WIRING SPACE PROVIDED BY THE RAISED COVER IS 2" x 3" x 3/4" = 4 1/2 CUBIC INCHES

Figure 10–44 The volume of raised covers that are marked with their cubic inch volume may be used when making box fill calculations.

SELECTING THE CORRECT SIZE BOX

In addition to the space taken by cable clamps, wiring devices, fixtures studs, etc., two other factors have an effect on proper box sizing. One is when all of the conductors are the same size, the second when the conductors are different sizes.

Same Size Conductors

Example: A nonmetallic box is marked as having a volume of 22.8 cubic inches. The box contains no fixtures stud or cable clamps. How many No. 14 AWG conductors are permitted in this box?

From *Table 370-16(b),* the volume requirement for a No. 14 AWG conductor is 2.00 cubic inches. Therefore, the maximum number of No. 14 AWG conductors permitted in this box is:

$$\frac{22.8}{2} = 11.4 \text{ (Round down to 11 conductors.)}$$

Example: A box contains two internal cable clamps. Two No. 12/2 AWG w/ground nonmetallic-sheathed cables enter the box. A receptacle will be installed in this box. Determine the correct size box.

Four No. 12 AWG conductors	4
Add one count for one or more equipment grounding conductors	1
Add one count for one or more cable clamps	1
Add two count for the receptacle	2
Total	8

Referring to *Table 310-16(a),* a 3" × 2" × 3½" deep device box is suitable.

Example: Determine the proper size metal box to use where two No. 14/3 w/ground and one No. 14/2 w/ground nonmetallic-sheathed cables enter the box. External Romex connectors are used. One receptacle is to be installed. The box will have a ¾" raised cover attached to it that is marked "4½ cubic inches" (see Figure 10–45).

Add the No. 14 AWG conductors: 2+3+3=8	
Add the equipment grounding conductor (count one only)	1
Add double count for the receptacle	2
Total	11

Table 370-16(b) indicates that we need 2.00 cubic inches of space for a No. 14 AWG conductor. The minimum cubic inch volume required for the above example is:

$$11 \times 2 = 22 \text{ cubic inches}$$

Checking *Table 370-16(a),* Figure 10–39, we find that a 4 × 1½ inch box has a capacity of 21.0 cubic inches, and by itself can hold ten No. 14 AWG conductors. To this we add the 4½ cubic inch volume provided by the raised plaster cover.

$$21.0 + 4½ = 25½ \text{ total cubic inches}$$

The box and plaster ring combined volume is more than adequate for the minimum capacity of

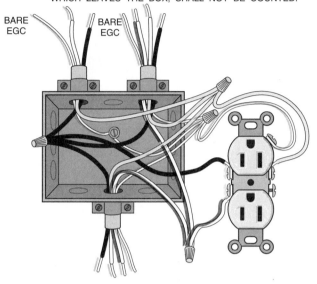

THE "PIGTAILS" CONNECTED TO THE RECEPTACLE NEED NOT BE COUNTED WHEN DETERMINING THE CORRECT BOX SIZE. THE CODE IN *SECTION 370-16(b)(1)* STATES, "A CONDUCTOR, NO PART OF WHICH LEAVES THE BOX, SHALL NOT BE COUNTED."

BARE EGC   BARE EGC

Figure 10–45   Determining size of box according to the number of conductors and wiring devices in the box.

22 cubic inches required for the conductors and receptacle.

Example: A wall box will contain one single-pole switch and one three-way switch. Nonmetallic-sheathed cables are routed so that two No. 14/2 w/ground and two No. 14/3 w/ground cables enter the box. The box has internal cable clamps. What size box is required?

Add the circuit conductors:
2 + 2 + 3 + 3 = 10
Add one count for one or more
equipment grounding conductors   1
Add one count for one or more
cable clamps   1
Add two count for each switch (2 + 2)   4
Total   16

Referring to *Table 370-16(a)*, two 3" × 2" × 3½" deep device boxes ganged together (9 + 9 = 18) are suitable for this example. A 4" square 1½" deep box with raised plaster ring is also large enough. The decision here it to make sure that the combined cubic inch volume equals not less than 16 × 2 = 32 cubic inches.

Different Size Conductors

Example: What is the minimum cubic inch volume required for a box that will contain two internal cable clamps, one switch, and one receptacle? Two nonmetallic-sheathed cables enter the box. There are two No. 14 AWG connected to the switch, and two No. 12 AWG conductors connected to the receptacle.

Two No. 14 AWG wires @
2 cubic inches per wire   = 4.00 cubic inches
Two No. 12 AWG wires @
2.25 cubic inches per wire   = 4.50 cubic inches
One cable clamp @
2.25 cubic inches   = 2.25 cubic inches
One switch @
2.00 cubic inches   = 2.00 cubic inches
One receptacle @
2.25 cubic inches   = 2.25 cubic inches
One count for the largest equip-
ment grounding conductor
@ 2.25 cubic inches   = 2.25 cubic inches
Total   = 17.25 cubic inches

Locate a box in the tables that has a volume of at least 17.25 cubic inches. When sectional boxes are ganged together, the total capacity is the total cubic inch volume of the assembled boxes, *Section 370-16(a)*. Two 3 × 2 × 2½ inch device boxes ganged together have a cubic inch volume of 12.5 + 12.5 = 25 cubic inch capacity. A 4" square × 1½" deep box has a volume of 21 cubic inches. To this box, attach a two gang device plaster ring that is marked 5.5 cubic inches. 21 + 5.5 = 26.5 cubic inches of capacity. Both of these possibilities result in more than ample space for this example.

CORRECT SIZE OF EQUIPMENT GROUNDING CONDUCTORS

The equipment grounding conductor in No. 14, No. 12, and No. 10 AWG nonmetallic-sheathed cables is the same size as the circuit conductors. In other words, the equipment grounding conductor in a No. 14/3 NMC is No. 14 AWG, etc.

Table 250-122 shows minimum size equipment grounding conductors based on the ampere rating of the overcurrent protective device.

COUNTING CONDUCTORS THAT RUN STRAIGHT THROUGH A BOX

A conductor that runs straight through a box is counted as one conductor, *Section 370-16(b)(1)*. Using a raceway such as electrical metallic tubing (EMT) makes it possible to "loop" conductors through the box. Looping conductors through a box is not possible when using cable (Figure 10–46).

Warning: Electricians and electrical inspectors have become very aware of the fact that GFCI receptacles, dimmers, and timers take up a lot of space. Where these devices are to be installed, make sure that you select boxes that will provide plenty of room to avoid pushing, jamming, and crowding the wires and wiring device into the box.

BOXES THAT ARE PART OF A RECESSED LIGHTING FIXTURE

Boxes that are an integral part of a "listed" recessed lighting fixture are handled differently from that of standard outlet boxes. The entire assembly of the

lighting fixture, the box, and the mounting brackets are covered by UL standards. This topic is covered in detail in Unit 11.

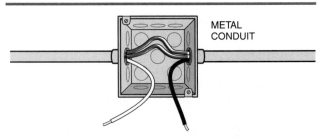

METAL CONDUIT

BY LOOPING THE CONDUCTORS THROUGH THE BOX, THE CONDUCTOR COUNT IN THIS ILLUSTRATION IS FOUR.

Figure 10–46   A conductor running straight through a box is counted as one conductor. This diagram has a conductor count of four. The metallic raceway provides the equipment ground, so additional equipment grounding conductors are not needed. If the same circuitry was done using two three-wire nonmetallic-sheathed cables, all of the circuit conductors would be counted plus one additional count for the equipment grounding conductor. The total count would be seven.

REVIEW QUESTIONS

1. You run a nonmetallic-sheathed cable through the outside wall of the house for an outdoor lighting fixture. In the spaces provided, mark T (True) or F (False) for the following statements.
   a) A box must be installed behind the lighting fixture.   _____
   b) It is not necessary to install a box behind the fixture. Just make up the splices in the canopy of the fixture, then fasten the fixture to the wall.   _____

2. What is the fundamental difference between boxes designed for use with nonmetallic-sheathed cable and boxes that are designed for use with armored cable?

   _____

   _____

3. What are the basic differences between standard device boxes and masonry device boxes?

   _____

   _____

4. The mounting holes in a device (switch) box are tapped for (a) No. ____ screws. The mounting holes in a outlet box are tapped for (b) No. ____ screws. The holes in metal boxes for attaching equipment grounding conductors are tapped for (c) No. ____ screws.

5. How can you identify a screw designed for terminating equipment grounding conductors?

   _____

   _____

6. Nonmetallic-sheathed cable must be secured within 12 inches of a device box. However, when using a nonmetallic device box no larger than 2¼ × 4 inches, and if the cable is stapled within 8 inches of the box, it _____ necessary to have a clamp in the box.

   a) is

   b) is not

7. In the spaces provided, mark T (True) or F (False) for the following statements.

   a) It is all right to run more than one nonmetallic-sheathed cable through one opening in a metallic outlet or device box. _____

   b) It is not all right to run more than one nonmetallic-sheathed cable through one opening in a nonmetallic outlet or device box. _____

   The answer to the above statements is found in *Section* _____.

8. In the spaces provided, mark T (True) or F (False) for the following statements.

   a) All unused openings other than nail holes in boxes must be closed. _____

   b) Boxes installed in noncombustible material are permitted to set back from the finish surface not more than ¼ inch. _____

   c) Boxes installed in combustible material such as wood paneling must be set flush with the finish surface. _____

   d) The open space around outlet and device boxes must not be more than ⅛ inch. _____

   e) Nonmetallic device boxes are suitable for attaching lighting fixtures to them. _____

   f) Outlet boxes that support ceiling paddle fans that weigh 35 pounds or less must be "listed" as being suitable for the support of ceiling fans, and will be marked *"Acceptable For Fan Support."*

   g) Ceiling-suspended (paddle) fans that weigh more that 35 pounds must be supported independently of the outlet box. _____

   h) The wall between dwellings in a two-family structure is required to be 2-hour fire rated. Electric wall boxes installed in this fire-rated wall must be separated by a minimum horizontal distance of 24 inches. This would prohibit mounting these boxes back-to-back in the same partition. _____

   i) When it becomes absolutely necessary to mount electrical boxes in the same partition of a fire-rated wall, an "intumescent" fire-resistant moldable putty-like material can be used to pack in and around the wall boxes. Should a fire occur, the intumescent material expands when heated. _____

9. From *Table 370-16(a)*, list the maximum number of conductors permitted in the following boxes. There are no cable clamps or fixture studs.

   a) No. 12s in a 4" × 1½" octagon box. _____

   b) No. 12s in a 4¹¹⁄₁₆" × 1½" square box. _____

   c) No. 12s in a 3" × 2" × 3½" device box. _____

   d) No. 14s in a 3" × 2" × 2½" device box. _____

   e) No. 10s in a 4" × 2⅛" square box. _____

   f) No. 8s in a 4" × 2⅛" square box. _____

10. Is it necessary to count fixture wires when counting the permitted number of conductors in a box according to *Section 370-16?*

    a) Yes

    b) No

11. *Table 370-16(a)* of the *NEC®* shows the maximum number of conductors permitted in a given size box that does not contain cable clamps, fixture studs, etc. In addition to counting the number of conductors that will be in the box, what is the additional volume that must be provided for the following items? Enter *single* or *double* volume allowance in the blank provided.

    a) One or more internal cable clamps: _____ volume allowance

    b) For a fixture stud: _____ volume allowance

    c) For one or more wiring devices on one yoke: _____ volume allowance

    d) For one or more equipment grounding conductors: _____ volume allowance

    e) A GFCI receptacle: _____ volume allowance

    f) A dimmer switch: _____ volume allowance

12. Two No. 12/2 w/ground and two No. 14/2 w/ground enter a box. The No. 12 conductors are properly spliced together with short No. 12 pigtails connected to a duplex receptacle. The No. 14 conductors are properly spliced together and are connected to a toggle switch. The receptacle and switch are on separate yokes. The equipment grounding conductors are terminated to No. 10-32 green, hexagon-shaped screws in the back of the box. Calculate the minimum cubic inch volume required for this box. The box contains two cable clamps.

13. Using the same size and number of conductors as in question 12, but using electrical metallic tubing (EMT), calculate the minimum cubic inch volume required for the box. There will be no separate equipment ground conductors, nor will there be any clamps in the box.

14. A conductor that runs straight through a box without a splice is counted as _____ conductors. What section of the Code supports your answer?

    a) one

    b) two

15. May the additional volume of raised covers or extension rings be used when making a box fill calculation? What section of the Code supports your answer?

    a) Yes

    b) No

Lighting Fixtures

OBJECTIVES

After studying this unit, you will be able to:

- Understand the importance of reading the label on fixtures.
- Understand lighting fixture terminology such as Type IC, Type NON-IC.
- Understand the color-coding of fixture and branch-circuit conductors.
- Have learned how to connect recessed fixtures, both prewired and nonprewired.
- Understand thermal insulation clearance requirements for recessed fixtures.
- Understand what a "fixture whip" is.
- Understand thermal protection for recessed lighting fixtures.
- Understand the basic Code rules for installing indoor and outdoor lighting fixtures.
- Describe a Class P ballast.
- Understand the importance of maximum wattage lamps for a given lighting fixture.

INTRODUCTION

Lighting fixtures come in two basic types: surface and recessed. Surface-type lighting fixtures fasten to an outlet box according to the instructions furnished with the lighting fixture. Recessed lighting fixtures demand more attention because of the heat generated and contained within the recessed fixture housing. Another concern is the installation of lighting fixtures in clothes closets. This unit addresses the key *National Electrical Code®* rules governing these concerns.

Residential lighting fixtures come in many varieties, a few of which are illustrated in Figure 11–1.

CODE REQUIREMENTS

No matter what type of fixtures are selected, they must be installed according to the Code. Most of the code rules for the installation of lighting fixtures are found in *Article 410* of the *National Electrical Code.®* Lighting fixtures are also referred to as "luminaires," the international term for a lighting fixture.

READ THE LABEL

The first thing you need to do when purchasing a lighting fixture is to make sure that the lighting fixture bears the Underwriters Laboratories "Listing" label. This ensures that the lighting fixture has been tested to the safety standards to which the fixture manufacturer must conform. Secondly, make sure that the lighting fixture you choose is a type designed and identified for the application where you intend to install it.

Figure 11–1   Some typical residential-type lighting fixtures. (*Courtesy of Progress Lighting*)

When a lighting fixture is "listed" or "labeled," *Section 110-3(b)* of the *NEC®* makes the requirements of the "listing" or "labeling" an enforceable part of code. Therefore, install the lighting fixture properly—according to the instructions furnished with the fixture—and to the requirements of the *National Electrical Code.®*

Many of the installation code requirements for lighting fixtures can be met simply by reading the label. The following list is made up of typical information that might be found on the label of a lighting fixture. You can readily understand why it is so important to read the label.

- For wall mount only.

- Ceiling mount only.

- Maximum lamp wattage.

- Type of lamp.

- Access from above ceiling required.

- Access from behind wall required.

- Suitable for air handling only.

- For chain or hook suspension only.

- Suitable for operation in ambient temperature not exceeding 140°F (60°C).

- Suitable for installation in poured concrete.

- For line volt-amperes, multiply lamp wattage by 1.25.

- Suitable for use in suspended ceilings.

- Suitable for use in noninsulated ceilings.

- Suitable for use in insulated ceilings.

- Suitable for damp locations.

- Suitable for wet locations.

- Suitable for use as a raceway.

- Suitable for mounting on low-density cellulose fiberboard.

- For supply connections, use wire rated for at least 194°F (90°C).

- Not for use in dwellings.

- Thermally protected.

- Type IC.

- Type NON-IC.

- Inherently protected.

MAXIMUM LAMP WATTAGE

One of the most common problems experienced with any type of lighting fixture is that of installing incandescent lamps having a wattage greater than the lighting fixture was designed for. The buildup of heat can damage the conductors, the lamp socket, as well as plastic lenses on the lighting fixture. Do not install lamps with a higher wattage than the maximum wattage marked on the lighting fixture.

GROUNDING LIGHTING FIXTURES

Lighting fixtures must be grounded according to the requirements of *Article 250* of the *NEC®*. Grounding is covered in Unit 12 of this book.

CONNECTING THE WIRES

It is important to connect the supply conductors to the lighting fixture wires correctly. Here are some of the basic rules for hooking up lighting fixtures.

- Be sure that the temperature rating of the supply conductors are suitable for the temperature of the lighting fixture. Nonmetallic-sheathed cable has 90°C (194°F) conductors. Fixtures are marked with the required supply conductor temperature ratings. In most instances, 90°C (194°F) conductors are satisfactory, but not always. Check the marking on the lighting fixture.

- Conductors within 3 inches (76 mm) of a fluorescent ballast must be rated at least 90°C (194°F).

- The white fixture wire is connected to the white supply *grounded* conductor.

- The black fixture wire is connected to the black supply *ungrounded* conductor. This might be a red conductor, depending on the circuitry.

- The bare (or green) equipment grounding conductor from the fixture is connected to the bare equipment grounding conductor from the supply. The bare EGC from the fixture may also be connected to a No. 10-32 grounding screw inside of the outlet box. Outlet boxes have a No. 10-32 hole tapped for this purpose. The equipment grounding screws are green in color, and are hexagon-shaped.

It might be tempting to twist the bare EGC from

the fixture around the No. 8-32 screws that attach the fixture mounting strap to the outlet box. This is a violation of *Section 110-14(a)*, which requires in part that the *"Connection of conductors to terminal parts shall ensure a thoroughly good connection . . ."*

In *Section 250-148(a)*, we find the statement that for metal boxes, *"a connection shall be made between the one or more equipment grounding conductors and a metal box by means of a grounding screw that shall be used for no other purpose, or a listed grounding device."*

The No. 8-32 are not terminals. They are for attaching the mounting strap.

Sheet-metal screws shall not be used to connect grounding conductors to enclosures, *Section 250-8*.

In residential installations, never connect an equipment grounding conductor to the grounded circuit conductor anywhere on the load side of the service equipment. *Section 250-142(b)* states that *"A grounded circuit conductor shall not be used for grounding noncurrent-carrying metal parts of equipment on the load side of the service disconnecting means or on the load side of a separately derived system disconnecting means or the overcurrent devices for a separately derived system not having a main disconnecting means."*

- Always make sure that the white supply *grounded* conductor is connected to the screw shell of a lampholder, *Sections 410-23* and *410-47*.

- Always make sure that the black supply *ungrounded* conductor is connected to the center contact of a lampholder.

TYPES OF LIGHTING FIXTURES

Lighting fixtures can be grouped into the following categories:

FLUORESCENT	INCANDESCENT
surface	surface
recessed	recessed
suspended ceiling	suspended ceiling

Wet, Damp, and Dry Locations

Lighting fixtures must be "listed" for the location where they will be installed.

Any outdoor location that is exposed to the weather is considered a wet location. *Section*

410-4(a) states that "*Fixtures installed in wet or damp locations shall be installed so water cannot enter or accumulate in the lampholders, wiring compartments, or other electrical parts.*" A shower location is considered to be a wet location. Lighting fixtures for use in wet locations must be marked "Suitable for Wet Locations."

Partially protected areas such as roofed open porches or areas under canopies are damp locations. Fixtures to be used in these locations must be marked "Suitable for Damp Locations" or "Suitable for Wet Locations."

Lampholders installed in wet or damp locations must be of the weatherproof type, *Section 410-49*.

Detailed definitions for wet, damp, and dry loca-tions can be found in Unit 8, and in *Article 100* of the *National Electrical Code.*®

Surface-Mounted Lighting Fixtures

Sections 410-15 and *410-16* cover the support-ing requirements for lighting fixtures.

Surface-mounted lighting fixtures are easy to install. It is simply a matter of following the manu-facturers instructions and diagrams that are fur-nished with the fixture. The lighting fixture is attached to the ceiling outlet box or wall outlet box using fixture studs, hickeys, bar straps, or fixture extensions.

Figure 11–2 is an assortment of hardware that might be needed to hang a lighting fixture. It is

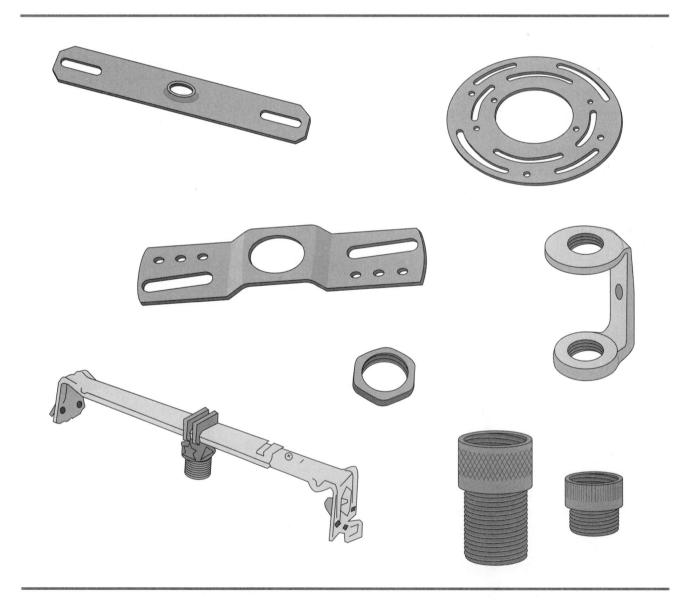

Figure 11–2 Illustrated are some of the hardware that may be needed to install a surface-mounted lighting fixture.

assumed that outlet boxes have been installed properly as discussed in Unit 10 so as not to have the lighting fixture fall down on someone's head. The maximum weight of a lighting fixture permitted to be supported directly from an outlet box is 50 pounds. For lighting fixtures weighing more than 50 pounds, support must be independent of the outlet box.

Recessed Lighting Fixtures

The Code requirements for the installation of recessed lighting fixtures are found in *Section 410-64* through *410-72* of the *NEC.*® The wiring of recessed lighting fixtures requires a little more attention than simple surface-mounted lighting fixtures.

Recessed lighting fixtures are available for both new work and remodel work. Remodel work recessed lighting fixtures can be installed from below an existing ceiling by cutting a hole, bringing power to the fixture, making the connections in the integral junction box on the lighting fixture, then installing the housing through the hole.

Recessed lighting fixtures as shown in Figure 11–3 have an inherent heat problem because they are recessed, are enclosed, in many instances are buried in thermal insulation, and may intentionally or unintentionally be overlamped with an improperly sized high-wattage lamp.

Recessed Lighting Fixture Trims

Recessed incandescent lighting fixtures are listed by Underwriters Laboratories with specific trims. Fixture/trim combinations are marked on the label. Do not use trims that are not listed for use with the particular recessed lighting fixture. Mismatching fixtures and trims is a violation of *Section 110-3(b)* of the *National Electrical Code,*® which states that *"Listed or labeled equipment shall be installed and used in accordance with any instructions included in the listing or labeling."*

Thermal Protection

To protect against overheating, Underwriters Laboratories requires that incandescent recessed lighting fixtures be equipped with an integral thermal protector. A thermal cutout is shown in Figure 11–4. This thermal protector will cycle on and off repeatedly until the heat problem is corrected. Thermal protection is also required by *Section 410-65(c)* of the *National Electrical Code.*® These lighting fixtures are marked that they are thermally protected.

Junction Box

Most recessed lighting fixtures come equipped with a junction box that is an integral part of the fixture housing. This is clearly shown in Figure 11–3. For most residential-type recessed lighting fixture installations, nonmetallic-sheathed cable or armored cable is run directly into this junction box. These cables have conductors rated 90°C. Check the marking on the lighting fixture for supply conductor temperature limitations.

Section 410-11 states that *"Branch-circuit wiring, other than 2-wire or multiwire branch circuits supplying power to fixtures connected together, shall not be passed through an outlet box that is an integral part of a fixture unless the fixture is identified for through-wiring."*

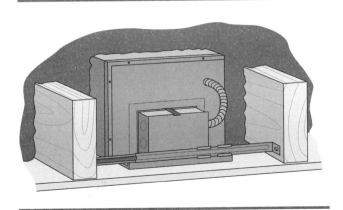

Figure 11–3 A typical recessed lighting fixture.

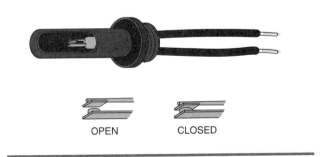

OPEN   CLOSED

Figure 11–4 A thermal cutout from a recessed lighting fixture. The contacts are normally closed, and open when excessive dangerous heat is generated within the fixture.

Section 410-31 states that *"Fixtures shall not be used as a raceway for circuit conductors."* Exceptions to this are:

- unless the fixture is marked for use as a raceway, or

- conductors may be run from fixture to fixture (daisy chained) with a recognized wiring method where there is more than one fixture, but again, only those conductors that supply the fixtures are permitted.

Some recessed lighting fixtures are marked "identified for through wiring," in which case branch circuit conductors in addition to the conductors that supply the lighting fixture are permitted to be run through the outlet box or wiring compartment on the fixture. A typical label will look like this:

"Maximum of _____ No. _____ AWG branch circuit conductors suitable for at least 90°C (194°F) permitted in a box (___ in: ___ out)."

The manufacturer of the recessed lighting fixture in conjunction with UL requirements will determine the maximum number, size, and temperature rating for the conductors entering and leaving the outlet box or wiring compartment, and will so mark the label.

Only those fixtures that have been tested for the extra heat generated by the additional branch circuit conductors will bear the above label marking.

Figure 11–5 shows three recessed lighting fixtures, each with an integral outlet box. If these fixtures are marked "identified for through-wiring,"

then it would be permitted to run conductors through the outlet boxes **in addition** to the conductors that supply the fixtures. If the "identified for through-wiring" marking is not found on the fixture, then it is a Code violation to run any conductors through the box, other than the conductors that supply the fixture itself.

Figure 11–6 highlights some of the common Code rules that apply to recessed lighting fixtures.

Type Non-IC and Type IC Recessed Lighting Fixtures

The term IC means *insulated ceiling.*

Section 410-66 points out the clearances needed for recessed lighting fixtures.

Type Non-IC must not be covered with thermal insulation. Insulation must be kept at least 3" (76 mm) from the sides of the fixture (Figure 11–7(A)).

Type IC recessed lighting fixtures are permitted to be completely buried in thermal insulation (Figure 11-7(B)).

Figure 11–7(C) is a more detailed cross-sectional drawing showing the clearances needed for Type Non-IC fixtures. Thermal insulation must be kept at least 3 inches (76 mm) from the sides of the fixture. Above the recessed lighting fixture, thermal insulation must be installed so as not to entrap heat in the cavity where the fixture is installed. Maintaining a 3-inch (76 mm) clearance above the fixture between the top of the fixture and the thermal insulation will meet this criterion. These requirements are found in *Section 410-66.*

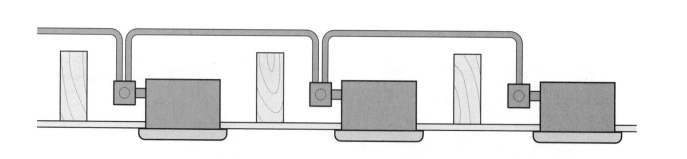

Figure 11–5 Three recessed lighting fixtures wired in "daisy chain" fashion. The supply conductors for the fixtures are the only conductors permitted to be run into and out of the outlet boxes on the fixtures. However, if the fixture is marked "identified for through-wiring," then it is permissible to run conductors through the outlet boxes *in addition* to the conductors that supply the fixtures.

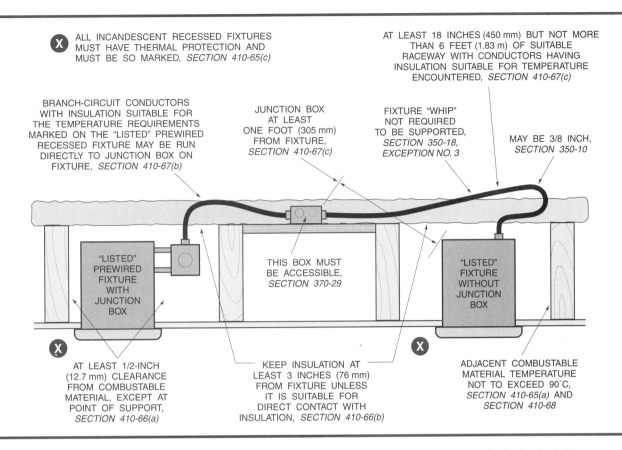

X ALL INCANDESCENT RECESSED FIXTURES MUST HAVE THERMAL PROTECTION AND MUST BE SO MARKED, *SECTION 410-65(c)*

AT LEAST 18 INCHES (450 mm) BUT NOT MORE THAN 6 FEET (1.83 m) OF SUITABLE RACEWAY WITH CONDUCTORS HAVING INSULATION SUITABLE FOR TEMPERATURE ENCOUNTERED, *SECTION 410-67(c)*

BRANCH-CIRCUIT CONDUCTORS WITH INSULATION SUITABLE FOR THE TEMPERATURE REQUIREMENTS MARKED ON THE "LISTED" PREWIRED RECESSED FIXTURE MAY BE RUN DIRECTLY TO JUNCTION BOX ON FIXTURE, *SECTION 410-67(b)*

JUNCTION BOX AT LEAST ONE FOOT (305 mm) FROM FIXTURE, *SECTION 410-67(c)*

FIXTURE "WHIP" NOT REQUIRED TO BE SUPPORTED, *SECTION 350-18, EXCEPTION NO. 3*

MAY BE 3/8 INCH, *SECTION 350-10*

"LISTED" PREWIRED FIXTURE WITH JUNCTION BOX

THIS BOX MUST BE ACCESSIBLE, *SECTION 370-29*

"LISTED" FIXTURE WITHOUT JUNCTION BOX

X

X

AT LEAST 1/2-INCH (12.7 mm) CLEARANCE FROM COMBUSTABLE MATERIAL, EXCEPT AT POINT OF SUPPORT, *SECTION 410-66(a)*

KEEP INSULATION AT LEAST 3 INCHES (76 mm) FROM FIXTURE UNLESS IT IS SUITABLE FOR DIRECT CONTACT WITH INSULATION, *SECTION 410-66(b)*

ADJACENT COMBUSTABLE MATERIAL TEMPERATURE NOT TO EXCEED 90°C, *SECTION 410-65(a)* AND *SECTION 410-68*

Figure 11–6 Clearance requirements and acceptable supply conductor wiring methods for installing recessed lighting fixtures.

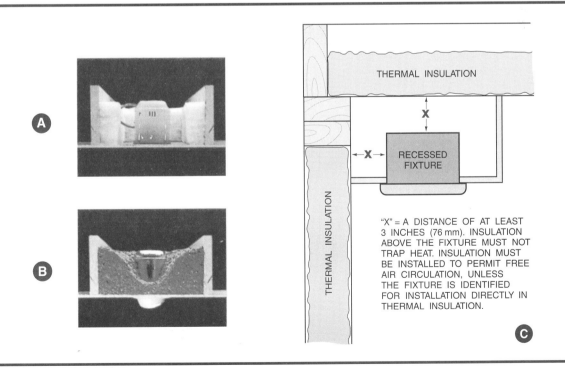

A

B

THERMAL INSULATION

X

RECESSED FIXTURE

X

THERMAL INSULATION

"X" = A DISTANCE OF AT LEAST 3 INCHES (76 mm). INSULATION ABOVE THE FIXTURE MUST NOT TRAP HEAT. INSULATION MUST BE INSTALLED TO PERMIT FREE AIR CIRCULATION, UNLESS THE FIXTURE IS IDENTIFIED FOR INSTALLATION DIRECTLY IN THERMAL INSULATION.

C

Figure 11–7 (A) A Type Non-IC recessed lighting fixture that requires 3-inch clearance from thermal insulation. (B) A Type IC recessed lighting fixture that may be completely covered with thermal insulation. (C) Clearances for recessed lighting fixture installed near thermal insulation. Any insulation above the fixture shall be so located as to not trap excessive heat in the cavity where the fixture is installed.

Fluorescent Lighting Fixtures

When marked "Recessed Fluorescent Fixture," they are intended for installation in cavities in ceilings and walls, and are to be wired according to *Section 410-64* of the *National Electrical Code.*® These fixtures may also be installed in suspended ceilings if they have the necessary mounting hardware.

When marked "Suspended Ceiling Fluorescent Fixture," they are intended only for installation in suspended ceilings where the acoustical tiles, lay-in panels, and suspended grid are not part of the actual building structure. This is the type of fixture commonly used in "dropped" ceilings in finished recreation rooms in homes.

Figure 11–8 shows a fluorescent "lay-in" lighting fixture suitable for use in a suspended ceiling grid. *Section 410-16(c)* requires that the framing members of the suspended ceiling be securely attached to the building structure, and that the fixture be secured to the framing members of the suspended ceiling.

Fixture Whips

Figure 11–9 shows how a "fixture whip" is used to connect a fixture.

When a recessed lighting fixture does not have a junction box on it, an outlet box must be installed at least one foot (305 mm) from the fixture. This box must be accessible. From this outlet box, a flexible raceway is run to the fixture. This flexible connection is referred to in the electrical trade as a "fixture whip." *Section 350-10(a)* permits ⅜-inch flexible metal conduit in lengths not to exceed 6 feet to make the connection between the junction box and the fixture with conductors that are suitable for the supply conductor temperature requirements marked on the fixture.

A flexible metal conduit "fixture whip" must:

- be not less than 18 inches (450 mm) nor longer than 6 feet (1.83 m), *Section 410-67(c)*.

- contain conductors that have a temperature rating as marked on the fixture.

- be used with connectors that are "listed" for grounding purposes.

IMPORTANT: TO PREVENT THE FIXTURE FROM INADVERTENTLY FALLING, *SECTION 410-16(c)* OF THE CODE REQUIRES THAT 1) SUSPENDED CEILING FRAMING MEMBERS THAT SUPPORT RECESSED FIXTURES MUST BE SECURELY FASTENED TO EACH OTHER, AND MUST BE SECURELY ATTACHED TO THE BUILDING STRUCTURE AT APPPROPRIATE INTERVALS, AND 2) RECESSED FIXTURES MUST BE SECURELY FASTENED TO THE SUSPENDED CEILING FRAMING MEMBERS BY BOLTS, SCREWS, RIVETS, OR SPECIAL "LISTED" CLIPS PROVIDED BY THE MANUFACTURER OF THE FIXTURE FOR THE PURPOSE OF ATTACHING THE FIXTURE TO THE FRAMING MEMBER.

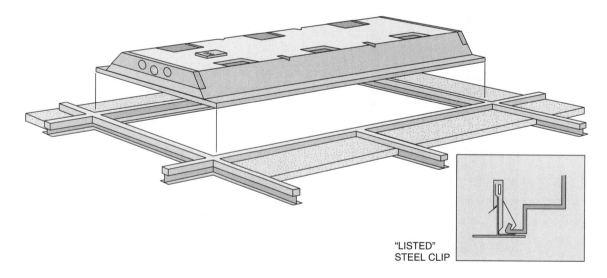

"LISTED" STEEL CLIP

Figure 11–8 A typical ceiling grid showing how a "lay-in" fixture is installed. Note the requirements of *Section 410–16(c)* for the supporting of the grid to the building structure, and the supporting of the fixture to the ceiling grid.

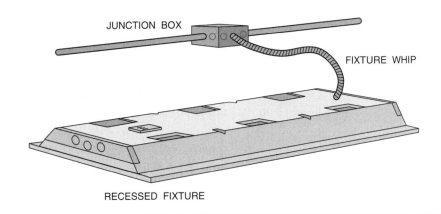

JUNCTION BOX

FIXTURE WHIP

RECESSED FIXTURE

Figure 11–9   This illustration shows how a "fixture whip" runs from the outlet box to the fluorescent lighting fixture.

- be protected by an overcurrent device not over 20 amperes.

A "fixture whip" can also be flexible nonmetallic conduit. If the fixture whip is flexible nonmetallic conduit, then a separate equipment grounding conductor must be installed through it to properly ground the fixture.

Section 333-7-(b)(3) permits Type AC (BX) cable to be used as a "fixture whip" when *"Not more than 6 feet (1.83 m) in length from an outlet for connections within an accessible ceiling to lighting fixtures or equipment."*

Still another method permitted by the *NEC*® for use as a "fixture whip" is found in *Section 336-18, Exception No. 3.* This section permits nonmetallic-sheathed cable to be used in *"Lengths not more than 4½ feet (1.37 m) from an outlet for connection within an accessible ceiling to lighting fixtures(s) or equipment."*

Fixture whips are exempt from the 12 inch and 4½ feet supporting requirements.

Fixture whips are exempt from the supporting requirements for flexible raceways, *Section 350-18, Exception No. 3.*

Underwriters Laboratories Standard 1570 covers recessed and suspended ceiling fixtures in detail.

Combustible Low-Density Ceiling Cellulose Fiberboard

Section 410-76(b) addresses combustible low-density cellulose fiberboard. This section states that fluorescent lighting fixtures that are surface

mounted on this material must be spaced at least 1½ inch (38 mm) from the fiberboard surface. Surface-mounted lighting fixtures that are permitted to be mounted directly on low-density fiberboard must be marked "Suitable for Surface Mounting on Low-Density Cellulose Fiberboard."

Figure 11–10 shows a surface-mounted fluorescent lighting fixture mounted on a low-density cellulose fiberboard ceiling.

The *Fine Print Note* to *Section 410-76(b)* explains combustible low-density cellulose fiberboard as sheets, panels, and tiles that have a density of 20 pounds per cubic foot or less that are formed of bonded plant fiber material. It does not include

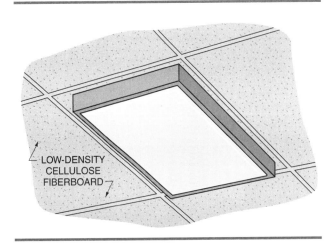

LOW-DENSITY CELLULOSE FIBERBOARD

Figure 11–10   When mounted on low-density ceiling fiberboard, surface-mounted fluorescent fixtures must be marked "Suitable for Surface Mounting on Low-Density Cellulose Fiberboard."

fiberboard that has a density of over 20 pounds per cubic foot or material that has been integrally treated with fire-retarding chemicals to meet specific standards. Solid or laminated wood does not come under the definition of "combustible low-density cellulose fiberboard."

This is a fire issue. Be sure to read the label on the fixture.

Lighting Fixtures in Clothes Closets

Clothing, boxes, and other combustible material normally stored in clothes closets are fire hazards. These items may ignite upon contact with the hot surface of an exposed lamp (bulb) or from the hot pieces of a broken lamp (bulb). Incandescent lamps have a hotter surface temperature than fluorescent lamps. Here again, we have a possible fire concern.

The *National Electrical Code®* does not require lighting fixtures in clothes closets. However, when lighting fixtures are installed in clothes closets, they must be of the proper type and they must be properly installed. *Section 410-8* of the *NEC®* gives very specific rules relative to the location and types of lighting fixtures permitted in clothes closets. The key point here is to understand exactly what the Code considers to be "storage space."

Figure 11–11 illustrates a typical clothes closet, and clearly shows the spaces that are defined as "storage space."

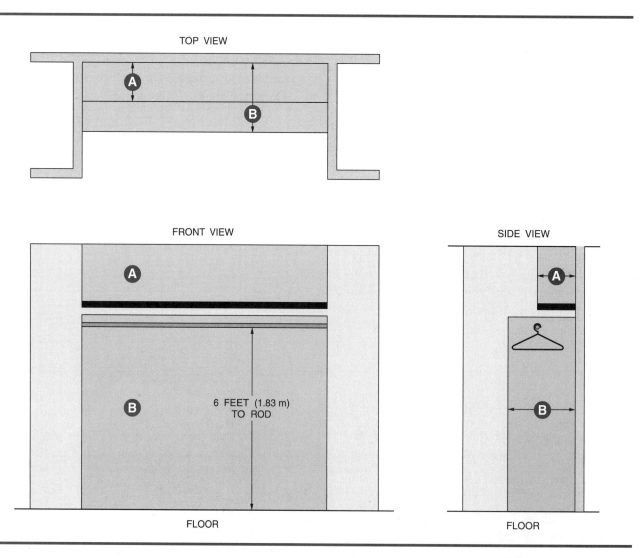

Figure 11–11 A typical clothes closet with one shelf and one rod. The shaded area defines "storage space." Dimension A is width of shelf of 12 inches from wall, whichever is greater. Dimension B is below rod, 24 inches from wall. See *Section 410-8(a)* of the *NEC®*

Figure 11–12 shows the types of lighting fixtures permitted in clothes closets.

Figure 11–13 shows the permitted locations for lighting fixtures in clothes closets.

Lighting Fixtures In Bathrooms

The choices are virtually unlimited for surface-mounted and recessed lighting fixtures for bathrooms. But there are some restrictions.

Section 410-4(d) prohibits cord-connected lighting fixtures, hanging lighting fixtures, lighting track, pendants, or ceiling-suspended (paddle) fans from being installed within a zone measuring 3 feet (914 mm) horizontally and 8 feet (2.44 m) vertically from the top of the bathtub rim. Recessed or surface-mounted lighting fixtures and recessed exhaust fans may be located within the restricted area.

Figure 11–14 shows a top view of the restricted area.

Figure 11–15 shows a side view of the restricted area.

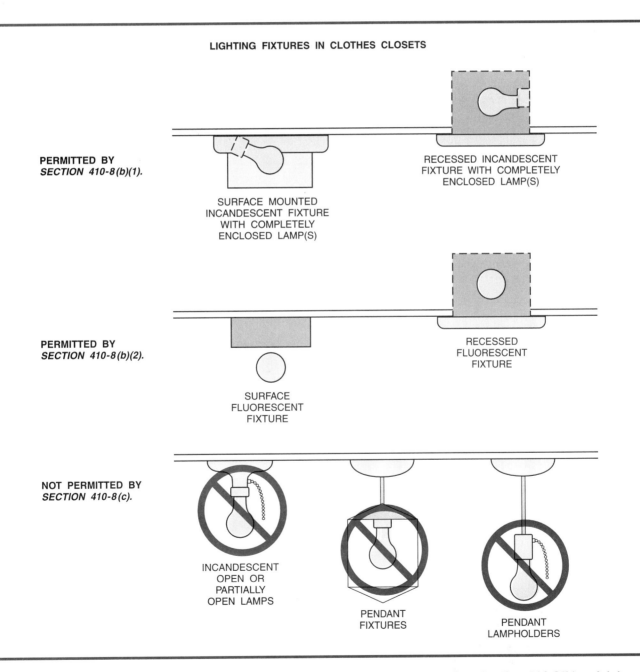

LIGHTING FIXTURES IN CLOTHES CLOSETS

PERMITTED BY
SECTION 410-8(b)(1).

SURFACE MOUNTED
INCANDESCENT FIXTURE
WITH COMPLETELY
ENCLOSED LAMP(S)

RECESSED INCANDESCENT
FIXTURE WITH COMPLETELY
ENCLOSED LAMP(S)

PERMITTED BY
SECTION 410-8(b)(2).

SURFACE
FLUORESCENT
FIXTURE

RECESSED
FLUORESCENT
FIXTURE

NOT PERMITTED BY
SECTION 410-8(c).

INCANDESCENT
OPEN OR
PARTIALLY
OPEN LAMPS

PENDANT
FIXTURES

PENDANT
LAMPHOLDERS

Figure 11–12   The types of lighting fixtures permitted in clothes closets. See *Section 410-8(b)* and *(c).*

LOCATION OF LIGHTING FIXTURES IN CLOTHES CLOSETS

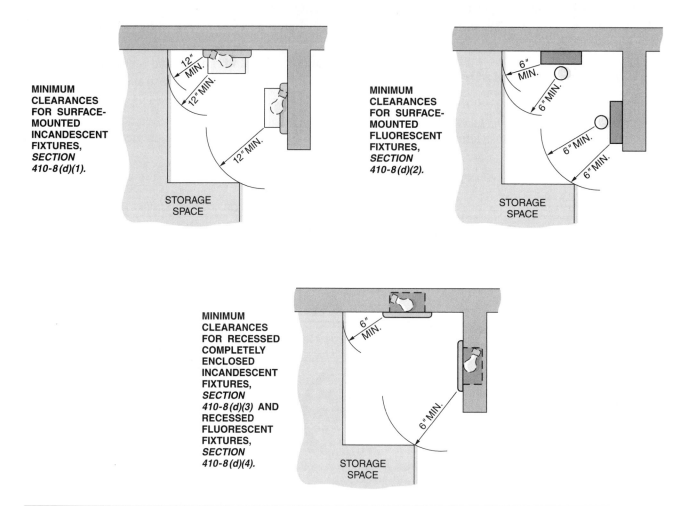

Figure 11–13 The permitted locations for installing lighting fixtures in clothes closets. See *Section 410-8(d)*.

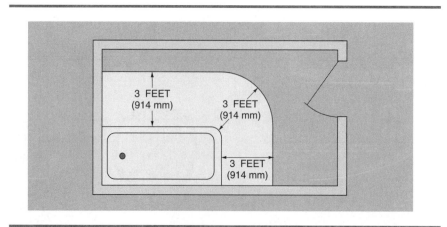

Figure 11–14 A top view of a tub and adjacent area. No part of cord-connected lighting fixtures, hanging fixtures, track lighting, pendants, or ceiling paddle fans are permitted in the shaded area. See *Section 410-4(d)*.

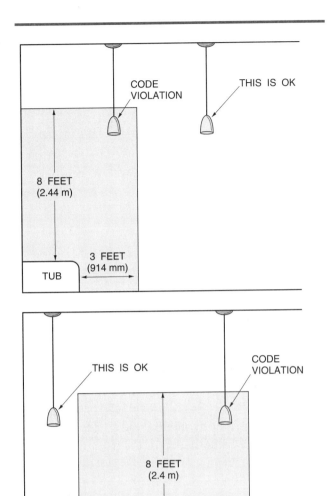

Figure 11–15 A side view of a tub and adjacent area. No part of cord-connected lighting fixtures, hanging fixtures, track lighting, pendants, or ceiling suspended (paddle) fans are permitted in the shaded area. See *Section 410-4(d)*.

Voltage Limitations

The maximum branch-circuit voltage allowed for residential lighting fixtures is 120 volts between conductors per *Section 210-6* of the *NEC.*

Section 410-80(b) makes a further restriction stating that for dwellings, no lighting equipment shall be used if it operates with an open-circuit voltage of over 1,000 volts. This really focuses on the use of neon lighting for decorative lighting purposes (Figure 11–16).

Class P Ballasts

Section 410-73(e) requires that all fluorescent ballasts installed indoors (except simple reactance-type ballasts) must have thermal protection built into the ballast. Thermally protected ballasts are called *Class P* ballasts. Should excessive heat develop, the thermal protector will cycle on and off until the problem is corrected.

The newer electronic ballasts are 25 to 40% more energy efficient than conventional magnetic (core and coil) ballasts.

Dimming Fluorescent Lamps

Dimming of fluorescent lamps requires special "dimming" ballasts. This was covered in Unit 5.

Track Lighting

Track lighting has become a popular form of lighting in homes.

Track lighting is covered in *Sections 410-100* through *410-105* of the *National Electrical Code.*

Figure 11–17 shows a typical track lighting installation.

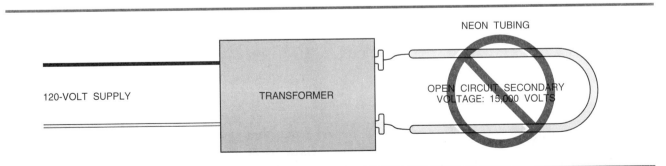

Figure 11–16 *Section 410-80(b)* prohibits the installation of lighting in or on residences where the open-circuit voltage exceeds 1,000 volts.

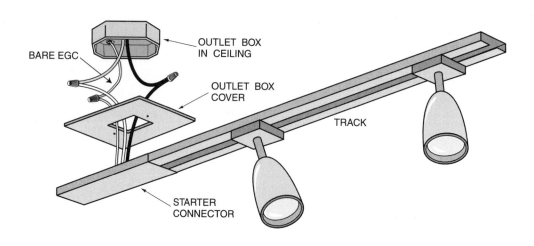

Figure 11–17   A typical lighting track and electrical connections.

Here is a summary of those requirements. Track lighting:

- shall be permanently installed and permanently connected to the branch circuit wiring.

- must be installed according to the manufacturer's instructions.

- that is 4 feet or less in length requires a minimum of two supports.

- that is longer than 4 feet must have at least one support for each 4 feet.

- shall use fittings and fixtures that are "listed" for the specific track.

- shall have not more load attached than the rating of the track.

- normally does not require any additional load calculations for the purpose of determining the branch-circuit sizing because track lighting load in homes is already included in the general purpose lighting load calculations. Most home-type track lighting does not have a great numbr of fixture heads attached, and generally would not present an overload situation. Add up the wattages of the lamps to verify that a possible overload condition might present itself.

- shall not be installed in wet or damp locations.

- shall not be run through walls or partitions.

- shall not be mounted less than 5 feet (1.52 m) above the floor unless it is protected from phys-

ical damage, or if it operates at an open-circuit voltage of less than 30 volts.

- shall not be installed in the zone 3 feet (914 mm) horizontally and 8 feet (2.44 m) vertically from the top of the bathrub rim.

Portable Track Lighting

Another form of track lighting is really nothing more that a cord- and plug-connected strip that can be fastened to a wall or ceiling, then "plugged in." Fixtures are snapped into the track according to the manufacturer's instructions. According to UL requirements, portable track lighting:

- shall not be longer than 4 feet.

- shall have a cord no longer than 9 feet.

- shall have integral switching in the lampholders.

- not have its length changed in the field.

Outdoor Lighting

There is virtually an unlimited array of outdoor lighting fixtures available for uplighting, downlighting, diffused lighting, moonlighting, shadow and texture lighting, accent lighting, silhouette lighting, and bounce lighting. These might be surface mounted or recessed. Earlier in this unit, we covered how to attach surface-mounted lighting fixtures to an outlet box, and the many requirements for installing recessed fixtures.

It is all right to run approved raceways or cables,

properly protected against physical damage, up the side of a tree to support outdoor lighting fixtures. *Section 410-16(h)* states that *"outdoor lighting fixtures and associated equipment shall be permitted to be supported by trees."* In some parts of the country, "treescape" lighting is very popular. Figure 11–18 shows the intent of *Section 410-16(h)*.

Do not support overhead conductors from trees or other vegetation, *Section 225-26*. This was covered in Unit 9.

The key is to make sure that the lighting fixtures are "listed" by Underwriters Laboratories, and that the wiring methods and mounting methods "Meet Code."

Although not a *National Electrical Code®* issue, you should check with your local electrical inspector and/or building official to see if there are any restrictions regarding outdoor lighting. In recent years, more and more complaints are coming from neighbors claiming that they are being bothered by "nuisance" lighting. The term nuisance includes glare, brightness, and light spillover. A number of communities have legislated strict outdoor lighting laws, specifying various restrictions such as the location, type, size, wattage, and/or foot-candles for outdoor lighting fixtures.

Figure 11–18 *Section 410-16(h)* allows lighting fixtures to be supported by a tree.

Landscape Lighting

Outdoor lighting can be 120 volts or low voltage.

In Unit 4 and in Unit 9 we discussed how boxes and conduit bodies are to be supported when they are attached to raceways that rise out of the ground.

Figure 11–19 shows a variety of outdoor lighting fixtures.

Figure 11–19 Typical outdoor lighting fixtures used for decorative purposes under shrubs and trees, in gardens, and for lighting paths and driveways. (*Courtesy Progress Lighting*).

Quite popular today are low-voltage landscape lighting systems. These systems come complete with a step-down transformer (120 volts to 15 volts), cords of the proper length, and the correct number of fixtures for the assembly. There are systems for "Indoor Use Only" and there are systems for "Outdoor Use Only."

Figure 11–20 illustrates a very common decorative outdoor low-voltage lighting system.

Cord-connected outdoor lighting is intended to be plugged into a receptacle that has a cover that is suitable for use in wet locations. This would also apply to decorative Christmas lighting. As explained in Unit 7, all outdoor receptacles must be GFCI protected. There is no exception for a receptacle that is used for decorative lighting.

Low-Voltage Lighting Systems

Article 411 of the *National Electrical Code®* is a very short article that covers lighting systems that operate at 30 volts or less. During a visit to a home center, an electrical distributor, or a lighting showroom you will no doubt see on display many low-voltage lighting systems.

The key thing to remember is that because these low-voltage lighting systems are designed as a "package," they must be installed according to the instructions furnished with the low-voltage lighting system, *Section 110-3(b),* and to the requirements of *Article 411.*

Low-voltage lighting systems must be "listed for the purpose," *Section 411-3.* The components of a low-voltage lighting system are matched. These

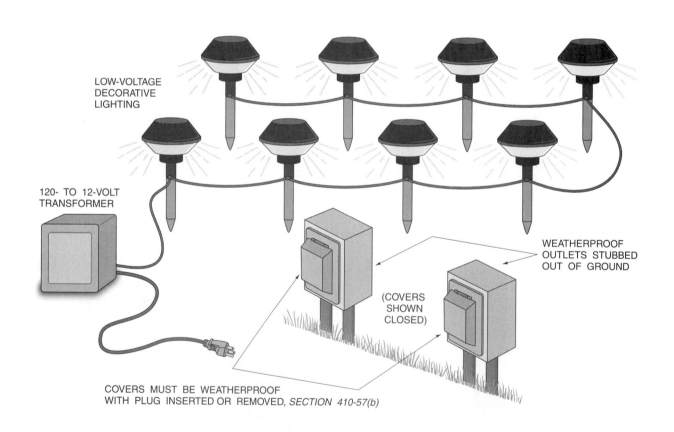

LOW-VOLTAGE DECORATIVE LIGHTING

120- TO 12-VOLT TRANSFORMER

WEATHERPROOF OUTLETS STUBBED OUT OF GROUND

(COVERS SHOWN CLOSED)

COVERS MUST BE WEATHERPROOF WITH PLUG INSERTED OR REMOVED, *SECTION 410-57(b)*

Figure 11–20 Low-voltage decorative outdoor lighting that can be plugged into weatherproof receptacle outlets that are stubbed out of the ground.

systems will contain the necessary transformer, conductors, and lighting fixtures.

The basic requirements for these systems follow.

They shall:

- operate at 30 volts rms (42.4 volts peak) or less.

- have one or more secondary circuits, each limited to 25 amperes maximum.

- be supplied through an isolated power supply.

- be "listed for the purpose."

- not have the wiring concealed in walls or other building partitions nor shall it be extended through a building unless the wiring is a conventional wiring method covered in *Chapter 3* of the *NEC®* such as nonmetallic-sheathed cable.

- not be installed within 10 feet (3.05 m) of pools, spas, or water fountains, unless specifically permitted in *Article 680.*

- have the secondary of the transformer isolated from the branch circuit through the use of an isolating transformer. This would be a two-winding transformer where primary and secondary are totally independent windings. This requirement is not as tough as for transformers that supply underwater swimming pool lighting, where transformers are required by *Section 680-5(a)* to have a grounded metal barrier between the primary and secondary windings.

- not have the secondary of the isolating transformer grounded.

- must be supplied by a branch circuit that does not exceed 20 amperes.

Voltage Drop Problems

There seems to be a mind-set that other than the fact that low-voltage lighting systems operate at a lower voltage than standard 120-volt circuits, no consideration is given to the much greater current draw for a given wattage at the lower voltage. This is a critical issue. When the length of the low-voltage wiring is long, larger conductors are necessary to compensate for voltage drop. If ignored, the lamps will not get enough voltage to burn properly (the lamps will burn dim), and the small conductors can reach dangerous temperatures that might cause a fire.

Example: A 480 watt load at 120 volts draws:

$$480 \div 120 = 4 \text{ amperes}$$

A 480 watt load at 12 volts draws:

$$480 \div 12 = 40 \text{ amperes}$$

You can readily see that for this 480 watt example, the conductors must be considerably larger for the 12 volt application to safely carry the current.

Pre-wired low-voltage lighting assemblies that bear the UL label have the correct size and length of conductors. Adding conductors in the field must be done according to the manufacturers' instructions.

Voltage drop calculations are covered in Unit 2.

RECAP

Figure 11–21 is a handy reference and recap of *National Electrical Code®* rules for lighting fixtures. As pointed out in this unit, always read the labels on the lighting fixture, the instructions furnished with the lighting fixture, and the *National Electrical Code®* to make sure that the wiring installation for the fixture is correct, and that the fixture is suitable for use where it is being installed.

Final Words

- Make sure that the premise wiring is in conformance to the *National Electrical Code®.* (Install the lighting fixture in conformance to the requirements of the *National Electrical Code®.*

- Make sure that the lighting fixture to be installed has been tested and "listed" by a recognized testing laboratory such as Underwriters Laboratories. Never install a lighting fixture that does not have a "listing" mark.

FIXTURES

- SEE *ARTICLE 410, NEC®*
- WILL BE MARKED WITH INSULATION TEMPERATURE RATING REQUIRED FOR SUPPLY CONDUCTORS IF OVER 60°C.
- READ THE LABEL AND INSTRUCTIONS FURNISHED WITH ALL FIXTURES.

FLUORESCENT

- IF BRANCH-CIRCUIT CONDUCTORS ARE WITHIN 3 INCHES (76.2 mm) OF BALLAST, USE 90°C. CONDUCTORS, *SECTION 410-33, NEC.®*
- DO NOT USE AS RACEWAY UNLESS PERMITTED, *SECTION 410-31, NEC.®*
- ALL FLUORESCENT BALLASTS INSTALLED INDOORS MUST BE CLASS P TYPE, *SECTION 410-73(e), NEC.®*

SURFACE

- NOT SUITABLE FOR MOUNTING WITHIN 1½ INCHES (38.1 mm) OF THE SURFACE OF LOW-DENSITY CEILING FIBERBOARD UNLESS MARKED "SUITABLE FOR SURFACE MOUNTING ON LOW-DENSITY CELLULOSE FIBERBOARD." SEE *SECTION 410-76(b), NEC.®*
- READ THE LABEL ON THE FIXTURE FOR SPECIAL REQUIREMENTS OR LIMITATIONS.

RECESSED

- SUITABLE FOR RECESSED INSTALLATION.
- MAY BE MOUNTED IN SUSPENDED CEILINGS IF PROVIDED WITH APPROPRIATE HARDWARE FOR MOUNTING TO OR IN SUSPENDED CEILINGS.
- READ FIXTURE LABEL FOR SPECIAL REQUIREMENTS OR LIMITATIONS.

SUSPENDED

- FOR INSTALLATION IN A SUSPENDED GRID ONLY WHERE THE LAY-IN TILES ARE NOT FASTENED IN PLACE, AND WHERE THE TIE WIRES, T BARS, CEILING TILES AND OTHER COMPONENTS DIRECTLY ASSOCIATED WITH THE GRID ARE NOT PART OF THE BUILDING STRUCTURE.
- THE FIXTURES ARE INTENDED TO BE MOUNTED IN THE CEILING OPENINGS.
- READ FIXTURE LABEL OR INSTRUCTIONS FURNISHED WITH THE FIXTURE FOR SPECIAL REQUIREMENTS OR LIMITATIONS.

INCANDESCENT

SURFACE

- FOR SURFACE MOUNTING ONLY.
- READ FIXTURE LABEL OR INSTRUCTIONS FURNISHED WITH THE FIXTURE FOR SPECIAL REQUIREMENTS OR LIMITATIONS.

RECESSED

TYPE IC
- ARE MARKED TYPE IC
- MAY BE INSTALLED IN INSULATED CEILINGS WHERE INSULATION AND OTHER COMBUSTIBLE MATERIALS MAY BE IN DIRECT CONTACT WITH AND OVER THE TOP OF THE FIXTURE.
- HAS INTEGRAL THERMAL PROTECTION THAT DEACTIVATES THE LAMP IF THE FIXTURE IS MIS-LAMPED.
- IS MARKED "NOTICE—THERMALLY PROTECTED FIXTURE. BLINKING LIGHT MAY INDICATE IMPROPER LAMP WATTAGE OR IMPROPER LAMP SIZE." MAY ALSO BE MARKED WITH OTHER CONDITIONS THAT WILL CAUSE OVERHEATING AND WILL RESULT IN THE LAMPS LINKING.
- MAY BE USED IN NONINSULATED CEILINGS.
- USUALLY ARE LOW-WATTAGE FIXTURES.
- READ FIXTURE LABEL OR INSTRUCTIONS FURNISHED WITH THE FIXTURE FOR SPECIAL REQUIREMENTS OR LIMITATIONS.

TYPE NON-IC
- FOR INSTALLATION IN UNINSULATED CEILINGS.
- IF INSTALLED IN AN INSULATED CEILING, KEEP INSULATION AT LEAST 3 INCHES (76.2 mm) FROM SIDES AND NOT PLACED OVER THE FIXTURE SUCH THAT IT WOULD ENTRAP THE HEAT PRODUCED BY THE FIXTURE.
- HAS AN INTEGRAL THERMAL PROTECTION THAT WILL DEACTIVATE THE LAMP IF INSULATION COVERS THE FIXTURE, RESULTING IN AN OVERHEATING SITUATION.
- UNLESS OTHERWISE MARKED, KEEP FIXTURE AT LEAST ½ INCH (12.7 mm) FROM COMBUSTIBLE MATERIAL (LIKE WOOD JOISTS) EXCEPT AT SUPPORT POINTS.
- READ FIXTURE LABEL OR INSTRUCTIONS FURNISHED WITH THE FIXTURE FOR SPECIAL REQUIREMENTS OR LIMITATIONS.

INHERENTLY PROTECTED
- IF MARKED INHERENTLY PROTECTED, THE FIXTURE IS SO DESIGNED THAT THE SURFACE TEMPERATURE WILL NOT EXCEED 90°C EVEN IF THE FIXTURE IS COVERED WITH INSULATION, IS MIS-LAMPED OR OVER-LAMPED, AN EXAMPLE MIGHT BE "DOUBLE-WALLED" CONSTRUCTION.
- THESE FIXTURES ARE NOT THERMALLY PROTECTED.
- READ FIXTURE LABEL OR INSTRUCTIONS FURNISHED WITH THE FIXTURE FOR SPECIAL REQUIREMENTS OR LIMITATIONS.

SUSPENDED

- FOR INSTALLATION IN A SUSPENDED GRID ONLY WHERE THE LAY-IN TILES ARE NOT FASTENED IN PLACE, AND WHERE THE TIE WIRES, T BARS, CEILING TILES, AND OTHER COMPONENTS DIRECTLY ASSOCIATED WITH THE GRID ARE NOT PART OF THE BUILDING STRUCTURE.
- THE FIXTURES ARE INTENDED TO BE MOUNTED IN CEILING OPENINGS.
- READ FIXTURE LABEL OR INSTRUCTIONS FURNISHED WITH THE FIXTURE FOR SPECIAL REQUIREMENTS OR LIMITATIONS.

Figure 11–21   A recap and handy reference for some of the more important Underwriters Laboratories and *National Electrical Code®* **requirements for lighting fixtures.**

REVIEW QUESTIONS

1. To ensure that a lighting fixture has been manufactured properly, and has been subjected to testing per specific Underwriters Laboratories standards, in the spaces provided, mark T (True) or F (False) for the following statements.

    a) Read the instructions in the box. The manufacturer will state in the instructions that the fixture has been UL Approved.   _____

    b) Look for the UL label on the fixture, which will confirm that the fixture has been "listed."   _____

    c) Ask the salesperson if the product is UL Listed. This is enough proof that the product has been "listed" and is safe.   _____

2. Lighting fixtures that bear the "listed" label may be installed in a manner that the electrician is accustomed to.

    a) True

    b) False

3. One of the most common problems, particularly with surface-mounted ceiling types, is the damage that heat can do to the insulation on the conductors in the box above the fixtures. The insulation on the wire dries out, bakes, crumbles off, or melts. This can allow the wires to short out or go to ground. This can cause a fire. The manufacturer provides a blanket of insulation in the canopy for surface-mounted fixtures. But there still can be a heat problem. In the spaces provided, mark T (True) or F (False) for the following statements.

    a) Read the labeling inside the fixture. It will show the maximum lamp wattage permitted in the fixture.   _____

    b) Install the largest wattage lamp that will physically fit in the fixture.   _____

    c) Remove the insulation blanket in the canopy if it makes it easier to splice the fixture wires to the branch circuit wires in the box.   _____

4. You are installing a lighting fixture. It has one white wire, one black wire, and one bare equipment grounding conductor. In the spaces provided, mark T (True) or F (False) for the following statements.

    a) The equipment grounding conductor may be wrapped around the No. 8-32 screw that secures the fixture mounting strap to the outlet box.   _____

    b) The equipment grounding conductor must be terminated inside the outlet box under a No. 10-32, green, hexagon-shaped grounding screw, or spliced to an equipment grounding conductor inside the outlet box.   _____

    c) Connect the bare equipment grounding conductor from the fixture to the white grounded circuit conductor in the outlet box.   _____

    d) Connect the bare equipment grounding conductor from the fixture to the outlet box using a sheet metal screw that fits tightly into the outlet box.   _____

5. Incandescent recessed lighting fixtures are required to have _____ protection.

6. You are installing six recessed incandescent lighting fixtures that have an integral junction box on them. All six fixtures will be controlled by one wall switch. You will run nonmetallic-sheathed cable from fixture to fixture. In the spaces provided, mark T (True) or F (False) for the following statements.

    a) You are permitted to run only those conductors that supply the fixtures through the outlet boxes on the recessed fixtures. There is no special marking on the fixtures indicating that they are "Suitable for through-wiring." _____

    b) You are permitted to run conductors, other than the conductors that supply the recessed fixtures, through the outlet boxes on the recessed fixtures. For example, you might want to carry the "hot" circuit through the outlet boxes on the fixtures to connect a receptacle on the other end of a string of recessed fixtures. There is no special marking on the fixtures indicating that they are "Suitable for through-wiring." _____

    c) You are permitted to run conductors, other than the conductors that supply the recessed fixtures, through the outlet boxes on the recessed fixtures. For example, you might want to carry the "hot" circuit through the outlet boxes on the fixtures to connect a receptacle on the other end of a string of recessed fixtures. The fixtures are marked "Suitable for through-wiring." _____

7. In the spaces provided, mark T (True) or F (False) for the following statements.

    a) Recessed lighting fixtures that are marked NON-IC must not be covered with thermal insulation. _____

    b) Thermal insulation must be kept at least 3 inches from the sides of recessed lighting fixtures that are marked NON-IC. _____

    c) Recessed lighting fixtures that are marked IC may be completely buried in thermal insulation. _____

    d) Above recessed fixtures marked NON-IC, thermal insulation must be installed so that heat will not be trapped in the cavity where the lighting fixture is installed. _____

    e) Fluorescent lighting fixtures that will be installed in a dropped ceiling in a recreation room must be "listed" and marked "Suspended Ceiling Fluorescent Fixture." _____

    f) Fluorescent lighting fixtures "listed" and marked "Suspended Ceiling Fluorescent Fixture" are suitable for use as a recessed fixture. _____

8. Recessed lighting fixtures that do not have an outlet box as an integral part of the fixture require special consideration. In the spaces provided, mark T (True) or F (False) for the following statements.

    a) Connect the fixture with conductors that have a temperature rating for the temperature that may be encountered. _____

    b) The fixture must be connected by using a "fixture whip," which must be at least 3 feet long. _____

    c) The fixture must be connected by using a "fixture whip," which must be at least 18 inches long but not more than 6 feet long. _____

d) The "fixture whip" must be run to a junction box that is located not less than 1 foot from the fixture. _____

e) ⅜" flexible metal conduit is acceptable for use as a "fixture whip." _____

f) When ⅜" flexible metal conduit is used as a "fixture whip," a separate equipment grounding conductor must also be pulled through the flex. _____

g) Type AC armored cable is suitable for use as a "fixture whip." _____

h) Type NMC cable is suitable for use as a "fixture whip." _____

9. Lighting fixtures in clothes closets can present a fire hazard. The *National Electrical Code®* recognized this potential fire hazard and instituted special requirements for clothes closets. Answer the following questions that pertain to lighting fixtures in clothes closets in homes.

a) Does the *NEC®* permit bare incandescent lamp fixtures such as porcelain keyless or porcelain pull-chain lampholders to be installed?_____

b) Does the *NEC®* permit bare fluorescent lamp fixtures to be installed? _____

c) Does the *NEC®* permit pendant fixtures or pendant lampholders to be installed? _____

d) What is the minimum clearance from the storage area to surface-mounted incandescent lighting fixtures? _____ inches.

e) What is the minimum clearance from the storage area to surface-mounted fluorescent lighting fixtures? _____ inches.

f) What is the minimum clearance from the storage area to recessed incandescent or fluorescent lighting fixtures? _____ inches.

g) Define the "storage area" of a clothes closet.

_____

_____

_____

_____

10. For safety reasons, what section of the *National Electrical Code®* prohibits installing cord-connected lighting fixtures, hanging lighting fixtures, lighting track, or ceiling-mounted paddle fans from being installed within a "zone" 3 feet horizontally and 8 feet vertically from the top of a bathtub rim.

Section _____.

11. Is it permissible to install a recessed exhaust fan above a bathtub?

a) Yes

b) No

12. Some upscale homes use a limited amount of neon lighting for accent lighting and similar uses. This is permitted by the *National Electrical Code®* as long as the open-circuit voltage does not exceed _____ volts. Select the correct answer.

a) 250

b) 600

c) 750

d) 1,000

13. Thermally protected fluorescent ballasts are referred to as:
    a) Type TPF
    b) Class P
    c) Self-protected

14. *Section* _____ of the *NEC®* permits lighting fixtures to be attached to and supported by trees.

15. When installing low-voltage outdoor decorative lighting, it is extremely important that the assembly be properly "listed."
    a) Outdoor Use Only
    b) Indoor Use Only

16. You are installing a decorative multilamp lighting fixture above the medicine chest located above the wash basin. For convenience, this lighting fixture has a single grounding-type receptacle on one end. What precautions are necessary when installing and connecting up this fixture?

    _____

    _____

    _____

Grounding and Bonding

OBJECTIVES

After studying this unit, you will be able to:

- **Understand the fundamentals of grounding and bonding an electric service.**
- **Understand the fundamentals of grounding equipment and appliances.**
- **Discuss the purpose served by ground rods.**
- **Know what a grounding electrode conductor is.**
- **Be able to size a grounding electrode conductor.**
- **Know what an equipment grounding conductor is.**
- **Be able to size equipment grounding conductors.**
- **Know the different types of accepted equipment grounding conductors.**
- **Understand some of the hazards associated with improper grounding.**

INTRODUCTION

Grounding is one of the most important and most misunderstood subjects in the *National Electrical Code.*® Proper grounding reduces the likelihood of a person receiving an electrical shock, and reduces the chances of a fire because of some failure or breakdown of the insulation on a "live" conductor. This unit discusses the logic of grounding, and the *National Electrical Code*® rules for correct grounding.

GROUND

Earlier in this book you learned about the *grounded conductor,* the *grounding conductor,* and *ground-fault circuit interrupters (GFCI).* Now, let us delve a little deeper into the subject of grounding. There are two distinct categories of grounding. The first category is the grounding of equipment. The second category is the grounding of the electrical system.

Most of the general grounding requirements are found in *Article 250* of the *National Electrical Code.*® We also find grounding requirements scat-

tered throughout the Code for specific items such as portable tools, appliances, motors, etc.

Definitions Relating to Grounding

To fully understand the subject of grounding, a good grasp of the meaning of certain words is essential. Here are the definitions of key words relating to the subject of grounding as listed in *Article 100* of the *NEC.*®

- ***Ground:*** *A conducting connection, whether intentional or accidental, between an electrical circuit or equipment and the earth, or to some conducting body that serves in place of the earth.*

- ***Grounded:*** *Connected to earth or to some conducting body that serves in place of the earth.*

- ***Grounded, Effectively:*** *Intentionally connected to earth through a ground connection or connections of sufficiently low impedance and having sufficient current-carrying capacity to prevent the buildup of voltages that may result in undue hazards to connected equipment or to persons.*

- *Grounded Conductor:* *A system or circuit conductor that is intentionally grounded. This is the white "neutral" conductor for branch circuits and feeders. For services, the grounded conductor is permitted to be bare.*

- *Grounding Conductor:* *A conductor used to connect equipment or the grounded circuit of a wiring system to a grounding electrode or electrodes.*

- *Grounding Conductor, Equipment:* *The conductor used to connect the noncurrent-carrying metal parts of equipment, raceways, and other enclosures to the system grounded conductor, the grounding electrode conductor, or both, at the service equipment or at the source of a separately derived system.*

  Section 250-118 recognizes that an equipment grounding conductor can be a copper or aluminum conductor, rigid metal conduit, electrical metal tubing (EMT), the metal armor of armored cable (BX), or flexible metal conduit (Greenfield).

- *Grounding Electrode Conductor:* *The conductor used to connect the grounding electrode to the equipment grounding conductor, to the grounded conductor, or to both, of the circuit at the service equipment or at the source of a separately derived system.*

  The grounding electrode conductor is permitted to be connected in the main service panel or in the meter base, *Section 250-24(a)(1)*. However, electric utilities usually frown on having the connection made in their sealed meter base. They do not like the possibility of people tampering with their "cash register."

Grounding electrode conductors are sized according to *Table 250-66*. Figure 12–1 shows sizes for grounding electrode conductors used for residential services. Refer to *Table 250-66* for service conductors larger than 600 kcmil.

Definitions Relating to Bonding

Closely associated with grounding is another term—*bonding*. Here are some definitions of key words relating to bonding as shown in *Article 100* of the *NEC*.

GROUNDING ELECTRODE CONDUCTORS

SIZE OF SERVICE-ENTRANCE CONDUCTOR*		SIZE OF GROUNDING ELECTRODE CONDUCTOR	
Copper	Aluminum or Copper-Clad Aluminum	Copper	Aluminum or Copper-Clad Aluminum
No. 2 or smaller	No. 1/0 or smaller	8	6
No. 1 or 1/0	No. 2/0 or 3/0	6	4
No. 2/0 or 3/0	No. 4/0 or 250 kcmil	4	2
Larger than No. 3/0 thru 350 kcmil	Larger than No. 250 kcmil thru 500 kcmil	2	1/0
Larger than 350 kcmil thru 600 kcmil	Larger than 500 kcmil thru 900 kcmil	1/0	3/0
Larger than 600 kcmil thru 1100 kcmil	Larger than 900 kcmil thru 1750 kcmil	2/0	4/0
Larger than 1100 kcmil	Larger than 1750 kcmil	3/0k	250 kcmil

*On large services, where service conductors are run in parallel, add the circular-mil area of the conductors that are run in parallel. Then refer back to the above using the total circular-mil area of the paralleled conductors.

Figure 12–1 Table showing grounding electrode conductor sizes for different sizes of service-entrance conductors typically used for residential services.

- *Bonding (Bonded):* *The permanent joining of metallic parts to form an electrically conductive path that will ensure electrical continuity and the capacity to conduct safely any current likely to be imposed.*

- *Bonding Jumper:* *A reliable conductor to ensure the required electrical conductivity between metal parts required to be electrically connected.*

- *Bonding Jumper, Equipment:* *The connection between two or more portions of the equipment grounding conductor.*

- *Bonding Jumper, Main:* *The connection between the grounded circuit conductor and the equipment grounding conductor at the service.*

Bonding equipment together does not necessarily mean that the equipment is properly grounded. Bonding must be done according to the *NEC* requirements.

Reasons to Ground and Bond

Section 250-2 is subdivided into four very important general requirements for grounding and bonding.

a. For those electrical systems that are required to be grounded, the connection to earth (the grounding connections) is done so to *"limit the voltage imposed by lightning, line surges, or unintentional contact with higher voltage lines and that will stabilize the voltage to earth during normal operation."*

b. Electrically conductive materials, such as metal raceways and boxes, appliances, lighting fixtures, electrical equipment, etc. are connected to earth (the grounding connections) *"so as to limit the voltage to ground on these materials."*

c. Electrically conductive materials (metal water piping, metal gas piping, structural steel.) *"that are likely to become energized shall be bonded as specified by this article to the supply system grounded conductor..."*

d. A grounding path or a bonding path is referred to in the *NEC®* as a "fault current path." The requirement is that *"The fault current path shall be permanent and electrically continuous, shall be capable of safely carrying the maximum fault likely to be imposed on it, and shall have sufficiently low impedance to facilitate the operation of overcurrent devices under fault conditions."*

Proper grounding and bonding means that all equipment, metal piping, and other metal surfaces such as air ducts will be at the same potential (voltage) so that should you come in contact with two different metallic objects, you will not receive a shock.

Overcurrent protective devices (fuses and/or circuit breakers) will operate fast when responding to ground faults, but only if the equipment is effectively grounded. Effective grounding occurs when a low-impedance (AC resistance) ground path is provided. Remember Ohm's law—in a given circuit, the lower the impedance, the higher the value of current. As the value of ground-fault current increases, there is an increase in the speed with which a fuse will open or a circuit breaker will trip off. This is called an *inverse time* relationship.

There will be less equipment and/or conductor damage when fault current is kept to a low value, and when the time the fault current is allowed to flow is kept to a minimum. The impedance of the circuit determines the *amount* of fault current that will flow. The speed of operation of a fuse or circuit breaker determines the length of *time* the fault current will flow. This time/current relationship holds true for line-to-line faults, or line-to-ground faults.

Grounding conductors and bonding conductors carry an insignificant—literally zero—amount of current under normal conditions. However, when a ground fault occurs, these conductors must be able to safely carry the fault current for the duration of time it takes the circuit breaker or fuse to open.

A grounding conductor that is too small might burn off under a fault condition, leaving the equipment "hot"—a real shock hazard.

House wiring involves proper grounding of the main service and other equipment such as appliances, outlet boxes, switch boxes, and lighting fixtures. Both issues are covered in this unit.

Grounding Electrode System

Part H of *Article 250* covers the requirements that establish a *grounding electrode system*. Simply stated, this means that everything metallic is intentionally "tied" together.

Here is how:

• The neutral of the service supplying the house is grounded by the utility somewhere on their lines, usually at the secondary of the transformer that supplies the house. See *Section 250-20(b)*.

• The service-entrance neutral conductor is again grounded at the main panel by running a grounding electrode conductor from the main panel to the grounding electrode, *Section 250-24(c)*. This is illustrated in Figures 12–2 and 12–3. In Figure 12–2, note that one option for the grounding electrode conductor is to run it from the ground rod to the inside of the meter base. As previously mentioned, many electric utilities do not allow this connection because should inspection of the grounding electrode conductor connection in the meter base be necessary, the meter seal would have to be broken.

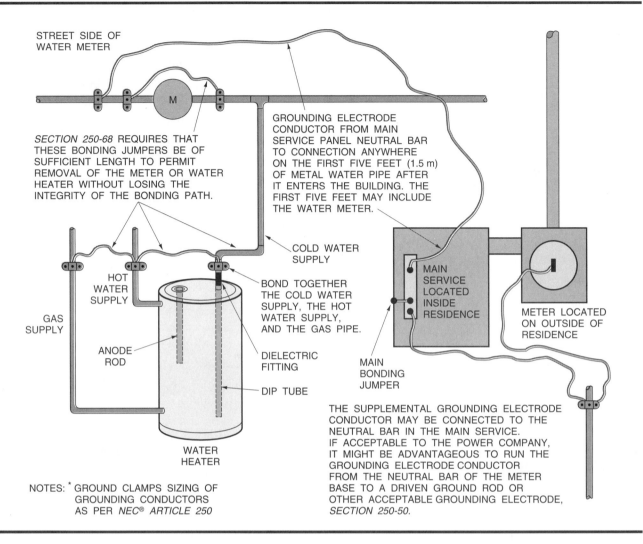

STREET SIDE OF
WATER METER

SECTION 250-68 REQUIRES THAT
THESE BONDING JUMPERS BE OF
SUFFICIENT LENGTH TO PERMIT
REMOVAL OF THE METER OR WATER
HEATER WITHOUT LOSING THE
INTEGRITY OF THE BONDING PATH.

GROUNDING ELECTRODE
CONDUCTOR FROM MAIN
SERVICE PANEL NEUTRAL BAR
TO CONNECTION ANYWHERE
ON THE FIRST FIVE FEET (1.5 m)
OF METAL WATER PIPE AFTER
IT ENTERS THE BUILDING. THE
FIRST FIVE FEET MAY INCLUDE
THE WATER METER.

COLD WATER
SUPPLY

HOT
WATER
SUPPLY

BOND TOGETHER
THE COLD WATER
SUPPLY, THE HOT
WATER SUPPLY,
AND THE GAS PIPE.

GAS
SUPPLY

ANODE
ROD

DIELECTRIC
FITTING

DIP TUBE

MAIN
SERVICE
LOCATED
INSIDE
RESIDENCE

METER LOCATED
ON OUTSIDE OF
RESIDENCE

MAIN
BONDING
JUMPER

WATER
HEATER

THE SUPPLEMENTAL GROUNDING ELECTRODE
CONDUCTOR MAY BE CONNECTED TO THE
NEUTRAL BAR IN THE MAIN SERVICE.
IF ACCEPTABLE TO THE POWER COMPANY,
IT MIGHT BE ADVANTAGEOUS TO RUN THE
GROUNDING ELECTRODE CONDUCTOR
FROM THE NEUTRAL BAR OF THE METER
BASE TO A DRIVEN GROUND ROD OR
OTHER ACCEPTABLE GROUNDING ELECTRODE,
SECTION 250-50.

NOTES: * GROUND CLAMPS SIZING OF
GROUNDING CONDUCTORS
AS PER *NEC® ARTICLE 250*

Figure 12–2 This diagram shows the main bonding jumper in the main panel, the grounding conductor electrode connected to the incoming metal water pipe, the bonding jumper around the water meter, the ground rod used as the required supplemental grounding electrode, the grounding electrode conductor attachment to the ground rod, and the bonding of the hot and cold water pipes.

A connection in the main service disconnect panel is much more accessible, and allows inspection without having to break the meter seal. Check this out locally before running the grounding electrode conductor to the meter base.

Figure 12–4 is a variation of Figure 12–3 in that a second subpanel has been added. Note that the neutral bus in the subpanel is not connected to the equipment grounding bus. It is only at the main service panel that the grounded neutral conductor, the equipment grounding conductors, the enclosure, and the grounding electrode conductor are permitted to be connected together.

• The neutral conductor is bonded to the main electrical panel, usually with a screw that connects between the neutral bar in the main electrical panel and the metal enclosure. This is referred to as the *main bonding jumper* and is shown in Figures 12–2, 12–3, and 12–4. See *Sections 250-24(a)(4), 250-28, 250-92, 250-102,* and *250-142.* This bonding is done only at the main electrical service panel, never at subpanels, *Section 250-142.*

• A grounding electrode conductor, *Sections 250-24(c), 250-62* through *250-70,* is run to a grounding electrode, usually the underground metal water pipe, where it enters the house,

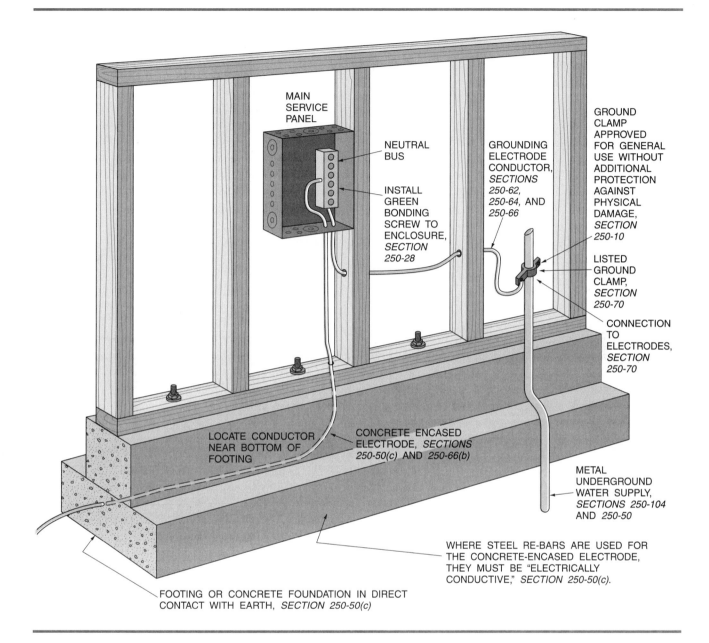

MAIN
SERVICE
PANEL

NEUTRAL
BUS

INSTALL
GREEN
BONDING
SCREW TO
ENCLOSURE,
SECTION
250-28

GROUNDING
ELECTRODE
CONDUCTOR,
SECTIONS
250-62,
250-64, AND
250-66

GROUND
CLAMP
APPROVED
FOR GENERAL
USE WITHOUT
ADDITIONAL
PROTECTION
AGAINST
PHYSICAL
DAMAGE,
SECTION
250-10

LISTED
GROUND
CLAMP,
SECTION
250-70

CONNECTION
TO
ELECTRODES,
SECTION
250-70

LOCATE CONDUCTOR
NEAR BOTTOM OF
FOOTING

CONCRETE ENCASED
ELECTRODE, *SECTIONS*
250-50(c) AND *250-66(b)*

METAL
UNDERGROUND
WATER SUPPLY,
SECTIONS 250-104
AND *250-50*

WHERE STEEL RE-BARS ARE USED FOR
THE CONCRETE-ENCASED ELECTRODE,
THEY MUST BE "ELECTRICALLY
CONDUCTIVE," *SECTION 250-50(c)*.

FOOTING OR CONCRETE FOUNDATION IN DIRECT
CONTACT WITH EARTH, *SECTION 250-50(c)*

Figure 12–3   This diagram shows the main panel neutral bus, the main bonding jumper (the bonding screw), the grounding electrode conductor connected to the incoming metal water pipe, and a supplemental concrete-encased grounding electrode in the concrete footing. The minimum size for the concrete-encased ground is No. 4 AWG bare copper conductor laid near the bottom of the footings, and encased by at least 2 inches of concrete. The minimum length for the concrete-encased electrode is 20 feet. See *Section 250-50(c)*.

Section 250-50. The underground metal water piping is then supplemented with a ground rod, *Section 250-50(a)(2).* Refer back to Figure 12–2.

- A bonding jumper is used between the hot and cold metal water pipes, *Section 250-80(b).* This is easily done at the water heater as illustrated in Figure 12–2. This is done because we do not want to rely on the plumbing piping to establish

ground to the hot water metal piping. We want to establish proper grounding of all metal piping according to acceptable practices as set forth in the *NEC.*® Some electrical inspectors will accept the many mixer water faucets as adequate to ground the hot water piping. Pipe dope makes this questionable. In the case of nonmetallic water piping, this is not an issue.

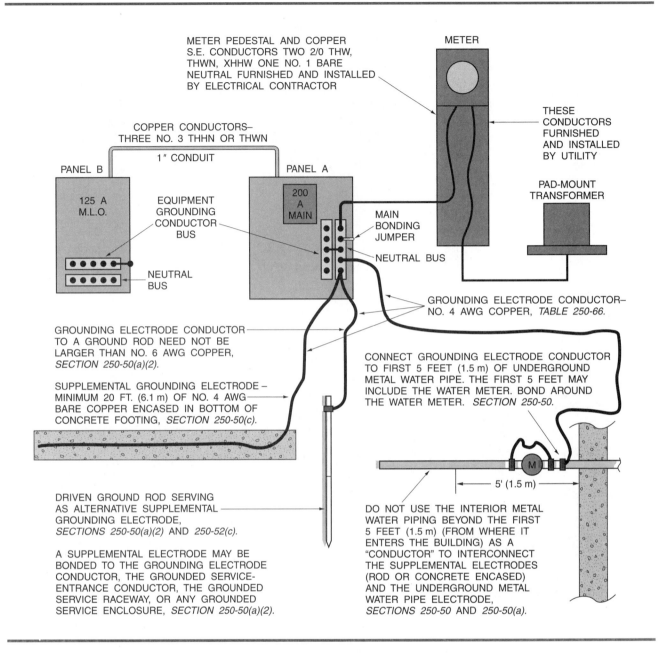

METER PEDESTAL AND COPPER
S.E. CONDUCTORS TWO 2/0 THW,
THWN, XHHW ONE NO. 1 BARE
NEUTRAL FURNISHED AND INSTALLED
BY ELECTRICAL CONTRACTOR

METER

THESE
CONDUCTORS
FURNISHED
AND INSTALLED
BY UTILITY

COPPER CONDUCTORS–
THREE NO. 3 THHN OR THWN

1″ CONDUIT

PANEL B

PANEL A

PAD-MOUNT
TRANSFORMER

125 A
M.L.O.

EQUIPMENT
GROUNDING
CONDUCTOR
BUS

200
A
MAIN

MAIN
BONDING
JUMPER

NEUTRAL
BUS

NEUTRAL BUS

GROUNDING ELECTRODE CONDUCTOR–
NO. 4 AWG COPPER, *TABLE 250-66.*

GROUNDING ELECTRODE CONDUCTOR
TO A GROUND ROD NEED NOT BE
LARGER THAN NO. 6 AWG COPPER,
SECTION 250-50(a)(2).

SUPPLEMENTAL GROUNDING ELECTRODE –
MINIMUM 20 FT. (6.1 m) OF NO. 4 AWG
BARE COPPER ENCASED IN BOTTOM OF
CONCRETE FOOTING, *SECTION 250-50(c).*

CONNECT GROUNDING ELECTRODE CONDUCTOR
TO FIRST 5 FEET (1.5 m) OF UNDERGROUND
METAL WATER PIPE. THE FIRST 5 FEET MAY
INCLUDE THE WATER METER. BOND AROUND
THE WATER METER. *SECTION 250-50.*

DRIVEN GROUND ROD SERVING
AS ALTERNATIVE SUPPLEMENTAL
GROUNDING ELECTRODE,
SECTIONS 250-50(a)(2) AND 250-52(c).

A SUPPLEMENTAL ELECTRODE MAY BE
BONDED TO THE GROUNDING ELECTRODE
CONDUCTOR, THE GROUNDED SERVICE-
ENTRANCE CONDUCTOR, THE GROUNDED
SERVICE RACEWAY, OR ANY GROUNDED
SERVICE ENCLOSURE, *SECTION 250-50(a)(2).*

5' (1.5 m)

DO NOT USE THE INTERIOR METAL
WATER PIPING BEYOND THE FIRST
5 FEET (1.5 m) (FROM WHERE IT
ENTERS THE BUILDING) AS A
"CONDUCTOR" TO INTERCONNECT
THE SUPPLEMENTAL ELECTRODES
(ROD OR CONCRETE ENCASED)
AND THE UNDERGROUND METAL
WATER PIPE ELECTRODE,
SECTIONS 250-50 AND 250-50(a).

Figure 12–4   This diagram shows a main panel and a subpanel. Note that the neutral bus and ground bus in the subpanel are not connected together. To do so would be a violation of *Section 250-142.*

- The metal gas pipe is also bonded to the grounding electrode system, *Section 250-104(b)*. The gas pipe does not serve as a grounding electrode, *Section 250-52(a)*. It is bonding so that all metal piping and surfaces are at the same potential.

Avoiding Electrical Shock

Take a look at the unlucky fellow in Figure 12–5. He is touching the water pipe and the gas

pipe at the same time. The "hot" conductor of a 20-ampere, 120-volt circuit, for whatever reason, comes in contact with the gas pipe.

Two scenarios are possible.

Scenario One: A bonding jumper (A) has been installed in conformance to the *National Electrical Code.* Let us say that the bonding jumper was No. 4 AWG copper meter base conductor and was 2 feet long. From *Table 8* in *Chapter 9* of the *NEC*, we can calculate the resistance of this bonding jumper. The DC resistance of a No. 4 AWG copper conductor is:

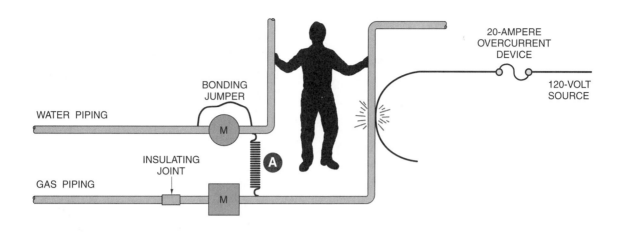

Figure 12–5   Here we have a person touching a water pipe and a gas pipe at the same time. The text explains two scenarios: one with proper bonding, the other with improper bonding.

- 0.308 ohms for 1,000 feet
- 0.000308 ohms for 1 foot
- 0.000616 ohms for 2 feet

We can now calculate the amount of current that would flow at 120 volts.

$$I = \frac{E}{R} = \frac{120}{0.000616} = 194,805 \text{ amperes}$$

This is a purely theoretical calculation that is shown for illustrative purposes. In an actual installation, the total resistance (impedance) of all parts of the entire circuit would be much higher than these simple calculations. The actual fault current flow would be much, much lower than as calculated above, yet would be of such a high value that a 20-ampere fuse or circuit breaker would open the circuit instantaneously. The individual would not be injured or killed. The low impedance path afforded by the bonding jumper saved his life, because both piping systems are at the same voltage potential.

Scenario Two: Let us see what happens when the bonding jumper (A) is not in place. The "live" conductor touches the gas pipe. The gas pipe now has 120 volts on it. The gas pipe is energized. It is "hot." The insulating joint in the gas pipe results in a very poor path to ground. Assume the resistance (impedance) to ground is 8 ohms.

$$I = \frac{E}{R} = \frac{120}{8} = 15 \text{ amperes}$$

A current of 15 amperes will not cause the 20-ampere fuse or circuit breaker to open.

The gas piping is now at the same potential as the 120-volt circuit.

If our unlucky fellow touches the gas pipe and the water pipe at the same time, he becomes a parallel path to ground. Because he is now a resistance (impedance) between the gas pipe and water pipe, current flows through his body. If the body resistance is 12,000 ohms, the current flowing through his body is:

$$I = \frac{E}{R} = \frac{120}{12,000} = 0.01 \text{ amperes (10 milliamperes)}$$

The total current flowing through the 20-ampere overcurrent device is:

$$15 + 0.01 = 15.01 \text{ amperes}$$

This amount of current will still not open the 20-ampere fuse or circuit breaker.

In Unit 7 we learned that a current of only 1,000 milliamperes for 0.03 seconds can cause ventricular fibrillation, which is usually fatal. The 10 milliamperes is in the range of where the person cannot let go. Without the bonding jumper in place, the stage has been set for a fatality.

Similar dangerous situations are present when a person touches a faulty appliance and a faucet at the same time, touching a faulty appliance and a grounded appliance at the same time, or touching a faulty tool and standing on the ground at the same time. These are just a few of the myriad of situations that set the stage for receiving an electrical shock.

You might want to review Unit 7 to refresh your memory about electrical shock hazards.

Section 250-2(d) states that for a grounding path to be effective, it must meet all three of the following issues. The path must:

- be permanent and electrically continuous,

- have the capacity to conduct safely any fault current likely to be imposed on it, and,

- have sufficiently low impedance to facilitate the operation of the circuit overcurrent protective device under fault conditions.

Section 250-90 states that "*bonding shall be provided where necessary to assure electrical continuity and the capacity to conduct safely any fault current likely to be imposed.*"

Section 250-96 states in part that the bonding of metal raceways, enclosures, fittings, and so forth, that serve as the grounding path "*shall be effectively bonded where necessary to assure electrical continuity and the capacity to conduct safely any fault current likely to be imposed on them.*"

The Grounding Electrode

Think of the earth as the *common denominator* to all electrical systems. The electric utility connects the neutral points and metal enclosures of their generators, transformers, substations, switchgear, steel structures, steel fences, and other equipment to "ground." Electricians are required by the *National Electrical Code®* to again connect the neutral of the main service and the main service enclosure to "ground."

Let us take a closer look at *National Electrical Code®* grounding requirements.

Section 250-50(a) states that metal underground water piping system 10 feet (3.05 m) or longer in direct contact with the earth is an acceptable grounding electrode. Because there is always the possibility that a metal water piping system might be interrupted with insulating joints and nonmetallic piping, *Section 250-50(a)(2)* requires that a metal water piping system must be *supplemented* by at least one additional grounding electrode.

For residential wiring in most parts of the country, underground metal water piping is generally used as the primary grounding electrode. Underground metal water piping must be supplemented by at least one additional grounding electrode according to *Section 250-50(a)(2)*, and this is usually a driven copper ground rod. In conformance to *Section 250-56*, this ground rod must have a resis-

tance to ground of 25 ohms or less. If the resistance to ground exceeds 25 ohms, then one additional ground rod must be driven. The grounding electrode conductor from the main service to the ground rod, and to the second ground rod where required, is sized according to *Table 250-66*.

Also, where the underground water piping system is nonmetallic, *Section 250-52* recognizes other acceptable grounding electrodes. For example, the grounding electrode may be a single driven ground rod having the resistance to ground of not over 25 ohms. If the resistance to ground of one driven ground rod exceeds 25 ohms, then a second ground rod is required. See *Section 250-56.*

Here again, you must check with the local electrical inspector to find out what is acceptable in your area.

Section 250-52 tells us that made electrodes should be installed below the permanent moisture level, if practical.

In *Section 250-52(c),* we find that a ground rod must:

- Be at least 8 feet (2.44 m) long.

- Be not less than ½ inch (12.7 mm) in diameter.

- Be driven so that at least 8 feet (2.44 m) of the rod is in direct contact with the earth. If rock bottom is encountered making it impossible to drive the rod vertically, then it may be driven at an angle that does not exceed 45 degrees from vertical, or buried in a trench that is at least 2½ feet (762 mm) deep.

- Be driven so that the top of the rod is flush or just below the surface of the ground. If the upper end of the rod is aboveground, the ground clamp and the grounding electrode conductor must be protected from physical damage.

- Have a resistance-to-ground of 25 ohms maximum. If when using a Megger (an instrument used for measuring resistance-to-ground), the resistance-to-ground is found to exceed 25 ohms, then one additional ground rod must be driven. Drive the second ground rod not less than 6 feet (1.83 m) from the first rod. See *Section 250-56.* Textbooks on ground rods suggest keeping the spacing of ground rods not less than the length of the rods. In other words, when driving two 8-foot rods, the spacing should be 8 feet apart.

• Some jurisdictions require that two ground rods be installed right away to supplement the water pipe ground. This eliminates the concern of whether or not one ground rod would meet the maximum 25-ohms-or-less-to-ground criterion as required by *Section 250-56.*

Another very popular supplemental electrode, particularly in areas where the soil is dry, is the so-called UFER ground, named after the gentleman who developed the concept. In the *National Electrical Code,*® this is referred to as a *concrete encased electrode.* In *Section 250-50(c)*, we find that this type of a concrete encased electrode must be at least 20 feet (6.1 m) of bare copper conductor not smaller than No. 4 AWG buried in concrete at least 2 inches (50.8 mm) thick that is in direct contact with the earth.

A second type of concrete encased electrode is at least 20 feet (6.1 m) of bare steel reinforcing bars, ½ inch (12.7 mm) minimum diameter.

A concrete encased electrode does not need a supplemental grounding electrode as does a metal underground water pipe electrode.

Concrete-encased grounding electrodes are shown in Figures 12–3 and 12–4.

There are other types of electrodes, such as a *ground ring, galvanized pipe,* and *plates,* but these are hardly, if ever, used for house wiring. These are discussed in *Section 250-50* and *250-52.*

In *Section 250-58,* we find that the same electrode must be used throughout a building. We would not use a ground rod only as the grounding electrode for the electric service, and leave the metal water piping completely separate and isolated. We must bond these together so as to create in essence, a single grounding electrode. When we properly bond everything together, we have created a *grounding electrode system* as previously discussed.

Ground Clamps

The ground bus in the main panel provides the necessary means to terminate the grounding electrode conductor in the panel.

To connect the grounding electrode conductor to a ground rod, the most common method for residential wiring is to use ground clamps. These clamps must be "listed" for use with the material of the grounding electrode, and if the clamp will be underground, it must be "listed" for direct contact with soil. Clamps and other connection methods are covered in *Section 250-70.*

Section 250-70 states that the use of solder is not acceptable because under high levels of fault current, the solder would melt, resulting in loss of the integrity of the grounding or bonding path.

Various types of ground clamps typically used for attaching a grounding electrode conductor to a ground rod or water pipe are shown in Figure 12–6.

Figure 12–7 shows a ground clamp of the type used to attach a grounding electrode conductor to a metal well casing.

Connecting Grounding and Grounded Conductors in Main Panel and Subpanel

According to the second paragraph of *Section 384-20,* equipment grounding conductors (these are the bare copper equipment grounding conductors

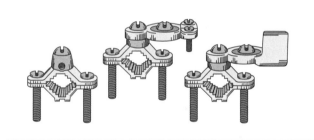

Figure 12–6 Three different types of ground clamps quite often used in residential applications for attaching the grounding electrode conductor to a ground rod or metal water pipe.

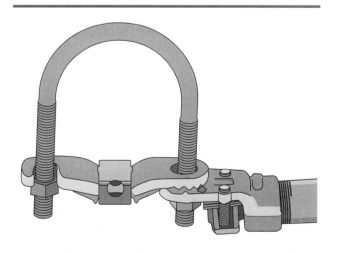

Figure 12–7 A large ground clamp used to attach a grounding electrode conductor to a metal well casing.

found in nonmetallic-sheathed cable) shall not be connected to the grounded conductor (this is the white neutral conductor) terminal bar (neutral bar) unless the bar is identified for the purpose, and is located where the grounded conductor is connected to the grounding electrode conductor at the service only.

In a typical main electrical panel, a green No. 10/32 bonding screw that is furnished with the panel is inserted through the ground bus terminal into a pretapped hole in the panel enclosure.

This screw becomes the main bonding jumper in the main panel. It bonds the neutral bar, the ground bus, the grounded neutral conductor, the grounding electrode conductor, and the panel enclosure together. This is discussed in *Section 250-28* and is illustrated in Figure 12–8, as well as in Figures 12–2 and 12–3.

Warning

Never bond the grounded conductor and the equipment grounding conductor together anywhere on the load side of the main service disconnect, *Section 250-142(b)*. If you are installing a subpanel, do not use the bonding screw that may be furnished with the panel. Throw it away.

Installing Grounding Electrode Conductors

In *Section 250-62,* we find that grounding electrode conductors:

* may be copper (the most commonly used), aluminum, or copper-clad aluminum

* may be solid or stranded

* may be bare or insulated

Section 250-64(b) states that a No. 4 AWG or larger copper or aluminum grounding electrode conductor may be securely fastened directly to the surface on which it is run, and it must be protected from *severe* physical damage.

A No. 6 AWG grounding electrode conductor may be securely fastened directly to the surface on which it is run without physical protection if not subject to physical damage. Note that the word *severe* is not present.

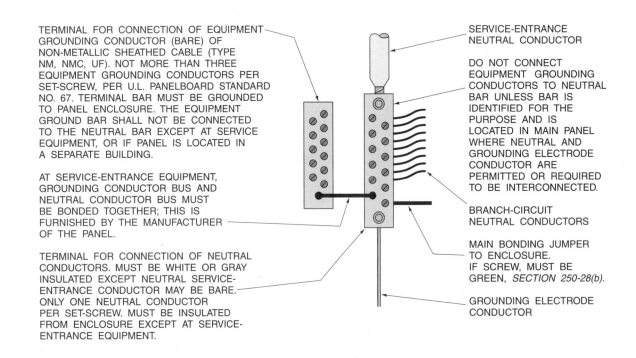

TERMINAL FOR CONNECTION OF EQUIPMENT GROUNDING CONDUCTOR (BARE) OF NON-METALLIC SHEATHED CABLE (TYPE NM, NMC, UF). NOT MORE THAN THREE EQUIPMENT GROUNDING CONDUCTORS PER SET-SCREW, PER U.L. PANELBOARD STANDARD NO. 67. TERMINAL BAR MUST BE GROUNDED TO PANEL ENCLOSURE. THE EQUIPMENT GROUND BAR SHALL NOT BE CONNECTED TO THE NEUTRAL BAR EXCEPT AT SERVICE EQUIPMENT, OR IF PANEL IS LOCATED IN A SEPARATE BUILDING.

AT SERVICE-ENTRANCE EQUIPMENT, GROUNDING CONDUCTOR BUS AND NEUTRAL CONDUCTOR BUS MUST BE BONDED TOGETHER; THIS IS FURNISHED BY THE MANUFACTURER OF THE PANEL.

TERMINAL FOR CONNECTION OF NEUTRAL CONDUCTORS. MUST BE WHITE OR GRAY INSULATED EXCEPT NEUTRAL SERVICE-ENTRANCE CONDUCTOR MAY BE BARE. ONLY ONE NEUTRAL CONDUCTOR PER SET-SCREW. MUST BE INSULATED FROM ENCLOSURE EXCEPT AT SERVICE-ENTRANCE EQUIPMENT.

SERVICE-ENTRANCE NEUTRAL CONDUCTOR

DO NOT CONNECT EQUIPMENT GROUNDING CONDUCTORS TO NEUTRAL BAR UNLESS BAR IS IDENTIFIED FOR THE PURPOSE AND IS LOCATED IN MAIN PANEL WHERE NEUTRAL AND GROUNDING ELECTRODE CONDUCTOR ARE PERMITTED OR REQUIRED TO BE INTERCONNECTED.

BRANCH-CIRCUIT NEUTRAL CONDUCTORS

MAIN BONDING JUMPER TO ENCLOSURE. IF SCREW, MUST BE GREEN, *SECTION 250-28(b)*.

GROUNDING ELECTRODE CONDUCTOR

Figure 12–8 This illustration shows the connections of the service neutral, the grounding electrode conductor, the branch-circuit neutrals, and the branch-circuit equipment grounding conductors, *Section 384–20*. At the main panel, not at any subpanels, a main bonding jumper ties the neutral bus and the ground bus together.

Figure 12–9 An armored ground cable. The conductor is copper protected by a metal armor similar to that on armored cable (BX).

Grounding electrode conductors smaller than No. 6 AWG must have physical protection. Physical protection is usually provided by using an armored ground conductor, Figure 12–9.

Section 250-64(c) states that grounding electrode conductors: must be in one continuous length from the main panel to the grounding electrodes. Splices are not permitted unless the splicing is done by exothermic welding or be an irreversible compression connector "listed" for the purpose.

A bare grounding electrode conductor can also be protected by installing it in a raceway, such as EMT.

In order to carry high values of ground-fault current, the ground path must have as low an imped-

ance as possible. We have already discussed *Section 250-2(d)* relative to proper grounding.

A metal raceway that encloses a grounding electrode conductor must be continuous from the main panel to the ground clamp—or made continuous by proper bonding. *Section 250-92(a)(3)* requires that the enclosing metal raceway must be bonded at *both ends*! At first thought, it might appear that simply installing the grounding electrode conductor in a metal raceway makes for a neat and workmanship-like installation. Bonding at one end only sets up a choke coil that results in a high impedance ground path. Past testing has shown that for a 300-ampere ground fault, approximately 295 amperes flow through the metal raceway, and only 5 amperes flow through the grounding electrode conductor inside of the metal raceway. Textbooks on grounding cover this issue.

Figures 12–10, 12–11, 12–12, and 12–13 are diagrams of proper and improper ways to install a grounding electrode conductor in a metal raceway. Only when the metal raceway is bonded at both ends will the installation "Meet Code."

If the protecting raceway is nonmetallic (PVC), then none of the above requirements for bonding are

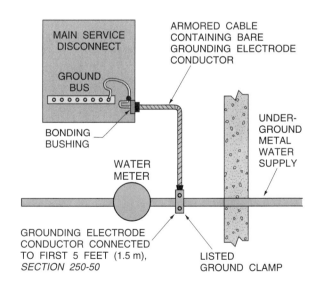

Figure 12–10 This installation of an armored ground cable "Meets Code" because a proper "listed" armored cable connector and a "listed" ground clamp have been used.

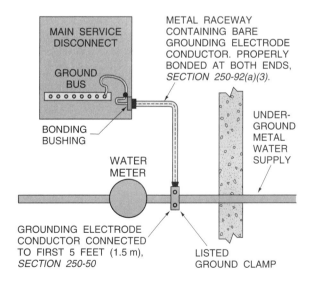

Figure 12–11 This installation of a metal raceway through which the grounding electrode conductor has been pulled "Meets Code" because a proper "listed" connector and ground clamp have been used.

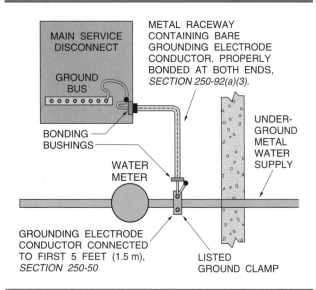

Figure 12–12 This installation of a metal raceway through which the grounding electrode has been pulled "Meets Code" because a proper "listed" connector has been used at the panel, a "listed" bonding type of fitting at the end of the raceway, and a "listed" ground clamp have been used.

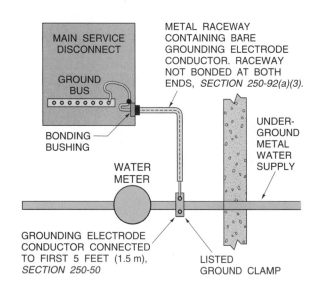

Figure 12–13 This installation of a metal raceway through which the grounding electrode conductor has been pulled **does not** "Meet Code" because the metal raceway is bonded on one end only.

necessary. Grounding and bonding requirements are irrelevant for nonmetallic wiring materials (Figure 12–14).

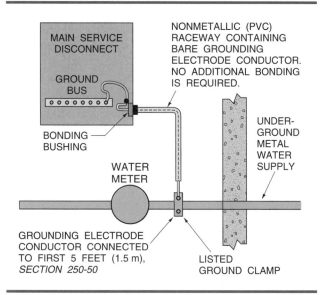

Figure 12–14 This installation of a nonmetallic (PVC) raceway through which the grounding electrode conductor has been pulled "Meets Code" because nonmetallic material cannot be grounded or bonded. Grounding and bonding requirements are irrelevant for nonmetallic wiring materials.

Sheet Metal Screws

Sheet metal screws are not permitted to be used as a means of terminating equipment grounding conductors or grounding electrode conductors, *Section 250-8.* Sheet metal screws do not have the same fine thread that No. 10-32 machine screws have. Sheet metal screws "force" themselves into a hole instead of threading themselves nicely into the pretapped matching holes in outlet boxes and panelboards. Furthermore, sheet metal screws have never been tested as to their ability to safely carry ground-fault current as required by *Section 250-2(d).*

Bonding

Bonding is very important at the main service equipment, around the water meter, and around a water heater. Consider the fact that there is really no overcurrent protection for the service conductors other than the electric utilities primary transformer fuses. Fault currents can easily reach 20,000 amperes at the main service of residential installations.

Should a fault occur on the *line side* of the main service panel, bonding conductors must be able to carry this amount of available fault current for a short period of time—the time it takes the utilities primary transformer fuse to open.

Should a fault occur on the *load side* of the main service panel, bonding conductors must be able to carry whatever the amount of available fault current for a short period of time—the time it takes to open the main fuses or circuit breaker.

Bonding is the "permanent joining of metallic parts to form an electrically conductive path that will assure electrical continuity and the capacity to conduct safely any current likely to be imposed."

Figure 12–15 shows an equipment bonding jumper.

Figure 12–16 shows a circuit bonding jumper.

For services, a bonding jumper must not be smaller than the grounding electrode conductor, *Sections 250-28(d)* and *250-102.* The size of a grounding electrode conductor is found in *Table 250-66.*

Section 250-92(a) lists the parts of service-entrance equipment that must be bonded. These

include the meter base, service raceways, service cable armor, and main disconnect. Where a metal raceway such as EMT is used for the service, a bonding bushing such as the type shown in Figure 12–17 must be used to accomplish bonding of the service raceway.

Sections 300-4(f) and *373-6(c)* state that when No. 4 AWG or larger conductors are installed in a raceway, an insulating bushing or equivalent must be used. Figure 12–18 shows one bushing made entirely of insulating material, and another that is a combination metal bushing with an insulated throat.

For service-entrance metal raceways, bonding is also required where there are concentric or eccentric knockouts, *Section 250-94.* These are combination knockouts (e.g., ½", ¾", 1", or 1", 1½", 1½", 2") that you pry out the knockout to the desired size of opening needed.

Bonding is also required where reducing washers are installed.

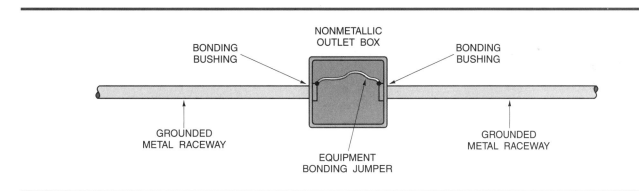

Figure 12–15   An *equipment* bonding jumper.

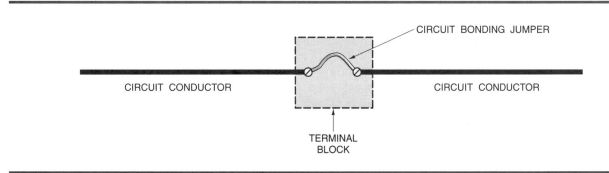

Figure 12–16   A *circuit* bonding jumper.

Figure 12–17 A bonding bushing that has an insulated throat, a lug for the grounding electrode conductor, and a set screw to lock the bushing to the conduit after it has been securely tightened.

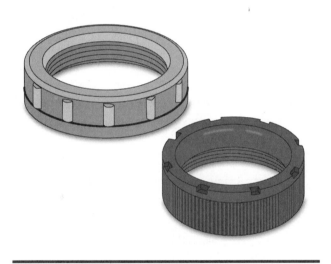

Figure 12–18 A bushing made entirely out of insulating material and a metal bushing that has an insulating throat. When a bushing is made totally of insulating material, a locknut also must be used, *Sections 300-4(f)* and *373-6(c)*.

SUMMARY OF SERVICE-ENTRANCE EQUIPMENT GROUNDING

Figure 12–19 is a montage of many of the code requirements for a typical residential electric service showing a main panel, meter, service-entrance conductors, service-drop conductors, grounding electrodes, grounding electrode conductors, bonding, bonding jumpers, clearances, etc. This diagram provides a good overview of key code issues for services.

- The system must be grounded so the maximum voltage-to-ground on the ungrounded conductors does not exceed 150 volts, *Section 250-20(b)(1)*.

- All grounding schemes shall be installed so that no objectionable currents will flow over the grounding conductors and other grounding paths, *Section 250-6(a)*.

- The grounding electrode conductor must be connected to the supply side of the service disconnecting means, *Section 250-24(a)(1)*. It must not be connected to any grounded circuit conductor on the load side of the service disconnect, *Section 250-24(a)(5)*.

- The neutral conductor must be grounded. This is generally done at the main panel, *Section 250-24(a)(1)*.

- Tie (bond) all metal piping together, *Sections 250-50, 250-52, 250-94,* and *250-104*.

- The main bonding jumper used to connect the equipment grounding conductor and the service disconnect enclosure to the grounded conductor shall not be spliced, *Section 250-28*.

- The grounding electrode conductor used to connect the grounded neutral conductor in the main panel to the grounding electrode must not be spliced, *Section 250-64(c)*.

- The grounding electrode conductor must be connected to the metal underground water pipe when the water pipe is 10 feet (3.05 m) long or more, including the well casing (if there is a well), *Section 250-50(a)*.

- Connect the grounding electrode conductor to the first 5 feet of metal underground water piping after it enters the building, *Section 250-50*. This will generally be ahead of the water meter, but does not have to be. A bonding jumper is required around the water meter if the meter is connected in the first 5 feet of piping. However, there are localities where the water meter is located in the street outside of the house, in which case bonding around the water meter would be the responsibility of the water utility. In all likelihood, there would be more than 10 feet of underground metal water piping serving as the grounding electrode, making the bonding around the street side water meter somewhat meaningless.

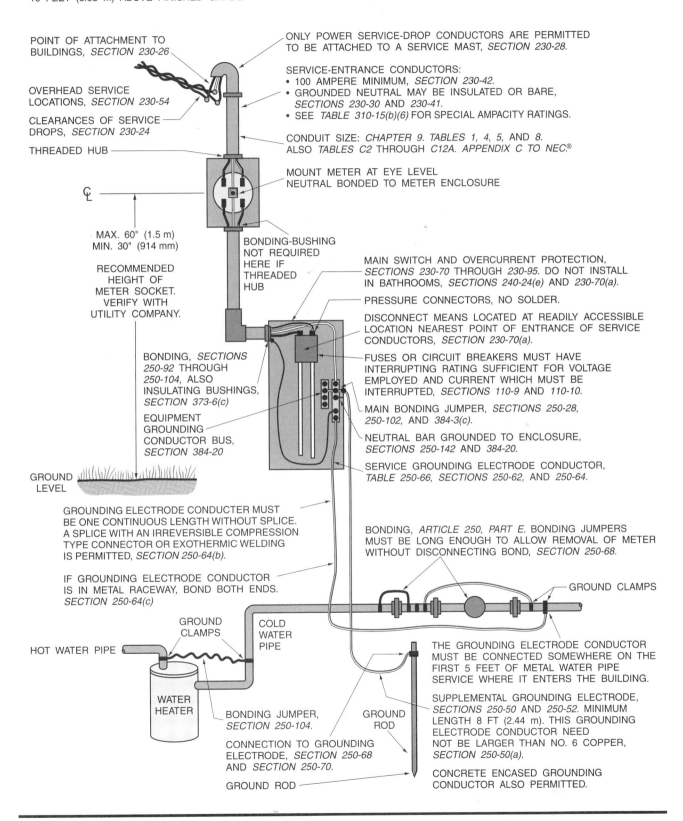

POINT OF ATTACHMENT TO BUILDINGS, *SECTION 230-26,* SHALL BE NOT LESS THAN 10 FEET (3.05 m) ABOVE FINISHED GRADE

POINT OF ATTACHMENT TO BUILDINGS, *SECTION 230-26*

OVERHEAD SERVICE LOCATIONS, *SECTION 230-54*

CLEARANCES OF SERVICE DROPS, *SECTION 230-24*

THREADED HUB

MAX. 60" (1.5 m)
MIN. 30" (914 mm)

RECOMMENDED HEIGHT OF METER SOCKET. VERIFY WITH UTILITY COMPANY.

ONLY POWER SERVICE-DROP CONDUCTORS ARE PERMITTED TO BE ATTACHED TO A SERVICE MAST, *SECTION 230-28.*

SERVICE-ENTRANCE CONDUCTORS:
• 100 AMPERE MINIMUM, *SECTION 230-42.*
• GROUNDED NEUTRAL MAY BE INSULATED OR BARE, *SECTIONS 230-30* AND *230-41.*
• SEE *TABLE 310-15(b)(6)* FOR SPECIAL AMPACITY RATINGS.

CONDUIT SIZE: *CHAPTER 9. TABLES 1, 4, 5,* AND *8.* ALSO *TABLES C2 THROUGH C12A. APPENDIX C TO NEC.®*

MOUNT METER AT EYE LEVEL NEUTRAL BONDED TO METER ENCLOSURE

BONDING-BUSHING NOT REQUIRED HERE IF THREADED HUB

MAIN SWITCH AND OVERCURRENT PROTECTION, *SECTIONS 230-70* THROUGH *230-95.* DO NOT INSTALL IN BATHROOMS, *SECTIONS 240-24(e)* AND *230-70(a).*

PRESSURE CONNECTORS, NO SOLDER.

DISCONNECT MEANS LOCATED AT READILY ACCESSIBLE LOCATION NEAREST POINT OF ENTRANCE OF SERVICE CONDUCTORS, *SECTION 230-70(a).*

FUSES OR CIRCUIT BREAKERS MUST HAVE INTERRUPTING RATING SUFFICIENT FOR VOLTAGE EMPLOYED AND CURRENT WHICH MUST BE INTERRUPTED, *SECTIONS 110-9* AND *110-10.*

BONDING, *SECTIONS 250-92* THROUGH *250-104,* ALSO INSULATING BUSHINGS, *SECTION 373-6(c)*

EQUIPMENT GROUNDING CONDUCTOR BUS, *SECTION 384-20*

MAIN BONDING JUMPER, *SECTIONS 250-28, 250-102,* AND *384-3(c).*

NEUTRAL BAR GROUNDED TO ENCLOSURE, *SECTIONS 250-142* AND *384-20.*

SERVICE GROUNDING ELECTRODE CONDUCTOR, *TABLE 250-66, SECTIONS 250-62,* AND *250-64.*

GROUND LEVEL

GROUNDING ELECTRODE CONDUCTOR MUST BE ONE CONTINUOUS LENGTH WITHOUT SPLICE. A SPLICE WITH AN IRREVERSIBLE COMPRESSION TYPE CONNECTOR OR EXOTHERMIC WELDING IS PERMITTED, *SECTION 250-64(b).*

IF GROUNDING ELECTRODE CONDUCTOR IS IN METAL RACEWAY, BOND BOTH ENDS. *SECTION 250-64(c)*

BONDING, *ARTICLE 250, PART E.* BONDING JUMPERS MUST BE LONG ENOUGH TO ALLOW REMOVAL OF METER WITHOUT DISCONNECTING BOND, *SECTION 250-68.*

GROUND CLAMPS

GROUND CLAMPS

COLD WATER PIPE

HOT WATER PIPE

WATER HEATER

BONDING JUMPER, *SECTION 250-104.*

CONNECTION TO GROUNDING ELECTRODE, *SECTION 250-68* AND *SECTION 250-70.*

GROUND ROD

GROUND ROD

THE GROUNDING ELECTRODE CONDUCTOR MUST BE CONNECTED SOMEWHERE ON THE FIRST 5 FEET OF METAL WATER PIPE SERVICE WHERE IT ENTERS THE BUILDING.

SUPPLEMENTAL GROUNDING ELECTRODE, *SECTIONS 250-50* AND *250-52.* MINIMUM LENGTH 8 FT (2.44 m). THIS GROUNDING ELECTRODE CONDUCTOR NEED NOT BE LARGER THAN NO. 6 COPPER, *SECTION 250-50(a).*

CONCRETE ENCASED GROUNDING CONDUCTOR ALSO PERMITTED.

Figure 12–19 This diagram shows most of the important *National Electrical Code*® rules for a typical residential electric service.

- Interior metal water piping shall not be used in interconnect electrodes and the grounding electrode conductor, *Section 250-50*. We do not want to rely on water piping to serve as part of the electrical system. Over time, water pipes may be disconnected, rerouted, and/or section of PVC nonmetallic piping might be installed, which, of course, would destroy the integrity of the ground path.

- A metal water pipe ground must be supplemented by another electrode, such as a bare conductor in the footing, *Section 250-50(a)(2)*, a grounding ring, *Section 250-50(d)*, the metal frame of a building if it is effectively grounded, *Section 250-50(b)*, a rod or pipe electrodes, *Section 250-52(c)*, or plate electrodes, *Section 250-52(d)*.

- The grounding electrode conductor shall be copper, aluminum, or copper-clad aluminum, *Section 250-62*.

- The grounding electrode conductor may be solid or stranded, insulated or uninsulated, covered or bare, and must not be spliced, *Section 250-62*.

- Bonding shall be provided around all insulating joints or section of the metal piping system that may be disconnected, *Section 250-68(b)*.

- The connection of the grounding electrode conductor to the grounding electrode shall be accessible, *Section 250-68(a)*. Accessibility to the connection is not required for "encased or buried connections to a concrete-encased, driven, or buried grounding electrode.

- The connection of the grounding electrode conductor to the grounding electrode may be buried using clamps listed for direct soil burial, *Section 250-70*.

- The grounding electrode conductor and bonding jumpers shall be connected tightly using proper lugs, connectors, clamps, exothermic welding, or other "listed" means. Solder connections are not permitted, *Section 250-70*.

EQUIPMENT GROUNDING

Once the main service panel has been properly grounded as discussed above, the next concern is to make sure that metallic outlet and switch boxes, lighting fixtures, and appliances are grounded in conformance to the requirements of the *National Electrical Code.*

Equipment grounding calls for an equipment grounding conductor that is large enough to carry sufficient current to cause the fuse or circuit breaker to open should a ground fault occur on a branch circuit or feeder. Equipment grounding conductors are sized according to the overcurrent device (fuse or circuit breaker) protecting the branch circuit or feeder. This is different than sizing grounding electrode conductors and main bonding jumpers at the main service, where there is no overcurrent protection ahead of the service conductors other than the utilities primary transformer fuse.

Sizing Equipment Grounding Conductors

Figure 12–20 shows sizes of equipment grounding conductors for different ratings of overcurrent devices (fuses and circuit breakers). We have included only the most common branch circuits and feeders used in residential wiring. For larger sizes, refer to *Table 250-122* in the *NEC.*

For nonmetallic-sheathed cable sizes No. 14, 12, and 10, the equipment grounding conductor is the same size as the circuit conductors.

The need for grounding these items is a safety issue that was discussed in Unit 7.

MINIMUM SIZE EQUIPMENT GROUNDING CONDUCTORS

Ampere Rating of Fuse or Circuit Breaker	Size of Equipment Grounding Conductor (Copper)	Size of Equipment Grounding Conductor (Aluminum or Copper-Clad Aluminum)
15	14	12
20	12	10
30	10	8
40	10	8
60	10	8
100	8	6

Figure 12–20 Size of equipment grounding conductors for different ampere rating overcurrent devices.

Warning

Never connect the white neutral *grounded* conductor to the equipment *grounding* conductor together anywhere on the load side of the main service panel. This requirement is found in *Section 250-142(b).*

Why? Because we want the return current in a circuit to return through the white grounded circuit conductor—not through questionable paths such as metal piping, metal ducts, aluminum foil, metal siding on a house, etc. As soon as we connect the grounded neutral conductor and the equipment grounding conductor together, we set up parallel paths where it becomes impossible to trace exactly where the current is flowing.

The only exception to this is at the main electrical service panel, where the incoming grounded neutral conductor, the grounding electrode conductor, all of the equipment grounding conductors from branch circuits, and the metal panel enclosure are all "tied" together.

There are still some who say "Why not ground equipment to the neutral? They all go to the same place, don't they?"

Let us analyze what might happen. Figure 12–21 shows a metal post lamp improperly grounded to the white grounded neutral conductor. Our unfortunate

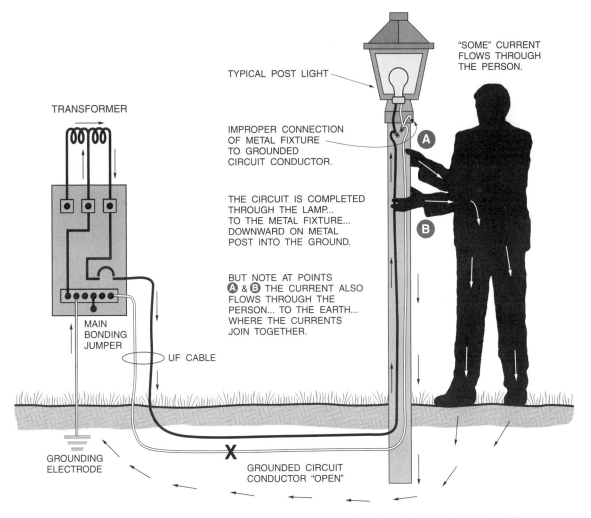

Figure 12–21 This sketch clearly illustrates the electric shock hazard associated with the Code violation practice of grounding metal objects to the grounded neutral circuit conductor. This is a violation of *Section 250-142(b).*

fellow touches the metal post. He now sets up a parallel return path with the metal post itself. The amount of current passing through him and the amount of current passing through the metal post is dependent upon the resistance (impedance) of each of the parallel paths. It is very difficult to know what the resistance values are. Suffice to say, the stage has been set for an electrocution.

Is the Earth a Good Ground?

The earth may or may not be a good ground. Is the soil moist, wet, dry, sandy, or rocky? There are too many variables. For safety sake, we cannot rely on using the earth as a ground!

Sections 250-2(d) and *250-54* of the *NEC®* tell us that *"The earth shall not be used as the sole equipment grounding conductor or fault current path."*

Figure 12–22 shows three installations of underground wiring to a post lamp.

- In (A), the installation does not "Meet Code" because the metal post in the ground serves as the only equipment ground for the post. This installation is relying solely on the earth as the return path for a ground fault. This is without question, a violation of *Section 250-2(d)* of the *NEC.®*

- In (B), the installation "Meets Code" because a separate equipment grounding conductor has been installed. Should the "hot" circuit conductor come in contact with the metal post, we know what the return path is.

- In (C), the installation does not "Meet Code" because the grounding of the post is accomplished by adding a ground rod. This installation is relying solely on the earth as the return path for a ground fault. This is without question, a violation of *Section 250-2(d)* of the *NEC.®*

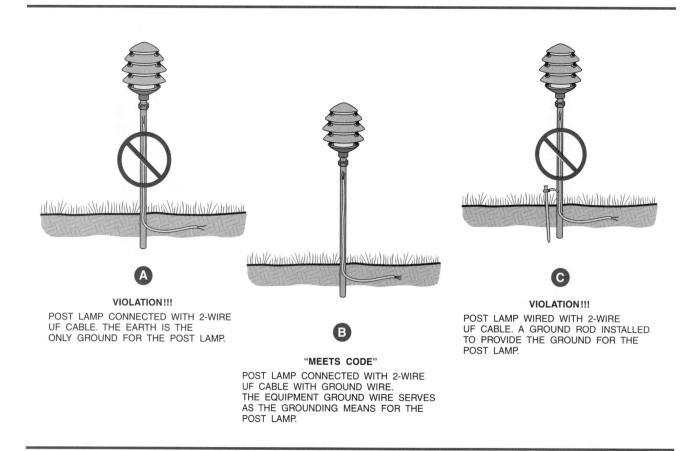

A

VIOLATION!!!
POST LAMP CONNECTED WITH 2-WIRE UF CABLE. THE EARTH IS THE ONLY GROUND FOR THE POST LAMP.

B

"MEETS CODE"
POST LAMP CONNECTED WITH 2-WIRE UF CABLE WITH GROUND WIRE. THE EQUIPMENT GROUND WIRE SERVES AS THE GROUNDING MEANS FOR THE POST LAMP.

C

VIOLATION!!!
POST LAMP WIRED WITH 2-WIRE UF CABLE. A GROUND ROD INSTALLED TO PROVIDE THE GROUND FOR THE POST LAMP.

Figure 12–22   These drawings illustrate the intent of *Section 250-2(d)* of the *NEC,®* which states that *"the earth shall not be used as the sole equipment grounding conductor or fault current path."*

Accomplishing Equipment Grounding

This subject is covered in *Part E* of *Article 250.* In some instances, the wiring method used provides the required equipment grounding. In other cases, a separate equipment grounding conductor is needed. For residential wiring, the following text discusses how equipment grounding is accomplished using various wiring methods.

Equipment Grounding When Using Nonmetallic-Sheathed Cable. When wiring with nonmetallic-sheathed cable, the equipment grounding conductor is one of the conductors in the cable. At each and every place where a nonmetallic-sheathed cable is terminated, the equipment grounding conductor is attached to the metal box, using No. 10-32 grounding screws or ground clips. Figures 12–23 and 12–24 clearly show the tapped hole in the back of a device box and an outlet box. This tapped hole is intended for a No. 10-32 grounding screw under which an equipment grounding conductor can be terminated. Figure 12–25 shows how to use a grounding clip in a device box to terminate an equipment grounding conductor.

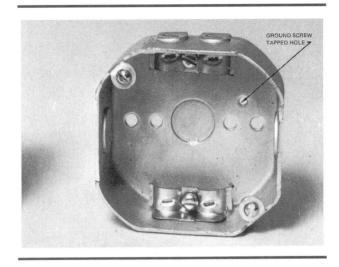

Figure 12–24   A metal outlet box with internal cable clamps for nonmetallic-sheathed cable. Note the tapped hole in the back of the box in which to insert a grounding screw, *Section 250-148(a).*

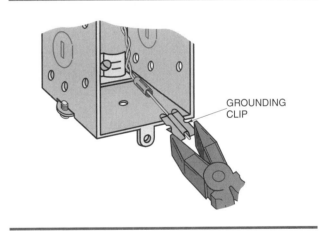

Figure 12–25   Attaching a grounding clip to a metal box for terminating an equipment grounding conductor, *Section 250-148(a).*

Figure 12–26 shows how an equipment grounding conductor of a nonmetallic-sheathed cable can be looped under an equipment grounding screw in the back of a device (switch) box, then brought out for connection to the equipment grounding terminal of a receptacle. Do not connect equipment grounding conductors in such a way that the continuity of the ground path would be interrupted by removing a receptacle or fixture, *Section 250-148.* Some key *NEC®* references are also indicated.

Figure 12–27 shows how multiple equipment grounding conductors can be spliced together with a

Figure 12–23   A metal device (switch) box with internal cable clamps for nonmetallic-sheathed cable. Note the tapped hole in the back of the box in which to insert a grounding screw, *Section 250-148(a).*

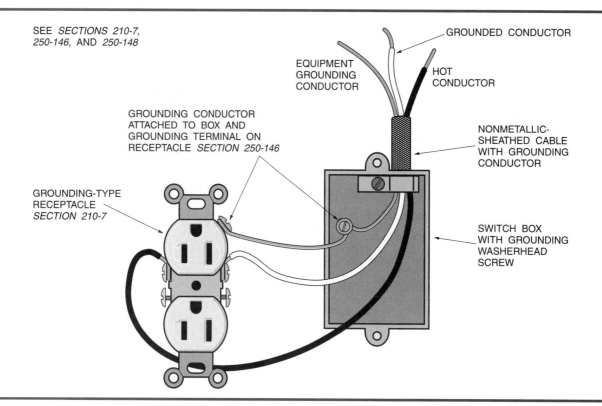

SEE *SECTIONS 210-7, 250-146*, AND *250-148*

GROUNDED CONDUCTOR

EQUIPMENT GROUNDING CONDUCTOR

HOT CONDUCTOR

GROUNDING CONDUCTOR ATTACHED TO BOX AND GROUNDING TERMINAL ON RECEPTACLE *SECTION 250-146*

NONMETALLIC-SHEATHED CABLE WITH GROUNDING CONDUCTOR

GROUNDING-TYPE RECEPTACLE *SECTION 210-7*

SWITCH BOX WITH GROUNDING WASHERHEAD SCREW

Figure 12–26   Detail showing connections for a receptacle. The wiring method is nonmetallic- sheathed cable with ground. The equipment grounding conductor is looped under the grounding screw in the metal box, then out to the grounding terminal on the receptacle. See *Sections 210–7, 250-146*, and *250-148*.

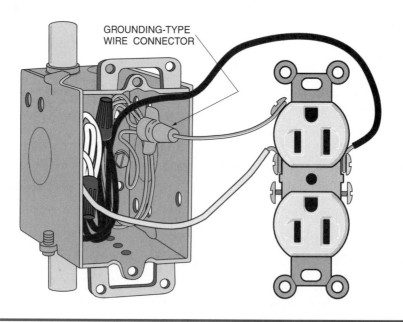

GROUNDING-TYPE WIRE CONNECTOR

Figure 12–27   An approved method of connecting equipment grounding conductors using a special wire conenctor. Should the receptacle ever be removed, the continuity of the ground path is still there. See *Sections 210–7, 250-146*, and *250-148*.

"listed" connector. From the connector, one short pigtail conductor is run to the equipment grounding screw in the back of the box. Another short pigtail conductor is brought out of the end of the connector and is terminated under the equipment grounding screw of the receptacle.

The equipment grounding conductor in nonmetallic-sheathed cable is sized in conformance to *Table 250-122*. The table shows the size of the equipment grounding conductors required for different branch-circuit ratings.

Equipment Grounding When Using Metal Raceways. When the wiring method is a metal raceway, such as electrical metallic tubing (EMT), the EMT itself serves as the equipment grounding conductor. This is illustrated in Figure 12–28.

Equipment Grounding When Using Armored Cable. When the wiring method is armored cable (BX), the metal armor plus the bonding strip inside of the armor serves as the equipment grounding conductor. This was discussed in detail in Unit 9.

Equipment Grounding When Using Nonmetallic Conduit or Tubing. When the wiring method is nonmetallic conduit or tubing, a separate equipment grounding conductor must be installed inside of the nonmetallic raceway. This is illustrated in Figure 12–29. The equipment grounding conductor is sized according to *Table 250-122*

When the wiring method is flexible metal conduit, the flex serves as the equipment grounding conductor under certain conditions previously discussed in Unit 9. In some instances, a separate equipment grounding conductor is required to be installed in the flex. The equipment grounding conductor is sized according to *Table 250-122*.

Equipment Grounding When Using Cords For Connecting Appliances. *Article 250, Part F* presents the requirements for the grounding of equipment.

Section 250-114 provides a list of specific residential cord- and plug-connected appliances that must be grounded. Most electrical appliances are furnished with a cord that contains an equipment grounding conductor. This equipment grounding conductor will be green, green with yellow stripes, or bare. Manufacturers of "listed" portable tools and appliances can eliminate the equipment grounding conductor in the attachment cord when they provide *double insulation* or equivalent. Double insulation means that the appliance or tool has two separate insulations between the "hot" conductor and the

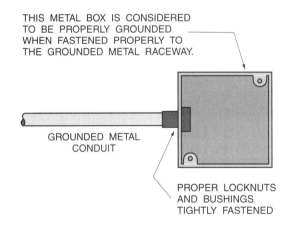

THIS METAL BOX IS CONSIDERED TO BE PROPERLY GROUNDED WHEN FASTENED PROPERLY TO THE GROUNDED METAL RACEWAY.

GROUNDED METAL CONDUIT

PROPER LOCKNUTS AND BUSHINGS TIGHTLY FASTENED

Figure 12–28 This illustration shows that a metal box is adequately grounded when fastened to a grounded metal raceway or armored cable (BX). These are acceptable grounding methods as listed in *Section 250-118.*

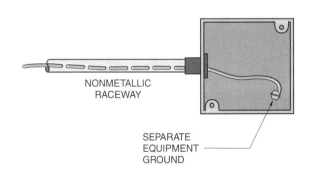

NONMETALLIC RACEWAY

SEPARATE EQUIPMENT GROUND

Figure 12–29 Grounding a metal box with a separate equipment grounding conductor that is run with the branch-circuit conductors in the nonmetallic raceway.

person using the appliance or tool. All double-insulated appliances and tools must be marked by the manufacturer to indicate that it is double insulated.

Section 250-138(a) indicates that for cord- and plug-connected equipment, an acceptable grounding means can be "*an equipment grounding conductor run with the power supply conductors in a cable assembly or flexible cord . . .*"

Detached Buildings

A detached garage with electric power must have at least one receptacle as required by *Section 210-52(g)*, and this receptacle must be GFCI protected, *Section 210-8(a)(2)*.

A detached tool shed or other similar structure is not required to have any electrical supply.

But if you decide to run electricity to a tool shed or other similar structure on the property, the wiring must be done with an acceptable wiring method, and must have their metal electrical boxes, motors, heaters, fixtures, or other electrical equipment grounded the same as if the equipment were located in the house.

Section 250-32 contains the grounding requirements for other buildings or structures on the property.

If only one branch circuit is run to an "outbuilding," the simplest way to provide proper and adequate grounding for the equipment in the outbuilding is to install an underground cable (Type UF) that contains an equipment grounding conductor in the cable. A grounded metallic raceway between the house and the outbuilding may also serve as the equipment grounding conductor. In this case, equipment grounding is provided by the separate equipment grounding conductor or the metal

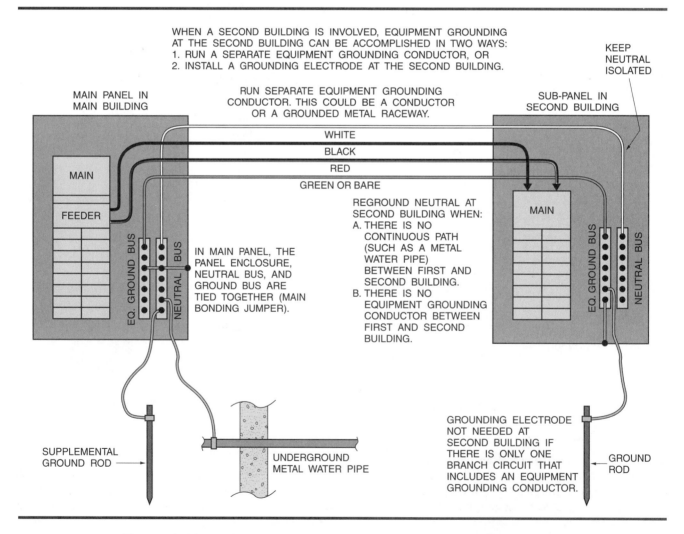

Figure 12–30   Equipment grounding at a second building. (*Section 250-32*)

raceway. These are acceptable grounding methods as listed in *Section 250-118.* A separate grounding electrode is not required, *Section 250-32(a), Exception.*

If a feeder to a second building supplies a panelboard that has more than one branch circuit, you can install a metal raceway between the two buildings, or you can install a separate equipment grounding conductor between the two buildings. A grounding electrode at the second building is required. The neutral bus and the equipment grounding bus are not connected together.

If there is no metal path such as a metal water pipe between the two buildings, and no equipment grounding conductor is run between the two buildings, a grounding electrode at the second building is required, and the neutral bus and the equipment ground bus are connected together.

These requirements are found in *Section 230-32(a)* and *(b),* and are illustrated in Figure 12–30.

REVIEW QUESTIONS

1. Match the following terms with the statement that defines the term.

   1. _____ Grounded conductor

   2. _____ Grounded

   3. _____ Equipment grounding conductor

   4. _____ Ground

   5. _____ Grounding conductor

   6. _____ Effectively grounded

   7. _____ Grounding electrode conductor

   a) Connection to earth or to some conducting body that serves in place of the earth

   b) The conductor used to connect the grounding electrode to the service equipment

   c) Intentionally connected to earth through a ground connection or connections of sufficiently low impedance and having sufficient current-carrying capacity to prevent the buildup of voltages that may result in undue hazards to connected equipment or to persons

   d) A conducting connection whether intentional or accidental, between an electrical circuit or equipment and the earth, or to some conducting body that serves in place of the earth

   e) A conductor used to connect equipment or the grounded circuit of a wiring system to a grounding electrode or electrodes

   f) The conductor used to connect the noncurrent-carrying metal parts of equipment, raceways, and other enclosures to the system grounded conductor, the grounding electrode conductor, or both, at the service equipment or at the source of a separately derived system

   g) A system or circuit conductor that is intentionally grounded

2. Grounding electrode conductors are sized according to *Table* _____ of the *NEC.*

3. Equipment grounding conductors are sized according to *Table* _____ of the *NEC.*

4. Match the following terms with the statement that defines the term.

1. _____ Bonding

2. _____ Bonding jumper

3. _____ Bonding jumper, circuit

4. _____ Bonding jumper, equipment

5. _____ Bonding jumper, main

a) A reliable conductor to ensure the required electrical conductivity between metal parts required to be electrically connected

b) The connection between two or more portions of the equipment grounding conductor

c) The connection between the grounded circuit conductor and the equipment grounding conductor at the service

d) The permanent joining of metallic parts to form an electrically conductive path that will ensure electrical continuity and the capacity to conduct safely any current likely to be imposed

e) The connection between portions of a conductor in a circuit to maintain required ampacity of the circuit

5. In your own words, explain why electrical systems are grounded.

_____

_____

_____

_____

_____

6. In your own words, explain why electrical equipment is grounded.

_____

_____

_____

_____

7. Where is the main bonding jumper located? What does it look like?

_____

_____

_____

_____

8. Mark in the spaces provided T (True) or F(False) for the following statements.

a) A grounding electrode conductor must not be smaller than the sizes indicated in *Table 250-66*. _____

b) A main bonding jumper must be installed at the main service panel. _____

c) A main bonding jumper must be installed at a subpanel to connect the neutral grounded conductor to the metal panel enclosure. The subpanel is supplied by a feeder from the main panel. _____

d) An underground metal underground water piping system 10 feet or longer in direct contact with the earth is an acceptable grounding electrode. _____

e) When an underground metal underground water piping system 10 feet or longer in direct contact with the earth is used as the grounding electrode, it must be supplemented by another electrode, such as an 8 foot long copper ground rod. _____

f) The connection of the grounding electrode conductor to metal water pipe must be connected somewhere on the first 5 feet of the metal water pipe where it enters the building. This is an underground metal water piping system that has more than 10 feet of the piping buried directly in the earth. _____

g) You have installed a 1½-inch metallic service raceway. The knockout in the main service panel has concentric knockouts. You punch (pry) out all of the concentric rings except the last one. You must use a bonding bushing or equivalent where this raceway enters the main service panel. _____

h) When the water service to a home is nonmetallic, the main service may be grounded by installing a concrete-encased electrode in the concrete footing. This is commonly referred to as a UFER ground. _____

i) The clamp used to connect a grounding electrode conductor to a driven ground rod must be visible above the ground. _____

j) To make sure that the cold and hot water metal piping is properly bonded, a bonding jumper can be installed between the cold and hot water pipes at the water heater. _____

k) An underground gas piping supply shall not be used as the grounding electrode. _____

l) Gas piping in a house must be bonded to the grounding electrode system. _____

9. The grounded white circuit conductor is permitted to be used as the equipment grounding conductor for a metal post light. What sections of the *NEC*® supports your answer?

    a) True

    b) False

    Sections _____ and _____.

10. There are a number of ways to meet the *NEC*® requirements for installing a grounding electrode conductor. In your area, what is the accepted method? _____

_____

_____

11. Grounding conductors and bonding conductors must be capable of carrying fault current for _____ periods of time.

    a) long

    b) short

12. The *NEC*® allows sheet metal screws to be used for terminating grounding conductors. What section of the *NEC*® supports your answer?

    a) True

    b) False

    Section _____

13. You are installing an equipment bonding jumper around the water meter and between the hot and cold water pipes on top of the water heater. These bonding jumpers shall not be _____ than the grounding electrode conductor that is required for the main electrical service.

14. A metal post light is installed in the front yard of a residence. Since the metal post light is set into the ground almost 2 feet, this is acceptable as suitable grounding of the metal post? Give the *National Electrical Code®* section reference number.

    a) True

    b) False

    Section _____

15. Mark T (True) or F (False) is the space provided for the following statements.

    a) All electrical outlet boxes and device boxes are metallic. Attach the equipment grounding conductor(s) to a No. 10-32, green, hexagon-shaped grounding screw in the back of the boxes. _____

    b) All electrical outlet boxes and device boxes are nonmetallic. The equipment grounding conductor must be connected to the equipment grounding screw on receptacles, and spliced to the equipment grounding conductor that is furnished with lighting fixtures, ceiling fans, etc. _____

    c) When more than one equipment grounding conductor enters a box, it is all right to twist them together without using a wire connector or other "listed" device. _____

    d) Electrical metallic tubing is an acceptable equipment grounding conductor. _____

    e) The metal armor of armored cable is an acceptable equipment grounding conductor. _____

    f) "Listed" flexible metal conduit (½") is an acceptable equipment grounding conductor if not longer than 6 feet, and the overcurrent device is not over 20 amperes. _____

    g) "Listed" flexible metal conduit (⅜") is an acceptable equipment grounding conductor if the flex is not longer than 6 feet, and the overcurrent device is not over 20 amperes. _____

    h) A grounding-type receptacle is installed in a metal device box that is flush with the finished surface of the wall. The equipment grounding conductor (EGC) from the nonmetallic-sheathed cable has been properly terminated in the box. It is not necessary to connect a separate EGC to the EGC screw on the receptacle. _____

16. Copper ground rods installed as a grounding electrode are commonly used. These rods must:

    a) be at least _____ inch in diameter.

    b) be driven to a depth of at least _____ feet unless solid rock is encountered, in which case the rod may be driven at a _____ degree angle, or it may be laid in a trench that is at least _____ feet deep.

    c) be separated by at least _____ feet when more that one rod is driven.

    d) have a resistance-to-ground of not over _____ ohms for one rod.

17. The list of cord- and plug-connected residential-type appliances that are required to be grounded is found in *Section* _____.

18. Electrical appliances that have "double insulation or equivalent" are not required to have a three-wire attachment cord, and as such, are not required to be grounded. What section of the *NEC®* supports your answer?

   a) True

   b) False

   Section _____

19. An electric range is protected with a 60-ampere breaker, supplied with No. 6 copper THW conductors. The branch-circuit conductors are installed in ENT. What size copper equipment grounding conductor is required? _____

20. Because of difficulty encountered in installing a branch circuit to an electric range, you are confronted with a situation where you must "fish" approximately 8 feet of flexible metal conduit through a crawl space and up through some cabinets to make the connection. Three No. 8 conductors are needed. The branch circuit is protected by a 40-ampere circuit breaker.

   a) What is the minimum size flexible metal conduit permitted for this installation? _____ inch.

   b) Is the flexible metal conduit acceptable as the equipment grounding means? _____ What section(s) of the *NEC®* support your answer?

   Sections _____, _____, _____

   c) Is a separate equipment grounding conductor needed?_____

   d) What is the minimum size equipment grounding conductor? _____

   e) What is the minimum size flexible metal conduit permitted for this installation? _____ inch.

21. Determine the minimum size grounding electrode conductors (copper) for the following residential 120/240 volt, single-phase services. The conductors are Type THW.

Ampere Rating of Service	SERVICE-ENTRANCE CONDUCTORS SIZED PER TABLE 310-16		SERVICE-ENTRANCE CONDUCTORS SIZED PER TABLE 310-15(B)(6)	
	Ungrounded Conductors (Copper)	Grounding Electrode Conductor (Copper)	Ungrounded Conductors (Copper)	Grounding Electrode Conductor (Copper)
100	No. 3		No. 4	
200	No. 3/0		No. 2/0	
400	Two No. 3/0 in parallel in two raceways.		Two No. 2/0 in parallel in two raceways.	

*Table 310-15(b)(6) is for 120/240-volt, 3-wire, single-phase residential services and feeders only. This table is not applicable to services and feeders other than residential.

UNIT 13

Appliances and Motors

OBJECTIVES

After studying this unit, you will be able to:

- Understand the important *National Electrical Code®* rules for appliances.
- Understand the basic *National Electrical Code®* rules for typical motor branch circuits.
- Be able to make simple appliance load calculations.
- Understand some of the special terminology used for air-conditioning and heat pumps.
- Understand when and where nonflexible cords are permitted to connect appliances.
- Have learned how to determine reduced neutral conductors for ranges, when permitted.
- Have a solid understanding of major concerns when connecting appliances, such as: 1) sizing the conductors, 2) sizing the branch-circuit short-circuit and ground-fault protection, 3) sizing the overload protection, 4) sizing and locating the disconnecting means, 5) selecting a wiring method, and 6) how to attain proper equipment grounding.

INTRODUCTION

We have come a long way. We have covered general purpose lighting branch circuits and small appliance branch circuits. It is now time to discuss specific kinds of electrical appliances found in homes.

DEFINITION OF AN APPLIANCE

You already know what an appliance is—a toaster, a microwave oven, a refrigerator, a range. But to understand the requirements of the *National Electrical Code®* relative to the installation of appliances, we need to "get technical."

The *NEC®* defines an *appliance* as: "*Utilization equipment, generally other than industrial, normally built in standardized sizes or types, that is installed or connected as a unit to perform one or more functions such as clothes washing, air conditioning, food mixing, deep frying, etc.*"

The *NEC®* defines *utilization equipment* as: "*Equipment that utilizes electric energy for electronic, electromechanical, chemical, heating, lighting, or similar purposes.*"

All appliances are furnished with installation instructions. These instructions are quite detailed, and describe such issues as the actual physical installation, minimum clearances from walls and cabinets, where the appliance is permitted to be used, the voltage, the current and/or wattage, the overcurrent protection requirements, grounding requirements, minimum branch-circuit rating requirements, and other restrictions or requirements as might be necessary for that particular appliance.

When planning for the electrical requirements for appliances, some key decisions must be made, such as:

- Branch circuit conductors: Appliances must be fed with conductors that are large enough to serve the appliance without overheating. Many times this information is found in the instructions furnished with the appliance. Other times calculations will have to be made.

- Branch-circuit short-circuit and ground-fault protection: Appliance branch circuits must be protected with fuses or circuit breakers that will guard against short circuits and ground faults.

- Overload protection: Appliances must be protected against overloads. In some appliances, overload protection is an integral part of the appliance. In other instances, separate overload protection must be provided.

- Disconnecting means: Appliances must be able to be disconnected from their electrical supply. The location of the disconnecting means is important.

- Wiring method: Appliances, unless identified for cord- and plug connection, must be hooked up with a recognized wiring method that is in accordance with the instructions furnished with the appliance, **and** in accordance with the *National Electrical Code,®* **and** in accordance with any local electrical codes. Wiring methods were discussed in Unit 9.

- Grounding for safety sake: All appliances, except *double insulated* appliances, must be grounded. We discussed grounding in Unit 12. Here is a quick review. Appliances may be grounded with:
  —A metal raceway such as EMT, flexible metal conduit, or armored cable
  —The equipment grounding conductor in a nonmetallic-sheathed cable
  —Nonmetallic type of raceway such as flexible nonmetallic conduit in which case a separate equipment grounding conductor must be installed in the flex
  —A cord- and plug connection where the cord has an equipment grounding conductor in it

GENERAL REQUIREMENTS FOR APPLIANCES

Here are some of the important Code references. *Article 422* in the *NEC®* contains most of the requirements for appliances, but there are many other specific requirements that may apply to a particular appliance.

- *Section 110-3(b)* states that: *"Listed or labeled equipment shall be installed and used in accordance with any instructions included in the listing or labeling."*

- *Section 240-3:* in general, protect the conductors at the conductor's ampacity. Conductor ampacities are found in *Table 310-16.* Where the conductor ampacity does not match a standard size fuse or circuit breaker, *Section 240-3(b)* allows the use of the next higher standard rating fuse or circuit breaker. This is only permitted where the fuse or circuit breaker does not exceed 800 amperes.

- *Section 250-114* requires that in homes, the following cord- and plug-connected appliances must be grounded unless they are *double insulated:* refrigerators, freezers, air conditioners, clothes washing, clothes dryers, food waste disposers, sump pumps, electrical aquarium equipment, handheld motor-operated tools, stationary and fixed motor-operated tools, light industrial motor-operated tools, motor-operated appliances of the following types: hedge clippers, lawnmowers, snow blowers, wet scrubbers, and portable hand lamps.

- *Section 400-7(a)(8):* Some appliances may be connected with a flexible cord. These are appliances that are designed for easy removal for maintenance and repair. These appliances must be intended or identified for flexible cord connection. According to *Table 400-4,* Types SP-3, SPE-3, and SPT-3 are all right for the appliances mentioned in *Section 422-16(b).* These are food waste disposers, built-in dishwashers, and trash compactors. Flexible cord- and plug connection is also permitted where necessary to prevent transmission of noise or vibration, *Section 422-16(a).*

- *Section 422-10(a):* The rating of the branch circuit must not be less than the rating marked on the appliance.

- *Section 422-10(a)*: Motor-operated appliances will, in most instances, be marked with their branch circuit requirements. If not marked, size the branch circuit as you would a regular motor, discussed later on in this unit.

- *Section 422-10(a)*: If a non motor-operated appliance is to be operated *continuously,* the branch-circuit rating shall be at least 125% of the rating marked on the appliances. The *NEC®* defines *continuous* as *"a load where the maximum current is expected to continue for three hours or more."* For residential applications, an electric water heater, electric snow melting cables, and similar electrical appliance loads could be considered to be *continuous*. It is somewhat of a judgment call. The Code rules for specific kinds of electrical appliances in most cases have already taken this issue consideration.

- *Section 422-11*: Protect appliance with an overcurrent device not larger than the ampere rating marked on the appliance.

- *Section 422-30*: Requires that all appliances must have a disconnecting means.

- *Section 422-31(a)*: For small appliances rated 300 volt-amperes or less, or ⅛ horsepower or less, permits the branch-circuit overcurrent device, usually a circuit breaker, to be the required disconnecting means.

- *Section 422-31(b)*: For appliances rated more than 300 volt-amperes, or greater than ⅛ horsepower, permits the branch-circuit switch or circuit breaker to be the disconnecting means if within sight of the appliance, or capable of being locked in the OFF position.

- *Section 422-34*: The disconnect switch or circuit breaker must clearly show when it is in the ON or OFF position.

- *Section 422-60*: The nameplate of appliances must show voltage and amperes, or voltage and wattage. The manufacturer of the appliance must locate the nameplate so as to be visible or easily accessible after the appliance is installed.

- *Section 422-62*: Appliances that also contain a motor must indicate the minimum supply conductors and the maximum size branch-circuit overcurrent protection. Exempt from this marking requirement are small appliances factory-equipped with a cord-and-plug assembly and appliances where the branch-circuit rating is 15 amperes or less.

- *Article 430*: This article covers the Code requirements for electric motors. It can be used if appliance consists of a motor only.

- *Article 440*: This article has the special requirements for air-conditioning equipment and heat pumps that contain a hermetic motor compressor.

CONDUCTOR ALLOWABLE AMPACITY

Sizing conductors for appliances is easy when you know what the current draw of the appliance is. Read the instruction manual and nameplate on the appliance for required circuit sizing and conductor sizing. It is quite difficult, if not impossible, to install the wiring for an appliance when you do not know what wattage or ampere rating it will be. To do so would be guessing, and in all probability, will not "Meet Code."

We discussed conductors in great detail in Unit 8. For quick reference, Figure 13–1 is *Table 310-16* from the *National Electrical Code®.* For the examples that follow, we will use copper conductors and the 60°C column for selecting conductor allowable ampacities.

Figure 13–2 shows the allowable ampacities for some of the more common cords used for residential applications. These ampacity values were excerpted from *Table 400-5(A)* of the *NEC®.*

We will now begin discussing some of the more common types of motors and appliances found in homes.

ELECTRIC MOTORS

Homes have many motors. Sump pumps, attic exhaust fans, swimming pool pump motors, saws, furnace fans, air-conditioning equipment are but a few examples. Some are cord- and plug-connected—others are "hard-wired." Some motors have integral overload protection—others require separate overload protection. Some have instructions specifying the minimum circuit and conductor sizing—others do not.

We have said it before, and will say it again, read the label!

Table 310-16. Allowable Ampacities of Insulated Conductors Rated 0 through 2000 Volts, 60°C through 90°C (140°F through 194°F) Not More than Three Current-Carrying Conductors in Raceway, Cable, or Earth (Directly Buried), Based on Ambient Temperature of 30°C (86°F)

Size	Temperature Rating of Conductor (See Table 310-13)						Size
	60°C (140°F)	75°C (167°F)	90°C (194°F)	60°C (140°F)	75°C (167°F)	90°C (194°F)	
AWG or kcmil	Types TW, UF	Types FEPW, RH, RHW, THHW, THW, THWN, XHHW, USE, ZW	Types TBS, SA, SIS, FEP, FEPB, MI, RHH, RHW-2, THHN, THHW, THW-2, THWN-2, USE-2, XHH, XHHW, XHHW-2, ZW-2	Types TW, UF	Types RH, RHW, THHW, THW, THWN, XHHW, USE	Types TBS, SA, SIS, THHN, THHW, THW-2, THWN-2, RHH, RHW-2, USE-2, XHH, XHHW, XHHW-2, ZW-2	AWG or kcmil
	COPPER			ALUMINUM OR COPPER-CLAD ALUMINUM			
18	—	—	14	—	—	—	—
16	—	—	18	—	—	—	—
14*	20	20	25	—	—	—	—
12*	25	25	30	20	20	25	12*
10*	30	35	40	25	30	35	10*
8	40	50	55	30	40	45	8
6	55	65	75	40	50	60	6
4	70	85	95	55	65	75	4
3	85	100	110	65	75	85	3
2	95	115	130	75	90	100	2
1	110	130	150	85	100	115	1
1/0	125	150	170	100	120	135	1/0
2/0	145	175	195	115	135	150	2/0
3/0	165	200	225	130	155	175	3/0
4/0	195	230	260	150	180	205	4/0
250	215	255	290	170	205	230	250
300	240	285	320	190	230	255	300
350	260	310	350	210	250	280	350
400	280	335	380	225	270	305	400
500	320	380	430	260	310	350	500
600	355	420	475	285	340	385	600
700	385	460	520	310	375	420	700
750	400	475	535	320	385	435	750
800	410	490	555	330	395	450	800
900	435	520	585	355	425	480	900
1000	455	545	615	375	445	500	1000
1250	495	590	665	405	485	545	1250
1500	520	625	705	435	520	585	1500
1750	545	650	735	455	545	615	1750
2000	560	665	750	470	560	630	2000
	CORRECTION FACTORS						
Ambient Temp. (°C)	For ambient temperatures other than 30°C (86°F), multiply the allowable ampacities shown above by the appropriate factor shown below.						Ambient Temp. (°F)
21–25	1.08	1.05	1.04	1.08	1.05	1.04	70–77
26–30	1.00	1.00	1.00	1.00	1.00	1.00	78–86
31–35	0.91	0.94	0.96	0.91	0.94	0.96	87–95
36–40	0.82	0.88	0.91	0.82	0.88	0.91	96–104
41–45	0.71	0.82	0.87	0.71	0.82	0.87	105–113
46–50	0.58	0.75	0.82	0.58	0.75	0.82	114–122
51–55	0.41	0.67	0.76	0.41	0.67	0.76	123–131
56–60	—	0.58	0.71	—	0.58	0.71	132–140
61–70	—	0.33	0.58	—	0.33	0.58	141–158
71–80	—	—	0.41	—	—	0.41	159–176

*See Section 240-3.

Reprinted with permission from NFPA 70-1999.

Figure 13-1 *Table 310-16* from the *National Electrical Code*® showing allowable ampacities for insulated conductors in raceways and cables. (*Courtesy National Fire Protection Association*)

ALLOWABLE AMPACITIES FOR FLEXIBLE CORDS

Conductor Size AWG Copper	Cords in Which Three Conductors Carry Current Example: black, white, red, plus equipment grounding conductor	Cords in Which Two Conductors Carry Current Example: black, white, plus equipment grounding conductor
14	15	18
12	20	25
10	25	30
8	35	40
6	45	55
4	60	70

Figure 13–2 Allowable ampacities for some of the more common flexible cords used for residential applicaitons. These values were taken from *Table 400-5(A), NEC.*® Refer to the *NEC*® table for other specific types and sizes.

Here are the basic Code requirements for a typical motor. Let us use a 240-volt water pump motor that has a full-load current rating (FLA) of 8 amperes for our example.

Conductors

The minimum ampacity for conductors supplying a motor is figured at 125% (1.25 times) of the motor's FLA, *Section 430-22(a)*.

Therefore: $1.25 \times 8 = 10$ amperes

We find in *Table 310-16* that a No. 14 AWG copper conductor is suitable for this load.

Branch-Circuit Short-Circuit and Ground-Fault Protection

Circuit Breakers. The maximum ampere rating of a circuit breaker according to *Table 430-152* is 250% (2.5 times) of the motor's FLA.

Therefore: $2.5 \times 8 = 20$ amperes

If the 250% sizing will not allow the motor to start because of extremely high starting current, *Section 430-52(c)(1), Exception No. 2(c)* permits a circuit breaker to be sized as large as 400% of the motor's FLA.

Nontime-Delay Fuses. The maximum ampere rating of a nontime-delay fuse according to *Table 430-152* is 300% (3 times) of the motor's FLA.

Therefore: $3 \times 8 = 24$ amperes

Section 430-52(c)(1), Exception No. 1 permits the use of the next standard size fuse, which according to *Section 240-6(a)* is a 25-ampere rating. In fact, you can size as large as 400% if the above ampere rating nontime-delay fuse will not allow the motor to start, *Section 430-52(c)(1), Exception No. 2a*. This is rarely a problem in residential applications.

Dual-Element, Time-Delay Fuses. The maximum ampere rating of a dual-element time-delay fuse according to *Table 430-152* is 175% (1.75 times) of the motor's FLA.

Therefore: $1.75 \times 8 = 14$ amperes

Section 430-52(c)(1), Exception No. 1 permits the use of the next standard size fuse, which according to *Section 240-6(a)* is a 15-ampere rating. In fact, you can size a dual-element, time-delay fuse as large as 225% (2.25 times) if the "next standard size" will not allow the motor to start, *Section 430-52(c)(1), Exception No. 2b*. This is not usually a problem in residential applications.

For practical applications, most dual-element time-delay fuses can be sized at 125% of the motor's FLA. This sizing also provides motor overload protection as required by *Section 430-32*.

Overload Protection

To protect against overheating, possible burnout of the motor, and fire, motor overload protection is required, *Section 430-32* of the *NEC®*. Here we find a number of percentages for motor overload protection as a multiple of the motor's FLA. The different percentages come from the fact that motors are grouped into more than one horsepower, one horsepower or less automatically started, and one horsepower or less nonautomatically started.

We also have the motor's marked temperature rise rating, and the *service-factor* of the motor to consider. A motor marked S.F. 1.0 cannot stand any overloads whatsoever for any appreciable time. A motor marked S.F. 1.15 may be loaded up to the horsepower obtained by multiplying the rated horsepower by the S.F. Do not multiply the full-load ampere rating on the motor's nameplate by the S.F. However, some motors may be marked S.F. amperes, which is the maximum current the motor can carry continuously without overloading it.

Motors that have integral overload protection meet these requirements. Motors that have integral overload protection can be recognized by a "reset button" somewhere on the motor.

For motors that do not have integral protection, separate overload protection is required. In most instances, this overload protection would be in time-delay fuses properly sized as discussed above, or a motor controller that contains *thermal overload elements*, quite often referred to as heaters.

Overload protection sizing varies from 115% (1.15 times) to as high as 170% (1.15 times). The most common sizing is 115% and 125%. You will have to read the nameplate on the motor before you can select the proper overload protection.

Here is a summary of requirements for motor overload protection. This table should cover almost all types of motor applications in homes. For more detail, refer to *Section 430-32*.

CONTINUOUS DUTY MOTORS	SEPARATE OVERLOAD DEVICE	INTEGRAL THERMAL PROTECTOR
Section 430-32(a): more than 1 horsepower.	–SF not less than 1.15 . . . . .125% –Temperature rise not over 40°C . . .125% –All other motors . . . . .115% *If these percentages do not allow the motor to start or to carry the load, OK to use next larger size, but do not exceed: –SF not less than 1.15 . . . . .140% –Temperature rise not over 40°C . . .140% –All other motors 130%	•Must be approved for use with the particular motor. –FLA not over 9 amperes . . . . .170% –FLA 9.1 thru 20 amperes . . . . . . .156% –FLA over 20 amperes 140%
Section 430-32(b): 1 horsepower or less, not automatically started	•If not permanently installed and within sight of the controller, OK for branch-circuit short-circuit protective device to be the overload protection. •If not in sight of the controller: overload protection is required: –SF not less than 1.15 . . . . .125% –Temperature rise not over 40°C . . .125% –All other motors . . . . .115% •If permanently installed: –SF not less than 1.15 . . . . .125% –Temperature rise notover40°C . . . .125% –All other motors 115%	•Must be approved for use with the particular motor
Section 430-32(b): 1 horsepower or less, automatically started	•Separate overload device: –SF not less than 1.15 . . . . .125% –Temperature rise not over 40°C . . .125% –All other motors . . . . .115% *If these percentages do not allow the motor to start or to carry the load, OK to use next larger size, but do not exceed: –SF not less than 1.15 . . . . .140% –Temperature rise not over 40°C . . .140% –All other motors 130%	•Must be approved for use with the particular motor

Disconnecting Means

A disconnect switch should be installed near and in sight of the motor, *Section 430-102(a)* and *(b)*. It makes common sense to have the disconnect near and in sight so as to be able to quickly turn off the switch if necessary. However, if the circuit breaker or disconnect switch is some distance away and can be individually locked in the OFF position, then the "in sight" requirement can be waived. Disconnect switches have provisions to attach a padlock, which will ensure that the switch is locked in the OFF position. Circuit breaker manufacturers supply "lock-off" devices that fit over the handle of a circuit breaker and lock the circuit breaker in the OFF position.

Section 430-107 requires that the disconnecting means be readily accessible, which means *"capable of being reached quickly for operation, renewal, or inspections, without requiring those to whom ready access is requisite to climb over or remove obstacles or to resort to portable ladders, chairs, etc."*

Section 430-109 recognizes that a motor disconnecting means may be a:

- Motor circuit switch, rated in horsepower. (This information is on the switch's label.)
- Circuit breaker
- General-use snap switch (e.g., a toggle switch) is all right for use on motors 2 hp or less, 300 volts or less, and the motor's FLA is not more than 80% of the ampere rating of the switch.
- Cord- and plug connection is all right to be the disconnecting means if the attachment plug cap and receptacle are hp-rated not less than the hp rating of the motor. The label on the attachment plug cap and the receptacle will show their hp rating. Examples of a cord-and plug-disconnecting means are the cord and plug on an overhead door unit, a refrigerator, an appliance, a sump, etc.

Box cover units, as shown in Figure 13–3, are often installed as the required disconnecting means. Type S fuses inserted into these box cover units can be sized to provide branch-circuit short-circuit and ground-fault protection. Sized closer to the motor's current rating, these fuses can also provide motor

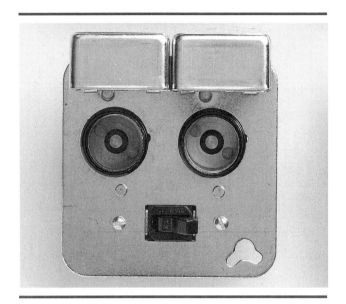

Figure 13–3 A box-cover unit that contains two fuseholders and two single-pole switches. These are widely used where disconnecting means is required. Many styles of box-cover units are available. (*Courtesy Bussman Division, Cooper Industries*)

overload protection. Box cover units are available for attachment to handy boxes, octagon outlet boxes, and square outlet boxes. They are available with fuseholder only, switch and fuseholder, receptacle and fuseholder, and pilot lamp and fuseholder. They are available with single-pole (for 120-volt applications) and two-pole switches (for 240-volt applications).

Fuses are discussed in Unit 17.

ELECTRIC CLOTHES DRYERS

Electric dryers are permitted to be cord- and plug-connected using Type SRD, SRDE, and SRDT cords as recognized in *Table 400-4.*

The cord set and receptacle must be four-wire. This was covered in detail in Unit 4.

Figure 13–4 shows a cord- and plug-connected dryer.

The minimum load requirements for an electric clothes dryer is 5,000 watts or the nameplate rating, whichever is larger, *Section 220-18.*

Let us take an actual nameplate rating of 5,700 watts.

$$I = \frac{W}{E} = \frac{5,700}{240} = 23.75 \text{ amperes}$$

Checking *Table 310-16,* we find in the 60°C column that a No. 12 AWG conductor has an ampacity

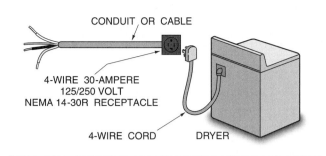

Figure 13–4 Illustration of an electric clothes dryer cord- and plug-connected. The cord set and receptacle must be 4–wire. This requirement was covered in Unit 4.

of 25 amperes. The maximum overcurrent protection for No. 12 conductors is 20 amperes, *Section 240-3(d).* Likewise, in *Table 310-16,* we find in the 60°C column that a No. 10 AWG conductor has an ampacity of 35 amperes, and in accordance with *Section 240-3(d),* must be protected by an overcurrent device not over 30 amperes.

Our selection for the dryer's branch-circuit conductors would be No. 10 AWG. See Figure 13–1, which is *Table 310-16* from the *NEC.*®

The branch circuit breaker or switch would be rated 30 amperes.

The receptacle and cord set would be rated 30 amperes.

You will find that most residential electric clothes dryers fall into this wattage and ampere range.

ELECTRIC WATER HEATERS

Electric water heaters are not permitted to be cord-and plug-connected. Since it is connected with water pipes, it does not meet the intent of having to been designed for easy removal for maintenance and repair, *Sections 400-7(a)(8)* and *422-16(a).*

Branch-circuit overcurrent protection must be sized as marked on the water heater. If not marked, the branch-circuit overcurrent protection shall not exceed 150% of the ampere rating of water heaters that draw more than 13.3 amperes. For small water heaters rated 13.3 amperes or less, the maximum branch-circuit overcurrent protection is 20 amperes. See *Section 422-11(e).*

For a single nonmotor-operated appliance, if the 150% sizing does not result in a standard size overcurrent device rating as listed in *Section 240-6(a),* then it is permitted to go to the next standard size, *Section 422-11(e)*.

This requirement applies to all single nonmotor-operated electric appliances

Let us take an example of a 3,000-watt, 240-volt electric water heater.

$$I = \frac{W}{E} = \frac{3,000}{240} = 12.5 \text{ amperes}$$

Section 422-13 requires that the branch circuit be rated not less than 125% of the nameplate rating of the water heater.

$$1.25 \times 12.5 = 15.625 \text{ amperes}$$

Let us take a look at the 60°C column of *Table 310-16*. The minimum size branch-circuit conductors would be No. 14 AWG that has an ampacity of 20 amperes. But *Section 240-3(d)* shows the maximum overcurrent protection for a No. 14 AWG conductor is 15 amperes.

We are looking for a branch-circuit rating of 15.625 amperes or greater. To meet this requirement, the branch circuit breaker or fuse would need to be rated 20 amperes, which is the next standard rating above the calculated 15.625 amperes.

Section 240-3(d) tells us that a 20-ampere overcurrent device will protect a No. 12 AWG conductor. A 20-ampere overcurrent device is too large to protect a No. 14 AWG conductor.

When making the load calculations for an electric water heater, be aware that some two-element electric water heaters are connected for simultaneous operation of the heating elements, and some are connected for "limited-demand," where only one heating element operates at any given time. The connections for the upper and lower thermostats determine this. Read the nameplate!

Electrical Wiring—Residential has a tremendous amount of technical information relative to different types of "utility-controlled" metering for water heaters (i.e., "time-of-use"), "customer-controlled" operation for water heaters, combination pressure/temperature safety relief valves, corrosion, heating element construction, recovery time charts for different capacity (in gallons) tanks and different heating element wattages, how water heater thermostats operate, a number of wiring diagrams showing the heating element and internal thermostat connections, and scalding hazards information.

ELECTRIC COUNTER-MOUNTED COOKTOPS, WALL-MOUNTED OVENS, RANGES (FREESTANDING)

Here is a summary of the basic circuit requirements for ranges, counter-mounted cooking units, and wall-mounted ovens.

Section 210-19(c)—for a branch circuit:

- The branch-circuit conductors must have an ampacity not less than the branch-circuit rating.

- The branch-circuit conductors' ampacity must not be less than the load being served.

- The branch-circuit rating must not be less than 40 amperes for ranges of 8¾ kW or more.

- For ranges 8¾ or more, the neutral may be 70% of the branch-circuit rating, but not smaller than No. 10 AWG.

- Tap conductors connecting the ranges, ovens, and cooking units to a 50-ampere branch circuit must be rated at least 20 amperes, be adequate for the load, and must not be any longer than necessary for servicing the appliance.

Section 220-19 —for a feeder:

The feeder demand for household ranges, cooktops, and ovens is computed by using the values found in *Table 220-19,* including the footnotes.

Section 220-22—for a feeder:

A feeder supplying household electric ranges, wall-mounted ovens, counter-mounted cooking units, the neutral load shall be considered as 70 percent of the load on the ungrounded conductors, as determined in accordance with *Table 220-19* for ranges. This is the value that is used when calculating the size of service-entrance conductors.

Section 422-11(b):

Branch-circuit requirements for household cooking appliances shall be permitted to be in accordance with *Table 220-19.*

Section 422-11(a):

This section states that appliance branch circuits must be protected in accordance with *Section 240-3*.

This section goes on to tell us that if the OCPD is marked on the appliance, the branch circuit OCPD shall not exceed the marked rating.

Sections 422-16(a) and 400-7(a)(8)

For ease in servicing, these appliances are

permitted to be cord-and plug-connected. The appliance must be identified for flexible cord connection, which would appear in the installation instructions. Permitted cords would be Types SRD, SRDE, and SRDT cords as recognized in *Table 400-4*.

Because these appliances are 120/240-volt-rated, the cord must be four-wire, consisting of one black, one white, one red, and one green or bare equipment grounding conductor. This was covered in detail in Unit 4.

These appliances are also permitted to be directly connected (hard-wired) using an acceptable flexible wiring method. Wiring methods were discussed in Unit 9. The disconnecting means would be the circuit breaker in the main panel. The disconnecting means must be in sight of the appliance, or be capable of being locked in the OFF position.

Section 422-32(b):

This section permits the receptacle to be under the range, but it must be accessible by pulling out and removing the bottom drawer of the range.

Section 422-11(f):

For ranges with demand over 60 amperes (i.e. some large double oven ranges), the power supply must be subdivided into two or more circuits each not over 50-amperes. This is done by the manufacturer of the appliance.

Section 422-11(e):

For nonmotor-operated appliances, the rating of the overcurrent protection shall:

- not exceed the rated current marked on the appliance. Not all appliances are marked with the maximum size overcurrent device.

- not exceed 150% of the appliance rated current if not marked, and the appliance is over 13.3 kW.

- not exceed 20 amperes if the appliance rated current is not marked, and the appliance is rated 13.3 kW or less.

Example: A counter-mounted cooking unit rated 7,450 watts and 120/240 volts. The current draw is:

$$I = \frac{W}{E} = \frac{7,450}{240} = 31.04 \text{ amperes}$$

Checking Figure 13–1 (*Table 310-16, NEC®*), we find that a No. 8 AWG conductor is minimum.

The neutral conductor cannot be reduced

because this cooking unit is less than 8¾ kW as stipulated as the minimum kW rating for application of the 70% rule. See *Section 210-19(c), Exception No. 2*.

The branch circuit breaker would be rated 40 amperes.

Example: A wall-mounted oven rated 6.6 kW @ 120/240 volts. The current draw is:

$$I = \frac{W}{E} = \frac{6,600}{240} = 27.5 \text{ amperes}$$

Checking Figure 13–1 (*Table 310-16, NEC®*), we find that a No. 10 AWG conductor is minimum.

The neutral conductor cannot be reduced because this cooking unit is less than 8¾ kW as stipulated as the minimum kW rating for application of the 70% rule. See *Section 210-19(c), Exception No. 2*.

The branch circuit breaker would be rated 30 amperes.

Example: A freestanding electric range marked 11.4 kW and 120/240 volts. The maximum current draw is:

$$I = \frac{W}{E} = \frac{11,400}{240} = 47.5 \text{ amperes}$$

However, the *National Electrical Code®* recognizes that all of the heating elements, lamps, timers, and so forth, of an electric range are unlikely to be turned on fully at the same time. The Code, in *Table 220-19*, recognizes this diversity, and permits a load demand factor to be used. For ranges rated not over 12 kW, the demand may be figured at 8 kW.

$$I = \frac{W}{E} = \frac{8,000}{240} = 33.3 \text{ amperes}$$

Checking Figure 13–1 (*Table 310-16, NEC®*), we find that a No. 8 AWG conductor is minimum.

The branch circuit breaker would be rated 40 amperes.

The neutral conductor may be reduced to 70% of the branch-circuit rating because this cooking unit is more than 8¾ kW as stipulated as the minimum kW rating for application of the 70% rule. See *Section 210-19(c), Exception No. 2*.

$$40 \times 0.70 = 28 \text{ amperes}$$

Checking Figure 13–1 (*Table 310-16, NEC®*), we find that a No. 10 AWG could be used for the neutral conductor. When the wiring method is nonmetallic-sheathed cable or armored cable, all conductors in the cable are the same size, and the equipment grounding conductor is sized per *Table 250-122*. This means that the opportunity to reduce the neutral conductor size for some electric ranges will only be possible for a raceway-type wiring method, such as EMT.

Example: An electric range rated 14,050 watts and 240 volts.

We need to look at *Note 1* to *Table 220-19*. For ranges rated over 12 kW but not over 27 kW, the demand load given in Column A must be increased by 5% for each kW or major fraction of a kW in excess of 12 kW. This is not as complicated as it sounds.

The demand is determined as follows:

2 kW (the kW over 12 kW) × 5% = 10%
8 kW (from *Column A, Table 220-19*) × 0.10 = 0.08 kW
Then: 8 kW + 0.08 kW = 8.8 kW (8,800 watts)

$$I = \frac{W}{E} = \frac{8,800}{240} = 36.7 \text{ amperes}$$

Checking Figure 13–1 (*Table 310-16, NEC®*), we find that a No. 8 AWG conductor is minimum.

The branch circuit breaker would be rated 40 amperes.

The neutral conductor may be reduced to 70% of the branch-circuit rating because this cooking unit is more than 8¾ kW as stipulated as the minimum kW rating for application of the 70% rule. See *Section 210-19(c), Exception No. 2.*

40 × 0.70 = 28 amperes

Checking Figure 13–1 (*Table 310-16, NEC®*), we find that a No. 10 AWG could be used for the neutral conductor. When the wiring method is nonmetallic-sheathed cable or armored cable, all conductors in the cable are the same size, and the equipment grounding conductor is sized per *Table 250-122*. This means that the opportunity to reduce the neutral conductor size for some electric ranges will only be possible for a raceway-type wiring method, such as EMT.

Hooking Up Ranges, Ovens, and Cooktops Using a Load Center

Sometimes the kitchen is located quite some distance from the main service panel. Instead of running separate branch circuits to each oven and counter mounted-cooktop, it might make economic sense to install a single feeder from the main service panel to a second load center (panel) near the kitchen. This is illustrated in Figure 13–5.

It is easy to make the calculation for sizing the feeder.

Checking *Note 4* to *Table 220-19,* we find that when counter-mounted cooking units and wall-mounted ovens are connected to one branch circuit, all that has to be done is to add up the nameplate wattages of the individual appliances, then consider this total to be equivalent to that of a single range of that wattage. It then becomes a simple matter to make the calculation as we did in the examples above for the 11.4 kW range and the 14,050 watt range.

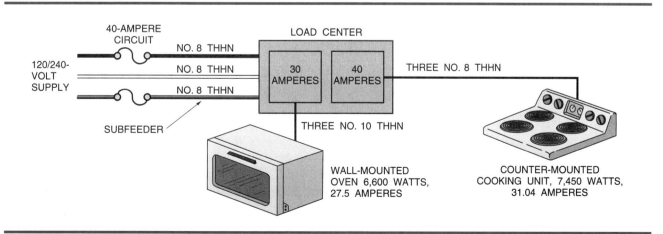

Figure 13–5 A wall-mounted oven and a counter-mounted cooking unit connected to a load center.

The branch-circuit conductors between the load center and the individual appliances are sized as discussed above.

The load center could also supply other branch circuits such as the required 20-ampere small appliance branch circuits, food waste disposer, or dishwasher in which case the feeder calculations would have to include these items. Unit 16 covers the procedure for including all of these loads into the computation.

Hooking Up Ranges, Ovens, and Cooktops Using "Taps"

Although rarely used, another way to hook up counter-mounted cooking units and wall-mounted ovens is to install a large branch circuit from the main service panel to a junction box or boxes, then connecting the appliances with "taps" that are sized for the individual appliance. Taps are nothing more than smaller size conductors connected to larger size conductors with no overcurrent protection where the "tap" is made. Figure 13–6 shows how this is done.

Because there is no overcurrent protection where the "tap" is made, certain restrictions apply. In accordance with *Section 210-19(c), Exception No. 1,* the "tap" conductors must:

- not be less than 20 amperes.
- be adequate for the appliance load to be served.

- not be longer than necessary for servicing the appliance.

The branch-circuit conductor calculation from the main panel is the same as we did for the load center option.

The branch-circuit conductors between the load center and the individual appliances are sized as discussed above.

A significant disadvantage to this installation is that to work on any of the appliances, shutting off the power off would de-energize all of the cooking appliances. A short circuit in any of the cooking appliances would trip the branch circuit breaker off, shutting off the power to all of the other cooking appliances.

The additional junction boxes, different cable sizes, additional cable connectors, conduit fittings, wire connectors, and labor might make this installation more costly than running individual branch circuits to each of the cooking appliances.

REFRIGERATORS

Residential refrigerators and freezers are cord- and plug-connected and meet all of the Underwriters Laboratories requirements relative to overcurrent protection for the internal wiring and hermetic motor compressor.

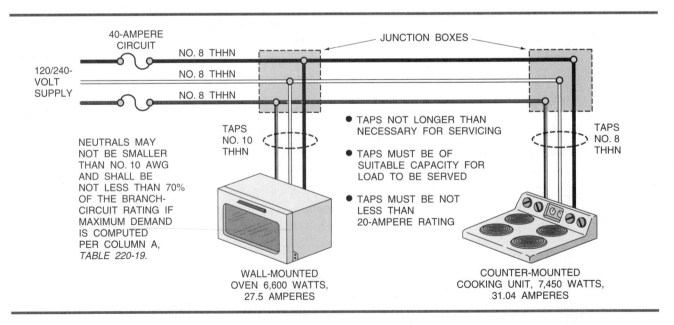

Figure 13–6   A wall-mounted oven and a counter-mounted cooking unit connected to one branch circuit.

"Listed" cord- and plug-connected appliances conform to all requirements of the UL standards.

The receptacle for a refrigerator in the kitchen or similar room in a home is permitted to be supplied by one of the two required 20-ampere small appliance branch circuits, *Section 210-52(b)*. However, *Exception No. 2* allows the receptacle for a refrigerator to be supplied by an individual branch circuit rated 15 or 20 amperes.

DISHWASHERS AND TRASH COMPACTORS

"Listed" cord- and plug-connected appliances conform to all requirements of the UL standards.

The nameplate and the installation instructions furnished with dishwashers and trash compactors provide the necessary minimum requirements for the branch circuit. A dishwasher has a motor, electric heater, and timer. It would be impossible in the field to determine what the maximum current draw would be since we do not know what electrical components might be on at the same time. So you are governed by the circuit requirements found on the nameplate and in the installation instructions.

Let us look at a typical dishwasher. It has a ⅓ hp motor rated 7.2 amperes and a 1,000-watt heating element for the drying cycle. For most dishwashers, the motor does not run during the drying cycle.

The current draw of the heating element is:

$$W = \frac{1,000}{120} = 8.33 \text{ amperes}$$

The connected load is:

Motor:	7.20 A
Heater:	8.33 A
Total:	15.53 A

Since the heater and the motor do not operate at the same time, the actual maximum current draw would be 8.33 amperes, easily served by No. 14 AWG conductors protected by a 15-ampere circuit breaker. If in doubt as to whether or not both motor and heater will be on at the same time, install a 20-ampere branch circuit using No. 12 AWG conductors.

It is best to connect each of these appliances to separate branch circuits, although it is sometimes possible to supply them with the same branch circuit. It depends on the ampere rating of the appliances. And you are playing a game trying to outguess what these ampere ratings might be, particularly if you are installing the branch circuits long before the appliances arrive.

Built-in dishwashers and trash compactors are permitted to be cord- and plug-connected with **almost** the same requirements as for food waste disposers. The differences are that the flexible cord must be 3 to 4 feet (0.910 to 1.22 m) in length, *Section 422-16(b)(2)*, and that the receptacle may be mounted in the same space as the built-in dishwasher and trash compactor, or in the adjacent space, which would generally would be under the sink, *Section 422-16(b)(2)(d)*.

Since residential dishwashers and compactors are rated 120 volts, the cord would be required to be a three-wire with grounding-type attachment plug cap.

Some local codes require that dishwashers and trash compactors be cord- and plug-connected for ease in disconnecting the appliance from the branch circuit.

FOOD WASTE DISPOSERS

Some local codes require that a food waste disposer be cord- and plug-connected.

UL lists food waste disposers both with and without a factory installed plug and cord.

A food waste disposer is permitted to be cord- and plug-connected using a flexible cord "identified for the purpose." If the cord is supplied with a "listed" food waste disposer, the cord is "identified for the purpose." If a cord is supplied and connected in the field, *Table 400-4* indicates that Types SP-3, SPE-3, and SPT-3 flexible cords are all right to use with food waste disposers, built-in dishwashers, and trash compactors. These flexible cord must have an equipment grounding conductor in it, and the cord must have a grounding-type attachment plug cap.

The flexible cord must not be longer than 3 feet (914 mm) nor shorter than 18 inches (457 mm), *Section 422-16(b)(1)(b)*.

Locate the three-wire grounding-type receptacle so as to avoid damage to the cord, *Section 422-16(b)(1)(c)*.

Do not connect the receptacle to any of the required 20-ampere small appliance branch circuit that serve the countertops, *Section 210-52(b)(2)*.

The receptacle is not required to be a GFCI-type because it does not serve the countertop surfaces. This was covered in Unit 7.

Locate the three-wire grounding-type receptacle so it will be accessible, *Section 422-16(b)(1)(d)*. This would probably be under the kitchen sink, but positioned so it will not be behind the water pipes, sink traps, or behind the food waste disposer itself. You want to be able to easily reach in and plug the attachment plug cap out of the receptacle.

A food waste disposer is a motor. The nameplate probably is marked with a horsepower rating, in amperes, and voltage. It has built-in thermal overload protection in conformance to UL standards.

Let us use an example of a food waste disposer marked ⅓ horsepower, 120 volts, 7.2 amperes.

The minimum ampacity for the conductors according to *Section 430-22(a)* is:

$$1.25 \times 7.2 = 9 \text{ amperes}$$

Checking Figure 13–1 (*Table 310-16, NEC®*), No. 14 AWG copper conductors could be used for the branch circuit.

The maximum branch circuit breaker is, according to *Section 430-52* and *Table 430-152:*

$$2.5 \times 7.2 = 18 \text{ amperes}$$

If supplied by a time-delay fused branch circuit, the maximum size time-delay fuse, according to *Section 430-52* and *Table 430-152:*

$$1.75 \times 7.2 = 12.6 \text{ amperes}$$

(All right to use the next standard size larger, which is 15 amperes, *Section 240-6(a)*.)

To turn the food waste disposer on and off, a wall switch mounted above the counter top next to the sink is standard, Figure 13–7. This also serves as the disconnecting means.

The disconnect means may also be a cord- and plug connection as illustrated in Figure 13–8.

In Figure 13–8, if the branch circuit supplying the wall receptacle was adequate, the receptacle could be a duplex, which could then also serve the dishwasher. Whether or not one 20 ampere branch circuit would be adequate to serve both appliances depends on the current ratings of both appliances. If you are not sure what the ampere ratings are, better to run two separate branch circuits. Better to be safe than sorry.

The branch circuit breaker, even though it is not in sight of the appliance serves as another disconnecting means if it is capable of being locked in the OFF position.

ELECTRIC BASEBOARD HEATERS

Although we are discussing electric baseboard heaters, *Article 424* of the *NEC®* also covers electric space heating cables, duct heaters, resistance-type boilers, and electric radiant heating panels. It is beyond the scope of this book to cover in detail all of the methods of electric heating.

Electric baseboard heaters are appliances. As with all appliances "listed" by Underwriters Laboratories, the appliances have met the safety requirements set by the standards. The premise wiring for hooking up the appliance must conform to the *NEC®*

Individual branch circuits are permitted to supply *fixed* electric baseboard heaters, *Section 424-3*. A baseboard heater fastened to the wall are *fixed* in place as opposed to a cord- and plug connection.

Branch circuits that supply two or more outlets for *fixed* electric space heating equipment must be rated 15, 20, or 30 amperes.

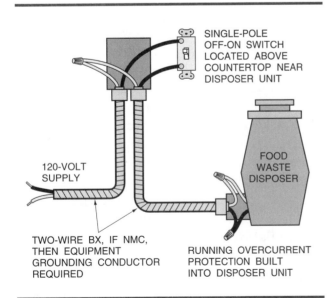

SINGLE-POLE
OFF-ON SWITCH
LOCATED ABOVE
COUNTERTOP NEAR
DISPOSER UNIT

120-VOLT
SUPPLY

FOOD
WASTE
DISPOSER

TWO-WIRE BX, IF NMC,
THEN EQUIPMENT
GROUNDING CONDUCTOR
REQUIRED

RUNNING OVERCURRENT
PROTECTION BUILT
INTO DISPOSER UNIT

Figure 13–7 Wiring for a food waste disposer operated by a separate wall switch located above countertop near the sink.

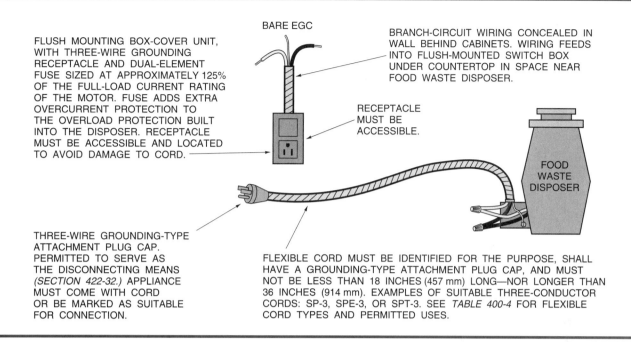

FLUSH MOUNTING BOX-COVER UNIT, WITH THREE-WIRE GROUNDING RECEPTACLE AND DUAL-ELEMENT FUSE SIZED AT APPROXIMATELY 125% OF THE FULL-LOAD CURRENT RATING OF THE MOTOR. FUSE ADDS EXTRA OVERCURRENT PROTECTION TO THE OVERLOAD PROTECTION BUILT INTO THE DISPOSER. RECEPTACLE MUST BE ACCESSIBLE AND LOCATED TO AVOID DAMAGE TO CORD.

BARE EGC

BRANCH-CIRCUIT WIRING CONCEALED IN WALL BEHIND CABINETS. WIRING FEEDS INTO FLUSH-MOUNTED SWITCH BOX UNDER COUNTERTOP IN SPACE NEAR FOOD WASTE DISPOSER.

RECEPTACLE MUST BE ACCESSIBLE.

FOOD WASTE DISPOSER

THREE-WIRE GROUNDING-TYPE ATTACHMENT PLUG CAP. PERMITTED TO SERVE AS THE DISCONNECTING MEANS *(SECTION 422-32.)* APPLIANCE MUST COME WITH CORD OR BE MARKED AS SUITABLE FOR CONNECTION.

FLEXIBLE CORD MUST BE IDENTIFIED FOR THE PURPOSE, SHALL HAVE A GROUNDING-TYPE ATTACHMENT PLUG CAP, AND MUST NOT BE LESS THAN 18 INCHES (457 mm) LONG—NOR LONGER THAN 36 INCHES (914 mm). EXAMPLES OF SUITABLE THREE-CONDUCTOR CORDS: SP-3, SPE-3, OR SPT-3. SEE *TABLE 400-4* FOR FLEXIBLE CORD TYPES AND PERMITTED USES.

Figure 13–8 Typical cord- and plug connection for a food waste disposer.

The branch-circuit conductors and the overcurrent protective device rating must be not less than 125% of the ampere rating of a fixed baseboard heating unit. If the 125% multiplier results in an ampere rating that does not match a standard rating fuse or circuit breaker, *Section 240-3(b)* permits the use of the next larger standard size.

Baseboard electric heating units for residential use are available in many different wattages. They are available in both 120-volt and 240-volt ratings. If possible, installing the 240-volt units is recommended because the current draw is only one-half that of the same wattage 120-volt unit.

You might want to review Unit 4 regarding the positioning of wall receptacles above electric baseboard heaters.

The disconnecting means is generally the branch circuit breaker that supplies the baseboard heating unit, and comes under the stipulation that the disconnecting means must be in sight (which is not practical in living areas of homes), or be capable of being locked in the OFF position. The term "within sight" appears many times in *Section 424-19.* And of course, the switch or circuit breaker must clearly indicate when in the ON or OFF position, *Section 424-21.* Code section numbers for this requirement were discussed above.

ELECTRIC SPACE HEATERS

These might be ceiling- or wall-mounted, recessed or surface-mounted. In any case, they are appliances insofar as the *NEC®* and UL are concerned. Installation instructions furnished with the heaters must be followed. The branch circuit requirements are stated in the installation instructions. The nameplate will show the voltage, wattage, and/or current rating. The calculation is easy. For example, let us consider a bathroom ceiling heater marked 1,500 watts, 120 volts.

$$I = \frac{W}{E} = \frac{1,500}{120} = 12.5 \text{ amperes}$$

Section 424-3(b) requires that the branch-circuit conductors and the overcurrent protection for fixed electric space heating shall not be less than 125% of the appliances' rating.

Therefore: $12.5 \times 1.25 = 15.625$ amperes

This requires a 20-ampere branch circuit and 20-ampere conductors.

ELECTRIC FURNACES

Central heating, such as an electric furnace or a gas furnace must be supplied by a separate branch circuit, *Section 422-12.*

The branch-circuit rating must be in conformance to the installation instructions and the information found on the nameplate.

An electric furnace is not permitted to be cord- and plug-connected. Because it is connected with hot and cold air ducts, and because of its physical size, a furnace does not meet the intent of having been designed for easy removal for maintenance and repair, *Sections 400-7(a)(8)* and *422-16.*

ANSI Standard Z21.47 prohibits gas furnaces to be cord- and plug-connected.

As with other appliances that we have already discussed, a "listed" electric furnace has already met the tough requirements of the Underwriters Laboratories standards. The premise wiring that is used to hook up an electric furnace must be done using a wiring method recognized in *Chapter 3* of the *National Electrical Code*® and/or local codes. We covered wiring methods in Unit 9.

The branch-circuit conductors and the overcurrent protective device must be not less than 125% of the total appliance load, which includes the motor and heating elements, *Section 424-3(b).* All of this would have already been taken into consideration when the product was "listed."

The 125% factor that is almost universally applied to major appliance loads is to provide a cushion—a bit of head room in determining the minimum branch circuit requirements.

The nameplate will indicate the maximum size and type of the branch-circuit overcurrent protective device. The overcurrent device must not exceed the value marked on the nameplate, and it must be of the type specified.

Figure 13–9 recaps a number of the very basic *National Electrical Code*® requirements for the branch-circuit wiring for an electric furnace.

AIR-CONDITIONING EQUIPMENT AND HEAT PUMPS

Air-conditioning equipment and heat pumps that have hermetic motor compressor(s) are somewhat different that an ordinary electric motor load. Hermetic motor compressors do not have a *full-load current* rating. They have a *rated-load ampere* rating.

Article 440 of the *NEC*® covers this type of equipment. For "packaged" air-conditioning and heat pump equipment, the electrician really needs to be concerned with the data on the nameplate.

Because hermetic motor compressors have characteristics that are different from standard electric motors, different terminology is marked on the nameplate. Here are the two items of most concern to the electrician.

Minimum Circuit Ampacity (MCA)

This is just what it says—the minimum circuit ampacity needed for the air-conditioning unit. The manufacturer has determined the current draw of the hermetic motor compressor, the fan, and any other loads that the equipment contains. The MCA value most often is some unusual value, like 26.4 amperes. The electrician need only to size the conductors equal to or greater than the MCA value. Whereas for standard electric motors, the conductors are sized based on 125% of the motor's FLA; this 125% factor has already been taken into consideration by the manufacturer of the end-use equipment. There is no need to apply the 125% factor again.

Maximum Overcurrent Protection (MOP):

MOP is the maximum rating of the branch-circuit overcurrent device. The MOP value is clearly marked on the nameplate. The nameplate and instructions furnished with the end-use equipment will also indicate whether the overcurrent device is permitted to be a nontime-delay fuse, a time-delay fuse, a standard type of circuit breaker, or a HACR circuit breaker. A HACR circuit breaker is one that has met the requirements for use with equipment that contains hermetic motor compressor(s). The letters stand for **H**eating, **A**ir-**C**onditioning, and **R**efrigeration. Use the size and type of overcurrent protection that the nameplate calls for, or you will void the appliance's warranty.

Some air-conditioning equipment is marked with the term *Branch Circuit Selection Current (BCSC)* instead of *Rated Load Amperes (RLA).* BCSC is greater than RLA, so the manufacturer of the end-use equipment would have used BCSC instead of RLA to determine the MCA.

Most air-conditioning equipment is marked with

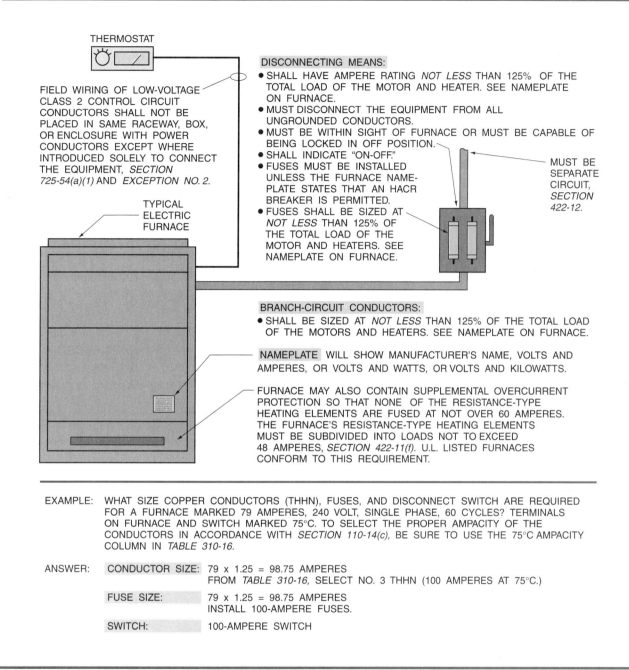

THERMOSTAT

FIELD WIRING OF LOW-VOLTAGE CLASS 2 CONTROL CIRCUIT CONDUCTORS SHALL NOT BE PLACED IN SAME RACEWAY, BOX, OR ENCLOSURE WITH POWER CONDUCTORS EXCEPT WHERE INTRODUCED SOLELY TO CONNECT THE EQUIPMENT, *SECTION 725-54(a)(1)* AND *EXCEPTION NO. 2*.

TYPICAL ELECTRIC FURNACE

DISCONNECTING MEANS:
- SHALL HAVE AMPERE RATING *NOT LESS* THAN 125% OF THE TOTAL LOAD OF THE MOTOR AND HEATER. SEE NAMEPLATE ON FURNACE.
- MUST DISCONNECT THE EQUIPMENT FROM ALL UNGROUNDED CONDUCTORS.
- MUST BE WITHIN SIGHT OF FURNACE OR MUST BE CAPABLE OF BEING LOCKED IN OFF POSITION.
- SHALL INDICATE "ON-OFF."
- FUSES MUST BE INSTALLED UNLESS THE FURNACE NAME-PLATE STATES THAT AN HACR BREAKER IS PERMITTED.
- FUSES SHALL BE SIZED AT *NOT LESS* THAN 125% OF THE TOTAL LOAD OF THE MOTOR AND HEATERS. SEE NAMEPLATE ON FURNACE.

MUST BE SEPARATE CIRCUIT, *SECTION 422-12*.

BRANCH-CIRCUIT CONDUCTORS:
- SHALL BE SIZED AT *NOT LESS* THAN 125% OF THE TOTAL LOAD OF THE MOTORS AND HEATERS. SEE NAMEPLATE ON FURNACE.

NAMEPLATE WILL SHOW MANUFACTURER'S NAME, VOLTS AND AMPERES, OR VOLTS AND WATTS, OR VOLTS AND KILOWATTS.

FURNACE MAY ALSO CONTAIN SUPPLEMENTAL OVERCURRENT PROTECTION SO THAT NONE OF THE RESISTANCE-TYPE HEATING ELEMENTS ARE FUSED AT NOT OVER 60 AMPERES. THE FURNACE'S RESISTANCE-TYPE HEATING ELEMENTS MUST BE SUBDIVIDED INTO LOADS NOT TO EXCEED 48 AMPERES, *SECTION 422-11(f)*. U.L. LISTED FURNACES CONFORM TO THIS REQUIREMENT.

EXAMPLE: WHAT SIZE COPPER CONDUCTORS (THHN), FUSES, AND DISCONNECT SWITCH ARE REQUIRED FOR A FURNACE MARKED 79 AMPERES, 240 VOLT, SINGLE PHASE, 60 CYCLES? TERMINALS ON FURNACE AND SWITCH MARKED 75°C. TO SELECT THE PROPER AMPACITY OF THE CONDUCTORS IN ACCORDANCE WITH *SECTION 110-14(c)*, BE SURE TO USE THE 75°C AMPACITY COLUMN IN *TABLE 310-16*.

ANSWER: CONDUCTOR SIZE: 79 x 1.25 = 98.75 AMPERES
FROM *TABLE 310-16*, SELECT NO. 3 THHN (100 AMPERES AT 75°C.)

FUSE SIZE: 79 x 1.25 = 98.75 AMPERES
INSTALL 100-AMPERE FUSES.

SWITCH: 100-AMPERE SWITCH

Figure 13–9 The basic *National Electrical Code®* requirements for hooking up an electric furnace. A "listed" electric furnace will have met all of the requirements set forth by Underwriters Laboratories for the appliance. The electrician generally only needs to install the proper size and type of branch-circuit wiring, including the disconnecting means. *Article 424* of the *NEC®* covers the installation of fixed electric space heating.

the term *Locked Rotor Current (LRC or LRA)*. For some HVAC equipment, you may have to convert the LRA value to an ampere rating in order to select a disconnecting means, *Section 440-12(b)(1). Table 430-151A* in the *National Electrical Code®* is a table

for making the conversion from LRA to horsepower for single-phase motors. You then select a disconnect switch that has the proper hp rating, and is capable of disconnecting the LRA of the equipment. Without getting into great detail, let us take an example of a 240-volt air-conditioning unit that is marked:

| Compressor LRA | 107 |
| Condensing Fan FLA | 1.5 |

The total LRA is 107 + (1.5 × 6) = 116 amperes

Since the small condensing fan is not marked with a LRA, we use the factor of six times in accordance with *Section 440-12(c)* as an approximation of the motor's LRA.

In *Table 430-151A,* the 116 LRA falls between 102 and 168 in the 230-volt column, which is in the 5-horsepower category. Checking manufacturers' catalogs, we find that a 30-ampere, 240-volt heavy-duty, single-phase disconnect switch has a 3-horsepower rating, and a 60-ampere, 240-volt heavy-duty, single-phase disconnect switch has a 10-horsepower rating. Therefore, our example requires a 60-ampere switch.

The disconnect must be in sight of the equipment and must be readily accessible, *Section 440-14.* The disconnect most popular for residential air-conditioning and heat pump units is a "pull-out."

This is illustrated in Figure 13–10. These are available in 30- and 60-ampere, 240-volt ratings in both fusible and nonfusible types. Another type is the "molded case switch" type, which is nothing more than a molded case circuit breaker with the tripping mechanism removed, and only the switching mechanism remains.

The disconnect is generally mounted on the side of the house instead of on the air conditioner itself for easy replacement of the air conditioner. The connection between the disconnect switch and the AC unit is flexible. See Unit 9 for acceptable wiring methods.

Figure 13–11 shows the basic requirements for a typical residential-type air conditioner or heat pump. The electrician must provide wiring based on the equipment's *Minimum Circuit Ampacity (MCA)* and branch-circuit overcurrent protection based on the equipment's *Maximum Overcurrent Protection (MOP).*

Electrical Wiring—Residential covers this air-conditioning equipment in much greater detail and should be referred to if you are involved in large air-conditioning equipment.

A

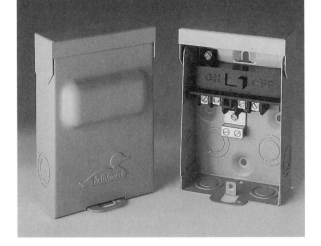

B

Figure 13–10   (A) is a fusible pullout disconnect (B) is a non-fusible pullout disconnect. These are available in 30- and 60-ampere ratings at 240 volts. (*Courtesy Midwest Electric Products*)

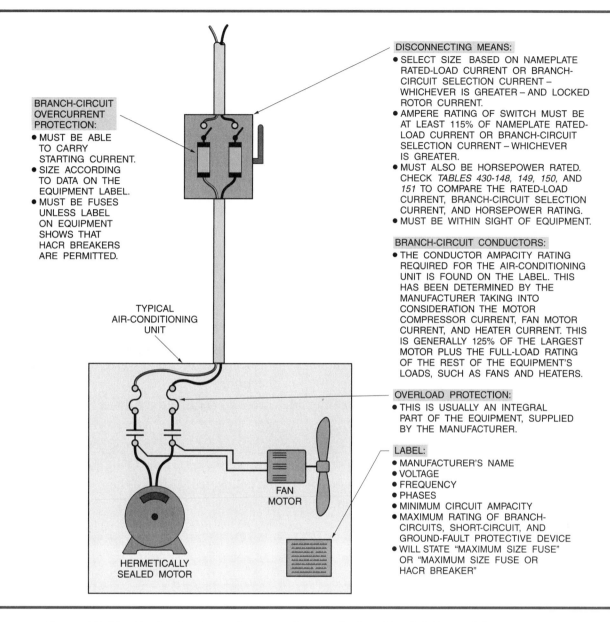

DISCONNECTING MEANS:
- SELECT SIZE BASED ON NAMEPLATE RATED-LOAD CURRENT OR BRANCH-CIRCUIT SELECTION CURRENT – WHICHEVER IS GREATER – AND LOCKED ROTOR CURRENT.
- AMPERE RATING OF SWITCH MUST BE AT LEAST 115% OF NAMEPLATE RATED-LOAD CURRENT OR BRANCH-CIRCUIT SELECTION CURRENT – WHICHEVER IS GREATER.
- MUST ALSO BE HORSEPOWER RATED. CHECK *TABLES 430-148, 149, 150, AND 151* TO COMPARE THE RATED-LOAD CURRENT, BRANCH-CIRCUIT SELECTION CURRENT, AND HORSEPOWER RATING.
- MUST BE WITHIN SIGHT OF EQUIPMENT.

BRANCH-CIRCUIT OVERCURRENT PROTECTION:
- MUST BE ABLE TO CARRY STARTING CURRENT.
- SIZE ACCORDING TO DATA ON THE EQUIPMENT LABEL.
- MUST BE FUSES UNLESS LABEL ON EQUIPMENT SHOWS THAT HACR BREAKERS ARE PERMITTED.

BRANCH-CIRCUIT CONDUCTORS:
- THE CONDUCTOR AMPACITY RATING REQUIRED FOR THE AIR-CONDITIONING UNIT IS FOUND ON THE LABEL. THIS HAS BEEN DETERMINED BY THE MANUFACTURER TAKING INTO CONSIDERATION THE MOTOR COMPRESSOR CURRENT, FAN MOTOR CURRENT, AND HEATER CURRENT. THIS IS GENERALLY 125% OF THE LARGEST MOTOR PLUS THE FULL-LOAD RATING OF THE REST OF THE EQUIPMENT'S LOADS, SUCH AS FANS AND HEATERS.

TYPICAL AIR-CONDITIONING UNIT

OVERLOAD PROTECTION:
- THIS IS USUALLY AN INTEGRAL PART OF THE EQUIPMENT, SUPPLIED BY THE MANUFACTURER.

LABEL:
- MANUFACTURER'S NAME
- VOLTAGE
- FREQUENCY
- PHASES
- MINIMUM CIRCUIT AMPACITY
- MAXIMUM RATING OF BRANCH-CIRCUITS, SHORT-CIRCUIT, AND GROUND-FAULT PROTECTIVE DEVICE
- WILL STATE "MAXIMUM SIZE FUSE" OR "MAXIMUM SIZE FUSE OR HACR BREAKER"

FAN MOTOR

HERMETICALLY SEALED MOTOR

Figure 13–11   Basic circuit requirements for a typical residential air conditioner or heat pump.

REVIEW QUESTIONS

1. In what Article of the *National Electrical Code®* are requirements found for appliances?

    Article _____

2. Instructions furnished with appliances _____ be followed when installing an appliance. What Code reference supports your supports your answer?

    a) should

    b) shall

    Section _____

3. Are all electrical appliances required to be grounded? What Code reference supports your supports your answer?

    a) Yes

    b) No

    Section _____

4. *Section 422-*_____ requires that an appliance branch circuit _____ be less than the marked rating on the appliance. Insert the Code section number and select the correct answer.

    a) shall

    b) shall not

5. *Section 400-7(a)(8)* permits flexible cords to be used for *"appliances where the fastening means and mechanical connections are specifically designed to permit ready removal for maintenance and repair, and the appliance is intended or identified for flexible cord connection."* The flexible cord must be *"equipped with an attachment plug and shall be energized from a receptacle outlet."* Section 422-16(b)(1) also permits food waste disposers to be cord- and plug-connected.

You have installed a receptacle outlet under the kitchen sink for plugging in the food waste disposer. According to the instructions furnished with the food waste disposer, the appliance is identified for cord- and plug-connection.

    a) From *Table 400-4,* what types of flexible cords are permitted for this application?
    Types _____, _____, _____.

    b) What is the maximum permitted length for the cord? _____ feet.

    c) May the receptacle under the kitchen sink be connected to one of the 20-ampere small appliance branch circuits?
    1) Yes
    2) No

    d) Does the receptacle under the sink have to be GFCI protected?
    1) Yes
    2) No

6. A home usually has a number of electrical motors, such as attic exhaust fans, water pumps, swimming pool water pumps, etc. In many cases, the motors are part of an appliance. In other instances, the motor becomes "a stand-alone" application. The conductors that supply an electric motor are generally sized at _____ percent of the motor's full-load current rating? The answer to this question is found in *Section* _____ of the *NEC.*®

7. For most typical motor branch-circuit short-circuit and ground-fault protection, indicate the maximum values for the various types of overcurrent protective devices.

    a) Circuit breaker: _____% of the motor's full-load current rating.

    b) Time-delay fuses: _____% of the motor's full-load current rating.

    c) Nontime-delay fuses: _____% of the motor's full-load current rating.

8. Overload protection for typical electric motors is based on _____% of the motor's full-load current rating. The answer to this question is found in *Section* _____ of the *NEC.*®

9. A disconnecting means for an electric motor must be located _____ _____ of the motor, or be capable of being _____ in the OFF position. The answer to this question is found in *Section* _____ of the *NEC*.®

10. A cord- and plug connection for a motor rated 2 horsepower or less is suitable as the disconnecting means for a motor.
    a) True
    b) False

11. From *Table 400-4,* what types of cords are permitted to connect electric ranges and dryers? _____, _____, _____.

12. Does the *NEC*® permit electric water heaters to be cord- and plug-connected?
    a) Yes
    b) No

    The answer to this question is found in *Sections* _____ and _____ of the *NEC*.®

13. Size THHN copper conductors and the maximum branch-circuit overcurrent protection for a 4,500-watt, 240-volt electric water heater. The nameplate does not indicate the size of overcurrent device. Show your work. Use the 60°C column of *Table 310-16* to select the conductors in conformance to *Section 110-14(c)*.

14. Size THHN copper conductors for a 6,000-watt, 120/240-volt electric surface-mounted cooktop. Size the branch circuit breaker. Show your work. Use the 60°C column of *Table 310-16* to select the conductors.

15. A freestanding electric range is rated 11.8 kW, 120/240 volts. The wiring method is nonmetallic-sheathed cable. Use the 60°C column of *Table 310-16* in conformance to *Section 110-14(c)*. Answer the following questions.
    a) According to *Column A, Table 220-19,* what is the maximum demand? _____ kW.
    b) What is the minimum size for the ungrounded conductors? No. _____AWG

    c) What is the minimum size neutral conductor? No. _____ AWG

d) What is the correct rating for the branch-circuit overcurrent protection?
_____amperes

16. When a kitchen is located a great distance from the main service panel, it sometimes makes sense to run a single "feeder" to a small load center near the kitchen instead of running individual branch circuits for a counter-mounted cooktop and for one or more wall-mounted ovens. The branch circuits for these appliances then become short branch circuits. In your own words, how would you make the calculations for this "feeder?"

_____

_____

_____

_____

_____

17. *Section 424-3(b)* requires that branch-circuit conductors for electric space heating such as baseboard heating units be sized at _____ _____% of the appliance's current rating. Select the correct answer and insert the correct percent value.

   a) not more than

   b) not less than

18. A central heating gas-fired residential-type furnace is permitted to be connected to the same branch circuit that supplies other loads in the same area. What section of the Code supports your answer?

   a) True

   b) False

   Section _____

19. Does the *National Electrical Code®* permit cord- and plug connection of an electric furnace? What section of the Code supports your answer?

   a) True

   b) False

   Section _____

20. *Section 424-3(b)* requires that the branch-circuit conductors and the overcurrent protective device for an electric furnace be sized at _____ _____% of the furnace's current rating. Select the correct answer and insert the correct percent value.

   a) not more than

   b) not less than

21. Air-conditioning units and heat pumps that contain a hermetic motor compressor must be treated differently from a standard electric motor. This is because a hermetic motor compressor has a unique rating called *rated load amperes,* whereas a standard electric motor has a *full-load ampere rating.* Because of this, the manufacturer subjects the unit to testing that result in the determination of two very important values that are marked on the nameplate. What are these terms?

_____

_____

22. The *NEC*® has requirements for the disconnect for AC and heat pump equipment. Mark T (True) or F (False) for the following statements.

    a) The disconnect must be within sight of the equipment.   _____

    b) The disconnect must be accessible.   _____

    c) The disconnect must be readily accessible.   _____

    d) The disconnect may be mounted on the outside wall next to the equipment.   _____

    e) The disconnect may be mounted on the outside wall behind the equipment.   _____

    f) The disconnect may be mounted on the side of the equipment.   _____

UNIT 14

Low-Voltage Wiring

OBJECTIVES

After studying this unit, you will be able to:

- **Understand the fundamentals of low-voltage circuits used in homes.**
- **Know how to hook up chimes.**
- **Have an appreciation of the Code requirements for Class 2 circuits.**
- **Have learned the basics for proper installation of telephone, cable television, fire alarm, and low-voltage remote control systems.**

INTRODUCTION

Up to this point in this book, we have been talking about power premise wiring for 120-volt and 240-volt circuits. Residential wiring also involves some low-voltage wiring, the installation requirements of which is not quite as severe as standard wiring methods.

A typical residence will have chimes, telephones, television (local or cable), thermostat wiring for the heating/cooling equipment, and possibly low-voltage control of lighting. Some homes will have a security system that includes smoke detectors, heat detectors, carbon monoxide detectors, alarms, and many other features.

The home might be wired in part as an "intelligent house," of which there are many systems available today.

Article 725 of the *National Electrical Code®* covers various types of remote-control, signaling, and power-limited circuits.

The *National Electrical Code®* describes a signaling circuit as "*Any electric circuit that energizes signaling equipment.*" These are fancy words for door chimes, doorbells, or buzzers.

Article 725 groups these circuits into three classes: Class 1, Class 2, and Class 3. Each class is covered by specific requirements. Since this book is all about house wiring, we will limit our discussion to that of Class 2 circuits.

As you study the following low-voltage material, you will note that there is much duplication and overlapping of the installation requirements. You also will note that the installation requirements are not as tough as the requirements for standard wiring methods.

CLASS 2 CIRCUITS

Class 2 circuits are defined in *Section 725-2* as "*The portion of the wiring system between the load side of a Class 2 power source and the connected equipment. Due to its power limitations, a Class 2 circuit considers safety from a fire initiation standpoint and provides acceptable protection from electric shock.*"

The most common Class 2 source is the secondary output from the low-voltage terminals of a chime transformer. The line-side (primary) is connected to a 120-volt supply. The load-side (secondary) supplies the necessary low voltage for the chime wiring. Another common Class 2 source is the transformer that is an integral part of a furnace. All of the low-voltage thermostat wiring is a Class 2 circuit.

Because all homes have door chimes, and it is a simple subject, let us talk about these first.

Chime transformers are "listed" by Underwriters Laboratories in two types.

- *Inherently limited:* these are designed to limit the short-circuit current to a maximum of 8 amperes. The impedance of the internal windings are such that they limit the short-circuit current to a value so low that there will be no overheating that could cause a fire. This *inherently limited* concept is the same for an electric clock. Should it stop completely, nothing will happen other than incorrect time.

  Open-circuit voltage is limited to 30 volts. Open-circuit voltage is the voltage across the secondary terminals of the transformer with no connected load.

- *Not inherently limited*: these are limited to 100 volt-amperes or less, and has overcurrent protection built into it. The open circuit voltage is limited to 30 volts.

    Should a short circuit occur in the bell-wiring conductors, the overload device inside of the transformer opens and closes repeatedly until the short is cleared.

    The transformer nameplate will be marked "Class 2 Transformer."

How Many Chimes Can Be Connected to One Transformer?

Chime transformers in homes have a secondary voltage of 16 volts.

Chime transformers are available with ratings of 5, 10, 15, 20, and 30 watts (VA).

Figure 14–1 shows a typical Class 2 chime transformer. Note the screw/bracket for securing the transformer through a ½" knockout of an outlet box.

The transformer must be adequate to serve one or more chimes. For more than one chime, do not install a transformer with a higher voltage rating. Install a transformer that has sufficient wattage rating. If you are adding a chime(s), take out the existing transformer and replace it with one having enough wattage capacity. Do not try to add another transformer and hook them up in parallel, as this is prohibited by *Section 725-41(b).*

The current required by different kinds of chimes varies. Figure 14–2 is a sample guideline for typical chimes. If still in doubt, you might want to consult the manufacturer of the chime transformer.

Figure 14–1 A Class 2 chime transformer of the type most commonly used in residential work. (*Courtesy of NuTone*)

TYPE OF CHIME	APPROXIMATE POWER CONSUMPTION
Standard two-note	10 watts
Repeating chime	10 watts
Internally lighted chime, two lamps	10 watts
Internally lighted chime, four lamps	15 watts
Chime and clock combination	15 watts
Motor-driven chime (more than two notes)	15 watts
Electronic chime	15 watts

Figure 14–2 Typical wattage ratings for various types of chimes. Consult the chime manufacturers' data if necessary.

Figure 14–3 is a wiring diagram for a two-note chime. The diagram also shows a second chime. A second chime is often needed in larger homes. Note that the low-voltage circuit is run from the transformer to the first chime. From this first chime, con-

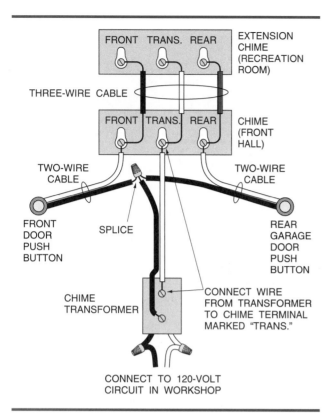

Figure 14–3   A wiring diagram for a two-note chime. A second chime also has been added.

ductors are run to each push button. Running the "live" circuit to the chime is a good idea should there be a need for the 16-volt "live" to be present at the chime location for such items as a chime with lights, or a multi-tone motor-operated chime. You might want to run a four-wire cable to the second chime. This is a decision you have to make. What is it that you expect the chimes to do?

Operation of Doorbell, Buzzers, and Chimes

Do not try to operate a buzzer (or bell) and a chime at same time from the same transformer. The buzzer (or bell) will cause the transformer to put out a fluctuating voltage, probably not allowing the

buzzer or chime to operate properly. In some instances, installing a larger wattage (more volt-amperes) rated transformer (not higher voltage) might work in this situation.

Conductors for Class 2 Circuits

The wire used for low-voltage bell and chime wiring, and for connecting thermostats must be suitable for use in Class 2 circuits. In the trade, you will here the words *bell-wire, thermostat wire,* and sometimes even the old term *annunciator wire.*

The low-voltage wiring for chimes, security systems, furnaces, air-conditioning equipment, heat pumps, and similar equipment that is derived from a small step-down transformer in the end-use equipment in almost in all cases classified as Class 2 wiring. The conductors are usually No. 18 or No. 20 AWG, and are marked as suitable for general use in Class 2 circuits. The marking is generally CL2. Security systems might call for the use of CL2X, which is limited for residential installations. Do not install these cables in hot or cold air ducts unless "listed" and marked for such use. These cables would be marked CL2P. Figure 14–4 is an illustration of a CL2P cable.

Simple "bell-wire" usually consists of solid copper conductors, has thermoplastic insulation, and is available in single-conductor, 2-wire, 3-wire, and 4-wire. This inexpensive low-voltage cable generally does not have an outer jacket over the insulated conductors. Bell-wire is usually No. 18 or 20 gauge, but might need to be larger for extremely long runs.

Because of the thinner insulation on the conductors, care must be taken when driving staples so as not to damage the conductors' insulation. Insulated staples are most often used. Staple guns are available that use rounded staples that straddle the cable instead of flattening and possibly short-circuiting the conductors.

Because these low-voltage conductors cannot

Figure 14–4   A typical low-voltage cable. This particular cable is marked CL2P. (*Courtesy GE Total Lighting Control*)

withstand physical abuse, do not run low-voltage wiring through the same holes in studs and joists as the nonmetallic-sheathed cable, armored cable, raceways, or other premise wiring methods.

SUMMARY OF *NEC*® REQUIREMENTS FOR CLASS 2 WIRING PER *Article 725*

Do not install Class 2 wiring in the same raceway or cable as light and power wiring, see *Section 725-54(a)(1)*.

Figure 14–5 shows low-voltage Class 2 conductors and 120-volt conductors in the same raceway. This Code violation is quite common where the low-voltage Class 2 conductors and the power supply to the outdoor condensing unit of a central air-conditioning system are pulled through the same raceway. This is absolutely not permitted. Some will say that it is OK to do this if the Class 2 wiring is done with 600-volt insulated conductors. Wrong!

You will see separate knockouts provided for Class 2 conductors on air-conditioning equipment, heat pumps, furnaces, and similar equipment.

Keep Class 2 wires at least 2 inches (50.8 mm) away from light or power wiring, see *Section 725-54(a)(3)*. Where the light and power wiring is installed in nonmetallic-sheathed cable, armored cable, or a raceway, you do not have to maintain this clearance. However, if the wiring is old knob-and-tube wiring, maintain the minimum 2-inch clearance.

Class 2 conductors may not enter the same enclosure, unless there is barrier in the enclosure that separates the Class 2 wiring from the light and power wiring, *Section 725-54(a)(1), Exception No. 1*.

Figure 14–6 shows a lighting panel and a cabinet where the power conductors are routed from the panelboard, through the nipples into the right-hand compartment of the cabinet to make connections to relays. The low-voltage Class 2 wiring is brought into the left-hand compartment of the cabinet. A barrier separates the Class 2 circuits and the power circuits. This arrangement is in conformance to *Section 725-54(a)(1), Exception No. 1*.

Section 725-54(a)(1), Exception Nos. 1 and *2* permit Class 2 and power conductors to come together where these conductors are introduced solely to connect the equipment. In boxes and equipment, power wires and Class 2 conductors must be separated by a barrier.

Do not support Class 2 conductors (or any other conductors for that matter) from raceways, *Sections 725-54(d)* and *300-11(b)*. The only time you are permitted to support any type of wiring from a raceway is if that raceway is for hooking up the electrical equipment, and the Class 2 conductors are solely for the connection of the equipment. This was illustrated in Figure 9–43.

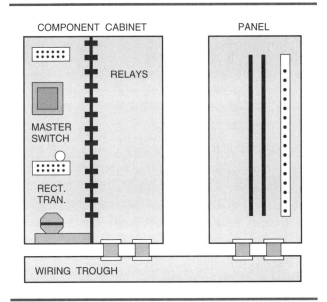

Figure 14–6 For complex low-voltage remote-control installations, a cabinet is usually installed near the electric panel. The 120-volt conductors are routed from the electric panel through short nipples, into the wiring trough, up through short nipples into the line-voltage compartment of the cabinet. The low-voltage Class 2 wiring is brought into the left-hand compartment of the cabinet. The barrier keeps the two systems separated.

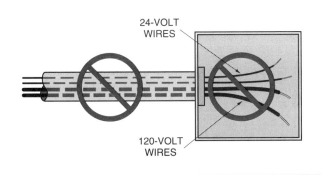

Figure 14–5 Class 2 low-voltage wiring *shall not* be run in the same raceway as the power conductors, *Section 725–54(a)(1)*.

Cables must be installed in a neat and workman-like manner, and must be supported from building structure, not just laid on top of a dropped ceiling or duct work. This is so the cables will not be damaged. This is a requirement of *Section 725-7*.

CABLE TELEVISION (CATV)

Article 820 in the *NEC®* covers Community Antenna Television, hence the term CATV. Cable companies install the CATV wiring outdoors, and generally run their coaxial cable through the wall of the house to some convenient point where the interior television wiring can be run to. Up to the point of entry, the cable company must conform to the requirements found in *Article 820* of the *NEC®* In addition to the *NEC®* most communities have their own local codes regulating the installation for cable television, and in most instances are more stringent that the *NEC®* Cable companies are well aware of local codes.

Here is a brief rundown on *Article 820*.

- The outer conductive shield (just under the outer jacket) of coaxial cables must be grounded as close to the point of entry as possible, *Section 820-33*.

- Coaxial cables shall not be run in the same raceways or boxes with electric light and power conductors, *Section 820-52(a)(1)(b)*.

- Coaxial cables shall not be supported to raceways that contain electrical light and power conductors, *Section 820-52(e)*.

- Where underground coaxial cables and electric service conductors are buried in the same trench, they must be separated by not less than 12 inches (305 mm). This separation is not needed if the service-lateral conductors are installed in a raceway or have a metal cable armor, *Section 820-11(b), Exception No. 1*.

- Where underground coaxial cables and branch circuit or feeder conductors are buried in the same trench, they must be separated by not less than 12 inches (305 mm). This separation is not needed if the branch circuit or feeder conductors are installed in a raceway, are metal-sheathed, are metal-clad, are Type UF, are Type USE, or if the coaxial cable has a metal armor or is installed in a raceway, *Section 820-11(b), Exception No. 2*.

- *Section 820-40* contains the rules for grounding a coaxial cable's metal shield. For proper grounding, the grounding conductor:

  a. must be insulated.
  b. must not be smaller than No. 14 AWG copper or other corrosion-resistant conductive material. Copper is the best choice.
  c. may be solid or stranded.
  d. must be run in a line as straight as practical.
  e. must be protected from physical damage.
  f. shall be connected to the nearest accessible location on one of the grounding points listed in *Section 820-40(b)*. This section is rather complex, so we will concentrate only on those items that ordinarily are found in homes. For residential installations, the following are generally used to ground the coaxial cable:
     — The building or structure grounding electrode system, *Section 250-50*.
     — The grounded interior metal water pipe, *Section 250-104(a)*.
     — The metal raceway that contains the service-entrance conductors.
     — The service equipment enclosure.
     — The grounding electrode conductor or its metal enclosure.
     — Do **not** ground the coaxial cable shield to a metal gas pipe. Metal gas piping is not an acceptable grounding electrode, *Section 250-52(a)*. Gas utilities frown on having their metal piping used to serve an electrical function. Years ago this may have been acceptable, but no longer.
  g. If none of the items in *f* are available, then ground to any of the electrodes per *Section 250-50*, such as metal underground water pipes, metal frame of the building, concrete encased ground, or a ground ring. These were discussed in Unit 12.
  h. If none of the items in *f* or *g* are available, (hardly unlikely in a house) then ground to an effectively grounded metal structure or driven ground rod or plate, *Section 250-50*. The most popular item to use is a driven ground rod as these are readily available at electric supply houses. Since most homes are not constructed of an effectively grounded metal structure, and metal plates are hard to

come by, items are rarely used in residential installations.

i. **Caution:** If you choose to drive a ground rod to establish the ground for the coaxial cable shield, then you must bond back to the main service grounding electrode with a bonding jumper not smaller than No. 6 AWG copper, *Section 820-40(d)*. This limits a difference of potential that could occur between the two electrodes.

Figure 14–7 shows a typical connection for grounding the shield of a coaxial cable.

Coaxial cables generally are 75 ohms as opposed to the old style 300-ohm cables rarely used today.

Standard switch boxes can be used where a television outlet is to be installed. All types of fittings, male and female connectors, splitters, amplifiers, couplers, wall plates, and tools for working with coaxial cables can be found in most electronic stores.

ANTENNAS

Article 810 of the Code covers antennas.

Whether installing a large outdoor "dish," or one of the smaller 18-inch digital satellite system (DDS) dishes," certain Code rules apply. Figure 14–8 shows an 18-inch digital satellite system (DDS) antenna and a larger 6-foot antenna.

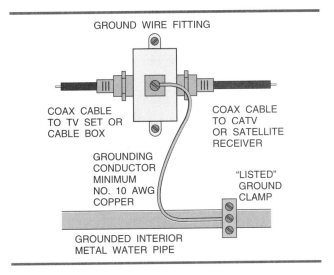

Figure 14–7 A typical connection for grounding the shield of a coaxial cable to a metal water pipe.

Figure 14–8 Photo (A) shows a digital satellite system 18-inch antenna, a receiver, and a remote control. (*Courtesy Thomson Consumer Electronics*); Photo (B) shows a large 6-foot satellite antenna.

Although instructions are supplied with antennas, the following key points regarding the installation of antennas and lead-in wire should be followed. In many cases, the rules are the same or similar as for coaxial cable discussed above.

1. Antennas and lead-in wires shall be securely supported.

2. Antennas and lead-in wires shall not be attached to the electric service mast.

3. Antennas and lead-in wires shall be kept away from all light and power conductors to avoid accidental contact with the light and power conductors.

4. Antennas and lead-in wires shall not be attached to any poles that carry light and power wires over 250 volts between conductors.

5. Lead-in wires shall securely attach to the antenna.

6. Outdoor antennas and lead-in conductors shall not cross over light and power wires.

7. Lead-in conductors shall be kept at least 2 feet (610 mm) away from open light and power conductors.

8. Where practicable, antenna conductors shall not be run under open light and power conductors.

9. On the outside of a building:
   a. Position and fasten lead-in wires so they cannot swing closer than 2 feet (610 mm) to light and power wire having not over 250 volts between conductors, and 10 feet (3.05 m) if over 250 volts between conductors.
   b. Keep lead-in conductors at least 6 feet (1.83 m) away from a lightning rod system, or bond together according to *Section 250-60*.

10. On the inside of a building:
    a. Keep antenna and lead-in wires at least 2 inches (51 mm) from open wiring, such as knob-and-tube wiring. This 2-inch distance is not required when the wiring method is raceway or cable.
    b. Keep lead-in wires out of electric boxes unless there is an effective, permanently installed barrier to separate the light and power wires from the lead-in wire.

11. Grounding:
    a. The grounding wire must be copper, aluminum, or copper-clad steel, bronze, or similar corrosion-resistant material.
    b. The grounding wire need not be insulated. It must be securely fastened in place; may be attached directly to a surface without the need for insulating supports; shall be protected from physical damage or be large enough to compensate for lack of protection; and shall be run in as straight a line as is practicable.
    c. The grounding conductor shall be connected to the nearest accessible location on: the building or structure grounding electrode (*Section 250-52*); or the grounded interior water pipe (*Section 250-104(a)*); or the metallic power service raceway; or the service equipment enclosure; or the grounding electrode conductor or its metal enclosure.
    d. If none of *c* is available, then ground to any one of the electrodes per *Section 250-50*, such as metal underground water pipe, metal frame of building, concrete-encased electrode, or ground ring.
    e. If neither *c* nor *d* are available, then ground to an effectively grounded metal structure or to any one of the electrodes per *Section 250-50*, such as a driven ground rod or pipe, or a metal plate.
    f. The grounding conductor may be run inside or outside of the building.
    g. The grounding conductor shall not be smaller than No. 10 AWG copper or No. 8 AWG aluminum.
    h. **Caution:** If any of the preceding rules results in one grounding electrode for the antenna and another grounding electrode for the electrical system, bond the two electrodes together with a bonding jumper not smaller than No. 6 AWG copper or equal. This will reduce the possibility of having a difference of voltage between the two systems.

TELEPHONE WIRING

Article 800 of the *National Electrical Code®* covers telephone wiring. This article goes much beyond that of simple telephone wiring in a home. We will discuss only that which pertains to a telephone wiring in a typical residence.

Since the deregulation of telephone companies, residential "Do-It-Yourself" telephone wiring in homes has become quite common. In new home construction, the interior telephone wiring is usually left up to the electrician.

The serving telephone company will install their telephone line(s) to a residence, and terminate at a protector device. This protector protects the system from hazardous voltage surges. The protector may be mounted either outside or inside the home.

Figure 14–9 shows a typical residential telephone installation.

The point where the telephone company ends and the homeowner's responsibility begins is called the *demarcation point.* The preferred demarcation point is outside of the home, generally near the electric meter where proper grounding and bonding can be done. However, the serving telephone company will make the decision of where to bring in their lines. In many communities, the electric utility and the telephone company have an arrangement whereby the electric utility will handle the burying of both the underground power conductors and the underground telephone cable in the same trench.

A *standard network interface,* referred to as SNI or just NI might be mounted indoors of the type shown in Figure 14–9 (A), or it might be a weatherproof type located outdoors, Figure 14–9(B).

Today, most telephone wiring is done using modular plugs and jacks.

Here are some of the basic requirements for interior telephone wiring:

• Grounding: This is generally is done by the telephone company. One method is shown in Figure 14–9 where the protector has been attached to the metal service raceway.

The grounding conductor:

—must be insulated and listed for the purpose.

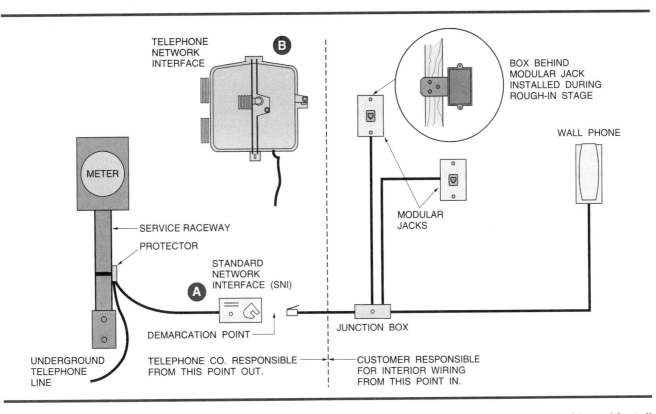

Figure 14–9 A typical residential telephone installation. The serving telephone company will provide and install the wiring up to the *demarcation point*. The homeowner is responsible for everything beyond the demarcation point. The *network interface* of the outdoor weatherproof type have become very popular in recent years as this reduces the need for telephone company workers to come into the house. Where the protector is attached to a metal service raceway, the required grounding is accomplished. Most telephone companies will provide, install, and service interior telephone wiring at additional cost.

—shall not be smaller than No. 14 AWG solid or stranded copper or other corrosion-resistant material.

—shall not be smaller than No. 14 AWG.

—shall be run in as straight as practicable to the grounding electrode.

—shall be protected against physical damage.

The grounding conductor must be connected to:

—the building or structure grounding electrode system, *Section 250-50*.

—the grounded interior metal water pipe, *Section 250-104(a)*.

—the metal raceway that contains the service-entrance conductors.

—the service equipment enclosure.

—the grounding electrode conductor or its metal enclosure.

—do **not** ground to a metal gas pipe. Metal gas piping is not an acceptable grounding electrode, *Section 250-52(a)*. Gas utilities frown on having their metal piping used to serve an electrical function. Years ago this may have been acceptable, but no longer.

Caution: If you choose to drive a ground rod to establish the ground for the coaxial cable shield, then you must bond back to the main service grounding electrode with a bonding jumper not smaller than No. 6 AWG copper, *Section 800-40(d)*. This limits a difference of potential that could occur between the two electrodes.

• Telephone wiring is not to be run in the same raceway with electrical wiring, *Section 800-52(c)*.

• Telephone wiring is not to be installed in the same outlet or junction boxes as electrical wiring, *Section 800-52(c)(1)*. Metal or nonmetallic boxes may be used.

• Telephone wires and cables must be kept at least 2 inches (50.8 mm) from light and power conductors. This separation is not required if the light and power conductors are in a raceway or cable such an nonmetallic-sheathed cable, armored cable, or EMT, *Section 800-(c)(2)*.

• Some telephone companies suggest that telephone wiring not be run in the same stud space as the electrical wiring.

• Some telephone companies recommend that telephone cables be kept a minimum of 12 inches away from electrical branch-circuit wiring where the telephone cable is run parallel to the branch circuit wiring. This guards against induction "noise" that might be picked up by the telephone lines.

• Keep the telephone cables away from hot water pipes, hot air ducts, and other heat sources that might harm the insulation.

• Do not support telephone cables from electrical raceways.

• For residential installations, types of "listed" cables permitted according to *Table 800-53* are:

—CMG: a general-purpose communications cable that is resistant to the spread of fire

—CM: a general-purpose communications cable that is resistant to the spread of fire

—CMX: is flame retardant and is all right for use in dwellings

Other types are available, but the above cables are commonly found in telephone stores, electronic stores, home centers, hardware stores, electrical distributors, and similar retail and wholesale outlets. Telephone cables are available in spools of long length for "roughing-in" the wiring, and in specific lengths (6, 9, 12, 25, or 50 feet) with modular plugs and jacks factory-attached. You will also find at these stores connectors, terminal blocks, jacks, plugs, faceplates, and other components needed to hooking up telephones in homes.

Figure 14–10 shows some of the many types of telephone cables available. The flat cords usually contain two conductors. The round cords have more than two conductors. Note the modular connectors on these cables.

Because of the thin insulation and jacket on telephone cables, care must be taken when driving staples so as not to damage the cables. Insulated staples are most often used. Staple guns are available that use rounded staples that nicely straddle the cable instead of flattening and possibly short-circuiting the conductors.

Telephone cables cannot withstand physical abuse. Do not run telephone cables through the same holes in studs and joists as the nonmetallic-sheathed cable, armored cable, raceways, or other premise wiring methods.

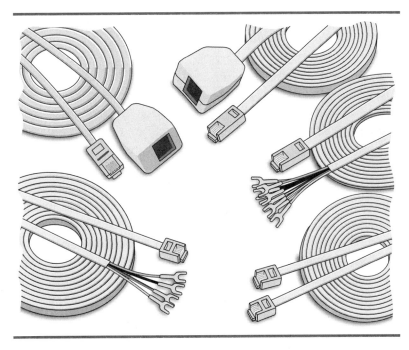

Figure 14–10   An assortment of telephone cords.

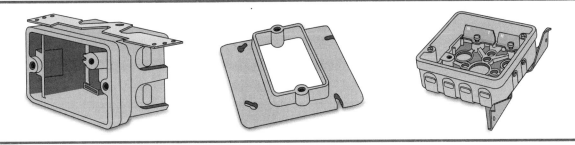

Figure 14–11   Types of boxes and a raised plaster ring that can be used for roughing in the wiring for residential telephone and television outlets.

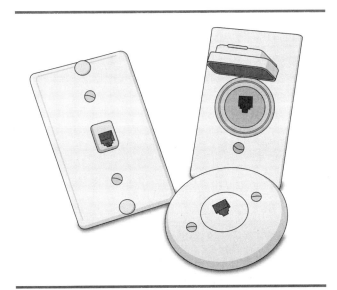

Figure 14–12   Typical modular telephone jacks in three different styles of faceplates.

For a typical residence, the electrician will rough-in wall boxes or plaster rings wherever a telephone outlet is wanted. These are illustrated in Figure 14–11.

Figure 14–12 shows modular telephone jacks as an integral part of three different styles of wall faceplates.

Telephone Cables

The size of the conductors in telephone cables is generally No. 24 AWG and No. 22 AWG. No. 24 AWG is all right for runs not over 200 feet in length. Use No. 22 AWG for runs not over 250 feet.

For residential installations, two-pair and three-pair telephone cables are generally adequate. However, for more involved systems, cables with additional pairs of conductors are available. The conductors in each pair are twisted together. Then the pairs of conductors are again twisted together.

The reason for the twisting is to reduce "cross-talk" problems, which occurs when signals traveling in one pair are picked up by the adjacent pair in the cable or cord. Cross-talk is virtually eliminated by using cords that have their conductors twisted at proper intervals as determined by telephone cable and cord manufacturers. Avoid the use of flat two-wire line cords in lengths over 7 feet. Flat one-line cords present no problem.

Figure 14–13 shows the color coding of the conductors used in telephone cables. Color coding establishes an easy way to keep track of what wires get connected to which terminal on telephone wiring devices and equipment. Telephone devices have their terminals marked with the letters G (for green), R (for red), Y (for yellow), and B (for black). Additional pairs of conductors can be used to serve additional telephone lines, faxes, security reporting back to a central station, speaker phones, dialers,

background music, etc. It all depends upon the complexity of the system.

Where To Run The Telephone Cables

The choice is yours! Much depends on the construction of the house and where the telephone outlets are to be located.

Figure 14–14 is a simple radial installation where a cable is run from each telephone outlet back to one point—a junction box located near the network interface. This junction box (terminal block inside) makes for the easy terminating of all of the individual conductors.

Figure 14–15 is a "loop" system where the cables are run from outlet to outlet. Should one of the cables become damaged, the circuit can be back-fed from the other end. This method is simple, with the only additional cable being the one that returns back to the junction box from the last outlet. All of the other outlets are connected end-to-end, sometimes referred to as "daisy chaining."

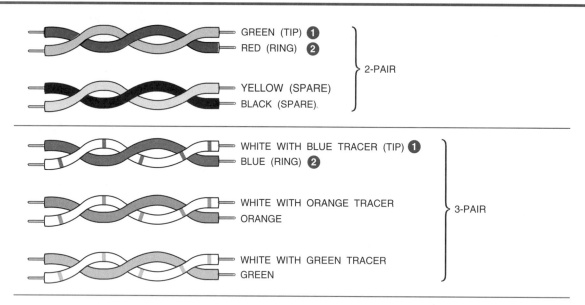

① *TIP* IS THE CONDUCTOR THAT IS CONNECTED TO THE TELEPHONE COMPANY'S "POSITIVE" TERMINAL. IT IS SIMILAR TO THE NEUTRAL CONDUCTOR OF A RESIDENTIAL WIRING CIRCUIT.

② *RING* IS THE CONDUCTOR THAT IS CONNECTED TO THE TELEPHONE COMPANY'S "NEGATIVE" TERMINAL. IT IS SIMILAR TO THE "HOT" UNGROUNDED CONDUCTOR OF A RESIDENTIAL WIRING CIRCUIT.

Figure 14–13 Color coding for a two-pair, four-conductor telephone cable and a three-pair, six-conductor telephone cable. The color coding for a two-pair stands alone. For three-pair or more cables, the color coding is white/blue, white/orange, white/green, white/brown, white/slate, red/blue, red/orange, red/green, red/brown, red/slate, black/blue, black/orange, black/green, black/brown, black/slate, yellow/blue, yellow/orange, yellow/green, yellow/brown, yellow/slate, violet/blue, violet/orange, violet/green, violet/brown, violet/slate.

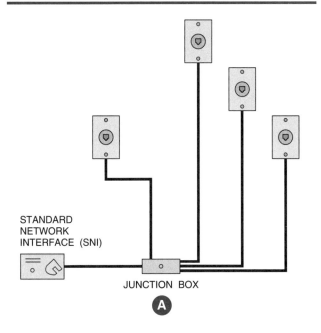

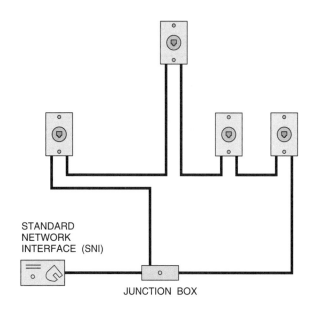

Figure 14–15 A "looped" system. The telephone cables are installed from the junction box, then from outlet-to-outlet, then back to the junction box. If something happens to one section of cable, the circuit can be fed from the other direction.

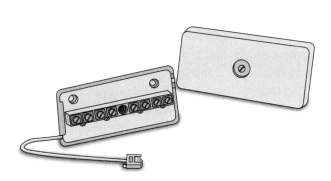

B JUNCTION BOX IT FEATURES A SHORT, PREWIRED CORD THAT PLUGS INTO A STANDARD NETWORK INTERFACE (SNI). IT ALLOWS EASY CONNECTION OF ADDITIONAL TELEPHONE EXTENSIONS.

Figure 14–14 A radial system. Individual telephone cables are run to each telephone outlet from a junction box.

How Many Telephones Can Be Connected To One Line?

This is usually not a problem in residence unless there are old style phones connected.

The newer electronic phones draw very little power. Therefore, many electronic phones can be connected to one line before running into problems. Old style phones that have an electromechanical ringer are another story. Generally, the maximum number of these old style phones is five. For those of you who want to get technical, telephones are rated

in RENs, which means *Ringer Equivalence Number.* An old style ringer was considered to be 1-REN. Consider five RENs the maximum.

Cordless telephones have a REN rating of 0.2 to 0.3 RENs.

Electronic phones have a REN rating of 0.4 to 0.7 RENs. It depends on the number of indicating lamps and the ringer.

Possible cutoff can occur when too many phones are connected to one line.

Safety

Open-circuit voltage between conductors of an idle pair of telephone conductors can range from 50 to 60 volts DC. The ringing voltage can reach 90 volts AC. Therefore, always work carefully with insulated tools and stay clear of bare terminals and grounded surfaces. Disconnect the interior telephone wiring by pulling out the modular connector from the network interface. Another hint is to take all phones off the hook, in which case the DC voltage level will dip, and there should be no AC ringing voltage delivered.

SECURITY SYSTEMS, HEAT AND SMOKE DETECTORS

Article 725 covers signal systems, which encompasses security systems.

You must recognize that at the time of this writing, the *National Electrical Code®* does not mandate that security systems and fire alarm systems be installed in homes. However, when such systems are installed, they then must conform to the requirements of the *NEC®* and whatever local codes might be in effect.

Article 760 of the *National Electrical Code®* covers fire alarm systems.

You will find in *Section 760-1, FPN No. 1* reference is made to *NFPA 72,* which is the *National Fire Alarm Code.* It is a consolidation of ten individual previous codes relating to fire alarm systems and contains complete information regarding the installation of fire alarm systems.

Individual smoke, heat, and carbon monoxide detectors are available with battery power only, or with both battery and 120-volt power. Some communities require that the combination battery and 120-volt devices be installed. We all have heard about deaths resulting from smoke inhalation and fire because the batteries in the detectors were either dead, or had been removed.

It is beyond the scope of this book to go into detail on the many types of security systems, heat, and smoke detectors available today. Here again, the complexity of the system will depend upon individual wants and needs. We will touch on the main concerns.

Manufacturers of "Do-It-Yourself" systems contain detailed installation instructions on where the detectors should be installed. These instructions are in conformance to the *National Electrical Code®* and the *National Fire Alarm Code.*

Never install "unlisted" equipment! Trying to compromise between price and safety is irrelevant when human lives are involved.

The main control panel for a home security system is usually a self-contained unit. A three-wire cord from the main control panel plugs into a 120-volt receptacle. From the main control panel, conductors are run to heat detectors, smoke detectors, alarms, keyboards, panic buttons, window and door entry detectors, window foil strips, telephone dialer to alert the provider of security, police, or fire department, floor mat detectors, outside alarms, outside strobe lights, etc. The possibilities of features available for these systems is endless.

The low-voltage wiring is Class 2 wiring because the transformer within the control panel is a Class 2 transformer. We have already discussed the installation requirements for Class 2 circuits.

Cables for fire alarm systems are types FPL (general use) and FPLP for use in ducts, *Section 760-61(c).*

Do not support fire alarm cables from electrical raceways, *Sections 760-54(c)* and *300-11(b).*

Install the cables in a neat and workmanship manner so that the cables will not be damaged, *Section 760-8.*

HOME AUTOMATION

Just about everyone has a remote control for their television set. "Intelligent" house wiring is a tremendous expansion and extension of remote manual and/or automatic control of just about anything electrical in the house.

In today's modern high-tech world, we find residences wired with conventional wiring methods, but in addition, many "bells and whistles" are added. Home automation systems have proliferated. Home centers, electrical distributors, electronic stores, and similar retail and wholesale outlets handle a variety of these systems for the control of general lighting, security lighting, energy management for HVAC equipment, entertainment, communication, etc. Many companies are involved in the developing and manufacturing of products for the intelligent home.

With the advent of Internet, the information highway, and the World Wide Web, the future possibilities are unlimited. Who knows how all of this equipment will be interconnected in the future!

You will hear about, you will read articles about, and possibly attend seminars about "intelligent" home automation wiring. Names such as ActiveHome, AmpOnQ, CEBus, X-10, Echelon, Elan, Enerlogic, Gemini, Greyfox Systems, Home Base, Honeywell Total-Home, IBM Home Director, IES Technologies, Leviton Intellisense, LightTouch, LonWorks, Lucent Technologies HomeStar, Lutrons RadioRA, Stargate, Smart-House, and Stanley Lightminder are some of the names associated with home automation systems.

These systems might be wireless (infrared), they might be wired with special cables such as coaxial, unshielded twisted pair, shielded twisted pair, and fiber-optic cables, or they might operate via carrier signals sent over the 120-volt conventional wiring in the house.

The Electronic Industries Association (EIA) and the Telecommunications Industry Association (TIA) have developed a series of five standards for commercial and residential telecommunications wiring.

These standards are broken down into five categories.

Category 1: Voice and low speed data
Category 2: Voice and low speed data
Category 3: Data networks up to l6-Mbps
Category 4: Data networks up to 20-Mbps
Category 5: Data networks up to l00-Mbps

The term M means *mega*, which is one million.
The letters *bps* mean bits per second.

A bit is the smallest unit of measurement in the binary system. The expression is derived from the term *binary digit*.

The term Mbps means *megabits per second.* For example, 20 Mbps means that twenty million (20,000,000) *bits* of data is being transmitted per second over copper or fiber-optic cable.

It takes eight bits to make one byte. A *byte* is the amount of data required to describe a single character of text.

Manual as well as remote control of electrical products such as "smart" appliances or appliance modules, receptacles, security systems, and security lighting is possible. Systems that operate with and through a personnel computer (PC) are available. Remote control is possible, even over telephone lines from virtually any place in the country or world for that matter.

Article 780 in the *National Electrical Code*® covers Closed Loop systems and is supplement to the general Code rules.

Underwriters Laboratories "lists" special cables that contain both power and low-voltage conductors under one outer jacket. These are for the wiring of intelligent houses. Figure 14–16 shows one type of hybrid cable.

What is important to understand is that all of the *National Electrical Code*® requirements for the number of receptacles, lighting outlets, small appli-

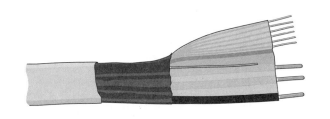

Figure 14–16 A typical hybrid cable that consists of power conductors and low-voltage control conductors. These cables are "listed," and must be used in conjunction with the proper boxes and fittings.

ance branch circuits, appliances (ranges, dryers, water heaters, etc.), grounding, and main service equipment are still applicable. Nothing in this respect has changed.

Home automation wiring merely adds features to a conventional electrical installation. Home automation is a rapidly changing industry, with new developments and inovations introduced so often that it is hard to stay on top of it.

For those interested in learning more about home automation, Internet Web Sites can be contacted such as:

(AmpOnQ) www.amponq.com
(Automated Home Technologies)
   www.autohometech.com
(Computer and Telecommunications Page)
   www.cmpcmm.com/cc/
(Electronic Industry Alliance) www.eia.org
(Fiber Optic Association, Inc.)
   www.world.std.com/~foa/
(Home Automation Association)
   www.homeautomation.org
(Home Automation Links) www.asihome.com
(Home Automation Magazine)
   www.electronichouse.com
(Home Automation Systems, Inc.)
   www.smarthome.com
(Home Controls, Inc.) www.homecontrols.com
(Home Team Network) www.hometeam.com/
(Home Toys) www.hometoys.com
(Intelligent Homes)
   http://web.cs.ualberta.ca/~wade/HyperHome/
   top.html
(Leviton Mfg. Co., Inc.) www.leviton.com

(Litetouch) www.litetouch.com

(Lucent Technologies)
www.lucent.com/netsys/homestar

(Lutron Electronics Co., Inc.) www.lutron.com

(Smart-House) www.smart-house.com

(Smart Home Pro) www.smarthomepro.com

(Technology Information Center)
www.tickit.com/,

(X-10 Pro) www.X10pro.com

One of the more important standards available from the Electonic Industry Alliance is the EIA/TIA-570 (a Residential and Light Commercial Telecommunications Wiring Standard). Their address is: 2500 Wilson Blvd., Arlington, VA 22201-3834.

Low-Voltage Remote-Control Systems

A system that has been around since before the advent of electronics is a simple low-voltage system that uses relays that in turn do the line voltage switching.

The system requires: a Class 2 transformer of the type illustrated in Figure 14–17. The transformer steps down the voltage from 120 volts to 24 volts. The transformer is rated no larger than 100 volt-amperes, and inherently limits the amount of current that can flow. Refer to *Table 12(a)* of *Chapter 9* of the *NEC*® for more detailed information relating to Class 2 power sources.

The cables that interconnect all of the components for this type of system are rated for Class 2 use as illustrated in Figure 14–4 previously shown. The conductors are usually No. 20 AWG copper conductors, but could be sized larger for extremely long runs.

Class 2 systems are not permitted to be run in the same raceway as light and power conductors, nor are they permitted to enter the same box, other than for the exception listed earlier in this unit in the Class 2 conductors section.

The system requires low-voltage relays of the type illustrated in Figure 14–18. The relay has three low-voltage leads: red for ON, black for OFF, and blue for COMMON. The two terminals are for the conductors that are in series with the load to be controlled, exactly like a standard single-pole toggle switch. Relays with four low-voltage leads can be used when a "pilot" contact is needed. Relays with five low-voltage leads are available when to totally separate "pilot" circuit is required. The four and five conductor relays are used to control other circuits, relays, master controls, pilot lights, and so forth.

The internal connections for these relays is shown in Figure 14–19. When 24 volts is applied to the red and blue leads, the switch will turn on. When 24 volts is applied to the black and blue leads, the switch will turn off.

These relays can be inserted through knockouts using a rubber grommet to quiet their operation. This is illustrated in Figure 14–20 where three low-voltage switches control a light. Note how the 120-volt wiring is separated from the 24-volt circuit.

Figure 14–17 A Class 2 transformer for 24-volt remote-control systems. (*Courtesy GE Total Lighting Control***)**

Figure 14–18 A low-voltage relay. (*Courtesy GE Total Lighting Control***)**

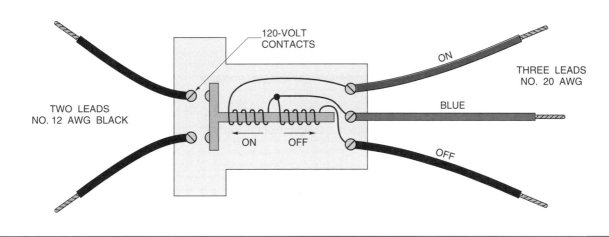

Figure 14–19 A cutaway view of a low-voltage relay showing the internal mechanism, the line-voltage leads, and the low-voltage leads.

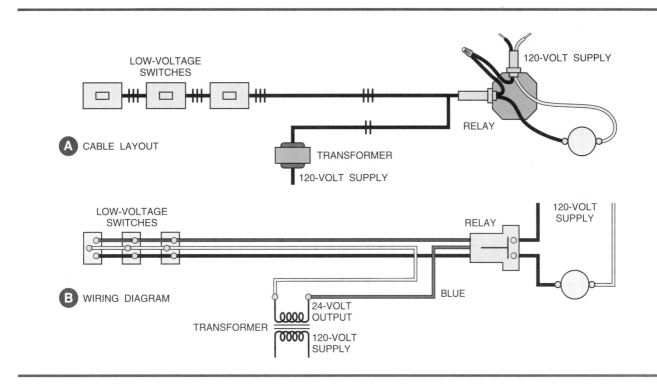

Figure 14–20 Typical cable layout and wiring diagram for a simple low-voltage remote-control installation.

Low-voltage switches are single-pole, double-throw momentary contact type. These switches have three low-voltage leads: red for ON, black for OFF, and white for COMMON. Figure 14–21 shows the details of a low-voltage switch.

Figure 14–22 shows an assortment of types of switches that are available for low-voltage systems.

Figure 14–23 shows a Master Selector Switch that has up to eight switches, pilot lights, or blanks.

Figure 14–24 shows a cable layout and a wiring diagram where each light is controlled by one switch. A Master Selector Switch has been added so that all three lights can be controlled at the Master Selector Switch location, such as in a bedroom for security reasons.

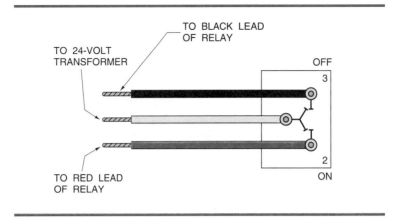

Figure 14–21 Internal wiring details of a low-voltage switch.

Figure 14–22 Types of low-voltage switches. From left to right: (A) switch with location light; (B) two unlighted switches; (C) unlighted key-operated switch, pilot light key-operated switch, blank filler, pilot light; (D) a key; and (E) a switch that is physically the same as the interchangeable line of switches. (*Courtesy GE Total Lighting Control*)

Figure 14–23 A master selector switch with room for up to eight switches, pilot lights, or blanks. (*Courtesy GE Total Lighting Control*)

For the more elaborate systems, Figure 14–6, shown previously in this unit, shows how relays can be mounted in a components cabinet located next to a lighting panel. A wiring trough is mounted below both cabinets with short conduit nipples connecting between the trough and the cabinets. The components cabinet has a barrier in the center, keeping the 24-volt circuitry in the left compartment, and the 120-volt circuitry in the right compartment. This makes the installation neat in appearance.

The options for these systems are almost unlimited.

Electrical distributors carry these systems in depth. Detailed installation instructions are furnished with the equipment.

Low-voltage lighting systems were discussed in Unit 11.

Final Words

No matter what the future holds relative to home automation, safety will still be the main concern. Remember the fundamental Code requirement in *Section 110-3(b)*: *"**Installation and Use:** listed or labeled equipment shall be installed and used in accordance with any instructions included in the listing or labeling."*

Whatever system you are installing, make sure that the equipment bears the Underwriters Laboratories label!

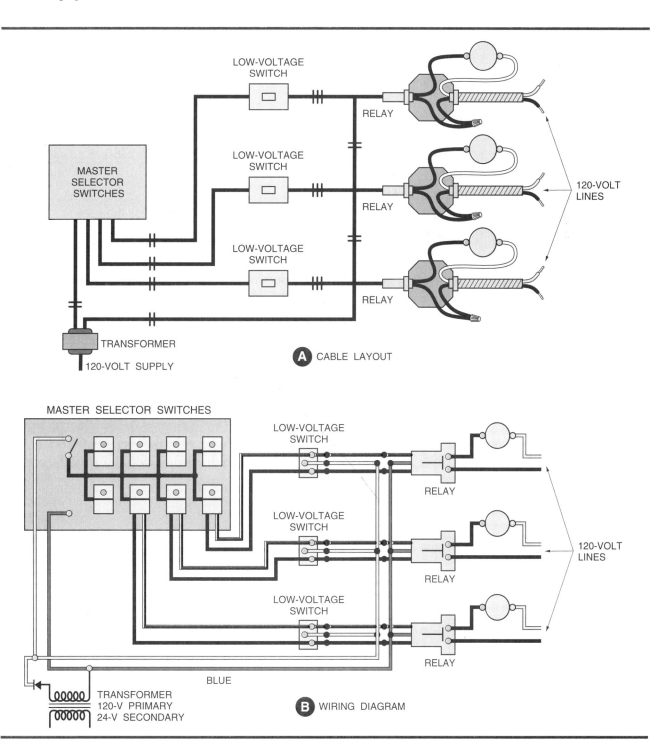

Figure 14–24 A cable layout and a wiring diagram for a low-voltage system. Each light is controlled by one switch. In addition, a master selector switch has been added so that all three lights can be controlled at the master selector switch location, such as in a bedroom for security reasons.

REVIEW QUESTIONS

1. Low-voltage chime transformer of the type used in house wiring are classified as:

    a) Class 1

    b) Class 2

    c) Class 3

2. Chime transformers have _____ power output.

    a) limited

    b) unlimited

3. A typical small residential chime is rated 16 volts, and has a power rating of approximately 10 watts. A second chime is added. Now neither of the chimes sounds loud enough. The way to solve the problem is: (Mark T (True) or F (False) in the space provided.)

    a) Replace the older transformer with one rated 32 volts.   _____

    b) Replace the older transformer with one rated 20 watts.   _____

    c) Buy another transformer exactly the same as the one already
       installed. Then hook them up in parallel.   _____

4. Installing Class 2 low-voltage chime wiring conductors, telephone cables, intercom conductors, security system cables, or coaxial TV cables in the same raceway as light and power conductors is: (Mark T (True) or F (False) in the space provided.)

    a) Permitted.   _____

    b) Permitted if the Class 2 conductors have a voltage rating of
       600 volts, the same as the conductors in the raceway for the
       light and power wiring.   _____

    c) Permitted if the TV cables are coaxial shielded type.   _____

5. Does the *National Electrical Code®* permit Class 2 low-voltage conductors, telephone cables, coaxial TV cables, and similar cables to be supported by electrical metallic tubing?

    a) Yes

    b) No

6. The metal conductive shield of a coaxial CATV cable _____ be grounded as close to the point of entry as possible. What section of the *NEC®* supports your answer?

    a) shall

    b) should

    Section _____.

7. Metal parts of a TV antenna, the telephone system, and the conductive metal shield of coaxial cable are required to be grounded. An acceptable ground is: (Mark T (True) or F (False) in the space provided.)

    a) The grounded interior metal water pipe   _____

    b) The metal service-entrance raceway   _____

    c) The service-entrance enclosure   _____

d) The metal gas pipe _____

e) A ground rod _____

8. What size conductors are generally used in residential-type telephone cables?

No. _____

9. Why are the conductors in two-wire and four-wire telephone cables twisted together?

_____

_____

10. What insulation colors are used for typical four-conductor telephone cable?

_____, _____, _____, _____.

11. Why is it not a good practice to run low-voltage wires and cables through the same holes in framing members as the nonmetallic-sheathed cables? _____

_____

12. Because there are so many types of equipment available today for security systems, heat and smoke detectors, automated house wiring systems, and so forth, the one thing that you can do to ensure that the equipment is safe is to make sure that the equipment is

_____

13. What is the approximate voltage used for low-voltage remote-control systems?

_____ volts.

14. The transformer used for Class 2 low-voltage systems shall not be rated larger than

_____ volt-amperes.

UNIT 15

Service-Entrance Equipment

OBJECTIVES

After studying this unit, you will be able to:

• Understand important terms used for describing service-entrance equipment.

• Understand the *National Electrical Code®* requirements for underground and overhead residential services.

• Have learned about the clearances needed for overhead service conductors.

• Have learned about panelboardboards and load centers.

• Know where to locate the main service disconnect.

• Be aware of certain cautions when installing electrical equipment near gas meters.

• Understand the meaning of the term "subpanelboard."

SETTING THE STAGE

All homes have a main electrical service. At the main electrical service panelboard, all of the individual branch circuits come together. The sizing (ampere rating) of the main service panelboard and service-entrance conductors depends upon the square footage of living space in the home and the rating of electrical appliances in the home.

DEFINITIONS

An electric service is required for all homes.

Before we begin our discussion about electric services, we need to understand the meaning of specific terms used in the *National Electrical Code®* that relate to services. Following are some definitions found in *Article 100* of the *NEC®*

• **Service:** *The conductors and equipment for delivering electric energy from the serving utility to the wiring system of the premises served.*

• **Service Cable:** *Service conductors made up in the form of a cable.*

• **Service Conductors:** *The conductors from the service point or other source of power to the service disconnecting means.*

• **Service Drop:** *The overhead service conductors from the last pole or other aerial support to and including the splices, if any, connecting to the service-entrance conductors at the building or other structure.*

• **Service-Entrance Conductors, Overhead System:** *The service conductors between the terminals of the service equipment and a point usually outside the building, clear of building walls, where joined by tap or splice to the service drop.*

• **Service-Entrance Conductors, Underground System:** *The service conductors between the terminals of the service equipment and the point of connection to the service lateral.*

• **Service Equipment:** *The necessary equipment, usually consisting of a circuit breaker(s) or* switch(es) and fuse(s), and their accessories, connected to the load end of service conductors to a building or other structure, or an otherwise designated area, and intended to constitute the main control and cutoff of the supply.

- **Service Lateral:** The underground service conductors between the street main, including any risers at a pole or other structure or from transformers, and the first point of connection to the service-entrance conductors in a terminal box or meter or other enclosure, inside or outside the building wall. Where there is no terminal box, meter, or other enclosure with adequate space, the point of connection shall be considered to be the point of entrance of the service conductors into the building.

- **Service Point:** The point of connection between the facilities of the serving utility and the premises wiring.

We discussed grounding and bonding in Unit 12, so there will be little need to repeat the topic in this unit unless absolutely necessary to emphasize a point.

Figure 15–1 is a simple sketch showing the components of a typical overhead service and a typical underground service.

The electric utility will "spot" the location where the meter is to be installed. The electrician

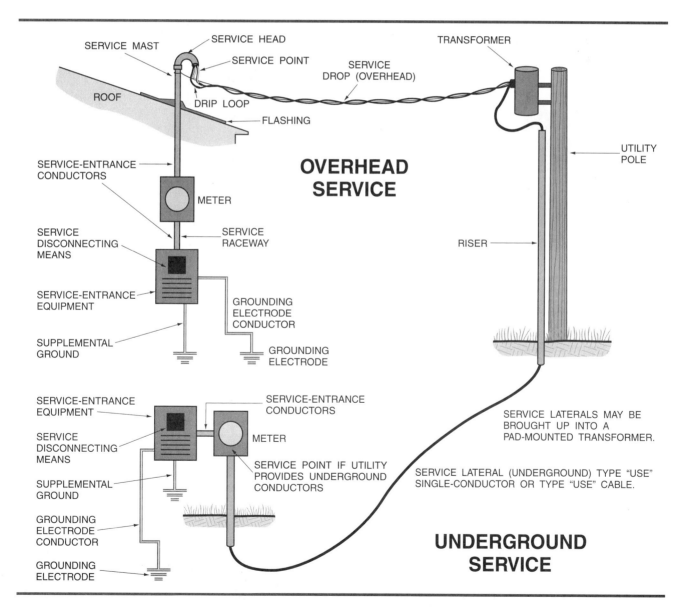

Figure 15–1   A simple sketch of an overhead service and an underground service.

installs the service equipment, including the meter base. The utility installs the watt-hour meter. In some areas, the electric utility furnishes the meter base. In most cases, the meter is mounted on the outside of the house, Figure 15–2, or it might be a pedestal type, Figure 15–3.

The pedestal-type is most often the choice when underground services are installed because it will house the service-lateral conductors from the utility and the service-entrance conductors from the meter to the main panelboard. In some localities, meter pedestals are located at the corner of the rear property line of the residence. A meter pedestal is really a trough-type enclosure. It is positioned so that the bottom end (the open end) is approximately 18 inches below grade. Extensions are available if needed.

PENETRATING AN OUTSIDE WALL

Sections 230-8, 300-5(g), and *300-7* require sealing raceways that pass from one area to another that have great temperature fluctuations. A service-entrance raceway is a good example of this. Figure 15–4 shows the sealing of a raceway passing through a basement or other outside wall. Without the seal, the passage of air back and forth through the raceway can, and will, result in condensation. The end result usually is the rusting of electrical equipment, or worse, can cause improper operation of the mechanism of circuit breakers, switches, and other equipment that has moving parts.

Figure 15–2   A typical residential-type meter base. The cover has been removed to show the internal "line" (on top) and "load" (on bottom) lugs. The neutral lugs are in the center. (*Courtesy of Milbank Manufacturing Co.*)

THE ELECTRIC UTILITY PROVIDES THE UNDERGROUND SERVICE LATERAL

Figure 15–3   Installing a meter "pedestal" allows for ease of installation of the underground service-lateral conductors from the electic utility and the service-entrance conductors that run from the meter to the main panel. (*Courtesy of Milbank Manufacturing Co.*)

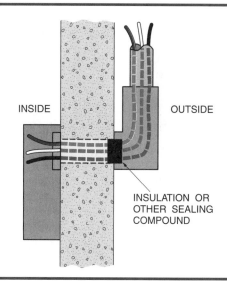

SECTION 300-5(g) AND *SECTION 300-7* STATE THAT WHEN RACEWAYS PASS THROUGH AREAS HAVING GREAT TEMPERATURE DIFFERENCES, SOME MEANS MUST BE PROVIDED TO PREVENT PASSAGE OF AIR BACK AND FORTH THROUGH THE RACEWAY. NOTE THAT OUTSIDE AIR IS DRAWN IN THROUGH THE CONDUIT WHENEVER A DOOR OPENS. COLD OUTSIDE AIR MEETING WARM INSIDE AIR CAUSES THE CONDENSATION OF MOISTURE. THIS CAN RESULT IN RUSTING AND CORROSION OF VITAL ELECTRICAL COMPONENTS. EQUIPMENT HAVING MOVING PARTS, SUCH AS CIRCUIT BREAKERS, SWITCHES, AND CONTROLLERS, IS ESPECIALLY AFFECTED BY MOISTURE. THE SLUGGISH ACTION OF THE MOVING PARTS IN THIS EQUIPMENT IS UNDESIRABLE.

INSULATION OR OTHER TYPE OF SEALING COMPOUND CAN BE INSERTED AS SHOWN TO PREVENT THE PASSAGE OF AIR.

SECTION 230-8 REQUIRES SEALS WHERE SERVICE RACEWAYS ENTER FROM AN UNDERGROUND DISTRIBUTION SYSTEM.

INSIDE

OUTSIDE

INSULATION OR OTHER SEALING COMPOUND

Figure 15–4 Sealing is required where a raceway passes from one area to another where there are extreme differences of temperature.

UNDERGROUND SERVICES

Underground services are by far the most popular method of supplying electrical power to homes. Many electric utility and telephone companies have agreements whereby the electric utility digs the trench, then buries the electric power and telephone underground cables in the same trench. The electric utility carries large reels of both the underground direct burial power cables and the underground direct burial telephone cables.

Underground services eliminate the "hassle" involved with conforming to the many *National Electrical Code®* requirements for overhead services. You will appreciate this statement as you study this unit.

Before installing the main service equipment, the electric utility must be contacted. They will determine when, where, and how they intend to hook up to the homeowner's service. They might prefer to "spot" the meter location on the side of the house rather than on the rear of the house. Why? Because raised wooden decks and concrete patios are usually in the back of a house. Patios and decks built at some later date would make it difficult to repair or replace the underground service-lateral conductors should there be problems. There is also the fact that the meter above or near a deck or patio is subject to physical damage.

Figure 15–5 shows a typical underground residential electric service. The two conduits that run to the meter base could have been a meter pedestal (trough) of the type shown in Figure 15–3. The *National Electrical Code®* defines the utilities underground cables as a *service lateral.* The pad-mount transformer usually is located on the property lot line in the rear of the home.

On occasion, there can be a conflict in the requirements for the underground service-lateral conductors. This is because an electrician doing the underground service lateral installation must conform to the *National Electrical Code®.* If the underground service-lateral conductors are furnished, installed, and maintained by the electric utility, the installation will be in conformance to the National Electric Safety Code. The utility will use the proper size and type of underground conductors, and they will bury them at the proper depth according to the NESC. But since the underground cables are on the homeowner's property, the electric utility will no doubt follow the code requirements enforced in your community. The electric utility will also make the line and load connection in the pedestal meter trough on the side of the house. They then install the watt-hour meter. They install a metal ring that secures the meter to the meter base. This ring is then locked in place with a seal that must be cut off

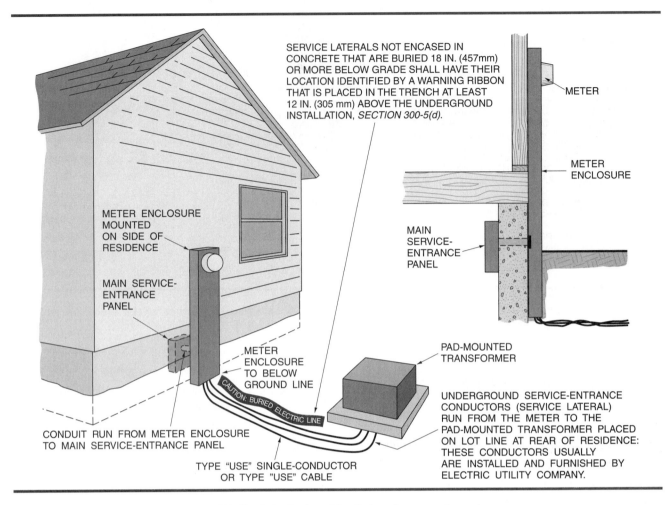

SERVICE LATERALS NOT ENCASED IN CONCRETE THAT ARE BURIED 18 IN. (457mm) OR MORE BELOW GRADE SHALL HAVE THEIR LOCATION IDENTIFIED BY A WARNING RIBBON THAT IS PLACED IN THE TRENCH AT LEAST 12 IN. (305 mm) ABOVE THE UNDERGROUND INSTALLATION, *SECTION 300-5(d).*

METER

METER ENCLOSURE

METER ENCLOSURE MOUNTED ON SIDE OF RESIDENCE

MAIN SERVICE-ENTRANCE PANEL

MAIN SERVICE-ENTRANCE PANEL

METER ENCLOSURE TO BELOW GROUND LINE

CAUTION: BURIED ELECTRIC LINE

PAD-MOUNTED TRANSFORMER

CONDUIT RUN FROM METER ENCLOSURE TO MAIN SERVICE-ENTRANCE PANEL

UNDERGROUND SERVICE-ENTRANCE CONDUCTORS (SERVICE LATERAL) RUN FROM THE METER TO THE PAD-MOUNTED TRANSFORMER PLACED ON LOT LINE AT REAR OF RESIDENCE: THESE CONDUCTORS USUALLY ARE INSTALLED AND FURNISHED BY ELECTRIC UTILITY COMPANY.

TYPE "USE" SINGLE-CONDUCTOR OR TYPE "USE" CABLE

Figure 15–5   A typical residential underground service.

before the ring and meter can be removed. This discourages tampering. Some meter pedestals have a ringless meter socket whereby the cover comes down over the watt-hour meter. This cover is then sealed to prevent tampering.

The *National Electrical Code®* exempts electric utilities from certain installations. These are clearly listed in *Section 90-2* of the *NEC.®* These exemptions are limited to electrical installations *"under the exclusive control of electric utilities for the purpose of communications, metering, generation, control, transformation, transmission, or distribution of electric energy."*

Where does the utility leave off? Where does the homeowner's responsibility begin? In most cases, this will be at the *Service Point,* defined in the *NEC®* as *"the point of connection between the facilities of the serving utility and the premises wiring."*

The *National Electrical Code®* rules for under-

ground service-lateral conductors are found in *Sections 230-30, 230-31, 230-32, 230-49* and *300-5.*

The conductor types used for underground services (Type USE) was discussed in Unit 8.

The conductor types between the meter base and the main panelboard need to be sized properly according to the *NEC.®* This will be discussed a little later on in this unit. The types of conductors suitable for wet and damp locations such as services was discussed in Unit 8.

MAST-TYPE (THROUGH-THE-ROOF) SERVICES

As with underground service, the electric utility must be contacted before starting the installation. They will "spot" the location of where they want the meter, and this will be a major determinant as to where the main service panelboard will be located in or on the house.

For overhead electrical services, a mast-type service is used. Mast services are used on buildings with low roofs, such as ranch-style homes, where the required clearance between ground level and the service-drop conductors cannot be maintained.

A service mast raceway is intended to support only power service-drop conductors. Do not attach or support television cables, telephone cables, or anything else to the mast. This requirement is found in *Section 230-28* of the *NEC.®* Figure 15–6 illustrates the intent of this requirement.

Where the service raceway is run through the roof, roof flashing and a neoprene seal fitting is used, Figure 15–7. For additional support of the

Figure 15–8 A "through-the-wall" support for a service mast.

mast, a "through-the-wall" support, Figure 15–8 might be needed.

Service Mast Support

The bending and pulling forces on the service mast increase with an increase in the length of the service drop.

Section 230-28 of the *NEC®* states that: "*Where a service mast is used for the support of service-drop conductors, it shall be of adequate strength or be supported by braces or guys to withstand safely the strain imposed by the service drop. Where raceway-type service masts are used, all raceway fittings shall be identified for use with service mast. Only power service-drop conductors shall be permitted to be attached to a service mast.*"

Electric utilities will provide you with detailed installation diagrams of their requirements showing exactly the size and type of mast required. They know the size and length of the service-drop conductor overhead span. The utility may also have clearance restrictions that are more stringent that those in the *National Electrical Code.®*

When intermediate metal conduit or rigid metal conduit is run through a roof and used as a service mast, secure fastening within 3 feet (914 mm) of the service head is not required, *Sections 345-12(a)* and *346-12(a)*. A roof mast kit provides adequate support when properly installed. However, a guy wire might be necessary if the service-drop conductors are long and/or heavy. You will know when and if a guy wire is needed by referring to the electric utilities' installation diagrams.

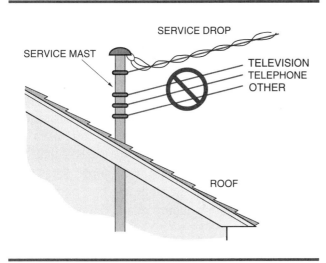

Figure 15–6 A service mast is for supporting service-drop conductors only—nothing else, *Section 230-28*.

Figure 15–7 A neoprene seal with metal flashing for use where the service raceway mast comes through the roof.

Clearances Over Roofs

Figure 15–9 shows the minimum clearances for service-drop conductors that pass above residential roofs. These requirements are listed in *Section 230-24(a)* of the *NEC®.*

Clearances Above Driveways, Sidewalks, and Roads

Clearance requirements aboveground for service-drop conductors for a typical residential electric service are found in *Section 230-24(b).* Minimum clearances from building openings, doors, porches, balconies, and similar locations are found in *Section 230-9.*

Figure 15–10 shows these minimum clearances.

Whether the overhead service-drop conductors are secured to a mast or to the side of the house, the point of attachment shall be such that the low portion of the sag in the service-drop conductors will meet the minimum clearance requirements mentioned above. The point of attachment of the service-drop conductors shall not be less than 10 feet (3.05 m) above finished grade, *Section 230-26.*

Clearances Over Swimming Pools

Minimum clearances for conductors above swimming pools, diving boards, platforms, and so forth, are covered in *Section 680-8* of the *NEC®.* This Code section contains a table and a diagram that nicely illustrates the required clearances. The main issue is the voltage-to-ground of the overhead conductors. Figure 15–11 is a condensed table that covers minimum service-drop conductor clearances over a typical residential pool. If higher voltage conductors are above the pool, consult *Table 680-8.*

Telephone and CATV cables that are owned, operated, and maintained by the respective utility are permitted to be over pools with a minimum of 10 feet (3.05 m) above the pool, diving board, platforms, and so forth, according to *Section 680-8.* More information on this can be found in Unit 18.

REAR OR SIDE OF HOUSE SERVICE

Some services are installed on the rear or side of a house so the service raceway does not need to penetrate the roof as does a mast-type service. The service raceway or service-entrance cable is run from the meter up the side of the house to a height that will provide the necessary clearances above sidewalks, driveways, roadways, and from windows and porches for the service-drop conductors and drip loop.

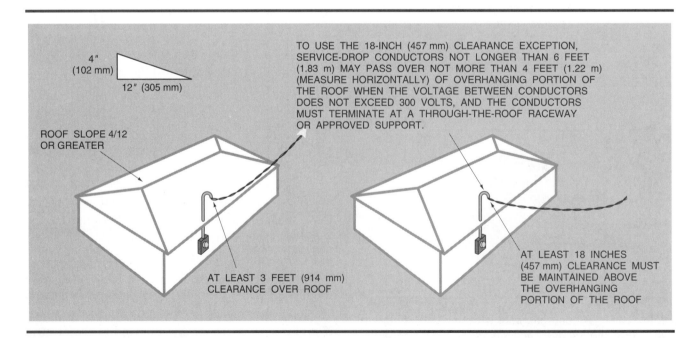

Figure 15–9 Clearance requirements for service-drop conductors passing over a residential roof in conformance to *Section 230-24(a)* of the *NEC®.*

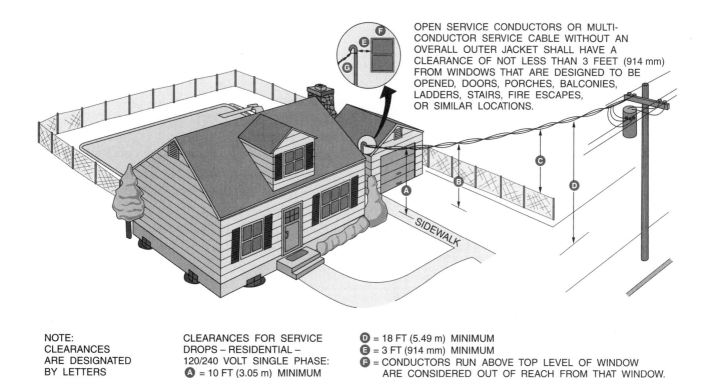

NOTE:
CLEARANCES
ARE DESIGNATED
BY LETTERS
A THROUGH G

CLEARANCES FOR SERVICE
DROPS – RESIDENTIAL –
120/240 VOLT SINGLE PHASE:
Ⓐ = 10 FT (3.05 m) MINIMUM
Ⓑ = 12 FT (3.66 m) MINIMUM
Ⓒ = 10 FT (3.05 m) MINIMUM

Ⓓ = 18 FT (5.49 m) MINIMUM
Ⓔ = 3 FT (914 mm) MINIMUM
Ⓕ = CONDUCTORS RUN ABOVE TOP LEVEL OF WINDOW
ARE CONSIDERED OUT OF REACH FROM THAT WINDOW.
3 FT (914 mm) CLEARANCE NOT REQUIRED.
Ⓖ = 10 FT (3.05 m) MIN. FROM DRIP LOOP TO FINISH GRADE

Figure 15–10 Minimum clearances for the service-drop conductors to a typical residence.

MINIMUM CLEARANCE	SERVICE-DROP CONDUCTORS, CABLED TOGETHER WITH A GROUNDED MESSENGER OR GROUNDED NEUTRAL CONDUCTOR. 0–750 VOLTS TO GROUND
Clearance in any direction to water level, edge of water surface, base of diving platform, or permanently-anchored raft.	22 feet (6.7 m)
Clearance in any direction to the diving platform or tower.	14 feet (4.27 m)
Horizontal limit of clearance measured from inside wall of the pool.	Distance extends to the outer edge of a diving platform or tower, but not less than 10 feet (3.05 m)

Figure 15–11 This table provides the minimum clearances for service-drop conductors above outdoor swimming pools.

PANELBOARDS

The term "panelboard" is defined and used throughout the *National Electrical Code*® and in Underwriters Laboratories Standards. Yet, when we check manufacturers' catalogs, we find that the term "load center" is used for residential-type panelboards. The term "load center" is not found in the Code. In the field, you will hear electricians call them *panelboards, panels or load centers.* For our purpose, they are one and the same.

Load centers are not as expensive as panelboards. The load center circuit breakers plug in as opposed to being bolted in as they are in panelboards. The depth of a load center enclosure is not as deep as that of a panelboard.

Panelboards offer more factory-installed options such as lighting contactors, wider gutter space, and feed-through lugs. Panelboards lend themselves to commercial and industrial installations, whereas load centers are designed to serve the residential market.

All load centers have a label that shows the ampere rating, voltage rating, and other required marking information. The manufacturer is required to provide this marking in conformance to *Section 384-13* of the Code and requirements of the UL Panelboard Standard No. 67. Panelboards also have a wiring diagram that clearly shows the panelboards bussing arrangement, the main lugs, main disconnect (if any), neutral lugs, and the provisions for terminating branch-circuit neutrals and equipment grounding conductors.

Figure 15–12 shows the inside of a typical 120/240-volt, single-phase, 3-wire panelboard. Clearly shown are the internal copper bus bars, the main lugs, and the neutral bus bars. This particular panelboard does not have a main disconnect, and is referred to as a *main lug only (MLO)* panelboard.

Figure 15–13 shows a panelboard that has a main circuit breaker.

In Unit 12, we discussed the *NEC*® requirements found in *Section 384-20* for the grounding and bonding of panelboards. Figure 12–8 in Unit 12 illustrated the bussing arrangement for the neutral and equipment grounding conductor connection in a panelboard.

Working Space

To provide for safe working conditions around electrical equipment, *Section 110-26* of the *NEC*® contains a number of rules that must be followed. For residential installations, here are some things to consider. In general, for electrical equipment:

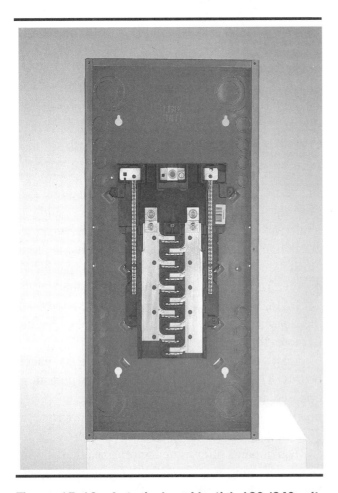

Figure 15–12 A typical residential 120/240-volt, single-phase, main lug-only panelboard. (*Courtesy Square D Company, Group Schneider*)

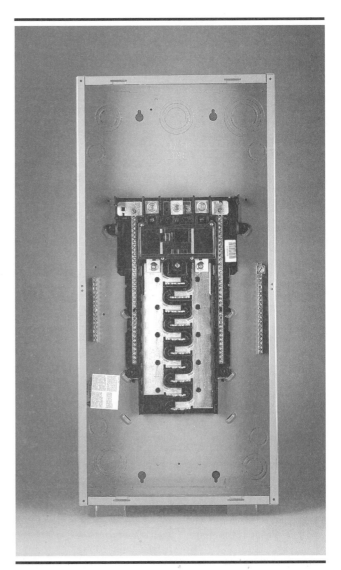

Figure 15–13 A typical residential 120/240-volt, single-phase panelboard with a main circuit breaker. (*Courtesy Square D Company, Group Schneider*)

- Allow a working space in front of the electrical equpment of not less than 30 inches wide where the electrical equipment is mounted. This area must be kept clear and not used for storage.

- Make sure that a working space of not less than 3 feet is kept clear in front of the electrical equipment. This area must be kept clear and not used for storage.

- Do not install electrical panelboards inside of cabinets, above shelving, washers, dryers, freezers, workbenches, etc.

- Do not install electrical panelboards above or close to sump pump holes.

- Make sure the hinged cover on the panelboard can be opened to a full 90 degrees.

- Keep the 30-inch wide space above the panelboard clear of any and all nonelectrical items (pipes, ducts, brackets, shelves, etc.). This "zone" is dedicated space intended for electrical equipment only, *Section 110-26(f)*.

Mounting Height

Disconnects and panelboards must be mounted so the center of the disconnect handle in its highest position is not higher than 6'7" (2.0 m), *Section 380-8*.

Headroom

Maintain not less than 6½ feet (1.98 m) of headroom around electrical panelboards, *Section 110-26(e)*. In the past, many service panelboards were installed in crawl spaces. This is no longer permitted.

Lighting

There is nothing worse than having to work on electrical equipment in the dark. Trying to find a blown fuse or a tripped circuit breaker in the dark is difficult.

Section 110-26(d) tells us that illumination must be provided for service equipment and panelboards. But the Code does not spell out how much illumination is required. In many cases, adjacent lighting in the area might be considered to be the required lighting. This becomes a judgment call on the part of the electrician and/or the electrical inspector. If in doubt, install some sort of lighting above and to the front of the panelboard.

MAIN SERVICE DISCONNECT LOCATION

The main disconnecting means will be a circuit breaker or fusible pullout. The main disconnect is usually an integral part of a residential panelboard that contains all of the required branch circuit breakers. Sometimes for larger size services, the main disconnect might, by necessity, be a separate disconnect.

The main service disconnecting means shall be installed at a readily accessible location outside or inside of the house so that the service conductors inside the building are as short as possible, *Section 230-70(a)*. Figure 15–14 illustrates this requirement.

Why this requirement? Think about it for a moment. There is no overcurrent protection ahead of the service-entrance conductors other than the utility's primary transformer fuse, and this fuse certainly does not provide the necessary overcurrent protection for the conductors as required by the *National Electrical Code.* The primary fuses for the utility transformer are sized for short-circuit protection of the transformer in accordance with the National Electrical Safety Code, not the *National Electrical Code.* These fuses have no relationship whatsoever to the current-carrying capabilities of the actual service-entrance conductors. These primary fuses could very well be many, many times the ampacity of the service-entrance conductors. Should a line-to-line or ground fault occur on the service conductors inside the dwelling, the only "line of defense" are the transformer's primary fuses.

If, for some reason, it is necessary to locate the main service equipment some distance inside the building, the way to do it is to install the conductors so they are "outside" of the building. *Section 230-6* tells us two ways that this can be done. For residential installations, this could be:

- Running the conduit under not less than 2 inches (50.8 mm) of concrete beneath the house. An example of this might be running the conduit under the garage floor to reach the main service panelboard that might be located on the inner house/garage wall.

- running the conduit within the building, but encased in at least 2 inches (50.8 mm) of concrete or brick.

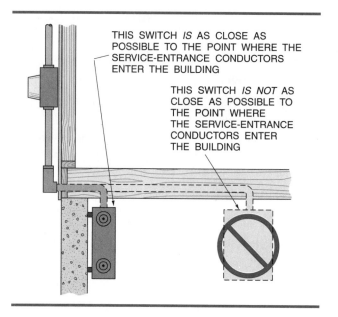

THIS SWITCH *IS* AS CLOSE AS POSSIBLE TO THE POINT WHERE THE SERVICE-ENTRANCE CONDUCTORS ENTER THE BUILDING

THIS SWITCH *IS NOT* AS CLOSE AS POSSIBLE TO THE POINT WHERE THE SERVICE-ENTRANCE CONDUCTORS ENTER THE BUILDING

Figure 15–14 Main service disconnect location should be as close as possible to where the service-entrance conductors enter the building.

Some other important requirements for main service disconnects and panelboards are that they:

- The service equipment must be marked "suitable for use as service equipment," *Section 230-66.*

- It must be readily accessible, *Section 240-24(a).*

- It shall not be installed in clothes closets or other areas close to easily ignitable material, *Section 240-24(d).*

- It shall not be installed in bathrooms, *Section 240-24(e).*

- It shall be as close as possible to the point where the service-entrance conductors enter the building, *Section 230-70(a).*

- It shall have a "legibly identified" circuit directory indicating what the circuits are for, *Sections 110-22* and *384-13.* The directory is furnished and attached to the inside of the panelboard by the manufacturer of the panelboard. If this directory is too small, then type a directory on a larger size card, and attach it to the inside of the panelboard cover.

Main Service Disconnects

Main service disconnects and panelboards shall consist of not more than six switches or six circuit breakers mounted in a single enclosure, or in a group of separate enclosures in the same location, *Section 230-71(a).* This permits the disconnection of all electrical equipment in the house with not more than six motions of the hand. Do not get confused with the six disconnect rule. It rarely comes into play in conventional residential wiring. It might come into play for extremely large upscale homes where the main service is large, like 600, 800, 1,000, or 1,200 amperes. The main service could consist of more than one panelboard, each having a main disconnect, which is unlikely for the typical home. We will keep our discussion focused on the needs for a typical home.

Some local electrical codes take exception to the six disconnect rule, and require a single main disconnect. Check this out with the local authority that has jurisdiction.

A lighting and appliance branch circuit panelboard shall not have more than two main circuit breakers or two sets of fuses for the panelboards overcurrent protection and disconnecting means. "Listed" panelboards meet this requirement. Panelboards that have two main disconnects are quite often called a "split-bus" panelboard, *Section 384-16(a).*

A lighting and appliance branch circuit panelboard shall not have more than 42 overcurrent devices in it, *Section 384-15*. A two-pole breaker is counted as two overcurrent devices. This is a *NEC®* requirement and a UL requirement. About the only way you could violate this requirement is to install a number of the so-called "twin" breakers. Twin breakers have two breakers in the case of a normal single breaker, allowing two breakers to take up the space of a single breaker. Obviously, you could end up with many more breakers than the panelboard was designed for. Read the label on the panelboard for information regarding the type, size, and number of breakers permitted.

Keep at least ¼-inch (6.35 mm) air space between the wall and the switch or panelboard where installed in wet or damp locations, *Section 373-2(a)*. Masonry in direct contact with the earth, such as a basement wall, is considered to be a wet location. Most disconnect switches, panelboards, meter sockets, and similar electrical equipment have raised mounting holes that provide the necessary clearance. Nonmetallic enclosure are not required to have the ¼-inch space.

SUB PANELBOARDS

Quite often a sub panelboard is located some distance from the main service panelboard. Let us say that the main service panelboard is located on the far corner of a house, and there is a concentration of electrical appliances (electric water heater, electric range and oven, and electric clothes dryer) at the opposite corner of the house. Rather than running many branch circuits, it might be advantageous to install a subpanelboard, using larger conductors to supply the subpanelboard. The appliance branch circuits are then connected to the subpanelboard.

Line losses (voltage and wattage) are considerably less than if the branch circuits had been run all the way back to the main panelboard. There can also be a significant cost of labor and material savings to install an extra panelboard as compared to the cost of running all of the branch circuit "homes runs" back to the main panelboard.

A sub panelboard is nothing more than a regular panelboard (load center) installed remote from the main panelboard. The difference is that you are **not permitted** to ground the neutral and the panelboard enclosure together as is done only at the main panelboard. If this is confusing, refer to Unit 12 for a refresher.

The conductors that run between the main panelboard and the sub panelboard are called feeders. For example, you might use a 100-ampere, two-pole circuit breaker in the main service panelboard to feed a subpanelboard. All the circuit breakers in the subpanelboard supply branch circuits.

The *NEC®* defines a *feeder* as: "*All circuit conductors between the service equipment, the source of a separately derived system, or other power supply source and the final branch-circuit overcurrent device.*"

The *NEC®* defines a *branch circuit* as: "*The circuit conductors between the final overcurrent device protecting the circuit and the outlet(s).*"

GAS EXPLOSION HAZARD

Although it might seem a bit out of place in an electrical book to talk about gas explosions, we must talk about it.

Often overlooked is a requirement in NFPA 54 (Fuel Gas Code), *Section 2.7.2(c)* that requires that gas meters be located at least 3 feet from sources of ignition. An electric meter or a disconnect switch are possible sources of ignition. Some utility regulations require a minimum of 3 feet clearance between electric metering equipment and gas meters and gas regulating equipment. It is better to be safe than sorry! Check this issue out with the local electrical inspector and/or the local electric utility before installing service-entrance equipment. Also consider the location of the air conditioner or heat pump disconnect on the outside of the house.

PANELBOARD RATINGS

In addition to panelboards (loadcenters) being marked with their voltage and current rating, they might also be marked with a "Series Rating." This is covered in detail in Unit 17.

REVIEW QUESTIONS

1. As we study the *National Electrical Code,*® there is terminology that is unique to the electrical industry that is necessary to know if one is to understand the Code. Match the following terms with the statement that defines the term.

1. _____ Service

2. _____ Service conductors

3. _____ Service drop

4. _____ Service entrance conductors, overhead system

5. _____ Service entrance conductors, underground system

6. _____ Service equipment

7. _____ Service lateral

8. _____ Service point

a) The service conductors between the terminals of the service equipment and a point usually outside the building, clear of building walls, where joined by tap or splice to the service drop.

b) The service conductors between the terminals of the service equipment and the point of connection to the service lateral

c) The overhead service conductors from the last pole or other aerial support to and including the splices, if any, connecting to the service-entrance conductors at the building or other structure

d) The underground service conductors between the street main, including any risers at a pole or other structure or from transformers, and the first point of connection to the service-entrance conductors in a terminal box or meter or other enclosure with adequate space, inside or outside the building wall. Where there is no terminal box, meter, or other enclosure with adequate space, the point of connection shall be considered to be the point of entrance of the service conductors into the building

e) The necessary equipment, usually consisting of a circuit breaker or switch and fuses, and their accessories, located near the point of entrance of supply conductors to a building of other structure, or an otherwise defined area, and intended to constitute the main control and means of cutoff of the supply

f) Service point is the point of connection between the facilities of the serving utility and the premises wiring

g) The conductors from the service point or other source of power to the service disconnecting means

h) The conductors and equipment for delivering energy from the electricity supply system to the wiring system of the premises served

2. Who generally determines where the meter location will be? _____

_____

3. In your area, what is the practice for installing the service equipment. Give the location of the main panel and meter, the ampere rating of the service, and what grounding electrode is used. Is service-entrance cable used? Are services required to be in a raceway? Give a brief description of a typical residential service in your area.

_____

_____

_____

4. When an outside wall is penetrated where the service-entrance conductors enter the building in a raceway, what must be done to prevent passage of air from the outside to the inside of the building? Give the *NEC®* section number that supports your answer.

_____

_____

_____

5. What are some of the advantages of underground services?

_____

_____

_____

6. A through-the-roof service mast consists of 2-inch rigid metal conduit. Because of the tremendous strength of this conduit, it is permissible to attach and support telephone and/or other cables to this mast. (CATV cables). What section of the *NEC®* supports your answer?
   a) True
   b) False
   Section _____

7. What is the minimum size ungrounded THWN copper conductors for the following 120/240-volt, single-phase residential services? Refer to *Table 310-15(b)(6)*.
   a) 100-ampere: No. _____ THWN
   b) 200-ampere: No. _____ THWN

8. What is the minimum size copper grounding electrode conductors for the following 120/240-volt, single-phase residential services? Take advantage of *Table 310-15(b)(6)*. See *Table 250-66*.
   a) 100-ampere: No. _____ grounding electrode conductor
   b) 200-ampere: No. _____ grounding electrode conductor

9. Show the minimum conductor clearances from grade level for the following 120/240-volt, single-phase residential service-entrance conductors.
   a) Above a residential driveway   _____
   b) Above a residential sidewalk   _____
   c) Above a roof that has not less than a 4/12 pitch   _____

    d) For the short distance where the service drop conductors
       pass over the overhang portion of a roof   _____

    e) Above a fence that can be climbed on   _____

    f) To the right or left of a window that can be opened   _____

    g) Above a window that can be opened   _____

    h) Service-drop conductors above the water level over a
       swimming pool   _____

10. What is the difference between a panelboard and a load center?

_____

_____

_____

11. Mark T (True) or F (False) for the following statements that relate to panelboards.

    a) Keep a working space having a width in front of the panel of
       at least 30 inches.   _____

    b) Keep a depth of at least 36 inches clear in front of the main panel.   _____

    c) Be sure that a hinged cover on a panel can be opened to at least
       90 degrees.   _____

    d) The minimum head room at an electrical panel is 6½ feet.   _____

    e) It is all right to install the service panel in a clothes closet.   _____

    f) It is all right to mount the main service panel below a drain pipe
       that runs across the wall.   _____

    g) Many new homes have 9-foot high ceilings in the basement.
       It is all right to mount the main service panel so that the center
       of the main breaker is 7½ feet above the floor.   _____

    h) It is necessary to provide lighting in the area where the main
       service panel is located.   _____

    i) A directory indicating all of the circuits must be filled out
       and attached to the main service panel.   _____

    j) It is all right to install the service panel in a bathroom.   _____

12. *Section 230-70(a)* states that the main service disconnecting means must be installed
at a readily accessible location, and that the service conductors inside of the dwelling
be kept as short as possible. Similar wording is found in *Section 240-24(a)*. In your
own words, explain why these requirements are in the *National Electrical Code.*®

_____

_____

_____

13. a) In order to shut off all of the electrical power in a home, what is the maximum
     number of disconnects permitted for a residential main electrical service?

    _____

    b) Does your local electrical code have any requirements regarding the number of
     main disconnects permitted? If yes, what are they?

_____

_____

14. In your area, are you permitted to use the so-called "twin" circuit breakers, where two breakers are enclosed in one housing that take up the space of a standard single circuit breaker?

    a) Yes

    b) No

15. In your area, are you permitted to install the so-called "split-bus" panelboards, where one of the upper circuit breakers is the main disconnect for the lower portion of the panelboard?

    a) Yes

    b) No

16. In your own words, why would you want to install a subpanel for a residential installation.

_____

_____

_____

17. The neutral conductor and the neutral bus in a subpanel must be grounded to the panel's metal enclosure. Give the Code reference that supports your answer.

    a) True

    b) False

    Section _____

18. One of the things often overlooked by electricians is the fact that there are requirements other than the *National Electrical Code®* that have a bearing on electrical equipment. For example, you are installing a disconnect and associated wiring for an air conditioner unit outside of the house. You notice that the gas company's meter and shutoff are located very close to where you intended to mount the disconnect switch. Can you give a Code reference that will influence where the disconnect should be mounted? Explain.

_____

_____

_____

_____

Service-Entrance Calculations

INTRODUCTION

In Unit 15, we discussed types of electrical services, service equipment, and the *National Electrical Code®* rules governing electrical services. But the question now becomes "What size service shall I install?" This unit covers this subject.

Residential wiring involves branch circuits for lighting, branch circuits for small appliances, and branch circuits for appliances. But we do not total up all of these to derive a total load to determine the size of the main service. There is much diversity. The *National Electrical Code®* recognizes this diversity, and over the years has developed load factors to be used to determine the minimum size for a main electrical service to a home.

SIZING SERVICE-ENTRANCE CONDUCTORS

The sample service-entrance calculations below are based on hypothetical loads and branch circuits. Once we know the final minimum ampacity for the service, the conductors are selected, and the raceway size is calculated.

Some municipalities have their own electrical codes for services that clearly state the minimum conductor size and raceway size. Check this out before installing a service! This text is based on the *National Electrical Code®.*

Much of the information regarding the size and rating of service-entrance conductors is covered in *Section 230-42* of the *NEC.®* After determining the minimum ampacity requirements for service conductors, the conductor size is then found in *Table 310-16* of the *NEC.®*

Section 310-15(b)(6) provides a special consideration for 120/240-volt, 3-wire, single-phase services. The neutral conductor is permitted to be smaller than the "hot" ungrounded conductors when the requirements of *Sections 215-2, 220-22,* and *230-42* are met. The sample calculations below show how these requirements are met. *Table 310-15(b)(6)* is a table that shows smaller size conductors for a given current value than *Table 310-16.* Figure 16–1 shows these special ampacity ratings.

The service neutral conductor serves two purposes. In addition to carrying the unbalanced

SPECIAL AMPACITY RATINGS FOR RESIDENTIAL 120/240-VOLT, 3-WIRE, SINGLE-PHASE SERVICE-ENTRANCE CONDUCTORS, SERVICE LATERAL CONDUCTORS, AND FEEDER CONDUCTORS THAT SERVE AS THE MAIN POWER FEEDER TO A DWELLING UNIT.

Copper Conductor AWG for Insulation of RH-RHH-RHW-THW-THWN-THHW-THHN-XHHW-USE	Aluminum or Copper-Clad Aluminum Conductors (AWG)	Allowable Special Ampacity
4	2	100
3	1	110
2	1/0	125
1	2/0	150
1/0	3/0	175
2/0	4/0	200
3/0	250 kcmil	225
4/0	300 kcmil	250
250 kcmil	350 kcmil	300
350 kcmil	500 kcmil	350
400 kcmil	600 kcmil	400

Note: Where the conductors will be exposed to temperatures in excess of 30°C (86°F), apply the correction factors found at the bottom of *Table 310-16*.

Figure 16–1 *Table 310-15(b)(6)* from the *NEC.* It is only for residential 120/240-volt, 3-wire, single-phase services that carry the main power to a dwelling. It cannot be used for feeders that run from the main panel to a subpanel.

current, the neutral conductor must also be capable of safely carrying any fault current that it might be called upon to carry, such as a line-to-ground fault in the service equipment. Sizing the neutral conductor only for the normal neutral unbalance current could result in a neutral conductor too small to safely carry fault currents. Under fault conditions, it could burn off, causing serious voltage problems, damage to equipment, and the creation of fire and shock hazards.

In Unit 8, we learned that conductors that are exposed to moisture (wet locations) such as service conductors must have the letter "W" in their type designation.

Minimum Size

The minimum size service for a single-family dwelling is 100 amperes per *Section 230-79(c)*.

Consider that a typical electric range is at least 8 kW, *Section 220-19*. Add to this the two required small appliance branch circuits, a laundry branch circuit, some lighting branch circuits, it is obvious that there are few, if any, single-family homes for

which a service smaller than 100 amperes can be installed.

For illustrative purposes, the sample calculation below is based on a residence that has 3,232 square feet of living area. It has many electrical appliances and many lighting and small appliance branch circuits. This example was developed to show the many types of electrical loads found in homes, and how load factors and demand factors are applied to these loads. Study this example.

The neutral conductor is permitted to be smaller than the "hot" ungrounded conductors because it is sized to carry unbalanced loads, subject to certain minimums. Straight 240-volt loads such as a 240-volt electric water heater or a 240-volt pump motor do not contribute to neutral loads, and may be omitted from the calculations for the "hot" ungrounded conductors.

At first, it may seem very confusing. However, it is a matter of "filling in the blanks," then do the math.

A blank form is provided in the back of this book on which you can make calculations for your own specific needs.

SINGLE-FAMILY DWELLING SERVICE-ENTRANCE CALCULATIONS

1. **General lighting load (*Section 220-3(a)*)**

   3,232 sq. ft. @ 3 VA per sq. ft. = 9,696 VA

2. **Minimum number of lighting branch circuits**

   9,696 VA ÷ 120 volts = 80.8 amperes
   Then:

   $$\frac{amperes}{15} = \frac{80.8}{15} = 5/387 \text{ (round up to 6)}$$
   15-ampere branch circuits

3. **Small appliance load (*Sections 210-11(c)(1) and 220-16(a)*)**

   (Minimum of two 20-ampere branch-circuits)

Kitchen	3
Laundry Area	1*
Workshop	2**

   6 branch circuits @ 1,500 VA each = 9,000 VA

4. **Laundry branch circuit (*Section 210-11(c)(2) and 220-16(b)*)**

   (Minimum of one 20-ampere branch circuit)
   1 branch circuit @ 1,500 VA = 1,500 VA

5. **Total general lighting, small appliance, and laundry load**

   Lines 1 + 3 + 4 = 20,196 VA

6. **Net computed general lighting, small appliance, and laundry loads (less ranges, ovens, "fastened-in-place" appliances). Apply demand factors from *Table 220-11*.**

   a. First 3,000 VA @ 100%   = 3,000 VA
   b. Line 5: 20,196 – 3,000 =
       17,196 @ 35%   = 6,019 VA
       Total a + b = 9,019 VA

---

* For this sample calculation, this branch circuit is in addition to the required minimum of one laundry branch circuit.
The intent is to serve a receptacle(s) in the laundry area for plugging in such things as an electric iron, steamer, or clothes press. This provides additional capacity to the laundry room and to the service.

** For this sample calculation, two branch circuits were arbitrarily added to the basic calculations to show how extra branch circuits can provide additional capacity for a workshop area and to the service.

7. **Electric range, wall-mounted ovens, counter mounted cooking units (*Table 220-19*)**

Wall-mounted oven:	= 7,450 VA
Counter-mounted cooking unit	= 6,600 VA
Total	14,050 VA
	(14kW)

   14 kW exceeds 12 kW by 2 kW
   2 kW × 5% = 10% increase, therefore:
   8 kW + 0.8 kW = 8.8 kW = 8,800 VA

8. **Electric clothes dryer (*Table 220-18*)**
   = 5,700 VA

9. **Electric furnace (*Section 220-15*)**
   Air conditioner, heat pump (*Article 440*)

Electric furnace:		= 13,000 VA
Air conditioner:	26.4 × 240	= 6,336 VA
(Enter largest value,		
Section 220-21)		= 13,000 VA

10. **Net computed general lighting, small appliance, laundry, ranges, ovens, cooktop units, HVAC**
    Line 6 + 7 + 8 + 9 = 36,519 VA

11. **List "fastened-in-place" appliances *in addition* to electric ranges, electric clothes dryers, electric space heaters, and air-conditioning equipment**

Appliance	VA Load
Water heater	= 3,000 VA
Dishwasher	= 1,000 VA
Motor: 7.2 × 120 = 864 VA	
Heater: 1,000 VA (watts)	
Enter largest value, *Section 220-21*	
Garage door opener: 5.8 × 120	= 696 VA
Food waste disposer: 7.2 × 120	= 864 VA
Water pump: 8 × 240	= 1,920 VA
Hydromassage tub: 10 × 120	= 1,200 VA
Attic exhaust fan: 5.8 × 120	= 696 VA
Heat/Vent/Lights: 1,500 × 2	= 3,000 VA
Freezer: 5.8 × 120	= 696 VA
Total	13,072 VA

12. **Apply 75% demand factor (*Section 220-17*) if four or more "fastened-in-place" appliances. If less than four, figure @ 100%.** Do not include electric ranges, electric clothes dryers, electric space heating, or air-conditioning equipment.

    Line 11: 13,072 × 0.75 = 9,804 VA

13. **Total computed load (lighting, small appliance, ranges, dryer, HVAC, "fastened-in-place" appliances)**

Line 10 + Line 12:
36,519 + 9,804 = 46,323 VA

14. **Add 25% of largest motor (*Section 220-14*)**

This is the water pump motor:
8 × 240 × 0.25 = 480 VA

15. **Grand total line 13 + line 14 = 46,803 VA**

16. **Minimum ampacity for ungrounded service-entrance conductors**

$$\text{Amperes} = \frac{\text{Line 15}}{240} = \frac{46,803}{240} = 195 \text{ amperes}$$

17. **Minimum ampacity for neutral service-entrance conductor, (*Sections 220-22* and *310-15(b)(6)*). Do not include straight 240-volt loads.**

a. Line 6   = 9,019 VA
b. Line 7: 8,800 × 0.70   = 6,160 VA
c. Line 8: 5,700 × 0.70   = 3,990 VA
d. Line 11: (Include only 120-volt loads.)

Freezer	696 VA
Food waste disposer	864 VA
Garage door opener	696 VA
Heat/Vent/Light	3,000 VA
Attic exhaust fan	696 VA
Hydromassage tub	1,200 VA
Dishwasher heater	1,000 VA
Total	8,152 VA

e. Line d total @ 75% Demand Factor:
8,152 × 0.75   = 6,114 VA
f. Add 25% of largest 120-volt motor.
This is the food waste disposer:
2 × 120 × 0.25   = 216 VA
           Total   25,499 VA

g. Total a + b + c + e + f   = 25,499 VA

$$\text{Amperes} = \frac{\text{volt} - \text{amperes}}{\text{volts}} =$$

$$\frac{25,499}{240} = 106.245 \text{ amperes}$$

18. **Ungrounded Conductor Size (copper) (*Table 310-15(b)(6)*)**   2/0 THWN

Note: For the 195 ampere computed load, these "hot" conductors could be No. 3/0 THW, THHW, THWN, XHHW, or THHN per *Table 310-16*, or a No. 2/0 (same types) using *Table 310-15(b)(6)*. This table may only be used for 120/240-volt, 3-wire, residential single-phase service-entrance conductors, service lateral conductors, and feeder conductors that serve as the main power feeder to a dwelling unit. We have assumed that all equipment is marked 75°C, and therefore, we read the ampacity of the THHW, XHHW, and THHN 90°C conductors from the 75° column per *Section 110-14(c)*.

19. **Neutral Conductor Size (copper) (*Section 220-22*)**   No. 2 AWG insulated or bare

Note: For the 106.245 ampere computed neutral load, a No. 2 AWG is adequate for the neutral per *Section 310-15(b)(6)*. This section permits the neutral conductor to be smaller than the ungrounded "hot" conductors if the requirement of *Sections 215-2, 220-22,* and *230-42* are met. When bare conductors are used with insulated conductors, the conductors ampacity is based on the ampacity of the other insulated conductors in the raceway, *Section 310-15(b)(3)*. The neutral conductor must not be smaller than the grounding electrode conductor, *Section 250-24(b)(1)*.

20. **Grounding Electrode Conductor Size (copper) (*Table 250-66*)**   No. 4 AWG

21. **Raceway Size**   1¼ inch

Obtain dimensional data from *Tables 1, 4, 5,* and *8, Chapter 9, NEC.*® For this example, use electrical metallic tubing.

CONDUCTOR SIZE BASED ON AMPACITY PER *TABLE 310-15(b)(6)*		CONDUCTOR SIZE BASED ON AMPACITY PER *TABLE 310-16*	
Two No. 2/0 THWN	0.2223 Sq. In.	Two No. 3/0 THWN	0.2679 Sq. In.
	0.2223 Sq. In.		0.2679 Sq. In.
One No. 2 bare	0.0670 Sq. In.	One No. 2 bare	0.0670 Sq. In.
Total	0.5116 Sq. In.	Total	0.6028 Sq. In.
EMT size	1¼ inch	EMT size	1½ inch

OPTIONAL CALCULATIONS

If the preceding calculations seem a bit too complicated, the *National Electrical Code® Section 220-30* shows a simple alternative method. This method may be used only when the service-entrance conductors have an ampacity of 100 amperes or more. Remember that the minimum size service for a one-family dwelling is 100 amperes, *Section 230-42(b)* and *230-79(c)*.

Section 220-30(a) tells us that we are permitted to calculate service-entrance conductors and feeder conductors (both phase conductors and the neutral conductor) using an optional method. *Section 220-30* addresses single-family dwellings. *Section 220-32* addresses multifamily dwellings.

Section 220-30(b) tells us to:

1. include 1500 volt-amperes for each 20-ampere small appliance circuit.

2. include 3 volt-amperes per square foot for lighting and general-use receptacles.

3. include the nameplate rating of appliances:
   - that are fastened in place, or
   - that are permanently connected, or
   - that are located to be connected to a specific circuit

   Appliances listed in *Section 220-30(b)(3)* are ranges, wall-mounted ovens, counter-mounted cooking units, clothes dryers, and water heaters.

4. include nameplate ampere rating or kVA for motors and all low-power-factor loads. The intent of the reference to low power factor is to address such loads as low-cost, low-power-factor fluorescent ballasts of the type that might be used in the recessed lay-in fluorescent fixtures in the recreation room. It is always recommended that high-power-factor ballasts be installed.

Section 220-30(c) tells us that for heating and air-conditioning loads, select the largest load from a list of six types of loads:

1. 100% of air conditioning and cooling.

2. 100% of heat pump compressors and supplemental heating unless both cannot operate at the same time.

3. 100% electric thermal storage heating if expected to be continuous at full nameplate value.

4. 65% of central electric space heating that also has supplemental heating in heat pumps where the compressor and supplemental heating will not operate at the same time.

5. 65% of electric space heating of less than four separately controlled units.

6. 40% of electric space heating if four or more separately controlled units.

Let us begin our optional calculation. We will use the same residence and the first sample calculation. It has an air conditioner and an electric furnace. For our purposes, we consider VA and watts to be the same.

Air-conditioner
$$4 \times 240 = 6,336 \text{ VA}$$

Electric furnace
$$13,000 \times 0.65 = 8,450 \text{ VA}$$

Select the electric furnace load for the calculations because it is the largest load. The air conditioner and electric furnace are *noncoincidental loads* as defined in *Section 220-21* because they would not operate at the same time.

We now add up all of the other loads:

General lighting load:
3,232 sq. ft. @ 3 VA per sq. ft.= 9,696 VA

Small appliance branch circuits:

Kitchen	3
Laundry (washer)	1
Laundry area	1
Workshop	2
	7 @ 1,500 VA each

	= 10,500 VA
Wall-mounted oven:	= 6,600 VA
Counter-mounted cooktop:	= 7,450 VA
Water heater:	= 3,000 VA
Clothes dryer:	= 5,700 VA
Dishwasher (maximum demand, heater only)	= 1,000 VA
Food waste disposer: 7.2×120	= 864 VA
Water pump motor: 8×240	= 1,920 VA
Garage door opener: 5.8×120	= 696 VA
Heat/Vent/Lights (2): 1500×2	= 3,000 VA
Attic exhaust fan: 5.8×120	= 696 VA
Hydromassage tub: 10×120	= 1,200 VA
Total other loads	= 52,322 VA

We now complete our optional calculations.

Electric furnace: $13,000 \times 0.65$ $= 8,450$ VA

Plus first 10 kVA of all other
 loads @ 100% $= 10,000$ VA

Plus remainder of all other loads
 @ 40% $52,322 - 10,000 = 42,322$
 Then, $42,322 \times 0.40$ $= \underline{16,929}$ VA
                   Total $= 35,379$ VA

$$\text{Amperes} = \frac{\text{volt} - \text{amperes}}{\text{volts}} = \frac{35,379}{240} = 147.4 \text{ amperes}$$

Checking *Table 310-15(b)(6)*, we find that No. 1 THW or THWN copper conductors could be used for this 147.4-ampere calculation.

Although permitted by the *NEC®*, it would probably be best to use the standard service-entrance calculations that resulted in the largest ampere value, which was 195 amperes. Some local codes do not permit the use of *Table 310-15(b)(6)*.

WATCHING OUT FOR HIGH TEMPERATURES

In Unit 8, we also learned how to apply correction factors for conductors that are subjected to high temperatures. In certain parts of the country such as the southwestern desert climates, extremely hot temperatures are common, particularly when exposed to direct sunlight. Figure 16–2 depicts a typical high temperature situation.

Check with local code officials to see if there are any special conductor sizing requirements because of the high temperatures.

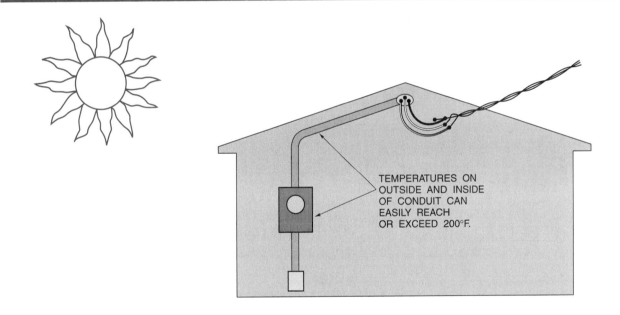

TEMPERATURES ON OUTSIDE AND INSIDE OF CONDUIT CAN EASILY REACH OR EXCEED 200°F.

Figure 16–2   An example of a high temperature location that might require derating the ampacity of the conductors.

REVIEW QUESTIONS

1. The basic requirements for sizing service-entrance conductors are found in *Section* _____ of the *National Electrical Code®.*

2. What is the minimum size 3-wire, 120/240-volt service for a one-family dwelling?

_____

3. Some areas of the country experience very high ambient temperatures, particularly in late afternoon when the sun has been shining on the house for hours. High temperatures can damage the insulation on conductors.

   Where subjected to these extremely high temperature conditions, services, feeders, and branch circuits must be _____ in conformance to the *C*_____ *F*_____ found at the bottom of *Table* _____.

4. Calculate the minimum size of copper service-entrance conductors and grounding electrode conductor required for a residence containing the following: floor area is 24' × 38' (7.3 m x 11.6 m). The dwelling will have:

   - a 12 kW, 120/240-volt electric range
   - a 5 kW, 120/240-volt electric clothes dryer consisting of a 4 kW, 240-volt heating element, a 120-volt motor and a 120-volt lamp that have a combined load of 1 kW
   - a 2,200 watt, 120-volt sauna heater
   - six individually controlled 2 kW, 240-volt baseboard electric heaters
   - a 12 ampere, 240-volt air conditioner
   - a 3 kW, 240-volt electric water heater
   - a 1,000 watt, 120-volt dishwasher
   - a 7.2 ampere, 120-volt food waste disposer
   - a 5.8 ampere, 120-volt attic exhaust fan

   Determine the sizes of the ungrounded "hot" conductors and the neutral conductor. Use Type THWN. Use *Table 310-15(b)(6)*. Also determine the correct size grounding electrode conductor.

   Do not forget to include the small appliance circuits and the laundry circuit.

   You may use the blank *Single-Family Dwelling Service-Entrance Calculation* form in the Appendix of this text, or use the form as a guide to make your calculations in the proper steps.

   a) Two No. _____ THWN ungrounded "hot" conductors

   b) One No. _____ THWN or bare neutral conductor

   c) One No. _____ Grounding electrode conductor

Overcurrent Protection, Short-Circuit Calculations

OBJECTIVES

After studying this unit, you will be able to:

- Understand the importance of proper overcurrent protection.
- Be able to describe an overload, overcurrent, short circuit, and ground fault.
- Have a better basic understanding of fuses and circuit breakers.
- Understand the meaning of the terms *interrupting rating, short-circuit current, available fault current, series-rated, fully-rated.*
- Understand that panelboards have a short-circuit rating.
- Be able to perform simple short-circuit calculations.

INTRODUCTION

Overcurrent protection, fuses, and circuit breakers, are the "safety valves" of an electrical circuit. Improperly sized, and you have set the stage for a disaster.

OVERCURRENT AND OVERLOAD

The *National Electrical Code®* provides us with a few definitions.

- *Overcurrent:* "Any current in excess of the rated current of equipment or the ampacity of a conductor. It may result from overload, short circuit, or ground fault."

- *Overload:* "Operation of equipment in excess of normal, full-load rating, or of a conductor is excess of rated ampacity that, when it persists for a sufficient length of time, would cause damage or dangerous overheating. A fault, such as a short circuit or ground fault, is not an overload."

Figure 17–1 illustrates a normal circuit where the current flowing is 10 amperes. The conductors and the overcurrent device are rated 15 amperes.

Figure 17–2 shows an overloaded circuit. The conductor and overcurrent device are rated 15 amperes. The 20 amperes of current is flowing though the "intended path." If allowed to continue for an extended period of time, the conductors will become overheated, the conductor's insulation will be damaged—a potential cause for fire.

Figure 17–3 shows a short-circuit condition where two conductors have come in contact with one another, resulting in a current flow that bypasses the connected load. Actually, the short circuit is in parallel with the connected load. The zero impedance of the short-circuit path results in an extremely high level of short-circuit current. In the real world, there would be considerably more impedance between the source and the point of the short circuit than the example shows. The impedances of the utility's generator, the transmission lines, the many transformers involved, service conductors, and branch-circuit conductors would reduce to fault current value tremendously. The impedance values in the example are for illustrative purposes only.

Ⓐ NORMAL CIRCUIT

THE CONDUCTORS CAN SAFELY CARRY THE CURRENT, THEY
DO NOT GET HOT. THE 15 AMPERE FUSES DO NOT OPEN.

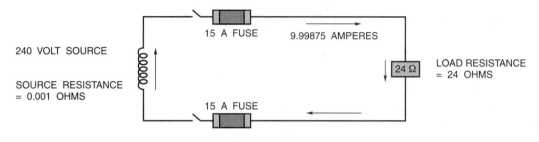

$$I \text{ (THROUGH CIRCUIT)} = \frac{E}{R} = \frac{240}{24 + 0.001 + 0.001 + 0.001} = \frac{240}{24.003} = 9.99875 \text{ AMPERES}$$

Figure 17–1   This is a normally loaded circuit.

Ⓑ OVERLOADED CIRCUIT

THE CONDUCTORS BEGIN TO GET HOT, BUT THE 15 AMPERE
FUSES WILL OPEN IN SOME PERIOD OF TIME BEFORE THE
CONDUCTORS ARE DAMAGED.

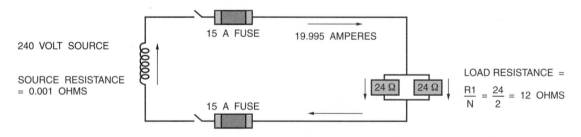

$$I \text{ (THROUGH CIRCUIT)} = \frac{E}{R} = \frac{240}{12 + 0.001 + 0.001 + 0.001} = \frac{240}{12.003} = 19.995 \text{ AMPERES}$$

Figure 17–2   This is an overloaded circuit.

Ⓒ SHORT CIRCUIT

THE CONDUCTORS GET EXTREMELY HOT. THE INSULATION WILL MELT OFF AND THE CONDUCTORS WILL MELT UNLESS THE FUSES OPEN IN A VERY SHORT (FAST) PERIOD OF TIME. CURRENT-LIMITING OVERCURRENT DEVICES WILL LIMIT THE AMOUNT OF "LET-THROUGH" CURRENT, AND WILL LIMIT THE TIME THE FAULT CURRENT IS PERMITTED TO FLOW.

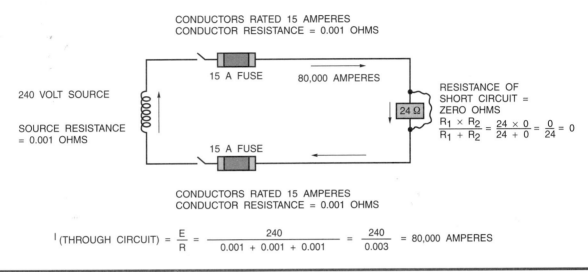

Figure 17–3   This circuit has a short circuit across the connected load.

A short circuit might be two "hot" conductors touching, or it might be a "hot" conductor and a "grounded" circuit conductor coming together. In either case, the current is outside of its "intended path." The heat of an arc is hotter than the temperature of the sun, and can certainly cause burns and/or a fire.

Figure 17–4 shows a "hot" conductor coming in contact with a grounded surface, such as a grounded metal raceway, metal water pipe, sheet metal, etc. The current flows outside of the "intended path." A ground fault of this type can be greater than normal, in which case the overcurrent device will open. A ground fault can also result in a current flow less than normal, in which case the overcurrent device may not open. Yet, the heat of the arcing at the point of the ground fault can cause a fire.

Ground-fault circuit interrupters (GFCIs) are an example of a device that protects against very low levels of ground fault. GFCIs were discussed in Unit 7.

Large services require *ground fault protection*

for equipment. This would not apply to a typical residential service. This subject is covered in Delmar's *Electrical Wiring — Commercial.*

Overcurrent Protection for Conductors

The *National Electrical Code*® covers most of the overcurrent protection requirements in *Article 240.* Overcurrent protection for specific appliances and motors was discussed in Unit 13.

Fuses and circuit breakers are sized by matching their ampere ratings to conductor ampacities and to the connected load currents. They sense overloads, short circuits, and ground-fault conditions, and protect the wiring and equipment from reaching dangerous temperatures.

We discussed some of the requirement for conductor protection in Unit 9. Here is a quick review.

Overcurrent devices are connected "in series" with the ungrounded "hot" conductor(s) of a circuit, *Section 240-20(a).*

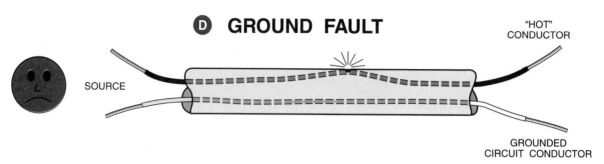

ⓓ GROUND FAULT

"HOT" CONDUCTOR

SOURCE

GROUNDED
CIRCUIT CONDUCTOR

"HOT" CONDUCTOR COMES IN CONTACT WITH METAL RACEWAY OR OTHER METAL OBJECT. IF THE RETURN GROUND PATH HAS LOW RESISTANCE (IMPEDANCE), THE OVERCURRENT DEVICE PROTECTING THE CIRCUIT WILL CLEAR THE FAULT. IF THE RETURN GROUND PATH HAS HIGH RESISTANCE (IMPEDANCE), THE OVERCURRENT DEVICE WILL NOT CLEAR THE FAULT. THE METAL OBJECT WILL THEN HAVE A VOLTAGE TO GROUND THE SAME AS THE "HOT" CONDUCTOR HAS TO GROUND. IN HOUSE WIRING, THIS VOLTAGE TO GROUND IS 120 VOLTS. PROPER GROUNDING AND GROUND-FAULT CIRCUIT INTERRUPTER PROTECTION IS DISCUSSED ELSEWHERE IN THIS TEXT. THE CALCULATION PROCEDURE FOR A GROUND FAULT IS THE SAME AS FOR A SHORT CIRCUIT, HOWEVER, THE VALUES OF "R" CAN VARY GREATLY BECAUSE OF THE UNKNOWN IMPEDANCE OF THE GROUND RETURN PATH. LOOSE LOCKNUTS, BUSHINGS, SET SCREWS ON CONNECTORS AND COUPLINGS, POOR TERMINATIONS, RUST, ETC. ALL CONTRIBUTE TO THE RESISTANCE OF THE RETURN GROUND PATH, MAKING IT EXTREMELY DIFFICULT TO DETERMINE THE ACTUAL GROUND-FAULT CURRENT VALUES.

Figure 17–4 In this circuit, the "hot" ungrounded conductor has come in contact with the grounded metal raceway. This is referred to as a ground fault.

Overcurrent devices are not permitted to be connected in series with the grounded conductor, *Section 240-22*. There are two exceptions to this rule:

1. where the grounded **and** ungrounded conductors open simultaneously, or

2. where used for motor overload protection, *Sections 430-36* and *430-37*.

Overcurrent protection for conductors generally matches the ampacity of the conductor, *Section 240-3*. Where the ampere rating of the overcurrent device does not exactly match the ampacity of the conductor, there are certain conditions where the Code allows us to use the next standard ampere rating, *Section 240-3(b)*. Checking *Section 240-3(d)*, we find that the maximum size overcurrent device for:

• No. 14 AWG copper conductors is 15 amperes.

• No. 12 AWG copper conductors is 20 amperes

• No. 10 AWG copper conductors is 30 amperes

The ampacities for these conductors are higher than the maximum permitted overcurrent protection. These higher ampacity values can be used to begin derating and correction factor calculations.

Overcurrent Protection for Appliances and Motors

We discussed overcurrent protection for appliances and motors in Unit 13.

TYPES OF OVERCURRENT PROTECTION

When a fuse opens or a circuit breaker trips, find out why! Locate and fix the problem. Then replace the fuse with the proper size and type, or reset the circuit breaker. Do not assume that the fuse is the proper size and type. It may not be. Do not continue to replace a blown fuse, or repeatedly reset a circuit breaker without solving the problem.

Plug Fuses, Fuseholders, and Adapters

The requirements for plug fuses are found in *Article 240*, Part E, specifically *Sections 240-50* through *240-54*.

Plug fuses are available in two types:

• Edison-base: their brass screw shell looks exactly like the shell of a light bulb.

• Type S base: their screw-in base looks much like the threads of a spark plug.

Plug fuses are available in two classes of time-current characteristics, non time-delay and time-delay.

Non time-delay:

• Has a single fuse element

• Does not have much time-delay to handle the momentary high inrush current of a motor

• Is best suited for circuits that do not have motors; an example of this are resistive loads such as an electric heater or incandescent lighting

Time-Delay:

• Has two fuse elements and is referred to as a *dual-element* fuse.

• The *overload element* provides more time-delay than an ordinary nontime-delay fuse.

• The *short-circuit element* opens quickly when a short circuit or ground fault occur.

• Is best suited for motor circuits. The momentary high starting current inrush of a motor will not cause the time-delay fuse to blow needlessly. Since most branch circuits have some motors, this is a good choice to use for general purpose branch circuits.

• Some time-delay fuses use a *loaded link* design whereby a slug of metal is placed on the short-circuit link to absorb heat developed during momentary overloads.

Plug fuses are for use on circuits 125 volts or less between conductors.

There is one exception. It is all right to use plug fuses on systems where the voltage to ground is not more than 150 volts. This is the case for the 120/240-volt system in a home, so you could use plug fuses in the disconnect switch for a 120/240-volt electric clothes dryer, a 240-volt electric water heater, or for a 240-volt motor.

Plug fuses have ampere ratings from 0 to 30 amperes. Actually, plug-type fuses are available in ratings $\frac{3}{10}$ through 30 amperes. Sizes $\frac{3}{10}$ through 14 amperes are usually of the time-delay type for overload protection of small motors. Ratings 15, 20, 25, and 30 amperes are standard branch-circuit ratings.

Plug fuses that are rated 15 amperes or less have a hexagon shape somewhere on the fuse, either the upper body of the fuse, or the window of the fuse.

Holders for plug fuses must have the screw shell connected to the load. The center contact of a plug fuseholder is the "hot" terminal.

Edison-base plug fuses may be used only to replace fuses in existing installations where there is no sign of overfusing or tampering. This is the old story of someone putting a penny behind the fuse, or putting in a fuse with an ampere rating much too large for the ampere rating of the conductors.

All new installations require Type S fuses.

Let us discuss Type S fuses. Type S fuses are "size limiting" in that a Type S adapter must first be inserted into an Edison-base fuseholder before a Type S fuse is screwed in. There are many ampere ratings of Type S fuses and adapters that provide a number of ampere ranges that prevent overfusing. These noninterchangeable ranges are accomplished by using different thread pitches and different sizes (thickness) of the bottom contact on the Type S fuse and on the adapter. Because Type S adapters are extremely difficult to remove without the proper tool, they are nontamperable. Figure 17–5 shows the ampere ranges of noninterchangeable combinations when using Type S fuses and adapters.

TYPE S FUSE INFORMATION

Type S Fuse Ampere Ratings	Type S Adapter Rating	Type S Fuse Ampere Ratings That Fit Into This Adapter
$\frac{3}{10}$, $\frac{1}{2}$, $\frac{8}{10}$, $\frac{4}{10}$, $\frac{6}{10}$, 1	1	1 ampere & smaller
$1\frac{1}{8}$, $1\frac{1}{4}$	$1\frac{1}{4}$	All smaller
$1\frac{4}{10}$, $1\frac{6}{10}$	$1\frac{6}{10}$	All smaller
$1\frac{8}{10}$, 2	2	$1\frac{8}{10}$, 2
$2\frac{1}{4}$, $3\frac{1}{2}$	$2\frac{1}{2}$	$1\frac{8}{10}$, 2, $2\frac{1}{4}$, $2\frac{1}{2}$
$2\frac{8}{10}$, $3\frac{2}{10}$	$3\frac{2}{10}$	$1\frac{8}{10}$, 2, $2\frac{1}{4}$, $2\frac{1}{2}$, $2\frac{8}{10}$, $3\frac{2}{10}$
$3\frac{1}{2}$, 4	4	$3\frac{1}{2}$, 4
$4\frac{1}{2}$, 5	5	$3\frac{1}{2}$, 4, $4\frac{1}{2}$, 5
$5\frac{6}{10}$, $6\frac{1}{4}$	$6\frac{1}{4}$	$3\frac{1}{2}$, 4, $4\frac{1}{2}$, 5, $5\frac{6}{10}$, $6\frac{1}{4}$
7, 8	8	7, 8
9, 10	10	7, 8, 9, 10
12, 14	14	7, 8, 9, 10, 12, 14
15	15	15
20	20	20
25	30	20, 25, 30
30	30	20, 25, 30

Figure 17–5 This table shows the many ampere rating combinations for Type S fuses and adapters.

Type S fuses sizes 0 through 14 amperes are time-delay type. Ratings 15, 20, 25, and 30 amperes are available in *nontime-delay* and *time-delay* types. The time-delay feature is advantageous for use on motor circuits, where the motor's momentary inrush would blow a nontime-delay fuse. Momentary overloads do not cause a time-delay fuse to blow. Branch circuits that supply motors and motor-driven appliances are good applications for time-delay fuses. Time-delay fuses are also referred to as *dual-element* fuses because they have a short-circuit element and an overload element.

Figure 17–6 shows an Edison-base dual-element time-delay plug fuse (marked T) and an Edison-base nontime-delay plug fuse (marked W).

Figure 17–7 shows a Type S fuse with its unique thread, and an adapter for a Type S fuse.

Figure 17–6 A 15-ampere Type W Edison base plug fuse and a 20-ampere Type T time-delay Edison base plug fuse. (*Courtesy of Cooper Bussmann, Inc.*)

Figure 17–7 A 15-ampere Type S fuse and a 15-ampere Type S adaptor. (*Courtesy Cooper Bussmann, Inc.*)

Cartridge Fuses

Cartridge fuses are tubular in shape in 0 through 30 amperes, and 35 through 60 amperes. They have brass ferrules on each end.

Cartridge fuses 110 through 600 amperes are tubular in shape, but have copper blades on each end. These are referred to as knife blade fuses.

There are a number of different case sizes of cartridge fuses. The maximum ampere rating in any given case size matches the ampere rating of disconnect switches and fuseholders. The following chart illustrates this.

DISCONNECT SWITCH AMPERE RATING	FUSE AMPERE RATINGS THAT FIT
30	0 through 30
60	35, 40, 50, 60
100	70, 80, 90, 100
200	110, 125, 150, 175, 200
400	225, 250, 300, 350, 400
600	450, 500, 600

Cartridge fuses are available in both time-delay dual-element and nontime-delay single-element types. On the outside, they look the same. But in the inside, the link construction is quite different.

Time-delay dual-element fuses are designed to handle temporary overloads and inrush surges caused by the starting of motors. They will hold approximately five times their ampere rating for about ten seconds, which should be enough time to get a motor started. This time versus current comparison is referred to as the fuse's time/current characteristic. Time/current characteristics vary with the ampere rating of the fuse. Manufacturers of fuses furnish time/current characteristic curves for all sizes and types of fuses.

Nontime-delay cartridge fuses are generally used in circuits where high inrush currents are not anticipated, such as resistive heating loads. Nontime-delay fuses are not the best choice for motor circuits because they must be sized large enough to handle motor starting inrush current. They will hold approximately three times their ampere ratings for about ten seconds. The time/current characteristic varies with ampere rating.

How to size fuses for specific loads is covered in Unit 13.

Figure 17–8 shows a 250-volt, 100-ampere Class R* dual-element, time-delay cartridge fuse. Also illustrated is a cutaway view of the same fuse showing the *overload element* and the *short-circuit elements*.

Ampere Ratings. Time-delay cartridge fuses are available in many fractional ampere ratings $\frac{1}{10}$ through 14 amperes for use on motor circuits. Ampere ratings available in the smaller sizes are

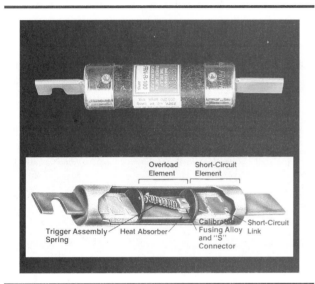

Figure 17–8 Cartridge-type dual-element fuse (A) is a 250-volt, 100-ampere fuse. The cutaway view in (B) shows the internal parts of the fuse. When this type of fuse has a rejection slot in the blades of 70–600 ampere sizes, or a rejection ring in the end ferrules of the 0–60 ampere sizes, they are U.L. Class RK1 or RK5 fuses. (*Courtesy Cooper Bussmann, Inc.*)

similar to those available for time-delay plug fuses as listed in Figure 17–5.

Nontime-delay cartridge fuses are available in many different ampere ratings, but not as many choices as time-delay cartridge fuses.

Standard ampere ratings for fuses are listed in *Section 240-6* of the *National Electrical Code.*® These ampere ratings are 15, 20, 25, 30, 35, 40, 50, 60, 70, 80, 90, 100, 110, 125, 150, 175, 200, 225, 250, 300, 350, 400, 450, 500, and 600. Ampere ratings above 600 amperes are Class T and Class L fuses.

Voltage Ratings. The voltage rating of cartridge fuses used in residential wiring is 250-volts.

In commercial wiring, 600-volt fuses are often used.

For most applications, a fuse may be used at or less than its voltage rating. For example, a 250-volt fuse will work all right on a 240-volt or 120-volt circuit. A 600-volt fuse will work all right on a 480-volt or 277-volt circuit. Consult fuse manufacturers technical data for exception to this general statement.

Class T Fuses

Figure 17–9 shows different sizes of Class T fuses. The 30 and 60 ampere ratings are ferrule type. Ampere ratings above 60 amperes are bolted into place.

---

* Class R, Class T, Class G, and Class L fuses are discussed in detail in *Electrical Wiring—Commercial.* These fuses have high interrupting ratings, and have rejection features that make it impossible to insert a fuse of one class into a fuseholder of another class.

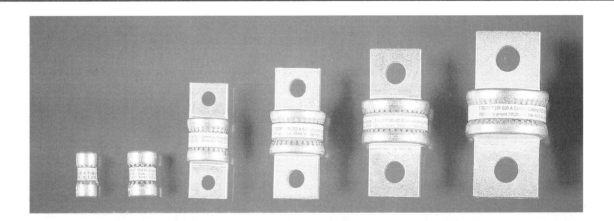

Figure 17–9 Class T fuses. Illustrated are 30-, 60-, 100-, 200-, 400-, and 600-ampere ratings. (*Courtesy Cooper Bussmann, Inc.*)

Class T fuses are rated 300 volts. For a given ampere rating, Class T fuses are considerably smaller than regular types of cartridge fuses.

Residential panelboards (load centers) used to be available with a fusible main pullout disconnect for standard cartridge type fuses and for Class T fuses. Many of these types of fusible main panelboards have been in use for years, and are performing very nicely. However, residential type fusible main load centers are no longer available. Panelboards (load centers) today are available with a circuit breaker main disconnect.

Class T fuses have a high interrupting rating (discussed later), and are current-limiting. Panelboard (load centers) that use Class T main fuses have a high interrupting rating. Class T fuses in combination with standard circuit breakers enable a panel to be "listed" as *Series Rated*.

Fuses Located In Extremely High or Extremely Low Ambient Temperatures

Fuses will carry **less** current than their ampere rating if located in extremely high ambient temperatures. The ampere rating of a fuse might have to be derated to 85% of its rating in an ambient temperature of 140°F. For example, a 60-ampere fuse in a switch located in the direct rays of the sun in the summer time might derate to: $60 \times 0.85 = 51$ amperes.

Fuses will carry **more** current than their ampere rating if located in extremely low ambient temperatures. The ampere rating of a fuse might have to be rerated to 110% of its rating in an ambient temperature of 32°F. For example, a 60-ampere fuse in a switch located outdoors where the ambient temperature is 32°F could carry $60 \times 1.1 = 66$ amperes

Fuse manufacturers provide rerating curves that show the effect of high and low ambient temperatures.

Fuse Reducers

Fuse reducers make it possible to install a smaller case size fuse than the fuse clips of a given disconnect switch's ampere rating. For example, if you had an existing 100-ampere disconnect switch, but wanted to install 60-ampere fuses, you would use fuse reducers. Fuse reducers are sold in pairs,

and slip over the end ferrules of the fuse. Figure 17–10 shows two sizes of fuse reducers. The top ones reduce from 100-ampere case size to 60-ampere case size. The bottom ones reduce from 200-ampere case size to 100-ampere case size.

Circuit Breakers

Sections 240-80 through *240-86* provide some of the basic requirements for circuit breakers.

Figure 17–11 shows a 15-ampere, single-pole circuit breaker used for 120-volt lighting and appli-

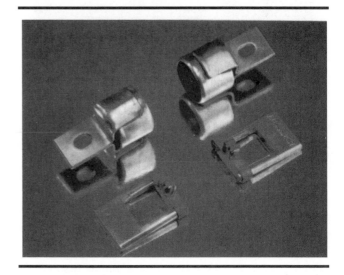

Figure 17–10 Fuse reducers. (*Courtesy Cooper Bussmann, Inc.*)

Figure 17–11 Typical molded case circuit breakers.

ance branch circuits. The two-pole circuit breaker type is used for a main, or for 240-volt branch circuits that supply electric ranges, clothes dryers, water heaters, and 240-volt motors.

Most residential-type circuit breakers are thermal/magnetic. On a continuous overload, a bimetallic element moves until the tripping mechanism is unlatched, allowing the circuit breaker to trip to its OFF position. When a short circuit occurs, a magnetic coil in the breaker quickly unlatches the tripping mechanism, allowing the circuit breaker to trip to its OFF position.

Circuit Breakers Located in Extremely High or Extremely Low Ambient Temperatures

If you see the words *ambient compensated* on a circuit breaker, it merely means that the tripping characteristics of the breaker will not change appreciably when subjected to above- or below-normal temperatures. Most circuit breakers are factory calibrated for use in up to 40°C (104°F) ambient. The label will be marked 40°C. In general, no derating is necessary when the ambient temperature is between 25°C and 40°C. Below 25°C, breakers will carry more than their rated current without tripping. Above 40°C, breakers will carry less than their rating current and may nuisance trip if this is not taken into consideration. You will have to consult the manufacturer of the circuit breaker to obtain re-rating curves for extreme ambient temperature applications.

Ampere Ratings. Standard ampere ratings for the common types of circuit breakers used in residential applications can be found in *Section 240-6* of the *National Electrical Code.* These ampere ratings are 15, 20, 25, 30, 35, 40, 50, 60, 70, 80, 90, 100, 110, 125, 150, 175, 200, 225, 250, 300, 350, 400, 450, 500, 600, 700, and 800. See *Section 240-6* for ampere ratings above 800.

Voltage Ratings. Residential circuit breakers are rated 120 volts for the single-pole type and 120/240 volts for the two-pole type. Two-pole circuit breakers are "common trip" so that both poles will open when the circuit breaker is shut off, or if a line-to-ground, a line-to-neutral, or a line-to-line fault occurs.

Handle Ties

The use of handle ties is somewhat limited. The *NEC®* has two tough rules on this.

1. Except where limited by *Section 210-4(b)*, individual single-pole circuit breakers, with or without approved handle ties, shall be permitted as the protection for the ungrounded conductor of multiwire branch circuits that serve only single-phase, line-to-neutral loads. An example of this might be a multiwire lighting branch circuit in a dwelling where a three-wire nonmetallic-sheathed cable is run to some point, then split into two 120-volt branch circuits. At the main panelboard, the conductors of this three-wire cable are connected to the two "hot" phases and to the neutral bus.

2. In grounded systems, individual single-pole circuit breakers with approved handle ties shall be permitted as the protection for each ungrounded conductor for line-to-line connected loads for single-phase circuits. An example of this might be using two single-pole circuit breakers with a handle tie for the 240-volt branch circuit for an electric water heater.

Other Code references relating to handle ties are *Sections 225-33(b), 230-71(b), 240-20(b),* and *305-4(e).*

Ground-Fault Circuit-Interrupter Circuit Breakers.

Circuit breakers are available with *ground-fault circuit-interrupter* features. This was discussed and illustrated in Unit 7.

HACR Circuit Breakers

Virtually all residential one-pole and two-pole circuit breakers are marked "HACR," which means they are suitable for use on branch circuits that serve **H**eating, **A**ir-**C**onditioning, and **R**efrigeration equipment. But to be sure, check the label.

The following are key points for circuit breakers:

• Shall be trip-free so that even if the handle is held in the ON position, the internal mechanism will trip to the OFF position.

- Shall indicate clearly whether they are on or off.

- Shall be nontamperable so that its tripping point cannot be readjusted.

- Shall have their ampere rating durably marked on the breaker. For small circuit breakers rated 100 amperes or less, the rating must be molded, stamped, or etched on the handle or other part of the breaker so as to be visible after the cover of the panelboard is installed.

- With an interrupting rating other than 5,000 amperes, shall have this rating marked on the breaker.

- when marked "SWD," are suitable to be used as a switch.

It is good practice to periodically turn a circuit breaker on and off to "exercise" its internal moving parts.

INTERRUPTING RATINGS

Section 110-9 of the *NEC®* states that all fuses and circuit breakers "*intended to interrupt current at fault levels shall have an interrupting rating sufficient for the nominal circuit voltage and the current that is available at the line terminals of the equipment.*"

Section 110-10 of the *NEC®* tells to "select and coordinate" overcurrent devices, total circuit impedance, and the short-circuit ratings of the equipment "*so as to clear a fault without extensive damage to the electrical components of the circuit.*"

The *National Electrical Code®* defines *interrupting rating* as "*The highest current at rated voltage that a device is intended to interrupt under standard test conditions.*" For simplicity we sometimes use the letters AIR, which mean *amperes interrupting rating*. The letters AIC mean *amperes interrupting capacity*. The terms AIR and AIC are oftentimes used interchangeably. There is a difference between the terms, but this subject is beyond the scope of this text.

Fuses must be marked with their interrupting rating if over 10,000 amperes.

Circuit breakers must be marked with their interrupting rating if over 5,000 amperes.

Overcurrent devices with inadequate interrupting ratings are, in effect, bombs waiting for a short circuit to trigger them into an explosion. Personal injury may result, and serious damage can be done to the electrical equipment. A fire could be started.

Fuses and circuit breakers can handle overloads quite nicely. However, should a major short circuit or ground fault occur, particularly in the main panelboard, these fuses or circuit breakers will be called upon to interrupt a considerable amount of current. This current is called *available fault current*. The amount of available fault current at any given point in an electric system depends upon a number of things, such as conductor size and length, transformer size, and the size and impedance of the transformer supplying the electrical system. Your electric utility is the best source for this information.

PANELBOARD (LOAD CENTER) SHORT-CIRCUIT RATING

For all practical purposes, the terms *panelboard* and *load center* mean the same thing. The *National Electrical Code®* defines a panelboard, but does not define a load center. In residential work, we freely use the term load center. In commercial and industrial work, we use the term panelboard, and would rarely, if ever, use the term load center. Circuit breakers for load centers are plug-in type. Circuit breakers for panelboards are usually the bolt-in type.

Panelboards usually are deeper, offer accessories such as relays, shunt trip devices, can be used with high interrupting rating circuit breakers, have higher short-circuit ratings than load centers, and are available in a number of different types of enclosures.

In Unit 15, we discussed service-entrance equipment such as panelboards (load centers). In Unit 16, we talked about how to calculate the minimum ampere rating for a service.

There is another issue often overlooked, and that is short-circuit rating of the panelboard.

The short-circuit rating of a panelboard is found on the nameplate of the equipment as required by *Section 240-83(c)*. This short-circuit rating has been established by testing the total assembly, which includes the breakers, fuses, disconnects, and bus bars in the panelboard.

If the service-entrance equipment is large, and the distance from the equipment is not too great, chances are that the *available fault current* at the line-side terminals of the equipment will exceed 10,000 amperes. Most residential-type circuit breakers have an interrupting rating of 10,000 amperes RMS symmetrical. Some of the larger ampere ratings for residential application have an interrupting rating of 22,000 amperes RMS symmetrical.

Panelboards (load centers) are marked with their short-circuit rating. A typical residential circuit breaker load center might be marked 22/10, in which case the panelboard has a 22,000 AIR main circuit breaker and 10,000 AIR branch circuit breakers. The panelboard is suitable to be connected to a system capable of delivering not over 22,000 of short-circuit current at the line-side terminals of the panelboard. Be very careful when installing new and replacement circuit breakers to make sure that their catalog number is listed as proper for use in the particular panelboard. This is particularly important for *series-rated* panelboards where the main circuit breaker and the branch circuit breaker work together under high level fault conditions.

CORRECT TYPE OF PANELBOARD TO USE

High values of available short circuit current are normally not a problem in single-family dwellings except in the case of large services. High available fault currents can be a problem in multi-family installations.

The following discussion covers the types of panelboards that might be considered for a given situation.

It has been determined that the available fault current at the location where a panelboard is to be installed there is approximately 20,800 amperes. Here are the possibilities.

- Install a *fully rated* panelboard that has main and all branch circuit breakers rated not less than 20,800 AIR. Circuit breakers are available that have an interrupting rating of 22,000 amperes. This is illustrated in Figure 17–12.

- Install a *series-rated* panelboard that has a 22,000 AIC main circuit breaker and standard 10,000 AIC branch circuit breakers. This panelboard is all right for use where the available fault-current does not exceed 22,000 amperes. The panelboard must be marked *Series Rated* for use where the available fault current does not exceed 22,000 amperes. This is shown in Figure 17–13.

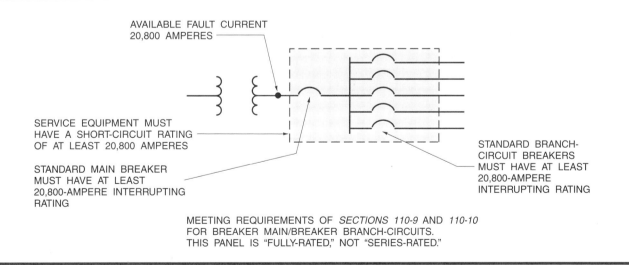

AVAILABLE FAULT CURRENT
20,800 AMPERES

SERVICE EQUIPMENT MUST HAVE A SHORT-CIRCUIT RATING OF AT LEAST 20,800 AMPERES

STANDARD MAIN BREAKER MUST HAVE AT LEAST 20,800-AMPERE INTERRUPTING RATING

STANDARD BRANCH-CIRCUIT BREAKERS MUST HAVE AT LEAST 20,800-AMPERE INTERRUPTING RATING

MEETING REQUIREMENTS OF *SECTIONS 110-9* AND *110-10* FOR BREAKER MAIN/BREAKER BRANCH-CIRCUITS. THIS PANEL IS "FULLY-RATED," NOT "SERIES-RATED."

Figure 17–12   A *fully rated* panel with a main circuit breaker and branch circuit breakers. All of the breakers in the panel are rated for the available fault current at the panel.

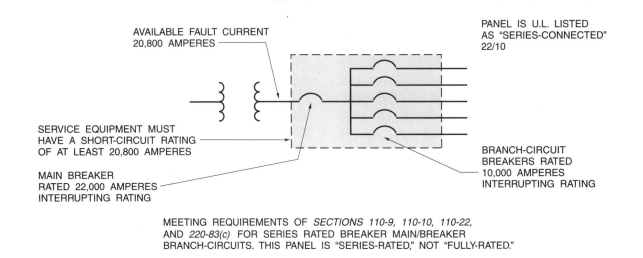

Figure 17–13 A *series-rated* panel with the main circuit breaker for not less than the available fault current at the panel. The interrupting rating of the branch circuit breakers is 10,000 amperes. The panel is marked "series rated."

- Install a *series rated* panelboard that has 100,000 ampere interrupting rating main Class T fuses and standard 10,000 interrupting rating branch circuit breakers. This panelboard is adequate for use where the available fault-current does not exceed 100,000 amperes, hardly likely in residential installations, but possible in apartment and condominium dwelling units. The panelboard must be marked *Series Rated* for use where the available fault-current will not

exceed 100,000 amperes. This is shown in Figure 17–14. New fusible panelboards for residential installations are virtually nonexistent today.

Series-Rated Panelboards

Section 110-22 states: "*Where circuit breakers or fuses are applied in compliance with the series combination ratings marked on the equipment by the manufacturer, the equipment enclosure(s) shall be*

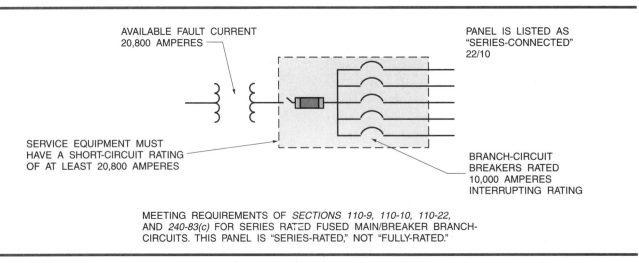

Figure 17–14 A *series-rated* panel with fusible main with an interrupting rating of not less than the available fault current at the panel. The interrupting rating of the branch circuit breakers is 10,000 amperes. The panel is marked "series rated." Most residential panels of this type usually have Class T fuses for the mains.

legibly marked in the field to indicate the equipment has been applied with a series combination rating. The marking shall be readily visible and state:

**Caution—Series Combination System
Rated _____ Amperes
Identified Replacement
Component Required.**

Section 240-86 states: "*When a circuit breaker is used on a circuit having an available fault current higher than its marked interrupting rating by being connected on the load side of an acceptable overcurrent-protective device having the higher rating, this additional series combination interrupting rating shall be marked on the end use equipment, such as switchboards and panelboards.*"

Section 240-86(b) limits the use of series-ratings when there is a significant amount of motor load. Motors that are running at the instant a short-circuit occurs act as generators, and contribute short-circuit current that is in addition to the fault current delivered by the utility. For residential installations, this is not a problem because motor loads do not constitute a big percentage of the connected load. In commercial and industrial applications, this could present a real problem. This is discussed in *Electrical Wiring—Commercial*.

CALCULATING AVAILABLE FAULT CURRENT

The terms *short-circuit current* and *available fault current* are synonymous. For all practical purposes, they mean the same thing.

Electric utilities will furnish you with available fault current at your main service.

A simple way to calculate fault current will give you a ball park figure that you can use to determine the minimum short-circuit rating needed for your main panelboard. You need to know the kVA rating and impedance of the transformer. Transformers installed by utilities, particularly the pad-mounted and pole-mounted types, have very low impedance values. As in Ohm's law, the lower the impedance, the higher the current.

This simple formula provides you with the approximate available fault current at the load side terminals of the transformer. It does not take into consideration the size and length of the conductors

between the transformer and the point of fault. It does provides you with a value of fault current to use when selecting service equipment.

Step 1.   Determine the transformer's normal full-load secondary current.

For single-phase transformers:

$$I = \frac{kVA \times 1000}{E}$$

where:

I = current, in amperes

kVA = kilovolt-amperes from nameplate of transformer

E = line-to-line voltage on secondary of transformer

Step 2. Using the percent impedance value given on the transformer nameplate, find a multiplier.

$$Multiplier = \frac{100}{percent\ impedance}$$

Step 3. The available line-to-line (L-L) short-circuit current (SCA) at the load side secondary terminals of the transformer is:

SCA = transformers normal full-load
secondary current × multiplier

Example: A single-phase transformer is rated 100 kVA. The secondary voltage is 120/240 volts. The impedance data from the nameplate is 1.5%. Find the available L-L short-circuit current at the secondary of this transformer.

Step 1.

$$I = \frac{KVA \times 1000}{E} = \frac{100 \times 1000}{240} = 416 \text{ amperes full load current}$$

Step 2.

$$Multiplier = \frac{100}{percent\ impedance} = \frac{100}{1.5} = 66.6$$

Step 3.   SCA = 416 x 66.6 = 27,706 amperes
available short-circuit current at the secondary
(L-L) terminals of the transformer

For simplicity, we could agree on an approximate available fault current value of 28,000 amperes.

Line-to-Neutral Fault Current

The line-to-neutral (L-N) available fault current for single-phase transformers is approximately 1½ times greater than the line-to-line (L-L) fault current.

For the preceding example, the approximate line-to-neutral (L-N) available short-circuit current is:

$$28,000 \times 1.5 = 42,000 \text{ amperes}$$

A Review of Interrupting Rating and Normal Ampere Rating

Circuit breakers and fuses are available with different interrupting ratings—10,000, 22,000, 35,000, 42,000, 65,000, 100,000, 200 000 amperes RMS symmetrical. Other interrupting ratings are available. Do not confuse interrupting ratings with the fuse or circuit breaker's normal ampere rating. Interrupting ratings have nothing to do with the fuse or circuit breaker's normal ampere rating. Interrupting ratings have to do with the ability to safely interrupt fault current. Ampere rating is the normal current-carrying ability, such as 15, 20, 30, 40, 50, or 100 amperes. Fuses and circuit breakers are first selected based on the required normal load requirements, then selected based upon the available fault current at different points in the system. The interrupting ratings of fuses and breakers in the main panelboard usually needs to be greater than those located a great distance from the transformer or main panelboard.

Available fault current decreases as the distance from the source increases. The major factors that limit available fault current are the impedance of the transformer, and the impedance of the conductors. The impedance of small size conductors is greater than the impedance of large size conductors. The impedance of aluminum conductors is greater than copper conductors of the same size. A long conductor has more impedance than a short conductor of the same size. A transformer with a low impedance will let through more fault-current than a transformer that has a higher impedance.

Other Ways to Calculate Fault Currents

There is a short-circuit calculation formula that takes into consideration the size, type, and length of conductors. The formula is a little more complicated than the example we showed previously. It is called the *Point-to-Point Method* and is found in *Electrical Wiring—Commercial*. For those of you who wish to dig deeper into this topic, you might want to obtain a copy of this text.

REVIEW QUESTIONS

1. Match the following statements to the correct definition.
   1. _____ An overcurrent
   2. _____ A short circuit
   3. _____ An overcurrent

   a) two circuit conductors touching one another
   b) an ungrounded (hot) conductor in contact with a grounded surface
   c) an excessive amount of current flowing in the conductor.

2. Very low levels of ground-fault current passing through a person can be fatal. To protect against this possibility, the *National Electrical Code*® requires that receptacles installed in bathrooms, kitchens, garages, basements, outdoors, and in crawl spaces be

   _____

3. In general, conductors must be protected against overcurrent according to the requirements of *Section* _____.

4. Fuses and circuit breakers are connected "in series" with the _____ conductor.
   a) ungrounded
   b) grounded
   This requirement is found in *Section* _____ of the *NEC*.®

5. For typical branch-circuit wiring, indicate the maximum ampere rating of the branch-circuit overcurrent device for the following copper conductors.

    a) No. 14 AWG   _____ amperes

    b) No. 12 AWG   _____ amperes

    c) No. 10 AWG   _____ amperes

6. Type S fuses are somewhat "size limiting" because Type S fuses have different threads for different ampere ratings. Mark the following statements T (True) or F (False).

    a) A 20-ampere Type S fuse will fit in a 15-ampere Type S adapter.   _____

    b) A 30-ampere Type S fuse will fit in a 15- or 20-ampere Type S adapter.   _____

    c) A 20-ampere Type S fuse will fit in a 15-ampere Type S adapter.   _____

    d) A 20-ampere Type S fuse will fit in a 30-ampere Type S adapter.   _____

    e) An 8-ampere Type S fuse will fit in a 6¼-ampere Type S adapter.   _____

7. When discussing fuses, the two basic characteristics that are available are _____-delay and _____-delay.

8. The standard type of plug fuse is referred to in the *National Electric Code*® as an _____ _____ fuse.

9. Disconnect switches are rated in amperes. List the standard ratings of disconnect switches through 600 amperes.

    _____

10. Standard ampere ratings of fuses and circuit breakers are found in *Section 240-6* of the *NEC*® Although there are many different ampere ratings for cartridge-type fuses, they all fall within a given "case size" so as to be able to fit into a given disconnect switch size. What are the "case sizes" for cartridge fuses?

    _____

11. You find an existing 60-ampere pullout disconnect in the main electrical panel. You wish to hook up a clothes dryer that calls for a 30-ampere time-delay fuse. How would you do this?

    _____

12. The proper fuse to use for a motor circuit is a _____ fuse. The proper fuse to use for a purely resistive load that does not have a high inrush current is a _____ fuse.

13. If you experience nuisance blowing fuses, or tripping of a circuit breaker, one possible cause is that they might be located in an extremely _____ temperature that could cause the fuse or breaker to operate at ampere values lower than their actual ampere rating.

14. Does the *NEC*® permit two single-pole circuit breakers to be used for a single-phase, line-to-neutral, multiwire branch circuit? What section of the Code supports your answer?

    a) Yes

    b) No

    Section _____

15. Does the *NEC*® permit two single-pole circuit breakers to be used for a straight 240-volt circuit?

    a) Yes

    b) No

16. Overcurrent devices such as fuses and circuit breakers not only must be capable of carrying and opening the circuit under overload or short-circuit conditions. These overcurrent devices might be called upon to open under severe short-circuit or ground-fault conditions. At a main panel, the available fault current can be very high. Therefore, in addition to selecting fuses and circuit breakers for their normal current rating and voltage rating, particular attention must be given to their *I*_____ *R*_____. What section of the *NEC*® supports your answer? *Sections* _____ and _____.

17. What does the term *interrupting rating* mean? _____

_____

18. We know that anything is only as good as its weakest link. We can apply this concept to electrical equipment. For example, a main service panel (load center) uses 10,000 AIC breakers, but the available fault current where the panel is located could reach fault currents in excess of 20,000 amperes. This is a violation of *Section* _____.

19. Mark T (True) or F (False) for the following statements.

    a) The panelboard has no marking showing its short-circuit rating. As long as you install breakers or fuses with adequate interrupting rating, the installation "Meets Code." _____

    b) The panelboard is "Series Rated" and is marked 22,000/10,000; therefore, is suitable for use where the available fault current does not exceed 22,000 amperes. _____

    c) The circuit breaker is marked "HACR" and is permitted to be used for the branch circuit supplying an air conditioner or heat pump. _____

    d) The panel label is marked with specific catalog numbers for the breakers permitted to be used in the panel, but since the electrical department of the home center has circuit breakers that have different part numbers, but are the same physical size that will fit into the panelboard, it is all right to use these breakers. _____

    e) Available fault current increases the closer you are to the transformer. _____

    f) Available fault current decreases the farther you are from the transformer. _____

20. The electric utility company has provided a letter to the electrical contractor stating that the available fault current at the line side of the main service-entrance equipment in a residence is 17,000 amperes RMS, line-to-line. In the space provided for each statement, write in "Meets Code" or "Violation" of *Section 110-9* of the *NEC*®

    a) Main circuit breaker is 10,000 amperes interrupting rating. Branch circuit breakers have a 10,000 amperes interrupting rating. The panel has no other marking relating to fault current. _____

b) Main current-limiting fuses have an interrupting rating of 200,000 amperes. Branch circuit breakers, 10,000 amperes interrupting rating. The panel is marked "Series Rated" for a maximum fault current of 22,000 amperes.   _____

c) Main circuit breaker is 22,000 amperes interrupting rating. Branch circuit breakers, 10,000 amperes interrupting rating. The panel is marked "Series Rated" for a maximum fault current of 22,000 amperes.   _____

21. Using the short-circuit calculations method shown in this unit, what is the approximate available short-circuit current at the secondary terminals of a 50 kVA, 120/240-volt, single-phase pad mount transformer? The transformer impedance is 1%. Show your work.

   a) line-to-line: _____ amperes available fault current

   b) line-to-neutral: _____ amperes available fault current

SINGLE-FAMILY DWELLING SERVICE-ENTRANCE CALCULATIONS

1. **General Lighting Load** (*Section 220-3(a)*)

   _____ sq. ft. @ 3 VA per sq. ft.　　　= _____ VA

2. **Minimum Number of Lighting Branch Circuits**

   $$\frac{\text{Line 1}}{120} = \frac{\underline{\hspace{2cm}}}{120}　　　　　= \underline{\hspace{2cm}} \text{ amperes}$$

   then: $\dfrac{\text{amperes}}{15} = \dfrac{\underline{\hspace{1.5cm}}}{15} = \underline{\hspace{2cm}}$ 15-ampere branch circuits

3. **Small Appliance Load** (*Sections 210-11(c)(1)* and *220-16(a)*)
   (Minimum of two 20-ampere branch circuits)

   _____ branch circuits @ 1,500 VA each　　= _____ VA

4. **Laundry Branch Circuit** (*Sections 210-11(c)(2)* and *220-16(b)*)
   (Minimum of one 20-ampere branch circuit)

   _____ branch circuit(s) @ 1,500 VA each　= _____ VA

5. **Total General Lighting, Small Appliance, and Laundry Load**
   Lines 1 + 3 + 4　　　　　　　　　　　　　　　= _____ VA

6. **Net Computed General Lighting, Small Appliance, and Laundry Loads (less ranges, ovens, and "fastened-in-place" appliances). Apply Demand Factors from** *Table 220-11.*

   a. First 3,000 VA @ 100%　　　　　　　　　= __3,000__ VA

   b. Line 5 _____ − 3,000 = _____ @ 35% = _____ VA

   　　　　　　　　　　　　　Total a + b = _____ VA

7. **Electric Range, Wall-Mounted Ovens, Counter-Mounted Cooking Units** (*Table 220-19*)　　　= _____ VA

8. **Electric Clothes Dryer** (*Table 220-18*)　　　= _____ VA

9. **Electric Furnace** (*Section 220-15*)
   Air Conditioner, Heat Pump (*Article 440*)
   (Enter largest value, *Section 220-21*)　　　= _____ VA

10. **Net Computed General Lighting, Small Appliance, Laundry, Ranges, Ovens, Cooktop Units, HVAC**
    Lines 6 + 7 + 8 + 9　　　　　　　　　　　= _____ VA

11. List "Fastened-in-Place" Appliances *in addition* to Electric Ranges, Electric Clothes Dryers, Electric Space Heating, and Air-Conditioning Equipment.

Appliance		VA Load	
Water heater:	=	_____	VA
Dishwasher:	=	_____	VA

    Motor: _____ VA

    Heater: _____ VA (watts)

    Enter largest value, *Section 220-21.*

Garage door opener:	=	_____	VA
Food waste disposer:	=	_____	VA
Water pump:	=	_____	VA
Gas-fired furnace:	=	_____	VA
Sump pump:	=	_____	VA
Other: _____	=	_____	VA
_____	=	_____	VA
_____	=	_____	VA
_____	=	_____	VA
Total	=	_____	VA

12. **Apply 75% Demand Factor (*Section 220-17*) if Four or More "Fastened-in-Place" Appliances. If Less Than Four, Figure @ 100%.** Do not include electric ranges, electric clothes dryers, electric space heating, or air-conditioning equipment.

    Line 11 Total: _____ × 0.75   = _____ VA

13. **Total Computed Load (Lighting, Small Appliance, Ranges, Dryer, HVAC, "Fastened-in-Place" Appliances)**

    Line 10 _____ + Line 12 _____   = _____ VA

14. **Add 25% of Largest Motor (*Section 220-14*)**

    _____ × 0.25   = _____ VA

15. **Grand Total Line 13 + Line 14.**   = _____ VA

16. **Minimum Ampacity for Ungrounded Service-Entrance Conductors**

$$\text{Amperes} = \frac{\text{Line 15}}{240} = \frac{\rule{2cm}{0.4pt}}{240} \qquad = \underline{\hspace{2cm}} \text{ amperes}$$

17. **Minimum Ampacity for Neutral Service-Entrance Conductor,** *Sections 220-22* and *310-15(b)(6)*. **Do Not Include Straight 240-Volt Loads.**

    a. Line 6:   = _____ VA

    b. Line 7: _____ @ 0.70   = _____ VA

    c. Line 8: _____ @ 0.70   = _____ VA

    d. Line 11: (Include only 120-volt loads.)

      _____   _____ VA

      _____   _____ VA

      _____   _____ VA

      _____   _____ VA

      _____   _____ VA

                Total   _____ VA

    e. Line d total @ 75% Demand Factor if four or more, otherwise use 100%

      _____ × 0.75   = _____ VA

    f. Add 25% of largest 120-volt motor.

      _____ × 0.25   = _____ VA

                        Total = _____ VA

    g. Total a + b + c + e + f.   = _____ VA

$$\text{Amperes} = \frac{\text{Line g}}{240} = \frac{\rule{3cm}{0.4pt}}{240} \qquad = \underline{\hspace{3cm}} \text{ amperes}$$

18. **Ungrounded Conductor Size (copper)** *(Table 310-15(b)(6))*   No. _____

    Note: *Table 310-15(b)(6)* may only be used for 120/240-volt, 3-wire, residential single-phase service-entrance conductors, service lateral conductors, and feeder conductors that serve as the main power feeder to a dwelling unit.

19. **Neutral Conductor Size (copper)** *(Section 220-22)*   No. _____

    Note: *Section 310-15(b)(6)* permits the neutral conductor to be smaller than the ungrounded "hot" conductors if the requirement of *Sections 215-2, 220-22,* and *230-42* are met. When bare conductors are used with insulated conductors, the conductors' ampacity is based on the ampacity of the other insulated conductors in the raceway, *Section 310-15(b)(3)*. The neutral conductor must not be smaller than the grounding electrode conductor, *Section 250-24(b)(1)*.

20. **Grounding Electrode Conductor Size (copper)** *(Table 250-66)*   No. _____

21. **Raceway Size.**   _____ inch

    Obtain dimensional data from *Tables 1, 4, 5,* and *8, Chapter 9, NEC®*.

International Association of Electrical Inspectors' Major Objectives

- Formulation of standards for safe installation and use of electrical materials, devices and appliances.
- Promotion of uniform understanding and application of the National Electrical Code®, other codes, and any adopted electrical codes in other countries.
- Promotion of uniform administrative ordinances and inspection methods.
- Collection and dissemination of information relative to the safe use of electricity.
- Representation of electrical inspectors in the electrical industry, nationally and internationally.
- Cooperation with national and international organizations in the further development of the electrical industry.
- Promotion of cooperation among inspectors, inspection departments of city, county and state at national and international levels, the electrical industry and the public.

Contact IAEI customer service department for information on our special membership categories - Section, National and International Member; Sustaining Member (Bronze, Silver, Gold or Platinum); and Inspection Agency Member.

MEMBERSHIP APPLICATION
for
Inspector Member
Associate Member
Student Member

Annual Membership Dues $40
Students $30

Services from the IAEI for you and your organization include all this and more:

- Subscription to *IAEI News* - Bimonthly magazine with latest news on Code changes
- Local Chapter or Division meetings - New products, programs, national experts answer questions on local problems
- Section meetings to promote education and cooperation
- Promote advancement of electrical industry
- Save TIME and MONEY by having the latest information

PLEASE PRINT

| Name - Last | First | M.I. | Chapter, where you live or work, if known |

| Title | | (Division, where appropriate) |

| Employer | | For Office Use Only | Section No. | Chapter No. | Division No. |

| Address of Applicant | | If previous member, give last membership number |

| City | State or Province | Zip or Postal Code | and last year of membership 19____ |

| Student applicants give school attending ** | Graduation Date / / |

Endorsed by R A Y C M U L L I N   Endorser's Membership Number 5 3 3 8 0 0

| Applicant's Signature | (Area Code) Telephone Number |

☐ Check ☐ Money Order ☐ Visa   Name on Card   Charge Card Number   Exp. Date
☐ MasterCard ☐ Amex ☐ Discover

☐ Inspector $40   ☐ Chief Electrical Inspector * $40   ☐ Associate $40   ☐ Student ** $30

Inspector Members and Chief Electrical Inspector Members MUST sign below:

☐ I, _____ , meet the qualifications for inspector member type as described below.

Inspector members must regularly make electrical inspections for preventing injury to persons or damage to property on behalf of a governmental agency, insurance agency, rating bureau, recognized testing laboratory or electric light and power company.

*Chief electrical inspector members must regularly provide technical supervision of electrical inspectors.

**Student member must be currently enrolled in an approved college, university, vocational technical school or trade school specializing in electrical training or approved electrical apprenticeship school.

Mail to: IAEI, P. O. Box 830848 Richardson, TX 75083-0848   For information call: (972) 235-1455 (8-5 CT)

Details: Plans, elevations, or sections that provide more specific information about a portion of a project component or element than smaller scale drawings.

Diagrams: Nonscaled views showing arrangements of special system components and connections not possible to clearly show in scaled views. A *schematic diagram* shows circuit components and their electrical connections without regard to actual physical locations. A *wiring diagram* shows circuit components and the actual electrical connections.

Dimmer: A switch with components that permits variable control of lighting intensity. Some dimmers have electronic components; others have core and coil (transformer) components.

Disconnecting Means*: A device, or group of devices, or other means by which the conductors of a circuit can be disconnected from their source of supply.

Drawings: A graphic representation of the work to be done. They show the relationship of the materials to each other, including sizes, shapes, locations, and connections. The drawings may include schematic diagrams showing such things as mechanical and electrical systems. They may also include schedules of structural elements, equipment, finishes, and other similar items.

Dry Niche Lighting Fixture*: A lighting fixture intended for installation in the wall of a pool or fountain in a niche that is sealed against the entry of pool water.

Dwelling Unit*: One or more rooms for the use of one or more persons as a housekeeping unit with space for eating, living, and sleeping, and permanent provisions for cooking and sanitation.

Elevations: Views of vertical planes, showing components in their vertical relationship, viewed perpendicularly from a selected vertical plane.

Feeder*: All circuit conductors between the service equipment or the source of a separately derived system and the final branch-circuit overcurrent device.

Fire Rating: The classification indicating in time (hours) the ability of a structure or component to withstand fire conditions.

Fine Print Note (FPN)*: Explanatory material is in the form of Fine Print Notes (FPN). Fine Print Notes are informational only, and are not enforceable as a requirement by the Code.

Flame Detector: A radiant energy-sensing fire detector that detects the radiant energy emitted by a flame.

Fully Rated System: Panelboards are marked with their short circuit rating in RMS symmetrical amperes. In a fully-rated system, the panelboard short circuit current rating will be equal to the lowest interrupting rating of any branch-circuit breaker or fuse installed. All devices installed shall have an interrupting rating greater than or equal to the specified available fault current. (See "Series-Rated System").

Fuse: An overcurrent protective device with a fusible link that operates and opens the circuit on an overcurrent condition.

Ground-Fault Protection of Equipment (GFPE)*: A system intended to provide protection of equipment from damaging line-to-ground fault currents by operating to cause a disconnecting means to open all ungrounded conductors of the faulted circuit. This protection is provided at current levels less than those required to protect conductors from damage through the operation of a supply circuit overcurrent device.

Ground*: A conducting connection, whether intentional or accidental, between an electrical circuit or equipment and the earth, or to some conducting body that serves in place of the earth.

Ground-Fault: A condition where an ungrounded conductor comes in contact with a grounded object such as but not limited to a grounded metal water pipe. A ground-fault is often referred to as a "line-to-ground" fault. See *Section 230-95* of the *NEC.*

Grounded*: Connected to earth or to some conducting body that serves in place of the earth.

Grounded Conductor*: A system or circuit conductor that is intentionally grounded.

Grounding Conductor*: A conductor used to connect equipment or the grounded circuit of a wiring system to a grounding electrode or electrodes.

Grounded, Effectively*: Intentionally connected to earth through a ground connection or connections of sufficiently low impedance and having sufficient current-carrying capacity to prevent the buildup of voltages that may result in undue hazards to connected equipment or to persons.

Ground-Fault Circuit-Interrupter (GFCI)*: A device intended for the protection of personnel that functions to de-energize a circuit or portion thereof within an established period of time when a current to ground exceeds some predetermined value that is less than that required to operate the overcurrent protective device of the supply circuit.

Grounding Conductor, Equipment*: The conductor used to connect the noncurrent-carrying metal parts of equipment, raceways, and other enclosures to the system grounded conductor, the grounding electrode conductor, or both, at the service equipment or at the source of a separately derived system.

Grounding Electrode Conductor*: The conductor used to connect the grounding electrode to the equipment grounding conductor, to the grounded conductor, or to both, of the circuit at the service equipment or at the source of a separately derived system.

HACR: A circuit breaker that has been tested and found suitable for use on heating, air-conditioning, and refrigeration equipment. The circuit breaker is marked with the letters HACR. Equipment that is suitable for use with HACR circuit breakers will be marked for use with HACR-type circuit breakers.

Heat Alarm: A single or multiple station alarm responsive to heat.

Hermetic Refrigerant Motor-Compressor: A combination consisting of a compressor and motor, both of which are enclosed in the same housing, with no external shaft or shaft seals; the motor operating in the refrigerant.

Household Fire Alarm System: A system of devices that produces an alarm signal in the household for the purpose of notifying the occupants of the presence of a fire so that they will evacuate the premises.

Identified*: (as applied to equipment) Recognizable as suitable for the specific purpose, function, use, environment, application, etc., where described in a particular Code requirement.

Identified* (conductor): The identified conductor is the insulated grounded conductor.

- For sizes No. 6 or smaller, the insulated grounded conductor shall be identified by a continuous white or natural gray outer finish or by three continuous white stripes on other than green insulation along its entire length.
- For sizes larger than No. 6, the insulated grounded conductor shall be identified either by a continuous white or natural gray outer finish or by three continuous white stripes on other than green insulation along its entire length, or at the time of installation by a distinctive white marking at its terminations.

Identified* (terminal): The identification of terminals to which a grounded conductor is to be connected shall be substantially white in color. The identification of other terminals shall be of a readily distinguishable different color.

IEC: The International Electrotechnical Commission is a worldwide standards organization. These standards differ from those of Underwriters Laboratories. Some electrical equipment might conform to a specific IEC Standard, but may not conform to the U.L. Standard for the same item.

Immersion Detection Circuit Interrupter: A device integral with grooming appliances that will shut off the appliance when the appliance is dropped in water.

In Sight*: Where this Code specifies that one equipment shall be "in sight from," "within sight from," or "within sight," etc., of another equipment, the specified equipment is to be visible and not more than 50 ft (15.24 m) from the other.

Inductive Load: A load that is made up of coiled or wound wire that creates a magnetic field when energized. Transformers, core and coil ballasts, motors, and solenoids are examples of inductive loads.

International Association of Electrical Inspectors: A not-for-profit and educational organization cooperating in the formulation and uniform application of standards for the safe installation and use of electricity, and collecting and disseminating information relative thereto. The IAEI is made up of electrical inspectors, electrical contractors, electrical apprentices, manufacturers, and governmental agencies.

Interrupting Rating*: The highest current at rated voltage that a device is intended to interrupt under standard test conditions.

Isolated Ground Receptacle: A grounding type device in which the equipment ground contact and terminal are electrically isolated from the receptacle mounting means.

Kilowatt: One thousand watts equals one kilowatt.

Kilowatt hour: One thousand watts of power consumed in one hour. Ten 100-watt lamps burning for ten hours is one kilowatt hour. Two 500-watt electric heaters operated for two hours is one kilowatt hour.

Labeled*: Equipment or materials to which has been attached a label, symbol, or other identifying mark of an organization that is acceptable to the authority having jurisdiction and concerned with product evaluation that maintains periodic inspection of production of labeled equipment or materials and by whose labeling the manufacturer indicates compliance with appropriate standards or performance in a specified manner.

Lighting Outlet*: An outlet intended for the direct connection of a lampholder, a lighting fixture, or a pendant cord terminating in a lampholder.

Listed*: Equipment, materials, or services included in a list published by an organization that is acceptable to the authority having jurisdiction and concerned with evaluation of products or services, that maintains periodic inspection of production of listed equipment or materials or periodic evaluation of services, and whose listing states that either the equipment, material, or service meets appropriate designated standards or has been tested and found suitable for use in a specified purpose.

Load: The electric power used by devices connected to an electrical system. Loads can be figured in amperes, volt-amperes, kilovolt-amperes, or kilowatts. Loads can be intermittent, continuous intermittent, periodic, short-time, or varying. See the definition of "Duty" in the *NEC.*

Load Center: A common name for residential panelboards. A load center may not be as deep as a panelboard, and generally does not contain relays or other accessories as are available for panelboards. Circuit breakers "plug-in" as opposed to "bolt-in" types used in panelboards. Manufacturers' catalogs will show both load centers and panelboards. The UL standards do not differentiate.

Locked Rotor Current: The steady-state current taken from the line with the rotor locked and with rated voltage and frequency applied to the motor.

Luminaire: A complete lighting unit consisting of lamp or lamps together with the parts designed to distribute the light, to position and protect the lamps, and to connect the lamps to the power supply. This is a lighting fixture.

Mandatory Rules: Required. The terms *shall* or *shall not* are used when a Code statement is mandatory.

Maximum Continuous Current (MCC): A value that is determined by the manufacturer of hermetic refrigerant motor compressors under high load (high refrigerant pressure) conditions. This value is established in the *NEC®* as being no greater than 156% of the marked rated amperes or the branch-circuit selection current.

Maximum Overcurrent Protection (MOP): A term used with equipment that has a hermetic motor compressor(s). MOP is the maximum ampere rating for the equipment's branch-circuit overcurrent protective device. The ampere rating is determined by the manufacturer, and is marked on the nameplate of the equipment. The nameplate will also indicate if the overcurrent device can be a HACR-type circuit breaker, a fuse, or either.

Minimum Circuit Ampacity (MCA): A term used with equipment that has a hermetic motor compressor(s). MCA is the minimum ampere rating value used to determine the proper size conductors, disconnecting means, and controllers. MCA is determined by the manufacturer, and is marked on the nameplate of the equipment.

National Electrical Code (*NEC®*): The electrical code published by the National Fire Protection Association. This Code provides for practical safeguarding of persons and property from hazards arising from the use of electricity. It does not become law until adopted by federal, state, and local laws and regulations. The *NEC®* is not intended as a design specification nor an instruction manual for untrained persons.

National Electrical Manufacturers Association (NEMA): NEMA is a trade organization made up of many manufacturers of electrical equipment. They develop and promote standards for electrical equipment.

National Fire Protection Association (NFPA): Located in Quincy, MA. The NFPA is an international Standards Making Organization dedicated to the protection of people from the ravages of fire and electric shock. The NFPA is responsible for developing and writing the *National Electrical Code*, The Sprinkler Code, The Life Safety Code, The National Fire Alarm Code, and over 295 other codes, standards, and recommended practices. The NFPA may be contacted at (800) 344-3555 or on their web site at www.nfpa.org.

Nationally Recognized Testing Laboratory (NRTL): The term used to define a testing laboratory that has been recognized by OSHA. For example, Underwriters Laboratories.

Neutral: In residential wiring, the neutral conductor is the grounded conductor in a circuit consisting of three or more conductors. There is no neutral conductor in a two-wire circuit, although electricians many times refer to the white grounded conductor as the neutral.

The voltages from the ungrounded conductors to the grounded conductor are of equal magnitude.

See the definition of "Multiwire Branch-Circuit" in the *NEC.*

No Niche Lighting Fixture*: A lighting fixture intended for installation above or below the water without a niche.

Noncoincidental Loads: Loads that are not likely to be on at the same time. Heating and cooling loads would not operate at the same time. See *Section 220-21* of the *NEC.*

Notations: Words found on plans to describe something.

Occupational Safety and Health Act (OSHA):   This is the Code of Federal Regulations, developed by the Occupational Safety and Health Administration, U.S. Department of Labor. The electrical regulations are covered in *Part 1910, Subpart S*. So as not to shorten and simplify the regulations, *Part 1920, Subpart S* contains only the most common performance requirements of the *NEC*® The *NEC*® must still be referred to, in conjunction with OSHA regulations.

Ohm: A unit of measure for electric resistance. An ohm is the amount of resistance that will allow one ampere to flow under a pressure of one volt.

Outlet*:   A point on the wiring system at which current is taken to supply utilization equipment.

Overcurrent*:   Any current in excess of the rated current of equipment or the ampacity of a conductor. It may result from overload, short circuit, or ground fault.

Overload*:   Operation of equipment in excess of normal, full-load rating, or of a conductor in excess of rated ampacity that, when it persists for a sufficient length of time, would cause damage or dangerous overheating. The current flow is contained in its normal intended path. A fault, such as a short circuit or ground fault, is not an overload.

Overcurrent Device: Also referred to as an overcurrent protective device. A form of protection that operates when current exceeds a predetermined value. Common forms of overcurrent devices are circuit breakers, fuses, and thermal overload elements found in motor controllers.

Panelboard*:   A single panel or group of panel units designed for assembly in the form of a single panel; including buses, automatic overcurrent devices, and equipped with or without switches for the control of light, heat, or power circuits; designed to be placed in a cabinet or cutout box placed in or against a wall or partition and accessible only from the front.

Permissive Rules: Allowed but not required. Terms such as *shall be permitted* or *shall not be required* are used when a Code statement is permissive.

Plans: Views of horizontal planes, showing components in their horizontal relationship. A set of construction drawings.

Power Supply: A source of electrical operating power including the circuits and terminations connecting it to the dependent system components.

Qualified Person*:   One familiar with the construction and operation of the equipment and the hazards involved.

Raceway*:   An enclosed channel of metal or non-metallic materials designed expressly for holding wires, cables, or busbars, with additional functions as permitted in this Code. Raceways include, but are not limited to, rigid metal conduit, rigid nonmetallic conduit, intermediate metal conduit, liquidtight flexible conduit, flexible metallic tubing, flexible metal conduit, electrical nonmetallic tubing, electrical metallic tubing, underfloor raceways, cellular concrete floor raceways, cellular metal floor raceways, surface raceways, wireways, and busways.

Rated-Load Current: The rated-load current for a hermetic refrigerant motor-compressor is the current resulting when the motor-compressor is operated at the rated load, rated voltage, and rated frequency of the equipment it serves.

Rate of Rise Detector: A device that responds when the temperature rises at a rate exceeding a predetermined value.

Receptacle*:   A receptacle is a contact device installed at the outlet for the connection of a single contact device. A single receptacle is a single contact device with no other contact device on the same yoke. A multiple receptacle is two or more contact devices on the same yoke.

Receptacle Outlet*:   An outlet where one or more receptacles are installed.

Resistive Load: An electric load that opposes the flow of current. A resistive load does not contain cores or coils of wire. Some examples of resistive loads are the electric heating elements in an electric range, ceiling heat cables, and electric baseboard heaters.

Root-Mean-Square (RMS):   The square root of the average of the square of the instantaneous values of current or voltage. For example, the RMS value of voltage line to neutral in a home is 120 volts. During each electrical cycle, the voltage rises from zero to a peak value ($120 \times 1.4142 = 169.7$ volts), back through zero to a negative peak value, then back to zero. The RMS value is 0.707 of the peak value ($169.7 \times 0.707 = 120$ volts). The root-mean-square value is what an electrician reads on an ammeter or voltmeter.

Schedules: Tables or charts that include data about materials, products, and equipment.

Sections: Views of vertical cuts through and perpendicular to components, showing their detailed arrangement.

Series-Rated System: Panelboards are marked with their short-circuit rating in RMS symmetrical amperes. A series-rated panelboard will be determined by the main circuit breaker or fuse, and branch-circuit breaker combination tested in accordance to UL *Standard 489*. The series-rating will be less than or equal to the interrupting rating of the main overcurrent device, and greater than the interrupting rating of the branch-circuit overcurrent devices. (See "Fully-Rated System.")

Service*: The conductors and equipment for delivering energy from the serving utility to the wiring system of the premises served.

Service Conductors*: The conductors from the service point to the service disconnecting means.

Service Drop*: The overhead service conductors from the last pole or other aerial support to and including the splices, if any, connecting to the service-entrance conductors at the building or other structure.

Service Equipment*: The necessary equipment, usually consisting of a circuit breaker or switch and fuses, and their accessories, located near the point of entrance of supply conductors to a building or other structure, or an otherwise defined area, and intended to constitute the main control and means of cutoff of the supply.

Service Lateral*: The underground service conductors between the street main, including any risers at a pole or other structure or from transformers, and the first point of connection to the service-entrance conductors in a terminal box or meter or other enclosure with adequate space, inside or outside the building wall. Where there is no terminal box, meter, or other enclosure with adequate space, the point of connection shall be considered to be the point of entrance of the service conductors into the building.

Service Point*: Service point is the point of connection between the facilities of the serving utility and the premises wiring.

Shall: Indicates a mandatory requirement.

Short Circuit: A connection between any two or more conductors of an electrical system in such a way as to significantly reduce the impedance of the circuit. The current flow is outside of its intended path, thus the term "short circuit." A short circuit is also referred to as a fault.

Should: Indicates a recommendation or that which is advised but not required.

Smoke Alarm: A single or multiple station alarm responsive to smoke.

Smoke Detector: A device that detects visible or invisible particles of combustion.

Specifications: Text setting forth details such as description, size, quality, performance, and workmanship, etc. Specifications that pertain to all of the construction trades involved might be subdivided into "General Conditions" and "Supplemental General Conditions." Further subdividing of the Specifications might be specific requirements for the various contractors such as electrical, plumbing, heating, and masonry, etc. Typically, the electrical specifications are found in Division 16.

Split Circuit Receptacle: A receptacle that can be connected to two branch-circuits. These receptacles may also be used so that one receptacle is live at all times, and the other receptacle is controlled by a switch. The terminals on these receptacles usually have breakaway tabs so the receptacle can be used either as a split circuit receptacle or as a standard receptacle.

Standard Network Interface (SNI): A device usually installed by the telephone company at the demarcation point where their service leaves off and the customers service takes over. This is similar to the service point for electrical systems.

Surge Suppressor: A device that limits peak voltage to a predetermined value when voltage spikes or surges appear on the connected line.

Switches*:

General-Use Snap Switch: A form of general-use switch constructed so that it can be installed in device boxes or on box covers, or otherwise used in conjunction with wiring systems recognized by this Code.

General-Use Switch: A switch intended for use in general distribution and branch-circuits. It is rated in amperes, and it is capable of interrupting its rated current at its rated voltage.

Motor-Circuit Switch: A switch, rated in horsepower, capable of interrupting the maximum operating overload current of a motor of the same horsepower rating as the switch at the rated voltage.

Symbols: A graphic representation that stands for or represents another thing. A symbol is a simple way to show such things as lighting outlets, switches and receptacles on an electrical Plan. The American Institute of Architects has developed a very comprehensive set of symbols that represent just about everything used by all building trades. When an item cannot be shown using a symbol, then a more detailed explanation using a notation or inclusion in the Specifications is necessary.

Terminal: A screw or a quick-connect device where a conductor(s) is intended to be connected.

Thermocouple: A pair of dissimilar conductors so joined at two points that an electromotive force is developed by the thermoelectric effects when the junctions are at different temperatures.

Thermopile: More than one thermocouple connected together. The connections may be series, parallel, or both.

Transient Voltage Surge Suppressors (TVSS): A device that clamps transient voltages by absorbing the major portion of the energy (joules) created by the surge, allowing only a small, safe amount of energy to enter the actual connected load.

UL: Underwriters Laboratories is an independent not-for-profit organization that develops standards, and tests electrical equipment to these standards.

UL Listed: Indicates that an item has been tested and approved to the standards established by UL for that particular item. The UL Listing Mark may appear in various forms, such as the letters UL in a circle. If the product is too small for the marking to be applied to the product, the marking must appear on the smallest unit container in which the product is packaged.

UL Recognized: Refers to a product that is incomplete in construction features or limited in performance capabilities. A "Recognized" product is intended to be used as a component part of equipment that has been "listed." A "Recognized" product must not be used by itself. A UL product may contain a number of components that have been "Recognized."

Ungrounded Conductor: The conductor of an electrical system that is not intentionally connected to ground. This conductor is referred to as the "hot" or "live" conductor.

Volt: The difference of electric potential between two points of a conductor carrying a constant current of one ampere, when the power dissipated between these points is equal to one watt. A voltage of one volt can push one ampere through a resistance of one ohm.

Voltage (of a circuit)*: The greatest root-mean-square (effective) difference of potential between any two conductors of the circuit concerned.

Voltage (nominal)*: A nominal value assigned to a circuit or system for the purpose of conveniently designating its voltage class (e.g., 120/240 volts, 480Y/277 volts, 600 volts).

The actual voltage at which a circuit operates can vary from the nominal within a range that permits satisfactory operation of equipment.

Voltage to Ground*: For grounded circuits, the voltage between the given conductor and that point or conductor of the circuit that is grounded; for ungrounded circuits, the greatest voltage between the given conductor and any other conductor of the circuit.

Voltage Drop: Also referred to as IR drop. Voltage drop is most commonly associated with conductors. A conductor has resistance. When current is flowing through the conductor, a voltage drop will be experienced across the conductor. Voltage drop across a conductor can be calculated using Ohm's Law: $E = IR$.

Volt-ampere: A unit of power determined by multiplying the voltage and current in a circuit. A 120-volt circuit carrying 1 ampere is 120 volt-amperes.

Watt: A measure of true power. A watt is the power required to do work at the rate of 1 joule per second. Wattage is determined by multiplying voltage times amperes times the power factor of the circuit: $W = E \times I \times PF$.

Watertight: Constructed so that moisture will not enter the enclosure under specified test conditions.

Weatherproof*: Constructed or protected so that exposure to the weather will not interfere with successful operation.

(FPN): Rainproof, raintight, or watertight equipment can fulfill the requirements for weatherproof where varying weather conditions other than wetness, such as snow, ice, dust, or temperature extremes, are not a factor.

Wet Niche Lighting Fixture*: A lighting fixture intended for installation in a forming shell mounted in a pool.

WEB SITES

The following list of Internet World Wide Web sites has been included for your convenience. These are current as of time of printing. Web sites are a moving target with continual changes. We will do our utmost to keep these web sites current. If you are aware of Web sites that should be added, deleted, or that are different than shown, please let us know. Most manufacturers will provide catalogs and technical literature upon request. You will be amazed at the amount of electrical equipment information that is available on these web sites.

Electrical Industry *(This search engine links to over 800 sites related to the electrical industry.)*
www.electricalsearch.com

ACCUBID	www.accubid.com
Advance Transformer Co.	www.advancetransformer.com
AFC Cable Systems	www.afcweb.com
Alflex	www.alflex.com
Allen-Bradley	www.ab.com
Allied Moulded Products, Inc.	www.alliedmoulded.com
Allied Tube & Conduit	www.industry.net/allied.tube.conduit
Alpha Wire	www.alphawire.com
American Lighting Association	www.americanlightingassoc.com
American Insulated Wire Corp.	www.aiwc.com
American National Standards Institute	www.ansi.org
American Society of Heating, Refrigerating, and Air-Conditioning Engineers	www.ashrae.org
American National Standards Institute	www.ansi.org
AMP	www.amp.com
Amprobe	www.amprobe.com
Angelo	www.angelobrothers.com
Appleton Electric Co.	www.appletonelec.com
Arrow Hart	www.crouse-hinds.com
Associated Builders & Contractors	www.abc.org
Automatic Switch Co. (ASCO)	www.asco.com
Baldor Motors & Drives	www.baldor.com
Belden	www.belden.com
Best Power	www.bestpower.com
BOCA	www.bocai.org
Brady	www.whbrady.com
Bridgeport	www.bpfittings.com
Broan	www.broan.com
Burndy	www.burndy.com
Bussmann	www.bussmann.com
CABO	www.cabo.org/
Canadian Standards Association	www.csa.ca
CEBus Industry Council	www.cebus.org
CEE News magazine	www.ceenews.com/
Certified Ballast Manufacturers	www.certbal.org
Chromalox	www.chromalox.com

Construction Specifications Institute	www.csinet.org
Continental Automated Building Association	www.caba.org
Cooper Hand Tools	www.coopertools.com
Cooper Power Systems	www.cooperpower.com
Copper Development Association Inc.	www.copper.org
Copper information	www.copper.org
Crouse-Hinds	www.crouse-hinds.com
Crouse-Hinds (Hazardous locations)	www.zonequest.com
Daniel Woodhead Company	www.danielwoodhead.com
Delmar Publishers	www.thomson.com/delmar/index.htm
Dual-Lite	www.dual-lite.com
Eagle Electric	www.eagle-electric.com
Eaton/Cutler-Hammer	www.cutlerhammer.eaton.com
EC&M magazine	www.ecmweb.com
Edwards	www.edwards-signals.com
Electrical Contractor magazine	www.ecmag.com
Erico, Inc.	www.erico.com
Estimation Inc.	www.estimation.com
Factory Mutual Insurance Co.	www.factorymutual.com
Federal Pacific Co.	www.electro-mechanical.com/fpc.html
Ferraz	www.ferraz.com
First Alert	www.FirstAlert.com
Fluke Corporation	www.fluke.com
F. W. Dodge	www.fwdodge.com
Gardner Bender	gardnerbender.com
GE Electrical Distribution & Control	www.ge.com/edc
GE Lighting	www.ge.com/lighting/business
Generac Power Systems	www.generac.com
GE Total Lighting Control	www.ge.com/tlc
Gould Shawmut	www.gouldshawmut.com
Greenlee Textron Inc.	www.greenlee.textron.com
Halo (Cooper Lighting)	www.cooperlighting.com
Heyco Molded Products, Inc.	www.heyco.com
Hilti	www.hilti.com
Home Automation Association	www.creator.hometeam.com/haa
HomeStar Wiring Systems	www.lucent/netsys/homestar
Home Toys (Home automation)	www.hometoys.com
Honeywell	www.Honeywell.com
Housing Urban Development	www.HUD.gov
Houston Wire & Cable Company	www.houwire.com
Hubbell Incorporated	www.hubbell.com
Hubbell Incorporated (Lighting)	www.hubbell-ltg.com
Hubbell Incorporated (Premise)	www.hubbell-premise.com
Hubbell Incorporated (Wiring)	www.hubbell-wiring.com
Hunt Dimmers	www.huntdimming.com
Husky	www.mphusky.com
ICBO	www.icbo.org
Ideal Industries, Inc.	www.idealindustries.com
ILSCO	www.ilsco.com

Independent Electrical Contractors	www.ieci.org
Institute of Electrical and Electronic Engineers, Inc.	www.ieee.org
inter.Light, Inc.	www.lightsearch.com
Intermatic	www.intermatic.com
International Association of Electrical Inspectors (IAEI)	www.iaei.com
Juno Lighting, Inc.	www.junolighting.com
Klein Tools	www.klein-tools.com
Kohler Generators	www.kohlergenerators.com
Leviton Manufacturing Co.	www.leviton.com
Lightolier	www.lightolier.com
Lithonia Lighting	www.lithonia.com
Littelfuse, Inc.	www.littelfuse.com
Lutron Electronics Co., Inc.	www.lutron.com
MagneTek Lighting Products	www.magnetek.com/ballast
Manhattan/CDT	www.manhattancdt.com
McGill	www.mcgillelectrical.com
Megger	www.avointl.com
Midwest Electric Products, Inc.	www.midwestelectric.com
Milbank Manufacturing Co.	www.milbankkmfg.com
Minerallac	www.minerallac.com
Motorola Lighting	www.motorola.com/aegg/mli
MYTECH	www.lightswitch.com
National Association of Electrical Distributors	www.naed.org
National Electrical Contractors Association (NECA)	www.necanet.org
National Electrical Manufacturers Association (NEMA)	www.nema.org
National Fire Protection Association	www.nfpa.org
National Joint Apprenticeship and Training Committee	www.njatc.org
NuTone	www.nutone.com
Occupational Safety & Health Act	www.osha.gov
Onan Corporation	www.onan.com
Optical Cable Corporation	www.occfiber.com
Osram Sylvania, Inc.	www.sylvania.com
Panasonic Lighting	www.panasonic.com/lighting
Panduit Corporation	www.panduit.com
Paragon Electric Company, Inc.	www.paragonelectric.com
Penn-Union	www.penn-union.com
Philips Lamps	www.ALTOlamp.com
Philips Lighting Company	www.lighting.philips.com/nam
Progress Lighting	www.progresslighting.com
RACO	www.racoinc.com
Russelectric	www.russelectric.com
SBCCI	www.sbcci.org
S&C Electric Company	www.sandc.com
Siemens-Furnas Controls	www.furnas.com

Simplex	www.simplexnet.com
Sola/Hevi Duty	www.sola-hevi-duty.com
Southwire Co.	www.southwire.com
Square D Company	www.squared.com
Sylvania	www.sylvania.com
Thomas & Betts Corporation	www.tnb.com
Trade Service	www.tra-ser.com
Underwriters Laboratories	www.ul.com
Underwriters Laboratories (Code Authority)	www.ul.com/auth/index.html
Underwriters Laboratories (Contact)	www.ul.com/contact.html
Waukesha Electric Systems	www.waukeshaelectric.com
Wheatland Tube Company	www.wheatland.com
Wiremold	www.wiremold.com
X-10 Pro	www.x10pro.com

CODE INDEX

Note: Page numbers in **bold type** reference non-text material.

INDEX

Note: Page numbers in **bold type** reference non-text material.